Supersymmetry and Trace Formulae
Chaos and Disorder

NATO ASI Series

Advanced Science Institutes Series

A series presenting the results of activities sponsored by the NATO Science Committee, which aims at the dissemination of advanced scientific and technological knowledge, with a view to strengthening links between scientific communities.

The series is published by an international board of publishers in conjunction with the NATO Scientific Affairs Division

A	**Life Sciences**	Kluwer Academic / Plenum Publishers
B	**Physics**	New York and London
C	**Mathematical and Physical Sciences**	Kluwer Academic Publishers Dordrecht, Boston, and London
D	**Behavioral and Social Sciences**	
E	**Applied Sciences**	
F	**Computer and Systems Sciences**	Springer-Verlag
G	**Ecological Sciences**	Berlin, Heidelberg, New York, London,
H	**Cell Biology**	Paris, Tokyo, Hong Kong, and Barcelona
I	**Global Environmental Change**	

PARTNERSHIP SUB-SERIES

1. Disarmament Technologies	Kluwer Academic Publishers
2. Environment	Springer-Verlag
3. High Technology	Kluwer Academic Publishers
4. Science and Technology Policy	Kluwer Academic Publishers
5. Computer Networking	Kluwer Academic Publishers

The Partnership Sub-Series incorporates activities undertaken in collaboration with NATO's Cooperation Partners, the countries of the CIS and Central and Eastern Europe, in Priority Areas of concern to those countries.

Recent Volumes in this Series:

Volume 368 — Confinement, Duality, and Nonperturbative Aspects of QCD
edited by Pierre van Baal

Volume 369 — Beam Shaping and Control with Nonlinear Optics
edited by F. Kajzar and R. Reinisch

Volume 370 — Supersymmetry and Trace Formulae: Chaos and Disorder
edited by Igor V. Lerner, Jonathan P. Keating, and David E. Khmelnitskii

Volume 371 — The Gap Symmetry and Fluctuations in High-T_C Superconductors
edited by Julien Bok, Guy Deutscher, Davor Pavuna, and Stuart A. Wolf

Series B: Physics

Supersymmetry and Trace Formulae
Chaos and Disorder

Edited by

Igor V. Lerner
University of Birmingham
Birmingham, United Kingdom

Jonathan P. Keating
University of Bristol
and Hewlett–Packard Laboratories
Bristol, United Kingdom

and

David E. Khmelnitskii
University of Cambridge
Cambridge, United Kingdom
and L. D. Landau Institute for Theoretical Physics
Moscow, Russia

Kluwer Academic / Plenum Publishers
New York, Boston, Dordrecht, London, Moscow
Published in cooperation with NATO Scientific Affairs Division

Proceedings of a NATO Advanced Study Institute on
Supersymmetry and Trace Formulae: Chaos and Disorder,
held September 8 – 20, 1997,
in Cambridge, United Kingdom

NATO-PCO-DATA BASE

The electronic index to the NATO ASI Series provides full bibliographical references (with keywords and/or abstracts) to about 50,000 contributions from international scientists published in all sections of the NATO ASI Series. Access to the NATO-PCO-DATA BASE is possible via a CD-ROM "NATO Science and Technology Disk" with user-friendly retrieval software in English, French, and German (©WTV GmbH and DATAWARE Technologies, Inc. 1989). The CD-ROM contains the AGARD Aerospace Database.

The CD-ROM can be ordered through any member of the Board of Publishers or through NATO-PCO, Overijse, Belgium.

Library of Congress Cataloging-in-Publication Data

Supersymmetry and trace formulae : chaos and disorder / edited by Igor
  V. Lerner, Jonathan P. Keating, and David E. Khmelnitskii.
        p.   cm. -- (NATO ASI series. Series B, Physics ; v. 370)
     "Proceedings of a NATO Advanced Study Institute on Supersymmetry
  and Trace Formulae: Chaos and Disorder, held September 8-20, 1997,
  in Cambridge, United Kingdom"--T.p. verso.
     "Published in cooperation with NATO Scientific Affairs Division."
  Includes bibliographical references and index.
  ISBN 0-306-45933-7
     1. Quantum chaos--Congresses.   2. Order-disorder models-
  -Congresses.   3. Supersymmetry--Congresses.   I. Lerner, Igor V.
  II. Keating, Jonathan P.   III. Khmelnitskii, David E.   IV. NATO
  Advanced Study Institute on Supersymmetry and Trace Formulae: Chaos
  and Disorder (1997 : Cambridge, England)   V. Series.
  QC174.17.C45S86   1998
  003'.857--dc21   98-42555
                                                             CIP

ISBN 0-306-45933-7

PREFACE

The motion of a particle in a random potential in two or more dimensions is chaotic, and the trajectories in deterministically chaotic systems are effectively random. It is therefore no surprise that there are links between the quantum properties of disordered systems and those of simple chaotic systems. The question is, how deep do the connections go? And to what extent do the mathematical techniques designed to understand one problem lead to new insights into the other?

The canonical problem in the theory of disordered mesoscopic systems is that of a particle moving in a random array of scatterers. The aim is to calculate the statistical properties of, for example, the quantum energy levels, wavefunctions, and conductance fluctuations by averaging over different arrays; that is, by averaging over an ensemble of different realizations of the random potential. In some regimes, corresponding to energy scales that are large compared to the mean level spacing, this can be done using diagrammatic perturbation theory. In others, where the discreteness of the quantum spectrum becomes important, such an approach fails. A more powerful method, developed by Efetov, involves representing correlation functions in terms of a supersymmetric nonlinear sigma-model. This applies over a wider range of energy scales, covering both the perturbative and non-perturbative regimes. It was proved using this method that energy level correlations in disordered systems coincide with those of random matrix theory when the dimensionless conductance tends to infinity. Building upon this, there has been considerable recent progress in developing non-perturbative techniques that encompass finite conductance corrections.

In Quantum Chaos, the aim is to understand the semiclassical asymptotics of the quantum properties of classically chaotic systems. In this case the main tool is Gutzwiller's trace formula, which links the quantum energy levels and eigenfunctions in a given system to the periodic orbits of the underlying classical dynamics. Here too it has been found in many examples that both spectral correlations on the scale of the mean level spacing and properties of the eigenstates are universal and well-modelled by random matrix theory, and it has been conjectured that generically this approximation becomes exact in the semiclassical limit. The problem is to prove this, and to describe deviations from the limit before it is reached. Approaches based on periodic orbit theory, which relate universal quantum correlations to classical ergodicity and link the non-universal deviations to features of the short-time dynamics, have proved very successful, but we are still far from a complete understanding.

Recently, there has been considerable interest in characterizing the similarities and differences between supersymmetric methods and those based on the trace formula in the hope that cross-fertilization will lead to further progress. Some of the key questions addressed include the following. Can one use an analogue of the nonlinear sigma-model to describe spectral correlations in a single deterministic system by averaging over the energy, or by introducing very weak disorder? Does ensemble averaging allow for the contribution of untypical realizations of the random potential? Is the approach to the random-matrix limit in ensembles of disordered systems related to the corresponding approach in a single deterministically chaotic system? How large is the exceptional set of strongly chaotic systems that do not to exhibit random-matrix statistics?

This volume is the product of a NATO Advanced Study Institute entitled "Supersymmetry and Trace Formulae: Chaos and Disorder", held at the Isaac Newton Institute in Cambridge UK from 8-19 September 1997. The lecture courses and seminars given there together formed a representative review of recent progress in both fields. It is our hope that the following papers, which appear in the order in which the corresponding lectures were presented, will provide a broad overview of the most topical ideas and the key problems. The ASI itself was part of a five month programme on *Disordered Systems and Quantum Chaos*. The aim was to bring together researchers in the two communities to focus on the questions raised above, and on the many others that link these fields. Thanks to the warm and stimulating environment at the Newton Institute, we believe that this was achieved.

Jon Keating
David Khmelnitskii
Igor Lerner

CONTENTS

Supersymmetry and Trace Formulae

Chaos and Disorder

PERIODIC ORBITS, SPECTRAL STATISTICS, AND THE RIEMANN ZEROS

J. P. Keating

School of Mathematics, University Walk,
Bristol BS8 1TW, U.K.,
and Basic Research Institute in the Mathematical Sciences,
Hewlett-Packard Laboratories Bristol, Filton Road,
Stoke Gifford, Bristol BS12 6QZ, U.K.

1. INTRODUCTION

My purpose in this article is to review the background to some recent developments in the semiclassical theory of spectral statistics. Specifically, I will concentrate on approaches based on the trace formula[1,2]; that is, on the link between quantum energy levels and classical periodic orbits. I will also review the closely related theory of the statistics of the zeros of the Riemann zeta function. My hope is to provide an introduction to the introductions of other papers in this volume on the same subjects, and with this in mind will discuss only in outline calculations to be described by them in greater detail.

The statistical properties we seek to understand concern fluctuations in the distribution of the quantum energy levels of a given system in the semiclassical limit. It has been conjectured that these fluctuations are, in this limit, universal, and depend only upon the chaotic nature and symmetries of the system's classical dynamics. For example, Berry and Tabor[3] proposed that the energy levels of classically integrable systems are generically uncorrelated in the semiclassical limit. They also suggested that for classically chaotic systems the levels might be correlated in the same way as the eigenvalues of random matrices. This was confirmed numerically for the distribution of level-spacings in a number of such systems[4,5] and put into the form of an explicit conjecture for all spectral statistics by Bohigas, Giannoni and Schmit[6], who also made detailed studies of several examples. The random matrix conjecture has led to extensive numerical investigations, the results of which may be found in the reviews by Berry[7] and Bohigas[8], to name but two.

One of the goals of the work to be reviewed here is to develop a theory that explains how universality arises in spectral statistics. This is complicated by the fact that some systems exhibit decidedly nonuniversal behaviour. For example, the cat maps are maximally chaotic, but their quantum spectra do not show any signs of being random-matrix correlated[9]. The same is also true for geodesic motion on compact surfaces of constant negative curvature associated with arithmetic groups[10]. Much more is known about this in the case of integrable systems. The harmonic oscillator is an obvious example for which the quantum levels are correlated[3]. Others include rectangular billiards in which the square of the aspect ratio is rational[3,11]. More surprising is the fact[12] that a large class of rectangular billiards for which this number is irrational have also been shown not to have a Poissonian limit, although another large class does. Any theory which hopes to explain universality must also account for these exceptional cases as well.

Another goal is to describe the non-universal deviations from the Poisson and random-matrix forms that occur before the semiclassical limit is reached (and which the conjecture implies disappear in the limit). Put another way, this would be a description of the asymptotic way in which the conjectured limits are approached.

Several theories have been developed to answer these questions. Of these, two are explicitly semiclassical, one being based on the trace formula and the other on field theory[13]. The trace formula was first used by Berry and Tabor[3] to show how the Poisson limit emerges for generic integrable systems. Hannay and Ozorio de Almeida[14] and Berry[15] then extended the approach to recover two-point random matrix correlations for chaotic systems. A key element of their work was the realization that in ergodic systems certain periodic orbit contributions (the *diagonal* terms - see Section 5) can be evaluated using a sum rule (now known as the *Hannay-Ozorio de Almeida sum rule*). Furthermore, Berry[15] also showed how the diagonal terms associated with short orbits can be used to describe some features of the nonuniversal approach to the random-matrix limit as $\hbar \to 0$. These methods were subsequently extended to include, for example, parametric correlations[16,17] and matrix element distributions[18,19]. They also generalize in a trivial way to quantum maps.

Going beyond the diagonal approximation means evaluating the *off-diagonal* terms (Section 6). To do this directly would require more knowledge about correlations between different periodic orbits than we possess at present. However, under certain assumptions, one can compute the off-diagonal contribution indirectly, by relating it to the diagonal terms[20]. This connection is very similar, but not identical to the one which exists between the perturbative and non-perturbative contributions to spectral correlation functions in disordered systems[21,22]. One of the aims of this article to discuss the similarities and differences.

Another aim is to review the links[23,24,25] between the theory of spectral statistics and the statistical distribution of the zeros of the Riemann zeta function (Section 7). The reason for doing this here is that, first, Montgomery[26] has conjectured that the Riemann zeros are correlated like the eigenvalues of matrices in the Gaussian Unitary Ensemble (GUE) of random matrices, and second, there exists a formula relating the zeros to the prime numbers that is the exact analogue of the trace formula. The problem of understanding the statistics of energy levels using periodic orbits is thus identical to that of understanding the statistics of the Riemann zeros in terms of the primes. For the zeros, the analogues of the diagonal periodic orbit terms can again be evaluated explicitly. Moreover, this is a case in which the off-diagonal contributions can be calculated directly as well[24,27,28,29], because we do, by virtue of certain celebrated conjectures due to Hardy and Littlewood[30], have a good understanding of the correlations that exist between the primes. The off-diagonal terms can of course also be calculated indirectly by relating them to the diagonal terms, as in the general semiclassical case, and the fact that these two independent approaches give the same answer represents an important test of the correctness of the indirect method.

2. PERIODIC ORBIT FORMULAE

The foundations of the theory of spectral statistics to be reviewed here rest on Gutzwiller's trace formula[1,2], which provides a semiclassical relationship between the density of states

$$d(E) = \sum_n \delta(E - E_n) \tag{1}$$

of a given quantum system and the periodic orbits of the underlying classical dynamics:

$$d(E) \sim \overline{d}(E) + \frac{1}{\pi\hbar}\mathrm{Re}\sum_{p}\sum_{n=1}^{\infty}\frac{T_p}{\left|\det\left(M_p^n - I\right)\right|^{\frac{1}{2}}}\exp\left(\frac{i}{\hbar}nS_p\right),$$

(2)

where $\overline{d}$ is the mean density; p labels primitive orbits and n their repetitions; T_p is the period of the pth orbit, S_p its action (defined here to include the Maslov index), and M_p the monodromy matrix that describes the flow linearized in its vicinity. Alternatively, this quantum-classical connection can be expressed as a formula for the spectral determinant

$$\Delta(E) = \det\left(E - \hat{H}\right),$$

(3)

which, when regularized to ensure convergence, has zeros at the energy levels (eigenvalues of the quantum Hamiltonian $\hat{H}$) E_n; for example, in systems with two degrees of freedom

$$\Delta(E) \sim B(E)\exp\left(-i\pi\overline{N}(E)\right)\prod_{p}\prod_{m=0}^{\infty}\left(1 - \frac{\exp\left(iS_p/\hbar\right)}{\left|\Lambda_p\right|^{\frac{1}{2}}\Lambda_p^m}\right)$$

(4)

where B is a function with no real zeros that is connected with the regularization, $\overline{N}$ is the mean of the eigenvalue counting function

$$N(E) = \int_0^E d(E')dE',$$

(5)

and Λ_p is the larger eigenvalue of M_p ($|\Lambda_p|>1$ in strongly chaotic systems - i.e., when all periodic orbits are isolated and unstable). Both equations, (2) and (4), have been written in a form appropriate for flows, but generalize immediately to quantum maps.

Loosely speaking, for most systems (2) and (4) are semiclassical approximations; that is, they represent the leading order asymptotics for d and Δ as $\hbar \to 0$. More precisely, this is the case for Δ, which is a smooth function, in the usual sense of the term. For the density, it is true in the sense that the positions of the singularities on the right-hand side of (2) are semiclassical approximations to the exact energy levels. This follows because these singularities coincide with the zeros of the function on the right-hand side of (4).

There are also systems for which both equations are exact equalities. One class of examples is provided by geodesic motion on compact surfaces of constant negative curvature. For these, (2) is known as the Selberg trace formula and the double product on the right-hand side of (4) is called the Selberg zeta function[31]. The quantum cat maps provide another class of examples[9]. In both of these cases, the classical dynamics is maximally chaotic, in the sense that the systems are Anosov.

When (2) and (4) are not exact, it has been found numerically that the approximation that they represent is good, in the sense that the positions of the singularities/zeros on the classical (right-hand) side are accurate approximations to the quantum energy levels on the scale of the mean level separation[32,33,34,35]. Specifically, it has been found in a wide class of examples, including integrable, chaotic, and mixed maps and flows in both two and three dimensions, that the root means square error in the difference between the positions of the zeros on the left and right-hand sides of (4) is a small fraction (typically less than 10%) of the mean level separation. Of course, one can in principle go further and achieve any desired accuracy by expanding to the appropriate order in $\hbar$ around the periodic orbits in the asymptotic evaluation of the path integral for the Green function that underlies the derivation of the trace formula.

The conclusion is that periodic orbit theory is able to reproduce a discrete spectrum which, on scales of the order of the mean level separation, represents a good approximation to the true quantum energy levels in systems (both maps and flows) whose classical dynamics is integrable, chaotic, or mixed; and that in principle the method can be extended to any desired order of accuracy.

3. SPECTRAL STATISTICS

In most physical problems one is less concerned with computing individual energy levels than with characterising their statistical distribution; that is, one wants to probe correlations in the spectrum of the quantum Hamiltonian in question. Studies of this problem have led to a number of interesting conjectures[3,6], which may be summarized as follows.

a. The statistical correlations in the spectrum of a single, typical, classically integrable system, measured on the scale of a fixed number of mean level spacings, are, in the semiclassical limit $\hbar \to 0$, Poissonian; that is, there are no correlations at all.

b. In typical classically chaotic systems, the corresponding correlations are, in the semiclassical limit, the same as those of the eigenvalues of random matrices. Specifically, the spectral correlations of time-reversal-symmetric systems correspond to those of random real symmetric matrices (the ensemble of which is denoted GOE), and the spectral correlations of non-time-reversal-symmetric systems correspond to those of random complex hermitian matrices (the ensemble of which is denoted GUE).

Measuring energy level correlations in a given system necessarily involves averaging some spectral function over a range of energy (or other suitable classical parameter). These conjectures then imply that for local statistics, the energy average is equivalent to an average over either the Poissonian ensemble, the GOE, or the GUE. Furthermore, since these ensemble averages know nothing about the details of the system in question, the conjectures imply that spectral statistics exhibit *universality*. It is, however, crucial to emphasize the there are exceptions (note the appearance of the word 'typical'). As already mentioned in the introduction, there are families of integrable systems whose spectral statistics are not Poissonian, and strongly chaotic systems whose spectral statistics are not random-matrix. Indeed, our current level of understanding is such that we cannot say what 'typical' (or equivalently 'generic') really means in this context, and so cannot make the conjectures precise in the mathematical sense (although in terms of physics they still have non-trivial content). These systems, whilst singular in some respects, serve to provide an important reminder that any proof of a conjecture like those discussed above must be based on more than just classical integrability or a measure of classical chaos, because these properties alone do not distinguish the exceptional cases.

The question then remains as to how to approach a deeper understanding of these issues. One can try to develop a theory based on averaging over a small family of quantum systems (i.e. much smaller than the spaces of random matrices), and such an idea will be described elsewhere in this volume[36]. Otherwise, one can stick to working with a single system (which is the essential element of the conjecture) and use the trace formula to relate the spectral statistics to the classical periodic orbits. More precisely, one may hope that the trace formula will provide a useful link between energy-level correlations and the statistical properties of classical orbits. The basis of this hope is the idea that if, as seems to be the case, the trace formula can be used to generate approximations to the energy levels in a given system that are accurate on the scale of the mean level separation, then it should also be able to describe the correlations between the levels on this scale as well. Put another way, if the approximate spectrum generated by the trace formula exhibits the same correlation statistics as the corresponding exact spectrum, then periodic orbits should be capable of describing these statistics.

This immediately suggests two questions: how can universality emerge from a trace formula that depends on periodic orbits which differ from system to system? And how does the individuality of these orbits influence spectral statistics? The answers lie in the different energy scales involved.

4. ENERGY SCALES

As a function of energy E, each periodic orbit contribution to the trace formula is locally periodic with period h/T_j, where T_j is the orbit period. This is to be compared with the mean level separation $\bar{d}^{-1}$, which Weyl's law implies is of the order of h^f in a system with f classical degrees of freedom. Thus the periodic orbits that contribute to correlations on the scale of $\bar{d}^{-1}$ have period $T \sim T_H = hd$, where the Heisenberg time T_H is of the order of h^{1-f}. Clearly as the semiclassical limit is approached the periods of the orbits that determine spectral correlations on universal scales tend to infinity when $f>1$. Conversely, a given periodic orbit contributes to spectral correlations on the energy scale T_H/T_j in units of the mean level spacing.

As an example, consider the two-point correlation function

$$R_2(x) = \frac{1}{\bar{d}^2}\left\langle d(E)d\left(E + x/\bar{d}\right)\right\rangle_E,$$

(6)

and its Fourier transform, the form factor

$$K(\tau) = \int_{-\infty}^{\infty}(R_2(x) - 1)\exp(2\pi i x \tau)dx.$$

(7)

The limit (x,τ)-fixed, $\hbar \to 0$ is dominated by semiclassically long orbits, while the limits $x \to \infty$ and $\tau \to 0$ with $\hbar$-fixed are dominated by short orbits.

We can now invoke the following information about periodic orbits. In ergodic systems, long periodic orbits are asymptotically, in the limit $T_j \to \infty$, uniformly dense on the energy shell (the surface of constant energy in phase space) when weighted by their stabilities; that is, the periodic orbits asymptotically approximate the invariant density. Mathematically, this is equivalent to the Hannay-Ozorio de Almeida sum rule[14]:

$$\frac{1}{\Delta T}\sum_{T \leq T_j \leq T + \Delta T}\frac{f(T_j)}{|M_j - I|} \sim \frac{f(T)}{T}$$

(8)

as $T \to \infty$ and $\Delta T \to 0$.

The basic idea underlying uniformity is that the periodic orbits in chaotic systems are dense and so can be used to describe phase space structures. It follows from Weyl's law that $\bar{d}$ is proportional to the volume of the energy shell, and so measuring spectral correlation lengths in units of the mean level spacing, $\bar{d}^{-1}$, corresponds to normalising this volume. Thus the key point is *that the long periodic orbits of all ergodic systems look the same in these units*. One can view this as saying that universality in quantum level statistics is related to the universality of long periodic orbits in ergodic systems. A further implication is that the way in which universality is approached in spectral statistics as $\hbar \to 0$ is related to the rate of approach to ergodic uniformity in the underlying classical dynamics in the long-time limit, information about which is encoded in the short periodic orbits.

5. THE DIAGONAL APPROXIMATION

Substituting the trace formula (2) into (6) gives a semiclassical approximation to the two-point correlation function:

$$
R_2(x) \approx 1 + \left\langle \frac{2}{T_H^2} \sum_p \sum_q \sum_{m=1}^{\infty} \sum_{n=1}^{\infty} \frac{T_p T_q}{\left| M_p^m - I \right|^{\frac{1}{2}} \left| M_q^n - I \right|^{\frac{1}{2}}} \right.
$$
$$
\left. \times \cos\left(\frac{mS_p - nS_q}{\hbar} + \frac{2\pi m T_p}{T_H} x \right) \right\rangle_E .
$$

$$(9)$$

This can be split up into diagonal contributions $R_2^{(d)}$, for which $mS_p = nS_q$, and off-diagonal contributions $R_2^{(off)}$, for which $mS_p \neq nS_q$, so that

$$
R_2(x) \approx 1 + R_2^{(d)}(x) + R_2^{(off)}(x).
\tag{10}
$$

Such a division is natural for two reasons. First, as functions of E, the diagonal terms are least oscillatory and so survive best the energy average in (9). Thus in appropriate regimes they may be expected to dominate the off-diagonal terms. Second, if the actions of different orbits (i.e. orbits not related by symmetry) are uncorrelated, then the off-diagonal contribution would vanish, being the average of a sum of terms with random phases. This observation will be of importance in the calculation of the off-diagonal contribution to be outlined in Section 6.

The diagonal contribution may be written in the form

$$
R_2^{(d)}(x) = \frac{2}{T_H^2} \sum_p \sum_{m=1}^{\infty} g_p \frac{T_p^2}{\left| M_p^m - I \right|} \cos\left(\frac{2\pi m T_p}{T_H} x \right),
\tag{11}
$$

where g_p is the action-multiplicity of the pth primitive orbit. The key question is how these multiplicities behave. When $x/T_H \to 0$ (as is the case in the limit $\hbar \to 0$ with x-fixed) the sum (11) is dominated by increasingly long orbits. For these, the multiplicity can, in typical systems, be replaced by its mean $\bar{g}$, which takes the values $\bar{g} = 1$ in non-time-reversal-symmetric systems and $\bar{g} = 2$ in the time-reversal-symmetric case, the difference being due to the symmetry between non-self-retracing orbits and their time-reversed twins. Then (11) can be evaluated using the Hannay-Ozorio de Almeida sum rule (8) to give

$$
R_2^{(d)}(x) \approx -\frac{\bar{g}}{2\pi^2 x^2},
\tag{12}
$$

which coincides precisely with the leading-order $x \to \infty$ asymptotics of the non-oscillatory (in x) contributions to the GUE (when $\bar{g} = 1$) and GOE (when $\bar{g} = 2$) results for the two-point correlation function of the eigenvalues of random matrices. This confirms the argument of the previous section that universality in spectral statistics, in this case represented by (12), is related to ergodic universality in the long periodic orbits of chaotic systems, here expressed through the Hannay-Ozorio de Almeida sum rule (8). Furthermore, it may be seen from (11) that when $x/T_H \to \infty$ the semiclassical approximation is governed by the short-time classical dynamics, and hence can be expressed in terms of the short periodic orbits, or equivalently, the decay of classical correlations.

It is also clear from the derivation of (12) that spectral universality is related to the behaviour of the multiplicities g_p for long orbits. Indeed, this is a centrally important issue, because in the strongly chaotic systems known to be exceptional in that their level statistics are not random-matrix, the mean multiplicity is not a constant but grows rapidly as a function of the period. Specifically, for geodesic motion on arithmetic surfaces of constant negative curvature and for the cat maps, both of which are fully ergodic, $\bar{g}$ grows with T as the square-root of the total number of periodic orbits of period T. Thus ergodicity alone is not enough to imply (12).

The distribution of orbit multiplities affects the non-universal regime too. As already argued, this regime is governed by the short periodic orbits for which, in time-reversal-symmetric systems, the multiplicties are typically erratic. Replacing g_p by its mean is then a highly questionable approximation.

The semiclassical structure of the two-point correlation function can be expressed succinctly by the following identity:

$$R_2^{(d)}(x) = \frac{2}{T_H^2}\,\mathrm{Re}\left[\frac{d^2}{ds^2}\ln Z(s) - a(s)\right]_{s=\frac{2\pi i x}{T_H}}, \tag{13}$$

where

$$\frac{1}{Z(s)} = \prod_p \prod_{m=0}^{\infty}\left(1 - \frac{\exp\left(sT_p\right)}{\left|\Lambda_p\right|\Lambda_p^m}\right)^{(m+1)g_p} \tag{14}$$

and

$$a(s) = \sum_p \sum_{m=0}^{\infty} g_p\,\frac{(m+1)T_p^2}{\left(\left|\Lambda_p\right|\Lambda_p^m \exp\left(-sT_p\right)-1\right)^2}. \tag{15}$$

If $g_p = 1$, Z corresponds to the Ruelle-type zeta-function associated with the spectral determinant of the Fobenius-Perron operator that generates the time evolution of phase-space densities in classical mechanics[37]. Ergodicity, or equivalently the Hannay-Ozorio de Almeida sum rule, implies that $Z(s)$ has a simple pole at $s=0$, and so (12) can be viewed as a direct consequence of this singularity. The non-universal asymptotic approach to the random-matrix limit is then related to the analytical structure of Z in the rest of the complex plane; for example, in the positions of the nearby singularities of $\ln Z$. In dynamical systems terms, this structure is precisely that which determines the decay of classical correlations, and hence the approach to ergodicity.

When $g_p \neq$ constant, Z is not exactly identifiable as a classical zeta function. In cases when $\bar{g} = 1$ it still has a simple pole at $s=0$, and so the universal limiting result (12) is unchanged. It is, however, not clear to what extent the analytical structure away from $s=0$ is affected by fluctuations in the mutiplicities. When $\bar{g} = 2$ the structure far away will almost certainly be different from the classical situation. In the exceptional cases, when the mean multiplicity increases with period, Z bears no obvious relation to a classical zeta function, and even the structure around $s=0$ may be changed, resulting in non-generic quantum spectral statistics.

The periodic orbit sum in (15) converges when $s=0$ and so

$$\frac{1}{T_H^2} a\left(\frac{2\pi i x}{T_H}\right) \to 0 \tag{16}$$

as $T_H \to \infty$. Hence a contributes to the non-universal approach to the universal limiting regime in R_2, but not to the limit itself. It thus plays the same role in (13) as the analytical structure of Z away from $s=0$.

The relationship between quantum spectral statistics and the Frobenius-Perron operator was first proposed by analogy with perturbative expressions for the spectral statistics in disordered systems[22], and a programme has been initiated to put it on a firmer footing using a nonlinear sigma-model for chaotic systems[38]. However, there are subtle differences from the results outlined above that would appear to warrant further investigation: the multiplicities g_p do not seem to play the same key role (the trace formula only leads to a classical zeta-function under the assumptions about g_p already stated), and the function a is absent from the formulae corresponding to (13). It is worth repeating that the multiplicities are essential to understanding the exceptional systems that are strongly chaotic (i.e. for which the classical zeta function has a simple pole at $s=0$, isolated by a gap from other singularities), but for which the spectral correlations are not random-matrix.

The picture for $K(\tau)$ is the same as the one painted above for R_2. The semiclassical formula for K, the fourier transform of (9), can be split into diagonal and off-diagonal contributions, where the diagonal contribution is given by

$$K^{(d)}(\tau) = \frac{1}{T_H^2} \sum_p \sum_{m=1}^{\infty} \frac{g_p T_p^2}{\left|M_p^m - I\right|} \delta\left(\tau - \frac{mT_p}{T_H}\right). \tag{17}$$

Replacing g_p by its mean $\bar{g}$, assuming this to be a constant, and evaluating the periodic orbit sum using the Hannay-Ozorio de Almeida sum rule gives

$$K^{(d)}(\tau) \approx \bar{g}\tau \tag{18}$$

in the limit $\tau T_H \to \infty$. When $\bar{g} = 1$ this coincides with the first term in the Taylor expansion of the GUE form-factor around $\tau = 0$, and when $\bar{g} = 2$ it coincides with the corresponding GOE result. This leads to the important conclusion that the off-diagonal terms do not contribute around $\tau = 0$ and hence that action correlations are negligible for $T \ll T_H$. In fact, when $\bar{g} = 1$ (18) coincides with the GUE form factor for $\tau \leq 1$ and so this conclusion holds for $T \leq T_H$.

In the regime $\tau T_H \to 0$, the spectral statistics are governed by the short-time classical dynamics and so are non-universal. The form factor can then either be expressed in terms of the individual short periodic orbits, via (17), or by the fourier transform of (13).

Finally, it is worth remarking that the analysis reviewed above relies only upon the existence of a trace formula and the Hannay-Ozorio de Almeida sum rule (8), which itself follows from classical ergodicity. The same approach thus also applies straightforwardly to maps and to integrable systems, where ergodicity on phase-space tori implies a sum rule that corresponds directly to (8). In the latter case, the results coincide exactly with the Poissonian expectation in the universal regime.

6. OFF-DIAGONAL CONTRIBUTIONS

The fact that $R_2(x) \neq R_2^{(d)}(x)$ for any of the random matrix ensembles suggests that if the conjectures reviewed in Section 3 are correct, and if the semiclassical approximation (9) is assumed accurate, then $R_2^{(off)}(x) \neq 0$. But as already discussed, if the periodic orbit contributions to (9) are uncorrelated, this would imply $R_2^{(off)}(x) = 0$. Hence there must be correlations. Unfortunately, we have no a priori knowledge of their origin; that is, there is at present no theory for them based purely on classical dynamics. One can, of course, work backwards by assuming that $R_2(x)$ is given precisely by the appropriate random matrix expressions, setting $R_2^{(off)}(x) = R_2(x) - R_2^{(d)}(x)$, with $R_2^{(d)}(x)$ given by (12), and then fourier-transforming with respect to $\hbar^{-1}$ to obtain a formula for a classical periodic orbit correlation function[39]. Numerical computations support the correctness of the result, but a derivation within classical mechanics is still lacking.

Since a direct evaluation of the sum over off-diagonal orbit pairs in (9) is, for generic systems, beyond our current horizon, we are forced to seek an indirect method of calculation. The basis of such an approach is suggested by the following observations. First, it follows from the general arguments presented in Section 4 that to resolve the quantum spectrum down to the scale of the mean level spacing requires orbits with periods up to the order of the Heisenberg time T_H. Orbits with periods longer than T_H determine spectral structure on scales shorter than the mean spacing, and so should not in principle be part of a theory of long-range statistics. Second, it was argued in Section 5 that, to a first approximation, orbits with periods less than T_H contribute to the semiclassical formula for $R_2(x)$ as if they are uncorrelated (the implication is that action correlations are needed to calculate the contributions from obits with periods larger than T_H). We conclude that a self-consistent semiclassical theory for spectral statistics should be based on orbits with periods up to the order of T_H, and should treat these as if they were uncorrelated.

Such a theory can be constructed as follows[20]. First, we use the fact that the zeros of the function on the right-hand side of (4) are semiclassical approximations to the exact energy levels. The restriction to orbits with periods T_p up to the order of T_H can then be made by truncating the p-product in (4) appropriately. The above arguments imply that the zeros of the truncated product remain good approximations to the energy levels on the scale of the mean level separation. It is a key assumption that this is the case.

The idea is then, essentially, to compute the correlations in the semiclassical spectrum obtained from the truncated product. However, this is complicated by the fact that unlike the exact energy levels, the approximations thus produced are not automatically real. Nevertheless, we can generate real approximations by using the real zeros of the real part of the right-hand side of (4) when truncated. This is semiclassically consistent, because hermiticity implies that the exact quantum spectral determinant is real when E is real and so the semiclassical approximation must be real to leading order in $\hbar$. Taking the real part thus corresponds to rearranging the higher orders. Basically, it can be thought of as imposing the functional equation, somewhat as in the derivation of the Riemann-Siegel lookalike formula[40,41].

The function B in (4) is itself real when E is real, and so the approximations e_n to the energy levels are the real zeros of

$$w(E) = \mathrm{Re}\left[\exp\left(-i\pi\overline{N}(E)\right)f(E)\right], \tag{19}$$

with

$$f(E) = \prod_{p}^{(T_H)} \prod_{m=0}^{\infty} \left(1 - \frac{\exp\left(iS_p/\hbar\right)}{\left|\Lambda_p\right|^{\frac{1}{2}} \Lambda_p^m}\right), \tag{20}$$

where the product is truncated smoothly to include orbits whose periods are less than or of the order of T_H. This is identical to truncating the p-sum in (2) in the same way, integrating as in (5) to obtain an approximation $n(E)$ to the spectral staircase (counting function), and defining e_n to be the solution of $n(E) = n + 1/2$.

The density associated with the semiclassical approximations e_n (defined as in (1) but with E_n replaced by e_n) can be written

$$\tilde{d}(E) = \frac{d}{dE} \sum_{k=-\infty}^{\infty} \frac{(-1)^k}{2\pi i k} \exp\left(2\pi i k \overline{N}(E)\right) \left(\frac{f^*(E)}{f(E)}\right)^k. \qquad (21)$$

Substituting this into (6) then gives an expression for the two-point correlation function of the semiclassical spectrum. The contribution from the $k=0$ term in (21) can be evaluated by again making the diagonal approximation[20]. The result has the same form as (13), but with the p-product in (14) and the p-sum in (15) truncated to include only orbits with T_p up to the order of T_H. When $x>>1$ the truncation can be ignored to leading order, and so the $k=0$ contribution reduces to the diagonal contribution of the previous section. This then gives (12) in the semiclassical limit, which, as already noted, coincides with the leading-order $x>>1$ asymptotics of the non-oscillatory terms in the corresponding RMT results.

Since the $k=0$ term in (21) corresponds to the diagonal contribution, the $k \neq 0$ terms must give the off-diagonal contribution. These may be evaluated by substituting (20) for f and then by interchanging the order of the energy average in (6) with the p-product in the ratio of the f-factors. It is this step that corresponds to assuming that the actions of different primitive orbits (of period less than or of the order of T_H) contribute as if they are uncorrelated.

It is crucial to note at this stage that it is primitive orbits whose actions are being treated as uncorrelated. The repetitions of a given primitive orbit are, of course, highly correlated - their actions are integer multiples of the primitive action - and it is a mistake to treat them otherwise.

Finally, the average over the m-products in the ratio of the f-factors can be evaluated exactly[20]. The result for the off-diagonal contribution to the two-point correlation function is that when $x>>1$ and, to take just one case, $g_p = 1$,

$$R_2^{(off)}(x) \approx \frac{2}{T_H^2} \left| \gamma^{-1} Z\left(\frac{2\pi i x}{T_H}\right) \right|^2 \mathrm{Re}\left[\exp(2\pi i x) \prod_p \chi_p\left(\frac{2\pi i x}{T_H}\right) \right], \qquad (22)$$

with

$$\chi_p(x) = {}_2\phi_1\left(\exp(-T_p x), \exp(-T_p x); \Lambda_p^{-1}; \Lambda_p^{-1}, |\Lambda_p^{-1}| \exp(T_p x)\right) \frac{|Z_p(0)|^2}{|Z_p(x)|^2}, \qquad (23)$$

where γ is the residue of the pole of $Z(s)$ at $s=0$, ${}_2\phi_1$ is the q-hypergeometric series, and Z_p is the p-th element of the product over primitive orbits in (14). Formally, the p-products in this expression should be truncated so as to include only those orbits with periods T_p up to the order of T_H, but when $x>>1$ there is no difference to leading order.

When $T_H \to \infty$, $\chi_p \to 1$ and (22) is dominated by the pole of Z at $s=0$. Hence in this limit (which corresponds to letting $\hbar \to 0$)

$$R_2^{(off)}(x) \to \frac{\cos(2\pi x)}{2\pi^2 x^2}. \qquad (24)$$

Combined with the result (12) for the diagonal terms, this then coincides precisely with the exact GUE expression when $\bar{g} = 1$. In the same way, when $\bar{g} = 2$ we recover the leading order $x >> 1$ asymptotics (rather than the exact form in this case) of the GOE two-point correlation function.

Treating all orbit actions as being uncorrelated, rather than just those of the primitive orbits, gives[20] a formula like (22) but with $\chi_p = 1$. This has the same limit as (22) when $T_H \rightarrow \infty$, but a different form for finite T_H. Remarkably, it coincides with the expression for the nonperturbative contribution to the two-point correlation function for disordered systems derived on the basis of the nonlinear σ-model[21]. It would be very interesting to understand the physical origins of the difference.

Given (22) and its analogue for time-reversal-symmetric systems, one can fourier-transform with respect to $\hbar^{-1}$ to obtain a formula for a classical periodic orbit action correlation function, as indicated at the start of this Section. The result takes the form of a sum over pairs of pseudo-orbits (linear combinations of periodic orbits) and generalizes that obtained by Argaman et al[39] in that it includes non-universal effects. It may appear paradoxical that one can obtain orbit correlations from an indirect calculation that ignores them; the point is that they are effectively included by the resummation that underlies the bootstrap formula (21). In the same way, fourier-transforming with respect to x gives an expression for the form-factor, also in terms of a sum of pseudo-orbit-pairs.

It is also worth pointing out that the calculation outlined above is based solely on the trace formula, and thus applies trivially to maps.

7. THE RIEMANN ZEROS

One way of testing the methods described in the previous two sections is to apply them to the zeros of the Riemann zeta function. This may be surprising upon first sight, because there is no proof of any link between the zeta function and a quantum system, but it is not difficult to see on the level of a mathematical 'toy-model'; that is, the analogy is mathematical, rather than physical.

The Riemann zeta-function is defined for Res>1 by a Dirichlet series

$$\zeta(s) = \sum_{n=1}^{\infty} \frac{1}{n^s},$$
(25)

or by an Euler product over the primes p,

$$\zeta(s) = \prod_{p} \left(1 - \frac{1}{p^s}\right)^{-1},$$
(26)

and then by analytic continuation to the rest of the complex plane[42,43]. It is a meromorphic function with a single simple pole at $s=1$, where it has unit residue, and ('trivial') zeros at $s=-2, -4, -6, \ldots$ The Riemann hypothesis (RH) is that all of the other ('non-trivial') zeros lie on the line Res=1/2. Put another way, the non-trivial zeros lie at points $s_n = 1/2 + iE_n$ where Im$E_n = 0$. Thus, assuming the hypothesis is correct, the set $\{E_n\}$ forms a real and discrete 'spectrum' which can be analysed statistically, in the same way as for energy levels. In fact, one can do this analysis even if the hypothesis is not correct, but for ease of presentation we will assume that it is.

A density of zeros can be defined exactly as in (1), and there is an explicit formula in terms of the primes:

$$d(E) = \overline{d}(E) - \frac{1}{\pi}\sum_{n=1}^{\infty}\frac{\Lambda(n)}{\sqrt{n}}\cos(E\log n), \tag{27}$$

where

$$\overline{d}(E) \approx \frac{1}{2\pi}\log\left(\frac{E}{2\pi}\right) \tag{28}$$

and

$$\Lambda(n) = \begin{cases} \log p & \text{if } n = p^k \\ 0 & \text{otherwise} \end{cases}. \tag{29}$$

The sum in (27) thus runs over primes and prime-powers, and is mathematically the exact analogue of the trace formula (2) (there are problems with its physical interpretation, but this does not affect its application here[25]).

The density can be substituted into (6) leading to a definition of the two-point correlation function of the Riemann zeros. Extensive numerical evidence[44], rigorous results[26,45], and heuristic calculations[20,24,28,29,46] all support the conjecture that this (and other statistics) tends to the corresponding GUE form in the limit $E \to \infty$. Mathematically, we are thus in the same position as when studying the spectral statistics of classically chaotic systems: there is a set of real numbers (assuming RH) whose limiting statistics is GUE and for which there is a trace formula. We can thus follow the analysis outlined in the previous two sections line by line.

The results are as follows, first the two-point correlation function can be expressed as

$$R_2(x) = 1 + \frac{1}{2\pi^2\overline{d}^2(E)}\sum_{m=1}^{\infty}\sum_{n=1}^{\infty}\frac{\Lambda(m)\Lambda(n)}{\sqrt{mn}}\left\langle\cos\left(E\log\left(\frac{m}{n}\right)+\frac{x}{\overline{d}(E)}\log m\right)\right\rangle_E, \tag{30}$$

which is the direct analogue of (9). This double sum can then be split up into diagonal ($m=n$) and off-diagonal ($m \neq n$) contributions. The diagonal part can be evaluated exactly and takes the form[20]

$$R_2^{(d)}(x) = \frac{1}{2\pi^2\overline{d}^2(E)}\operatorname{Re}\left[\frac{d^2}{ds^2}\log\zeta(s) - \sum_{p}\frac{\log^2 p}{(p^s-1)^2}\right]_{s=1+ix\,/\,\overline{d}(E)} \tag{31}$$

which corresponds to (13). In the limit $E \to \infty$ (and hence from (28) $\overline{d} \to \infty$), (31) is dominated by the pole of the zeta function at $s=1$, and so

$$R_2^{(d)}(x) \to -\frac{1}{2\pi^2 x^2}. \tag{32}$$

The off-diagonal contribution can be calculated using the methods outlined in Section 6: one can define approximations to the zeros by truncating the product (26) near to the 'Heisenberg prime' when $\log p = 2\pi d(E)$, and then calculate the two-point correlation function of these by assuming that the logs of the primes included are statistically uncorrelated. This leads to the result[20]

$$R_2^{(off)}(x) = \frac{1}{2\pi^2 \bar{d}^2(E)} \left| \zeta\left(1 + \frac{ix}{\bar{d}(E)}\right) \right|^2 \text{Re}\left[\exp(2\pi i x)\prod_p \left(1 - \frac{\left(p^{ix/\bar{d}(E)} - 1\right)^2}{(p-1)^2} \right) \right] \qquad (33)$$

when x>>1, which is the direct analogue of (22). In the limit $E \to \infty$ we again recover the limit (24) due to the pole of the zeta function. Combined with (32) this coincides precisely with the exact GUE expression for R_2.

The importance of the Riemann zeta-function is that it is the one example where we can evaluate the off-diagonal contribution directly from (30) without any reference to the indirect methods reviewed in Section 6. This is because we possess enough information about the correlations between the phases in the off-diagonal terms. The main steps in this calculation are as follows. First, the off-diagonal contribution is clearly given by

$$R_2^{(off)}(x) = \frac{1}{2\pi^2 \bar{d}^2(E)} \sum_{m \neq n} \frac{\Lambda(m)\Lambda(n)}{\sqrt{mn}} \left\langle \cos\left(E\log\left(\frac{m}{n}\right) + \frac{x}{\bar{d}(E)}\log m \right) \right\rangle_E . \qquad (34)$$

Because of the E-average, only terms with $m \approx n$ contribute. Hence writing $m=n+k$ and expanding the term multiplied by E (the large parameter) to first order in k,

$$R_2^{(off)}(x) = \frac{1}{2\pi^2 \bar{d}^2(E)} \sum_n \sum_{k \neq 0} \frac{\Lambda(n)\Lambda(n+k)}{n} \left\langle \cos\left(\frac{Ek}{n} + \frac{x}{\bar{d}(E)}\log m \right) \right\rangle_E . \qquad (35)$$

The statistical information corresponding to the periodic orbit action correlations discussed at the beginning of Section 6 is now provided by the Hardy-Littlewood conjecture[30], namely that in the limit $N \to \infty$

$$\frac{1}{N} \sum_{n<N} \Lambda(n)\Lambda(n+k) \to \alpha(k), \qquad (36)$$

where

$$\alpha(k) = \sum_{\substack{(r,q)=1 \\ r<q}} \exp\left(-2\pi i \frac{r}{q}k\right)\left(\frac{\mu(q)}{\phi(q)}\right)^2 , \qquad (37)$$

in which the sum (known as a singular series) runs over all rationals, $\mu(q)$ is the Möbius function, and $\phi(q)$ is Euler's totient function. Substituting this into (35), all sums can be performed analytically[20] to give (33) without the assumption that x>>1. This then provides an important check on the correctness of the steps involved in the indirect method outlined in Section 6.

Once again, one can reverse the direction of the calculation and, taking (33) as given by the indirect approach, fourier-transform with respect to E to derive (37); that is, one can recover the Hardy-Littlewood prime correlations from a result derived by ignoring them. This seemingly self-contradictory approach is the direct analogue of the calculation of the periodic orbit action correlation function mentioned at the end of Section 6. Moreover, fourier-transforming (35) with respect to x leads to an expression for the form-factor of the Riemann-zeros in terms of the singular series. This plays the role of the sum over pairs of pseudo-orbits in the corresponding semiclassical result.

8. CONCLUSIONS

It is, perhaps, too soon to try to draw conclusions about the recent developments reviewed here. It is clear that the trace formula provides a method to calculate quantum statistics that is simple and physically transparent. It is also general, in that it applies equally well to a range of related problems. For example, the methods outlined transfer directly[20] to quantum maps, to matrix elements and wavefunctions (for which there is also a trace formula), to parametric statistics, to integrable systems, and to higher-order correlation functions. Moreover, they also apply to the zeros of number-theoretical zeta functions, such as the Riemann zeta-function, where they can be checked against alternative methods of calculation and against numerical computations. Preliminary results support their correctness.

The weaknesses of the semiclassical method described here are that it is hard to see how it can be made rigorous (e.g. how to estimate the errors in the various approximations), and that at this stage it is not known how to go beyond leading-order in x in the calculation of correlation functions. An advantage is that it applies to systems that are non-generic with respect to their spectral statistics. Another advantage is that it applies to mixed systems; for example, it describes the rich critical behaviour associated with orbit bifurcations[47].

At this stage, a pressing problem seems to be to understand the relationship between the results obtained using the trace formula and those derived using supersymmetric methods. Are the differences in the non-universal regime a clue as to the relationship between the two methods? Do they suggest that the supersymmetric approach is best geared to describing the universal regime and the trace formula the non-universal behaviour? Do they point to the effects of the various uncontrolled approximations? Or do they mean that supersymmetric techniques are not yet applicable to individual systems? It is expected that these, and other related questions, will be at the centre of the next round of developments in this rapidly evolving area.

ACKNOWLEDGEMENTS

All of the recent results reviewed here were obtained in collaboration with Eugene Bogomolny. I have enjoyed too many conversations on this subject to list all of those who have influenced the views expressed. I am grateful to the Isaac Newton Institute for making many of them possible.

REFERENCES

1. Gutzwiller, M.C. *J. Math. Phys.* **12**, 343-358 (1971).
2. Gutzwiller, M.C. *Chaos in classical and quantum mechanics* (Springer, New York, 1990).
3. Berry, M.V. & Tabor, M. *Proc. R. Soc. Lond.* A **356**, 375-394 (1977).
4. MacDonald, S.W. & Kaufman, A.N. *Phys. Rev. Lett.* **42**, 1189-1191 (1979).
5. Berry, M.V. *Ann. Phys.* **131**, 163-216 (1981).
6. Bohigas, O., Giannoni, M.J. & Schmit, C. *Phys. Rev. Lett.* **52**, 1-4 (1984).
7. Berry, M.V. *Proc. R. Soc. Lond.* A**413**, 183-198 (1987).
8. Bohigas, O. *Les Houches Lecture Series* vol. 52 eds. Giannoni, M.J., Voros, A., & Zinn-Justin, J. (Amsterdam: North-Holland), 89-199 (1991).
9. Keating, J.P. *Nonlinearity* **4**, 309-341 (1991).
10. Bogomolny, E.B., Georgeot, B., Giannoni, M.J. & Schmit, C. *Physics Reports* **291**, 220-324 (1997).
11. Connors, R.D. & Keating, J.P. *J. Phys. A* **30**, 1817-1830 (1997).
12. Sarnak, P. *Curr. Dev. Math.* 84-115 (1997).
13. Simons, B.D. in this volume.
14. Hannay, J.H. & Ozorio de Almeida, A.M. *J. Phys. A* **17**, 3429-3440 (1984).
15. Berry, M.V. *Proc. R. Soc. Lond.* A**400**, 229-251 (1985).

16. Goldberg, J., Smilansky, U., Berry, M.V., Schweizer, W., Wunner, G., Zeller, G. *Nonlinearity* **4** 1-14 (1991).
17. Berry, M.V. & Keating, J.P. *J. Phys. A* **27**, 6091-6106 (1994).
18. Wilkinson, M. *J. Phys. A* **21**, 1173-1190 (1988).
19. Eckhardt, B., Fishman, S., Keating, J., Agam, O., Main, J. & Müller, K. *Phys. Rev. E* **52**, 5893-5903 (1995).
20. Bogomolny, E.B. & Keating, J.P. *Phys. Rev. Lett.* **77**, 1472-1475 (1996).
21. Andreev, A.V. & Altshuler, B.L. *Phys. Rev. Lett.* **75**, 902-905 (1995).
22. Agam, O., Altshuler, B.L. & Andreev, A.V. *Phys. Rev. Lett.* **75**, 4389-4392 (1995).
23. Berry, M.V. in *Quantum chaos and statistical nuclear physics* eds. Seligman, T.H. & Nishioka, H., 1-17 (1986).
24. Keating, J.P. in *Quantum Chaos* eds. Casati, G., Guarneri, I. & Smilansky, U., 145-185 (North-Holland, Amsterdam) (1993).
25. Berry, M.V. & Keating, J.P. in this volume,
26. Montgomery, H.L. *Proc. Symp. Pure Math.* **24**, 181-193 (1973).
27. Keating, J.P. in *Quantum Chaos* eds. Cerdeira, H.A., Ramaswamy, R., Gutzwiller, M.C. & Casati, G., 280-290 (World Scientific, Singapore) (1993).
28. Bogomolny, E.B. & Keating, J.P. *Nonlinearity* **8**, 1115-1131 (1995).
29. Bogomolny, E.B. & Keating, J.P. *Nonlinearity* **9**, 911-935 (1996).
30. Hardy, G.H., & Littlewood, J.E. *Acta Math.* **44**, 1-70 (1923).
31. Balazs, N.L. & Voros, A. *Physics Reports* **143**, 109-240 (1986).
32. Sieber, M. & Steiner, F. *Phys. Rev. Lett.* **67**, 1941-1944 (1991).
33. Tanner, G., Scherer, P., Bogomolny, E.B., Eckhardt, B. & Wintgen, D. *Phys. Rev. Lett.* **67**, 2410-2413 (1991).
34. Keating, J.P. & Sieber, M. *Proc. R. Soc. Lond.* A**447**, 413-437 (1994).
35. Schomerus, H. & Haake, F. *Phys. Rev. Lett.* **79**, 1022-1025 (1997).
36. Zirnbauer, M.R. in this volume.
37. Cvitanovic, P. & Eckhardt, B. *J.Phys.A* **24**, L237-L241 (1991).
38. Andreev, A.V., Agam, O., Simons, B.D. & Altshuler, B.L. *Phys. Rev. Lett.* **76**, 3947-3950 (1996).
39. Argaman, N., Dittes, F.M., Doron, E., Keating, J.P., Kitaev, Ya., Sieber, M. & Smilansky, U. *Phys. Rev. Lett.* **71**, 4326-4329 (1993).
40. Keating, J.P. *Proc. R. Soc. Lond.* A**436**, 99-108 (1992).
41. Berry, M.V. & Keating, J.P. *Proc. R. Soc. Lond.* A**437**, 151-173 (1992).
42. Edwards, H.M. *Riemann's Zeta Function* (Academic Press, New York and London, 1974).
43. Titchmarsh, E.C. *The theory of the Riemann zeta-function* (Clarendon Press, Oxford, 1986).
44. Odlyzko, A.M. *Math. of Comp.* **48**, 273-308 (1987).
45. Rudnick, Z. & Sarnak, P. *Duke Math. J.* **81**, 269-322 (1996).
46. Berry, M.V. *Nonlinearity* **1**, 399-407 (1988).
47. Berry, M.V., Keating, J.P. & Prado, S.D. 'Orbit bifurcations and spectral statistics' *J. Phys. A* in the press (1998).

QUANTUM CHAOS: LESSONS FROM DISORDERED METALS

A. Altland, C. R. Offer and B. D. Simons

Cavendish Laboratory, Madingley Road, Cambridge, CB3 0HE, UK

INTRODUCTION

The quantum description of systems which are chaotic in their classical limit is the subject of "Quantum Chaos". A wide variety of physical systems fall into this category. Amongst those most commonly studied are the neutron resonances of complex atomic nuclei[1], Rydberg atoms in strong magnetic fields[2], and electrons in semiconducting nanostructures (known as "quantum dots"[3]). In contrast to integrable systems, eigenstates of quantum chaotic structures are characterized solely by their energy rather than by a set of quantum numbers. Therefore, a useful description of chaotic systems is a statistical one. It is the development of a framework in which the statistical properties of general quantum chaotic systems can be studied which forms the focus of these lectures.

The study of the quantum properties of systems which are chaotic in their classical limit has been conducted largely along two parallel lines. The first approach has been developed within the framework of a semi-classical approximation based on the Feynman path integral. This method, known in the literature as "Periodic orbit theory", expresses the spectral quantum statistical properties of *clean chaotic* systems in terms of infinite sums over classical periodic trajectories — the "Gutzwiller Trace Formula"[4]. (Here the attribute 'clean' means that systems without stochastic disorder are considered.) This theory has proved successful in describing the large scale structure of the spectrum, but its success in describing long-time universal behavior has been limited.

At the same time, a complementary approach based on methods of quantum field theory has been developed to study the average properties of an ensemble of *disordered* systems such as weakly disordered conductors[5]. As well as offering a description of the short-time diffusive dynamics, by surrendering information specific to each given system the field theoretic approach is capable of describing the longest time scales, where the quantum behavior displays a remarkable degree of universality.

Despite the fact that these two approaches account for the same qualitative behavior, it was only recently that a formal correspondence between them was established. The goal of these lectures will be to introduce and explore a third approach[6, 7] aimed at describing the quantum statistical properties of *individual* chaotic structures over a wide range of energy scales, encompassing both the universal and non-universal regimes.

This third approach, which relies on *energy averaging*, generates a field theory in

Supersymmetry and Trace Formulae: Chaos and Disorder
Edited by Lerner *et al.*, Kluwer Academic / Plenum Publishers, New York, 1999

which the effective action is associated with classical flow in phase space[7]. However, the basic ingredients of this approach are not individual periodic orbits but global modes of the time evolution operator of the underlying classical system. These results show that the statistical quantum properties of the system are intimately related to the *irreversible* classical chaotic dynamics or, more precisely, to the Perron-Frobenius modes, in which a classical probability density relaxes into the ergodic distribution. We will discuss how both random matrix theory, and the "diagonal approximation" of the periodic orbit theory emerge naturally as limits of the theory.

However, in arriving at the general field theoretic description we will review qualitative aspects of quantum coherence phenomena in disordered and mesoscopic structures. The lecture notes are organized as follows. In the section entitled "Spectral Statistics: A Brief History" we briefly review some of the milestones encountered in the development of a theory of quantum chaos. In the section "Coherence Effects in Disordered Conductors" we will investigate mesoscopic quantum coherence phenomena in weakly disordered metallic systems. Our emphasis will be placed on energy level correlations, weak localization phenomena, and universal mesoscopic fluctuations. This discussion will provide a useful platform on which to explore quantum coherence phenomena in non-random or ballistic chaotic structures. It will also reveal the pitfalls of a purely perturbative analysis. These qualitative ideas are put on a rigorous footing in the following section, "Supersymmetry Method". Using a field theoretic analysis, quantum statistical properties of weakly disordered conductors will be cast in the form of a functional field integral, involving an effective action which takes the form of a non-linear σ-model. Both perturbative and non-perturbative aspects of the theory, as well as the connection to random matrix theory, will be discussed. The field theoretic approach is not, however, restricted to random systems. In the section entitled "Quantum Chaos: The Ballistic σ-Model", we develop a general field theoretic approach to study quantum statistical properties of ballistic chaotic structures. Finally, in the closing section we summarize current understanding and offer an outlook on the outstanding problems.

SPECTRAL STATISTICS: A BRIEF HISTORY

Our discussion begins with a brief summary of the early progress in the field. The study of spectral correlations in chaotic quantum systems has a long history that dates back to the investigation of complex many-body quantum systems[9].

Wigner Surmise

Guided by the very complexity of the system, a statistical ansatz was introduced by Wigner to explain strong level correlations observed in resonance spectra of complex nuclei[10, 1]. Taking the matrix elements to be statistically uncorrelated Wigner obtained, within a two-level approximation, an estimate of the distribution of neighboring level spacings,

$$P(S) = \langle \delta \left(S - (E_{i+1} - E_i) \right) \rangle \sim \frac{\pi}{2} \frac{S}{\Delta} \exp \left[-\frac{\pi S^2}{4\Delta^2} \right], \tag{1}$$

where $\Delta = \langle E_{i+1} - E_i \rangle$ denotes the average level spacing. Surrendering information specific to each individual nucleus, Wigner was able to quantitatively account for the characteristic level repulsion observed in experiment. Forty years on, similar considerations are now given to the ground state properties of artificial atoms, or quantum dots[3] (with so far only limited success!).

Gor'kov-Eliashberg Theory

While successfully studied and applied to the physics of complex nuclei, it was not until almost twenty years later that similar ideas were applied to the study of the quantum properties of weakly disordered metallic grains. By identifying spectral correlations of the grains with those of random Hermitian matrix ensembles, Gor'kov and Eliashberg investigated the dielectric response of metallic grains embedded in an insulating matrix[8]. Although often neglected, this application anticipated applications of random matrix theory to the field of mesoscopic physics by two decades.

Bohigas-Giannoni-Schmidt Conjecture

Until 1981, the application of random matrix theory had been limited to the study of complex many-body or disordered quantum systems. An extensive numerical survey by Bohigas, Giannoni and Schmidt[11] lead them to the following conjecture:

- The spectral properties of quantum systems which are chaotic in their classical limit (even those with few degrees of freedom) are UNIVERSAL, independent of material properties of the system, and coincide with those of random matrix ensembles.

This conjecture presented a milestone in the study of quantum chaos. It provided a common thread which united a whole class of qualitatively different systems. So far, very few exceptions to this rule have been identified. And for those that have, it is usually simple to understand why they lie outside the present scheme. However, despite more than ten years of intense study, and the accumulation of a wealth of supporting data, this conjecture still awaits a rigorous mathematical proof.

Supersymmetry: Efetov's Non-linear σ-Model

Belonging to a statistical ensemble, average properties of disordered quantum systems are more susceptible to analytical investigation than individual systems. Providing the statistical average of an individual system (for example, over a range of energy levels) is statistically equivalent to average over an ensemble of similar systems, analytical results can be obtained systematically and rigorously.

Relying on such an *ergodicity hypothesis*, Efetov developed a field theoretic approach to the study of level correlations in weakly disordered metallic grains[5]. Building upon the seminal work of Wegner[12] and others[13, 14, 15, 16, 17, 18] on the phenomenon of Anderson localization, Efetov provided a rigorous mathematical foundation for the random matrix hypothesis introduced by Gor'kov and Eliashberg. However, perhaps more importantly, Efetov's approach provided a systematic way in which non-universal properties reflecting the intrinsic diffusive dynamics of the system could be studied.

Recent studies by Muzykantskii and Khmel'nitskii[6], and later by Andreev *et al.*[7] revealed that the supersymmetry approach introduced by Efetov is, in fact, not limited to random systems, but can find an extension to the wider class of chaotic structures. It is to this development that we turn towards the end of these lectures.

Random Matrix Theory

In the years that followed Wigner's statistical hypothesis, the general properties of random matrix ensembles was put on a firm footing by a number of mathematicians and physicists[19, 20]. In particular, a classification of three statistical (circular) ensembles was introduced by Dyson[21]. Focusing on the ensemble of Gaussian distributed random matrices,

$$P(H)dH \propto \exp\left[-\frac{N\beta}{\lambda^2}\mathrm{tr}H^2\right]dH,\tag{2}$$

three principle universality classes were identified according to whether the $N \times N$ matrix H is constrained to be real symmetric ($\beta = 1$, Orthogonal), complex Hermitian ($\beta = 2$, Unitary), or real quaternion ($\beta = 4$, Symplectic). Hamiltonians invariant under time-reversal belong to the orthogonal ensemble, while those which are not belong to the unitary ensemble. Time-reversal invariant systems with half-integer spin and broken rotational symmetry belong to the third symplectic ensemble.

In passing we note that more recent analyses, based on a general mathematical classification scheme for symmetric spaces, indicate that *ten* rather than just three fundamental symmetry classes exist [22]. By now all additional classes have been identified in physics (viz. in mesoscopic superconductor/normal metal systems and in models of random Dirac fermions).

Expressed in the basis of eigenstates $H = U^\dagger \Lambda U$, where Λ denotes the matrix of eigenvalues, the probability distribution takes the form

$$P(H)dH = J(\{E_i\})\exp\left[-\frac{\beta N}{\lambda^2}\sum_i^N E_i^2\right]\prod_l^N dE_l dU$$

where the invariant measure

$$J(\{E_i\}) = \prod_{i<j}|E_i - E_j|^\beta.\tag{3}$$

reveals the characteristic repulsion of the energy levels.

Spectral properties of the ensemble are characterized by correlations of the density of states (DoS),

$$\nu(E) = \mathrm{tr}\,\delta(E - H).\tag{4}$$

While the average DoS depends explicitly on the form of the probability distribution, statistical properties of the fluctuations $\delta\nu = \nu - \langle\nu\rangle$ are universal, independent of the support of the spectrum. For the Gaussian ensembles, the average DoS takes the form of a semi-circle,

$$\langle\nu(E)\rangle = \frac{N}{\pi\lambda}\left[1 - \left(\frac{E}{2\lambda}\right)^2\right]^{1/2},\tag{5}$$

while the dimensionless two-point correlator of DoS fluctuations

$$R_2(\Omega) \equiv \Delta^2\langle\nu(E + \Omega/2)\nu(E - \Omega/2)\rangle - 1,\tag{6}$$

takes a universal form. For the unitary ($\beta = 2$) ensemble

$$R_2(\Omega) \overset{N\to\infty}{=} -\frac{\sin^2(\pi\Omega/\Delta)}{(\pi\Omega/\Delta)^2}.\tag{7}$$

General properties of all three random matrix ensembles are summarized by Mehta[19].

Although compelling in its simplicity, the conjecture of Bohigas, Giannoni and Schmidt begs the question: Which properties are universal and, perhaps more importantly, *which are not?* Empirically, the study of spectral statistics typically shows level correlations to be universal on energy scales comparable with the average energy level spacing Δ (see Fig. 1). However, universal correlations are typically observed to break down at some non-universal energy scale $E_c = \hbar/t_c$.

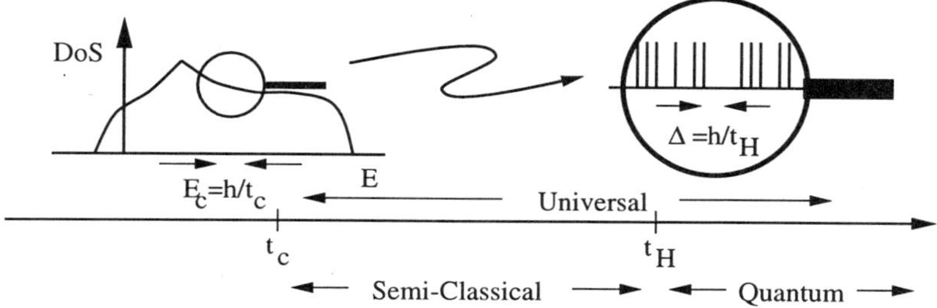

Figure 1. Time scales over which response functions such as $R_2(\Omega)$ are universal.

- What, in general, sets the range of universal correlations?

- How is chaos reflected in the statistics of the non-universal interval?

To address these questions and identify others we can gain useful insight by studying statistical properties of weakly disordered conductors.

COHERENCE EFFECTS IN DISORDERED CONDUCTORS

To guide our intuition in the technical analysis that follows we will begin with a survey of quantum coherence phenomena in weakly disordered conductors. To do so, we will focus on the quantum time evolution of a spreading wavepacket in a background of weakly scattering random impurities. We will see a history in which the particle explores several quite distinct regimes.

To be concrete, let us consider a particle confined to a d-dimensional region of size L, and subject to a random impurity potential. The corresponding Hamiltonian takes the form

$$\hat{H} = \frac{\hat{\mathbf{p}}^2}{2m} + \hat{V},$$

(8)

where $\hat{\mathbf{p}} = -i\hbar\partial$ and, for simplicity, the random potential is taken to be Gaussian δ-correlated white-noise with zero average and a correlator given by

$$\langle V(\mathbf{r})V(\mathbf{r}')\rangle = \frac{\hbar}{2\pi\nu\tau}\delta^d(\mathbf{r} - \mathbf{r}').$$

(9)

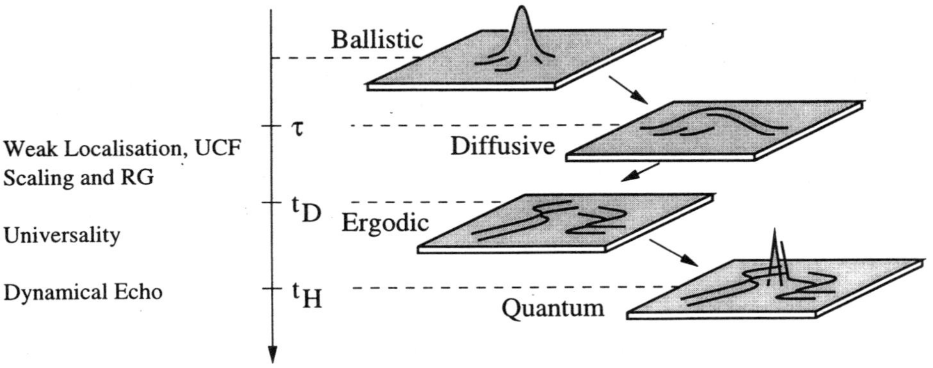

Figure 2. Times scales separating regimes of dynamics.

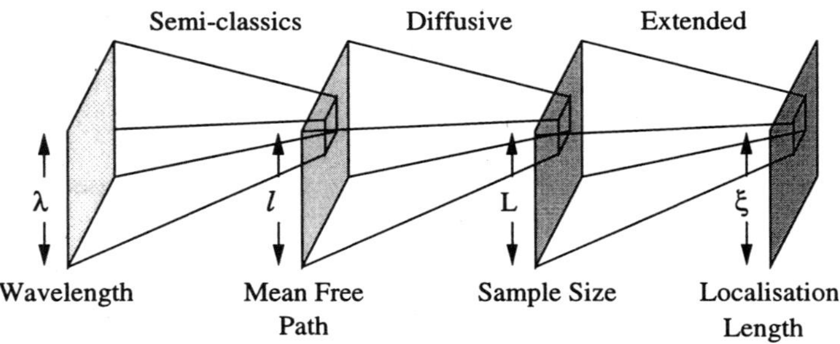

Figure 3. Length scales over which the properties of disordered conductors compare to general chaotic structures.

To stay within the regime in which properties of the disordered conductor bare comparison with those of general chaotic structures we will limit our discussion to length scales in which the dynamics is diffusive (see Fig. 3). Specifically, we will focus on the limit in which the wavelength of the particle λ_F is much smaller than the mean free path $\ell = v_F \tau$ (or, equivalently, $E_F \tau \gg \hbar$), wherein a semi-classical analysis applies. Moreover, we will consider the sample size L to be greatly in excess of the mean free path, while being much smaller than the localization radius ξ of any eigenstates which are localized.

A wavepacket, located in the vicinity of $\mathbf{r} = 0$ at a time $t = 0$, can be expressed at subsequent times as

$$\psi(\mathbf{r}, t) = \sum_\nu \psi_\nu^*(0)\psi_\nu(\mathbf{r})e^{-iE_\nu t/\hbar}, \tag{10}$$

where ψ_ν represents the stationary eigenstates of the time-independent Hamiltonian, and E_ν denotes the eigenenergy. Strictly speaking, the sum over eigenstates should be regarded as being truncated at some bandwidth E_B around an energy $E_F \gg E_B \gg \hbar/\tau$ so that the wavepacket is approximately localized in space and energy. The corresponding probability of finding the particle at a distance $\mathbf{r}$ after a time $t > 0$ can be expressed in terms of the propagator or (retarded) Green function as

$$\rho(\mathbf{r}, t) \equiv |\psi(\mathbf{r}, t)|^2 = |iG(0, \mathbf{r}; t)|^2. \tag{11}$$

On time scales in excess of the mean free scattering time τ the initial *ballistic* evolution of the wavepacket gives way to *diffusive* dynamics.

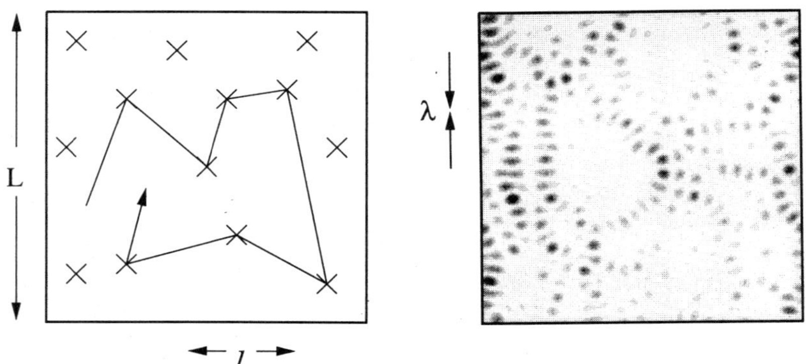

Figure 4. Typical wavefunction of a δ-correlated impurity potential.

Diffusive Regime: $\tau < t < t_D = L^2/D$

Expressed as a Feynman path integral, the probability *amplitude* of propagation takes the form

$$G(0,\mathbf{r};t) = \int_{\mathbf{r}(0)=0,\mathbf{r}(t)=\mathbf{r}} Dr\, \exp\left[-\frac{i}{\hbar}\int_0^t dt'\left(\frac{1}{2}m\dot{\mathbf{r}}^2 - V[\mathbf{r}(t')]\right)\right]. \tag{12}$$

Typically, as a result of the random impurity potential $V(\mathbf{r})$, trajectories arriving at $\mathbf{r}$ after a time t acquire different amplitudes and phases (see Fig. 5):

$$G(0,\mathbf{r};t) = \sum_i \sqrt{\rho_i}e^{-i\varphi_i}, \tag{13}$$

where $\varphi_i = S_i/\hbar$ represents the total action associated with the trajectory i. Focusing on just two possible paths each having a probability amplitude $G_{1,2} = \sqrt{\rho_{1,2}}\exp[-i\varphi_{1,2}]$, the total probability density is given by

$$\rho \equiv |G_1 + G_2|^2 = \rho_1 + \rho_2 + 2\sqrt{\rho_1\rho_2}\cos\left(\varphi_1 - \varphi_2\right), \tag{14}$$

where the first two terms represent the sum of classical probabilities, and the last term describes the interference between the two trajectories.

Typically, different paths have lengths that differ substantially implying a statistical independence of φ_1 and φ_2 ($|\varphi_1 - \varphi_2| \gg 2\pi$). Therefore, the average over the random distribution of phases reproduces the classical result $\langle\rho\rangle = \rho_1 + \rho_2$. Applied to the Green function itself, this random phase cancelation implies a decay of the ensemble average on length scales comparable to the mean free path ℓ. Indeed, for a δ-correlated white noise impurity potential, it is a straightforward (and useful) exercise to show that, within diagrammatic perturbation theory,

$$\langle G(0,\mathbf{r};t)\rangle = G_0(0,\mathbf{r};t)e^{-t/\tau}, \tag{15}$$

where G_0 denotes the free particle Green function.

By contrast, the "diagonal" phase cancelation of each Feynman amplitude generates long-range correlations of the average probability density. In the leading approximation, this leads to a diffusive relaxation of the average probability density,

$$\langle\rho(\mathbf{r},t)\rangle = \frac{1}{(2\pi Dt)^{d/2}}\exp\left[-\frac{\mathbf{r}^2}{2Dt}\right], \tag{16}$$

where $D = v^2\tau/d$ denotes the bare classical diffusion constant. Within diagrammatic perturbation theory this mechanism of density relaxation, known as a *diffuson mode*, can be expressed as a ladder series (see Fig. 6a),

$$\langle\rho(\mathbf{q},\Omega)\rangle = \frac{1}{-i\Omega + \hbar D\mathbf{q}^2}. \tag{17}$$

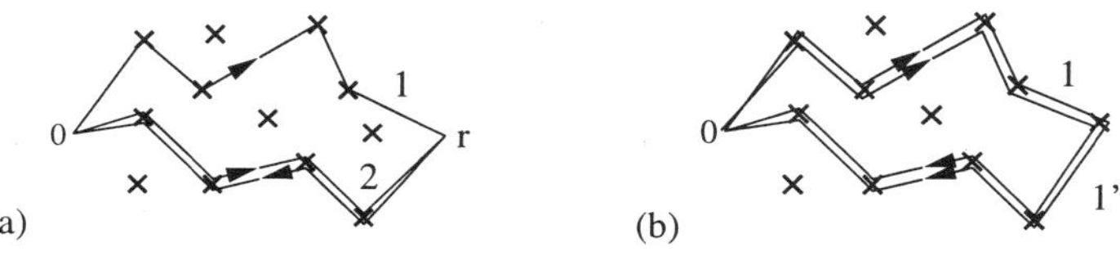

Figure 5. Schematic representation of different Feynman amplitudes contributing to the transmission probability.

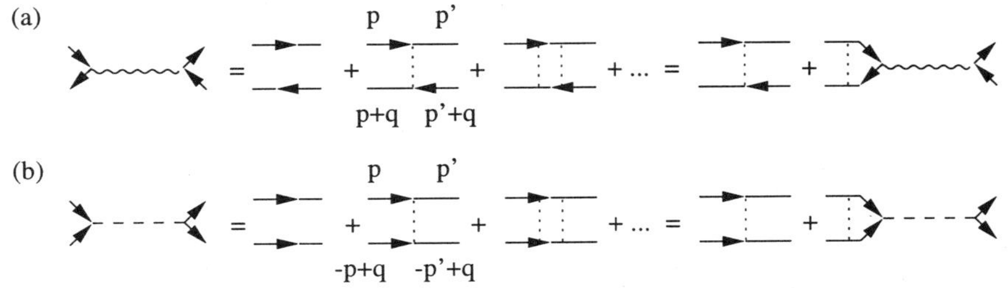

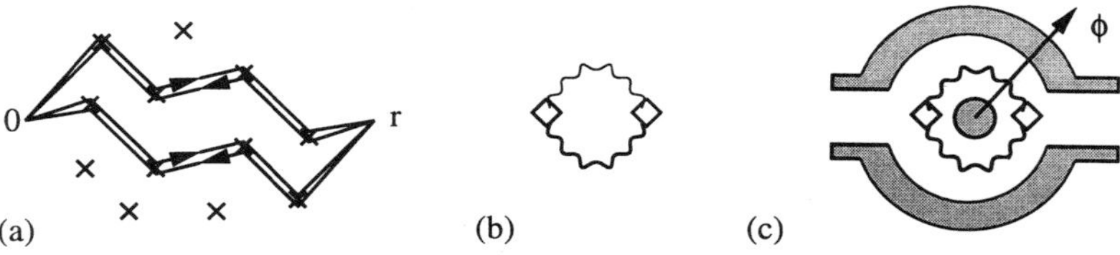

Figure 6. The diagrammatic representation of the dominant "diagonal" contribution to the probability amplitude arising from the interference of (a) overlapping trajectories (diffusons), and (b) the interference of paths with their time-reversed counterpart (Cooperons).

A second and quantum mechanism of density relaxation can be identified by studying the return probability. In the absence of an external magnetic field, the coherent superposition of a path (1) with its time-reversed counterpart (1′) leads to an enhancement of the average return probability (see Fig. 5b),

$$\rho = 2 \times 2\rho_1.$$

We will see below that this tendency to "weakly localize" the electron has important manifestations in quantum transport properties. Diagrammatically, contributions from the interference of time-reversed paths, known as *Cooperon* modes, correspond to a perturbative series involving maximally crossed impurity lines (see Fig. 6b). Together, the diffuson and Cooperon modes provide the elements from which a consistent theory of quantum coherence phenomena in disordered conductors can be assembled.

Fluctuations. While random phase cancelation removes mechanisms of quantum interference from the average probability density, such effects are manifest in *fluctuations*. Taking Eq. (14) for the probability density, its average fluctuation is shown to differ from the classical value by an interference term,

$$\left\langle \rho^2 \right\rangle = \langle \rho \rangle^2 + 2\rho_1\rho_2,$$

a signature of non-locality. This result, which relies on the non-vanishing average $\langle \cos^2(\varphi_1 - \varphi_2) \rangle = 1/2$, can be interpreted as an interaction involving both types of diffusion mode as shown in Fig. 7.

Aharonov-Bohm hc/e Oscillations. Both mechanisms of quantum interference shown in Fig. 7 are manifest in the transport properties of a metallic ring threaded by a magnetic flux, $\varphi = 2\pi\phi/\phi_0$, where $\phi_0 = hc/e$ denotes the flux quantum. In the

Figure 7. (a) Trajectories contributing to quantum interference corrections to fluctuations in the probability density together with (b) the diagrammatic interpretation. In the Aharonov-Bohm geometry (c) such contributions generate hc/e oscillations of the autocorrelation of conductance.

Aharonov-Bohm geometry (see Fig. 7) Feynman trajectories which pass anti-clockwise around the ring experience an additional phase shift relative to those that move clockwise. As a result, a modulation, periodic in hc/e, appears in the autocorrelator of the transmission probability, $\langle \rho(0)\rho(\varphi)\rangle$.

An analogous situation arises if a uniform magnetic field is applied to the disordered metal. In this case, an additional phase difference of $\delta\phi = \phi_1(H) - \phi_2(H)$ is induced between paths (1) and (2) in Fig. 5a. Thus, while leaving the fluctuation of the transmission probability unchanged, the autocorrelation, sensitive to the *relative* flux, decays on the scale of $H_c = \phi_0/A$, where A denotes the area of the sample:

$$\langle \rho(0)\rho(H)\rangle - \langle \rho\rangle^2 = \begin{cases} 2\rho_1\rho_2 & \phi \ll \phi_0, \\ 0 & \phi \sim \phi_0. \end{cases}$$

Magnetoconductance oscillations of this kind were observed experimentally by Webb et al.[23]

Weak Localization. As discussed above, the interference of paths with their time-reversed counterpart generates an enhanced return probability. This effect leads to a *quantum renormalization* of the classical diffusion constant and the phenomenon of weak localization (for reviews see Refs. 24, 25, 26, 27, 28, 29). An estimate of the quantum correction to the diffusion constant can be obtained by calculating the change in transmission probability along a diffusing path due to redundant excursions (see Fig. 8a). Requiring that the particle return to a volume comparable to the wavelength, we obtain[24]

$$\frac{\delta D}{D} = \frac{\delta P}{P} \approx -\frac{v\lambda^2}{a^{3-d}}\int_\tau^{t_D} \frac{dt}{(Dt)^{d/2}} = -\frac{1}{\hbar\nu_d D}\frac{2}{2-d}\left(L^{2-d} - \ell^{2-d}\right), \tag{18}$$

where $P(t) = (2\pi Dt)^{d/2}$ represents the return probability, and the d-dimensional DoS $\nu_d = \nu a^{3-d}$ derives from the three dimensional DoS, $\nu = 1/\hbar\lambda^2 v$, with a denoting the transverse width. Applying the Einstein relation (or *Drude formula*),

$$\sigma = e^2\nu_d D, \tag{19}$$

this estimate of the weak localization correction (18) is in accord with the diagrammatic calculation showing that, in two dimensions[30],

$$\delta\sigma = -\frac{e^2}{\pi^2\hbar}\ln\left(\frac{L}{\ell}\right). \tag{20}$$

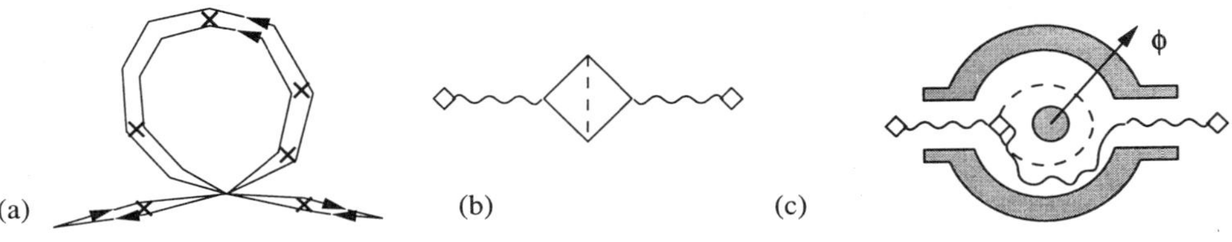

Figure 8. (a) Trajectories contributing to the quantum weak localization correction to the transmission probability. (b) Diagrammatic representation of the weak localization contribution. The junction between the two diffuson modes and the Cooperon is known as a "Hikami box". (c) Diagram responsible for $hc/2e$ oscillations of the conductance in the Aharonov-Bohm geometry.

Anomalous Magnetoconductance. The sensitivity of the Cooperon mode to an external magnetic field has a remarkable effect on the conductance of a mesoscopic sample. While classical considerations would suggest an effect when $\Omega_c \tau \gtrsim 1$, where $\Omega_c = eB/mc$ represents the cyclotron frequency, an enhancement of the conductance is already observed at much lower field strengths, $B \sim \varphi_0/L^2$. This effect is easily understood as the phase breaking of the Cooperon mode by the external magnetic field. Its effect is easily estimated by including a factor to account for the typical flux enclosed by the random walk,

$$\frac{\delta D}{D} = \frac{\delta \sigma}{\sigma} \approx -\frac{v\lambda^2}{a^{3-d}} \int_\tau^{t_D} \frac{dt}{(Dt)^{d/2}} \left[1 - \cos\left(\frac{Dt^2}{L_\varphi}\right) \right],$$

where $L_\varphi = \sqrt{\varphi_0/B}$ denotes the phase coherence length. This leads to an anomalous magnetoconductance observed in experiment.

Aharonov-Bohm $hc/2e$ **oscillations.** The sensitivity of the weak localization correction to an external magnetic field also has a dramatic effect on the conductance of a ring in the Aharonov-Bohm geometry (see Fig. 8). The application of a magnetic flux imparts a relative phase into the time-reversed paths encircling the ring. As a result oscillations in the *non-averaged* conductance appear with a period of $hc/2e$. This effect, predicted by Altshuler and Aronov[31] was first observed by Sharvin and Sharvin[32].

Scaling Theory of Localization. In fact, Eq. (18) represents just the first term in a perturbative series which leads to a *quantum renormalization* of the diffusion constant D, and is in accord with a one-parameter scaling theory of localization*. Beginning with the pioneering work of Thouless[33], the scaling theory of localization as proposed by Abrahams, Anderson, Licciardello, and Ramakrishnan[34] represented a breakthrough in understanding. Its main elements are summarized below.

The bare conductance of a metallic sample of dimension L can be expressed through the Einstein relation as the ratio of characteristic energy scales,

$$G = \sigma L^{d-2} = \frac{e^2}{\hbar} \nu L^d \frac{\hbar D}{L^2} = \frac{e^2}{\hbar} g, \qquad g = \frac{E_c}{\Delta}. \tag{21}$$

According to the one-parameter scaling theory of localization, the variation of the dimensionless conductance with system size obeys the Gell-Mann Low equation

$$\frac{d \ln g}{d \ln L} = \beta(g). \tag{22}$$

In particular, it is independent of the microscopic properties of the sample, such as the bare conductance and ℓ/λ.

This result has profound consequences on the localization properties of disordered conductors. For an ohmic conductor the scaling function takes a constant value $\beta = d - 2$, while deep in the insulating regime, where states are exponentially localized, $\beta(g) \approx \ln g$. A smooth interpolation between these limits suggests that below two-dimensions all states are localized, while above there is a critical conductance, g_*, above which states are extended. The unstable fixed point is associated with the Anderson localization transition.

---

*Although the leading quantum weak localization correction relies on the Cooperon channel, higher order corrections involving just diffuson modes can also operate.

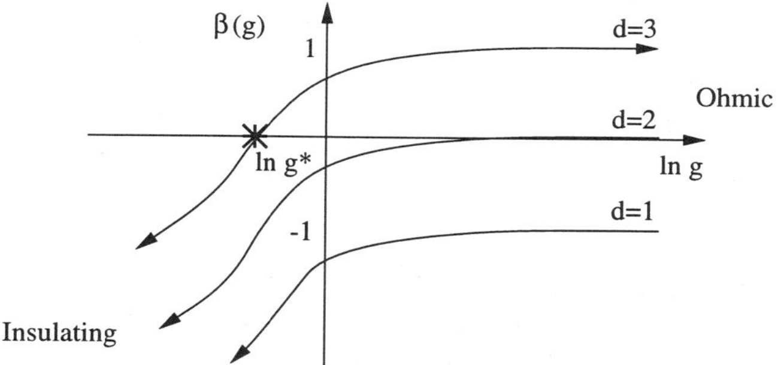

Figure 9. Scaling function of conductance, $\beta(g)$, in dimensions $d = 1$, 2 and 3. A localization transition is predicted in dimensions greater than two.

The situation in two-dimensions is more delicate. Localization properties depend sensitively on the asymptotic approach of the β function to the metallic limit. However, taking the first quantum correction from Eq. (20), we obtain $\beta(g) = -1/\pi^2 g$ which is consistent with localization of all states in two-dimensions. While the one-parameter scaling theory of localization has yet to find a rigorous mathematical basis, it nevertheless represents a milestone in phenomenology.

Level Statistics. Similar arguments can be applied to the two-point correlator of DoS fluctuations, R_2 (6). By making use of the identity

$$\nu(E) = \frac{1}{\pi}\text{tr Im } \hat{G}^-(E), \qquad \hat{G}^- = \frac{1}{E - \hat{H} - i0} \tag{23}$$

the fluctuation of the DoS, R_2, can be expressed as a correlation of Green functions. The dominant contribution arises from the interference of Feynman trajectories topologically equivalent to those shown in Fig. 5. In the language of diagrammatics, this contribution arises from the exchange of two diffusion ladders between two closed loops (see Fig. 10), and gives the celebrated result of Altshuler and Shklovskii[35]

$$R_2(\Omega) = \frac{1}{2\pi^2}\text{Re}\sum_{\mathbf{q}} \frac{\Delta^2}{(-i\Omega + \hbar D\mathbf{q}^2)^2}. \tag{24}$$

In systems invariant under time-reversal, a second mechanism arising from the exchange of Cooperon ladders is accounted by a further factor of two.

Therefore, far from being universal, Eq. (24) shows that, at least on energy scales Ω in excess of the "Thouless energy" (or the inverse transport time),

$$E_c \equiv \frac{\hbar}{t_D} = \frac{\hbar D}{L^2}, \tag{25}$$

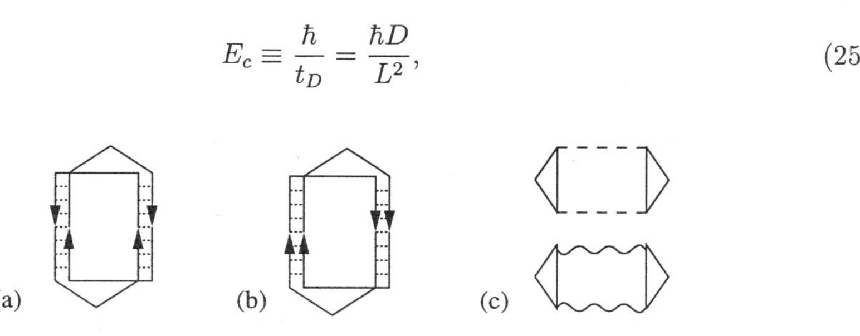

Figure 10. Diagrams showing the leading contribution to the two-point autocorrelator of DoS fluctuations.

level statistics depend explicitly on material properties such as geometry and morphology. Yet, as we will see, Eq. (24) establishes an important connection between disordered and quantum chaotic structures as well as having important physical consequences.

Having strayed far, let us return to the spreading of the wavepacket. In the disordered conductor, we have seen that the dominant mechanism of density relaxation is diffusion. Mechanisms of quantum interference generate quantum corrections and give rise to strong mesoscopic, or sample-to-sample fluctuations. Quantum renormalization of the diffusion constant serves to weakly localize the electron. However, if the impurity potential is not strong enough to bring about complete localization, the wavepacket continues to spread. Beyond the typical transport time $t_D = L^2/D$ the wavepacket is distributed approximately uniformly throughout the sample and further evolution of the wavepacket is, in this sense, *ergodic*. Here the properties of the system become *universal*, independent of the individual features of the system.

Ergodic Regime: $t_D < t < t_H = \hbar/\Delta$

Universality of the response in the ergodic regime is manifest in both spectral and transport properties. Turning first to spectral properties of the system, we return to the perturbative analysis of the two-point correlator of DoS fluctuations, R_2 above.

Universal Level Statistics. Taking the energy $\Omega \ll E_c$, the expression for R_2 (24) can be approximated by the lowest spatial mode $\mathbf{q} = 0$,

$$R_2(\Omega) \simeq -\frac{1}{2\pi^2} \left(\frac{\Delta}{\Omega}\right)^2. \tag{26}$$

In this limit fluctuations of the DoS become universal, independent of detailed properties of the system. Moreover, Eq. (26) is a manifestation of rigidity in the spectrum. Taking $N(\Omega)$ to be the number of levels in a band of width Ω, Eq. (26) shows fluctuations in N to increase only logarithmically with Ω,

$$\left\langle N^2(\Omega)\right\rangle_c \approx \frac{1}{\pi^2} \ln \left\langle N(\Omega)\right\rangle. \tag{27}$$

This compares with a random or Poissonian distribution of energy levels in which fluctuations of N grow as $\sqrt{\langle N \rangle}$.

Universal Conductance Fluctuations. Spectral rigidity (27) provided a simple physical understanding of the universality of conductance fluctuations observed in experiment[36]: It has been found that the dimensionless conductance $g = G\hbar/e^2$ exhibits fluctuations of $O(1)$ no matter how large the average conductance $\langle g \rangle$ (G: dimensionful conductance). To understand this phenomenon we note that Eq. (21) shows the dimensionless conductance to be equivalent to the number of levels inside a window of energy E_c. Thus, while the average conductance

$$\langle G \rangle = \frac{e^2}{\hbar} \langle N(E_c)\rangle = \frac{e^2}{\hbar} g, \tag{28}$$

is large in a good metal ($g \gg 1$), Eq. (27) implies that characteristic fluctuations are small,

$$\sqrt{\langle G^2 \rangle_c} \sim O(e^2/\hbar), \tag{29}$$

and universal[37, 38] (independent of the average conductance).

Breakdown of Perturbation Theory. Eq. (26) illustrates a further point concerning the domain of validity of perturbation theory. At energy scales $\Omega \ll \Delta$, or equivalently at time scales $t \gg t_H = \hbar/\Delta$, response functions obtained with diagrammatic perturbation theory show unphysical IR divergences. Perturbation theory relies not only the semi-classical approximation, $\lambda/\ell \ll 1$, but also that (c.f. $\delta D/D$)

$$\left(\frac{\Delta}{E_c}\right)^{d/2} \left(\frac{\Delta}{\Omega}\right)^{(2-d)/2} \ll 1. \tag{30}$$

In the ergodic regime, where the effective dimensionality is zero, this condition is not met. From this we learn that, in general, one cannot expect "$\hbar \to 0$" to uniquely define what is meant by "semi-classics".

Quantum Regime: $t > t_H = \hbar/\Delta$

Within the quantum regime a separate approach must be sought that does not rely on the parameter (30) of the perturbation theory. Fortunately, at least for non-interacting electronic systems, such an approach exists. The supersymmetry method finds its origin in the pioneering work of Wegner[12]. Motivated by the scaling ideas which were being developed in parallel, Wegner introduced a description of weakly disordered conductors which was based on a field theory of nonlinear σ-model type. The approach allowed for the analysis of the perturbative sector of the theory, including a description of diffusion modes, higher order interference processes and a more rigorous formulation of scaling arguments.

Subsequent investigation provided a microscopic justification for the effective action proposed by Wegner. The crucial step came in a development by Efetov[39]. Employing an approach based on a supersymmetric formalism, Efetov was able to circumvent formal difficulties of the existing theory and thereby explore the non-perturbative regime. As a result, Efetov was able to explicitly confirm the conjecture employed by Gor'kov and Eliashberg: On time scales in excess of the typical transport or diffusion time, average spectral and transport properties of weakly disordered conductors coincide with those of random matrix ensembles. Since then, considerable effort has been directed towards the investigation of the universal limit of the theory. These include universal statistical properties of wavefunctions as well as spectra too numerous to report here.

So within the quantum regime, the dynamics of the spreading wavepacket are almost featureless, with the density relaxing into the uniform zero-mode configuration. The exception concerns the Heisenberg time. As a consequence of the rigidity of the spectrum, an approximately coherent superposition of the wavepacket generates an "echo" of the particle at $t_H = \hbar/\Delta$[40].

Ballistic Chaotic Structures

Does an analogous scenario describe the quantum time evolution of a wavepacket in, say, an irregular cavity ("quantum billiard") without impurities? In such systems there too exists some ergodic time after which properties of the system become universal. Our aim is to explore the dynamics at shorter time scales where the unstable nature of the classical dynamics is reflected in the quantum evolution:

- What plays the role of diffusion in describing the low lying relaxational degrees of freedom in general chaotic quantum systems?

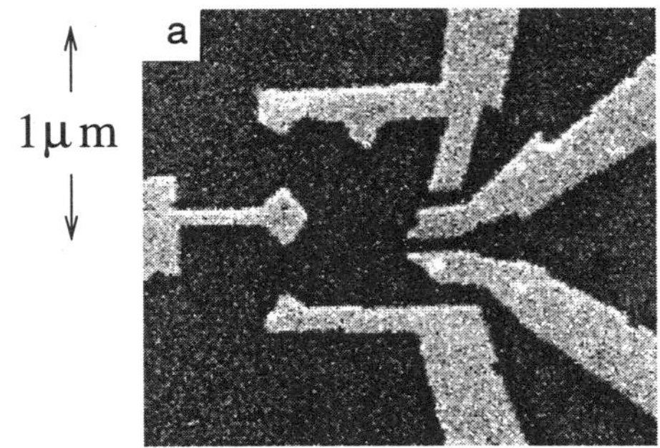
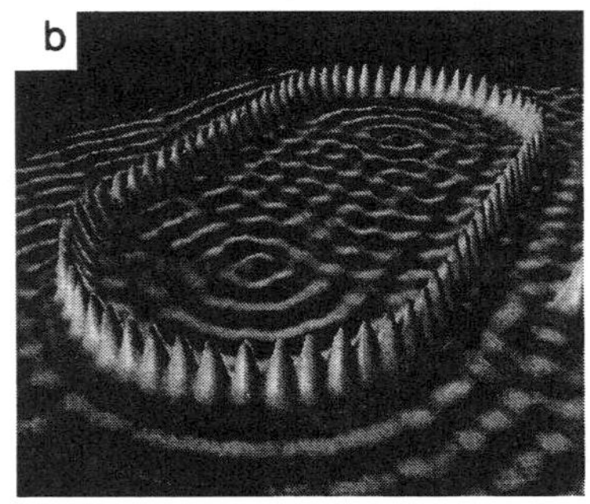

Figure 11. (a) Electron micrograph of a typical ballistic quantum dot taken from Ref. 3. Electrons are confined vertically to the ground state of a quantum well located at a GaAs/AlGaAs interface, and form a two-dimensional electron gas (2DEG). Metallic depletion gates confine electrons laterally. A mean free path and coherence length on the order of 10 microns ensures that the carriers are coherent and ballistic. (b) STM image taken from Ref. 41 of the surface electron density of a quantum stadium corral made from Iron atoms deposited on Copper (111).

- What is the analogue of *weak localization* and how are such quantum coherence effects manifest in experiment?

- What determines the domain of universality?

To address the first of these questions we begin with a semi-classical analysis based on the Feynman path integral.

Semi-classics and the Trace formula

The semiclassical analysis begins with Gutzwiller's trace formula, which expresses the dimensionless DoS as a sum over the classical periodic orbits[4, 42, 43, 44]

$$\frac{\nu(E)}{\nu} = 1 + \mathrm{Re}\frac{\Delta}{\pi\hbar}\sum_p T_p \sum_{r=1}^{\infty} \frac{e^{iS_p(E)r/\hbar - i\nu_p r}}{|\det(M_p^r - \mathbf{I})|^{1/2}}, \tag{31}$$

where p labels a primitive orbit with a period T_p, action $S_p(E)$, and Maslov phase ν_p. The sum over r accounts for repetitions of the classical periodic trajectory. Here M_p denotes the monodromy matrix associated with the linearized dynamics on the Poincarè section perpendicular to the orbit p.

Substituting Eq. (31) into Eq. (6), the two-point correlator of DoS fluctuations is associated with a double sum over periodic orbits. A random phase cancelation of long trajectories identifies the "diagonal contribution" as dominant. Taking this contribution alone, and expanding $S_p(E + \Omega) \simeq S_p(E) + T_p\Omega$, we obtain

$$R_2(\Omega) = \mathrm{Re}\frac{\Delta^2}{2\pi^2\hbar^2}\sum_p T_p^2 \sum_{r=1}^{\infty} \frac{e^{i\Omega T_p r/\hbar}}{|\det(M_p^r - \mathbf{I})|}. \tag{32}$$

The traditional way to deal with the above sum is to approximate it by an integral:

$$\sum_p \frac{f(T_p)}{|\det(M_p - \mathbf{I})|} \to \int_0^{\infty} \frac{dt}{t} f(t)$$

for any sufficiently smooth function $f(t)$. This approximation, known as the Hannay and Ozorio de Almeida sum rule[45], holds in the limit $t \to \infty$ where long periodic orbits which explore the whole energy shell uniformly are considered. In employing it for the calculation of $R_2(\Omega)$, the time t should be restricted to the regime where it is much larger than the shortest periodic orbits but still smaller than the Heisenberg time t_H. The result associated with it is therefore the universal one (26) which holds as long as $\Omega \gg \Delta$[43, 44]. Below we present a more careful treatment of the sum (32) that keeps the non-universal part of $R_2(\Omega)$.

To identify the physical operator to which the diagonal sum (32) corresponds we first consider the propagator of classical phase space density, the Liouville operator. Formally, the latter is defined by

$$\rho(\mathbf{x}, t) = \hat{\mathcal{L}}^t \rho(\mathbf{x}, 0) = \int d\mathbf{x}' \delta\left(\mathbf{x} - \mathbf{u}(\mathbf{x}'; t)\right) \rho(\mathbf{x}', 0), \tag{33}$$

where $\mathbf{x}'$ and $\mathbf{x}$ are phase space vectors representing coordinates and momenta, and $\mathbf{u}(\mathbf{x}; t)$ denotes the solution of the classical equations of motion of a particle starting at a position $\mathbf{x}$ after a time t.

From this definition, it is straightforward to obtain the following expression for the trace[46],

$$\mathrm{tr}\mathcal{L}^t = \sum_p T_p \sum_r \frac{\delta\left(t - rT_p\right)}{\left|\det(M_p^r - \mathbf{I})\right|}. \tag{34}$$

Comparing this expression to Eq. (32), and *neglecting repetitions*, i.e. the contribution of short periodic orbits that are traversed repeatedly, we obtain[47]

$$R_2(\Omega) = \frac{\Delta^2}{2\pi^2\hbar} \mathrm{Re} \frac{\partial}{\partial(i\Omega)} \int_0^\infty dt e^{i\Omega_+ t/\hbar} \mathrm{tr}\mathcal{L}^t = \frac{1}{2\pi^2} \mathrm{Re} \sum_\mu \frac{\Delta^2}{(-i\Omega_+ + \hbar\gamma_\mu)^2}, \tag{35}$$

where the second equality is based on $\mathrm{tr}\, \mathcal{L}^t = \sum_\mu \exp(t\gamma_\mu)$, $\{\gamma_\mu\}$ being the eigenvalues of (the logarithm of) the classical evolution operator. This result compares to Eq. (24) (with $\gamma_\mu \to D\mathbf{q}^2$), and identifies the modes of density relaxation as the eigenfunctions of the classical evolution operator.

But do the modes of the Liouville operator relax? What role is played by the repetitions neglected above? And how can we account for weak localization and corrections non-perturbative in Δ/Ω? These questions can not be answered with the framework of the diagonal approximation. Recent attempts to go beyond this approximation within the framework of periodic theory have met with some limited success[48]. However, these studies have, as yet, failed to identify weak localization corrections. Instead, we will employ a different approach based on a non-perturbative field theory. To introduce these ideas, we will first apply this approach to study quantum coherence phenomena in weakly disordered conductors.

SUPERSYMMETRY METHOD

Although the ideas presented in the previous section can be straightforwardly interpreted within the framework of a diagrammatic perturbation theory, the analysis of low energy, infrared phenomena relies on a non-perturbative description. The latter is provided by the above mentioned field theoretic approach introduced by Efetov. The purpose of the present section is to introduce this so-called supersymmetry method approach and demonstrate its application to the study of quantum coherence phenomena described in the previous section. The presentation will be by necessity concise. For a more comprehensive discussion the reader is referred to Ref. 5. Later we will show how an analogous approach can be developed for structures which are chaotic but not random.

Field Integral

The starting point of the field theoretic approach is the representation of the Green function as a functional field integral. Focusing on the advanced function, we have

$$G^-(\mathbf{r}, \mathbf{r}'; E) \equiv \left\langle \mathbf{r}' \left| \frac{1}{E - \hat{H} - i0} \right| \mathbf{r} \right\rangle \equiv \sum_\nu \frac{\psi_\nu(\mathbf{r}')^* \psi_\nu(\mathbf{r})}{E - E_\nu - i0} = \frac{i}{\mathcal{Z}} \int DS\, S(\mathbf{r}) S^*(\mathbf{r}') e^{-\mathcal{L}[S]},$$

where the integration measure $DS = \Pi d(\mathrm{Re}S) d(\mathrm{Im}S)/(2\pi)$ is taken over all space points, $\mathcal{Z} = \int DS \exp(-\mathcal{L}[S])$ denotes the constant of normalization, and

$$\mathcal{L}[S] = i \int S^* \left(E - \hat{H} - i0\right) S.$$

Expressed in this form, the presence of the normalization factor leads to unphysical vacuum loops in the ensemble average. Accountancy of the latter can, at least in principle, be treated within the "replica formalism" in which one exploits the analytic continuation[49, 50]

$$\ln \mathcal{Z} = \lim_{n \to 0} \frac{\mathcal{Z}^n - 1}{n}.$$

However, experience has shown considerable difficulty in correctly implementing this procedure in the non-perturbative sector of the theory[51]. Fortunately, at least for systems which are non-interacting, a second and more reliable approach can be developed.

The "supersymmetry method" exploits properties of Grassmann algebra to implement a scheme of "book-keeping". While a comprehensive introduction to superalgebra can be found in Ref. 52, essential properties of Grassmann variables are summarized in Ref. 5. Crucially, by exploiting the properties of Gaussian integrals over complex commuting or bosonic (B) variables S, and anticommuting or fermionic (F) variables χ,

$$\int e^{-S^* M S} d[S^*] d[S] = \frac{1}{\det M}, \qquad \int e^{-\bar{\chi} M \chi} d[\bar{\chi}] d[\chi] = \det M$$

the Green function can be expressed as a functional field integral in which the normalization, $\mathcal{Z}$, is by construction unity. Specifically, we can write

$$G^-(\mathbf{r}, \mathbf{r}'; E) = i \int D\psi\, S(\mathbf{r}) S^*(\mathbf{r}') e^{-\mathcal{L}[\psi]}, \qquad \mathcal{L}[\psi] = i \int \psi^\dagger \left(E - \hat{H} - i0\right) \psi, \quad (36)$$

where $\psi^T = (S, \chi)$ denotes a two-component superfield. The measure of the functional integral takes the form $D\psi = DS\, D\chi$, where $D\chi = \Pi\, 2d\chi^* d\chi$.

Finally, instead of representing each vertex separately, it is often more convenient to introduce a "source term" into the effective action from which arbitrary correlators can be constructed. Defining

$$\mathcal{Z}[J] = \int D\psi\, e^{-\mathcal{L}[\psi]} \exp\left[-i \int \left(J^\dagger \psi + \psi^\dagger J\right)\right],$$

the Green function can be expressed as a functional derivative over the vector superfields

$$G^-(\mathbf{r}, \mathbf{r}'; E) = -\frac{i}{2} \mathrm{tr} \left[\frac{\delta}{\delta J^\dagger(\mathbf{r})} \otimes \frac{\delta}{\delta J(\mathbf{r}')}\right]_{\mathrm{bf}} \mathcal{Z}\Big|_{J=0}.$$

By doubling the field space, this representation allows *two-point* correlators or response functions to be expressed in terms of the generating function,

$$\mathcal{Z} = \int D\psi\, e^{-\mathcal{L}[\psi]-\mathcal{L}_J[\psi]}, \tag{37}$$

$$\mathcal{L}[\psi] = i\int \bar{\psi}\left(E - \frac{\Omega_+}{2}\sigma_3^{\mathrm{ar}} - \hat{H}\right)\psi, \qquad \mathcal{L}_J[\psi] = i\int\left(\bar{J}\psi + \bar{\psi}J\right) \tag{38}$$

where $\Omega_+ \equiv \Omega + i0$ denotes the frequency source, and the field integral involves four-component superfields

$$\psi = \begin{pmatrix}\psi_-\\\psi_+\end{pmatrix}_{\mathrm{ar}}, \qquad \psi_\mp = \begin{pmatrix}S_\mp\\\chi_\mp\end{pmatrix}_{\mathrm{bf}}, \qquad \bar{\psi} = \psi^\dagger L, \qquad L = \begin{pmatrix}\sigma_3^{\mathrm{ar}} & 0\\0 & \mathbf{I}^{\mathrm{ar}}\end{pmatrix}_{\mathrm{bf}}.$$

Pauli matrices[†] σ_3^{bf} and σ_3^{ar} respectively break the symmetry between boson/fermion and advanced/retarded degrees of freedom of Q. Convergence of $\mathcal{Z}$ determines the form of the metric L in the BB sector. While the choice in the fermionic sector seems arbitrary, a consideration of the saddle-point manifold of the σ-model determines the form above[53]. This implies a group structure in the bosonic sector which is non-compact, while the fermionic sector is compact.[‡]

Finally, our preliminary discussion in the previous section classified two modes of density relaxation in the disordered metal, diffusons and Cooperons. Both will be identified in the analysis that follows. Accordingly, anticipating the structure of the saddle-point action, it is convenient to manipulate the action into the form

$$\mathcal{L}[\Psi] = \frac{1}{2}\left(\mathcal{L}[\psi] + \mathcal{L}[\psi]^T\right)$$

where the dimension of the vector space is doubled to include complex or time-reversed components,

$$\Psi = \begin{pmatrix}\psi\\\psi^*\end{pmatrix}.$$

The elements of the newly defined supervector Ψ are not independent, but fulfill the "time-reversal" symmetry relation

$$\Psi^\dagger = (C\Psi)^T, \qquad C = \begin{pmatrix}\sigma_1^{\mathrm{tr}} & 0\\0 & i\sigma_2^{\mathrm{tr}}\end{pmatrix}_{\mathrm{bf}}.$$

Regarding ψ as analogous to the wavefunction, the transformation $\psi \to \psi^*$, $\hat{H} \to \hat{H}^T$ is the analogue of the quantum mechanical time-reversal operation. In fact, any discrete symmetry of the microscopic Hamiltonian doubles the number of low-lying modes of density relaxation. In each case, it is convenient to double the vector space[54].

Altogether, after doubling the space, we obtain the action

$$\mathcal{L}[\Psi] = \frac{i}{2}\int \bar{\Psi}\left(E - \frac{\Omega_+}{2}\sigma_3^{\mathrm{ar}} - \hat{H}\right)\Psi, \qquad \mathcal{L}_J[\Psi] = \frac{i}{2}\int\left(\bar{J}\Psi + \bar{\Psi}J\right)$$

[†]
$$\sigma_1 = \begin{pmatrix}0 & 1\\1 & 0\end{pmatrix}, \qquad \sigma_2 = \begin{pmatrix}0 & -i\\i & 0\end{pmatrix}, \qquad \sigma_3 = \begin{pmatrix}1 & 0\\0 & -1\end{pmatrix}.$$

[‡]Formally, the difference between the bosonic and fermionic sectors introduces a factor $\det(-1)$, taken over space points, into $\mathcal{Z}$. Since this factor is identically cancelled at a later stage we will, for convenience, incorporate it into a redefinition of the integration measure.

In this notation, advanced and retarded Green functions are obtained from the generating function by the operation,

$$\hat{G}^{\mp}(E \mp \Omega/2) = -\frac{i}{2}\mathrm{tr}\left[\frac{\delta}{\delta \bar{J}_{\mp}} \otimes \frac{\delta}{\delta J_{\mp}}\right]_{\mathrm{bf,tr}} \mathcal{Z}\Big|_{J=0}. \tag{39}$$

To proceed we will examine the Hamiltonian (8) describing a particle propagating in a weakly disordered impurity potential. Linear in the potential, the ensemble average over the Gaussian δ-correlated distribution (9), is straightforward.

$$\left\langle \exp\left[\frac{i}{2}\int \bar{\Psi} V \Psi\right]\right\rangle = \exp\left[-\int \frac{\hbar}{16\pi\nu\tau}\left(\bar{\Psi}\Psi\right)^2\right]. \tag{40}$$

The result is an effective quartic interaction of the fields. An expansion of the total action in the interaction recovers diagrammatic perturbation theory. However, our preliminary survey of two-dimensional systems below identifies the dominant contributions arising from mechanisms of quantum coherence as coming from IR or long-range fluctuations of the density. This suggests the validity of a mean-field decomposition of the action in which the important fluctuations are long-ranged.

The situation is comparable to the BCS mechanism of superconductivity. There the formation of the Cooper pair condensate is associated with the growth of the mean-field order parameter described by a saddle-point or gap equation. Massless Goldstone fluctuations of the order parameter around the saddle-point are described by a Ginzburg-Landau type action. In the present case we will find that the Goldstone modes associated with fluctuations of the mean field represent the diffusion modes.

In the absence of the symmetry breaking sources, Ω and J, the action is invariant under pseudounitary rotations of the fields in superspace,

$$\psi \longrightarrow U\psi, \qquad U^{\dagger}LU = L, \qquad U \in SU(2,2/4). \tag{41}$$

Hubbard-Stratonovich Decoupling

To establish a useful mean-field decoupling of the interaction requires the identification of the low-lying modes of the theory. A diagrammatic analysis of the theory identifies two relevant channels: the diffuson and Cooperon. Both are represented diagrammatically in Fig. 12.

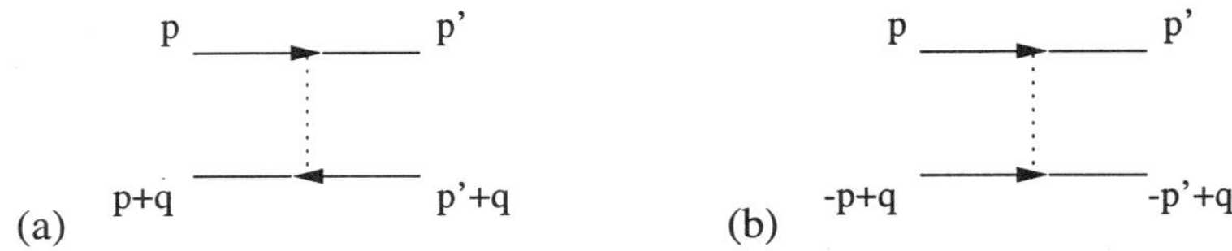

Figure 12. Diagrammatic representation of the terms in the Hubbard-Stratonovich decoupling that lead to (a) the diffusion mode, and (b) the Cooperon mode.

These processes are accounted for by contributions to the action from

$$\int \left(\bar{\Psi}\Psi\right)^2 \approx \sum_{\mathbf{p},\mathbf{p}'} \sum_{|\mathbf{q}|\ll \ell^{-1}} \left[\left(\bar{\psi}_{-\mathbf{p}'}\psi_{\mathbf{p}}\right)\left(\bar{\psi}_{-\mathbf{p}-\mathbf{q}}\psi_{\mathbf{p}'+\mathbf{q}}\right) + \left(\bar{\psi}_{-\mathbf{p}'}\psi_{\mathbf{p}}\right)\left(\bar{\psi}_{\mathbf{p}'-\mathbf{q}}\psi_{-\mathbf{p}+\mathbf{q}}\right)\right]$$

where the first and second terms generate the contributions from the diffuson and Cooperon respectively, and the summation over $\mathbf{q}$ is limited to low momentum transfer. Introducing the Hubbard-Stratonovich transformation,

$$\exp\left[\int \frac{\hbar}{16\pi\nu\tau}\mathrm{str}\left(\Psi \otimes \bar{\Psi}\right)^2\right] = \int DQ \exp\left[-\frac{\hbar}{4\tau}\int\left(\bar{\Psi}Q\Psi + \frac{\pi\nu}{2}\mathrm{str}Q^2\right)\right], \tag{42}$$

the contributions to the diffuson and Cooperon degrees of freedom are accounted for by *slow* fluctuations of the 8×8 supermatrix fields, Q.[§] Here the supertrace operation is defined by $\mathrm{str}M = \mathrm{tr}M_{FF} - \mathrm{tr}M_{BB}$. The supermatrix fields Q have an algebraic structure which reflects that of the dyadic product $\Psi \otimes \bar{\Psi}$,

$$Q = C L Q^T L C^T,$$

where the transpose of a supermatrix is defined by

$$M^T = \begin{pmatrix} M_{BB}^T & M_{FB}^T \\ -M_{BF}^T & M_{FF}^T \end{pmatrix}.$$

Altogether we obtain the following expression for the average

$$\langle \mathcal{Z} \rangle = \int D\psi D Q e^{-\mathcal{L}_J} \exp\left[-\int \left(\frac{\pi \hbar \nu}{8\tau} \mathrm{str} Q^2 + \frac{i}{2} \bar{\Psi} \hat{\mathcal{G}}^{-1} \Psi \right) \right],$$

where

$$\hat{\mathcal{G}}^{-1} = E - \frac{\Omega_+}{2} \sigma_3^{\mathrm{ar}} - \frac{\hat{p}^2}{2m} - \frac{i\hbar}{2\tau} Q, \tag{43}$$

defines the supermatrix Green function. Integrating over the fields Ψ, we obtain

$$\langle \mathcal{Z} \rangle = \int D Q \exp\left[-\int \mathrm{str} \left(\frac{\pi \hbar \nu}{8\tau} Q^2 - \frac{1}{2} \ln \hat{\mathcal{G}}^{-1} \right) + \int \int \frac{i}{2} \bar{J} \hat{\mathcal{G}} J \right].$$

A legitimate mean-field decomposition justifies treating the action within a saddle-point approximation. We will proceed by identifying a manifold of degeneracy associated with the saddle-point. Fluctuations around this degenerate manifold will generate an effective action for the diffusion modes.

Saddle-Point Equation

A variation of the action with respect to Q generates the saddle-point equation

$$Q_{\mathrm{sp}}(\mathbf{r}) = -\frac{i}{\pi \nu} \mathcal{G}(\mathbf{r}, \mathbf{r}). \tag{44}$$

From this equation it is possible to interpret the saddle-point solution of Q as representing the self-consistent Born approximation to the self-energy. Taking the symmetry breaking sources Ω and J to be vanishingly small, the ansatz that the saddle-point solution Q_{sp} is constant, independent of position, and takes non-zero matrix elements only on the diagonal, implies

$$\mathcal{G}(\mathbf{r}, \mathbf{r}) = \int d\mathbf{p} \left[E - \frac{\mathbf{p}^2}{2m} - \frac{i\hbar}{2\tau} Q_{\mathrm{sp}} \right]^{-1} = \nu \int d\xi \left[E - \xi - \frac{i\hbar}{2\tau} Q_{\mathrm{sp}} \right]^{-1} = i\pi\nu\,\mathrm{sgn}(Q_{\mathrm{sp}}).$$

From this we deduce that the elements of Q_{sp} take values of ± 1. However, signs cannot be assigned arbitrarily and analyticity of the Green function in each sector demands

---

[§] Note that as written, the Hubbard-Stratonovich transformation is not exact. If *all* degrees of freedom of Q (fast and slow) are taken into account, the decoupling (42) involves an overcounting by a factor of 2. This is because the saddle-point corresponding to the Cooperonic sector ($[Q, \sigma_3^{\mathrm{tr}}]$) can be found in fast fluctuations of the diffuson section, and *vice versa*

$Q_{\rm sp} = \sigma_3^{\rm ar}$. Moreover, the invariance of the action under pseudounitary rotations implies a degenerate manifold of solutions

$$Q_{\rm sp} = T^{-1}\sigma_3^{\rm ar}T, \qquad T = LU, \qquad Q_{\rm sp}^2 = \mathbf{I}. \tag{45}$$

Dividing out rotations that leave $\sigma_3^{\rm ar}$ invariant, the degeneracy of the manifold is specified by the factor space $SU(2,2/4)/SU(2/2) \otimes SU(2/2)$.

Substituting Eq. (45) into Eq. (43), we obtain the following expression for the supermatrix Green function at the saddle-point:

$$\mathcal{G}_{\rm sp}(\mathbf{r}, \mathbf{r}') = {\rm Re}\left\langle G^-(\mathbf{r}, \mathbf{r}')\right\rangle + i\pi\nu f_d(|\mathbf{r} - \mathbf{r}'|)Q_{\rm sp}, \tag{46}$$

where the average Green function, or "Friedel function"

$$f_d(r) \equiv \frac{{\rm Im}\left\langle G^-(\mathbf{r}, 0)\right\rangle}{{\rm Im}\left\langle G^-(0, 0)\right\rangle} = \Gamma(d/2)\left(\frac{2}{k_F r}\right)^{d/2-1} J_{d/2-1}(k_F r)e^{-r/2\ell}, \tag{47}$$

decays on a scale comparable to the mean free path. Note that a non-zero contribution to ${\rm Re}\langle G^-\rangle$ can be accommodated by a shift of the chemical potential (and can therefore be set to zero).

Non-linear σ-model

The degeneracy of the non-linear manifold implies the existence of Goldstone modes which can be obtained by expansion of the action near the extremum. To classify the fluctuations let us consider small deviations $\delta Q(\mathbf{r}) = Q(\mathbf{r}) - Q_{\rm sp}$. Neglecting symmetry breaking sources, an expansion of the effective action to quadratic order yields

$$\langle \mathcal{Z} \rangle = \int DQ e^{-S^{(2)}[Q]}$$

with

$$S^{(2)}[Q] = -\int d\mathbf{r}d\mathbf{r}'\frac{\hbar^2}{16\tau^2}{\rm str}\left[\mathcal{G}_{\rm sp}(\mathbf{r}, \mathbf{r}')\delta Q(\mathbf{r}')\mathcal{G}_{\rm sp}(\mathbf{r}', \mathbf{r})\delta Q(\mathbf{r}) - \frac{2\pi\nu\tau}{\hbar}\delta^d(\mathbf{r} - \mathbf{r}')\delta Q(\mathbf{r})\delta Q(\mathbf{r}')\right]$$

Depending on whether or not they change the eigenvalues of the supermatrix Q, the modes of fluctuation can be classified as longitudinal, δQ_l, or transverse, δQ_t,

$$[Q, \delta Q_l]_- = 0, \qquad [Q, \delta Q_t]_+ = 0.$$

Expressed in terms of these modes, the first term separates into two contributions, $S_l + S_t$, where

$$S_{\substack{l \\ t}}[\delta Q] = \pm\left(\frac{\pi\hbar\nu}{4\tau}\right)^2 \int d\mathbf{r}d\mathbf{r}' f_d^2(|\mathbf{r} - \mathbf{r}'|){\rm str}\left[\delta Q_{\substack{l \\ t}}(\mathbf{r})\delta Q_{\substack{l \\ t}}(\mathbf{r}')\right].$$

Since the Q-matrices vary slowly over the length scale ℓ at which the average Green function decays, this facilitates a gradient expansion. Using the identity

$$\int d\mathbf{r} f_d^2(|\mathbf{r}|) = \frac{2\tau}{\pi\hbar\nu},$$

we find that, in the leading approximation, the action of the transverse fluctuations vanishes, while the action for the longitudinal fluctuations takes the form

$$S_l^{(2)} = \int \frac{\pi\hbar\nu}{4\tau}{\rm str}\left[\delta Q_l\right]^2. \tag{48}$$

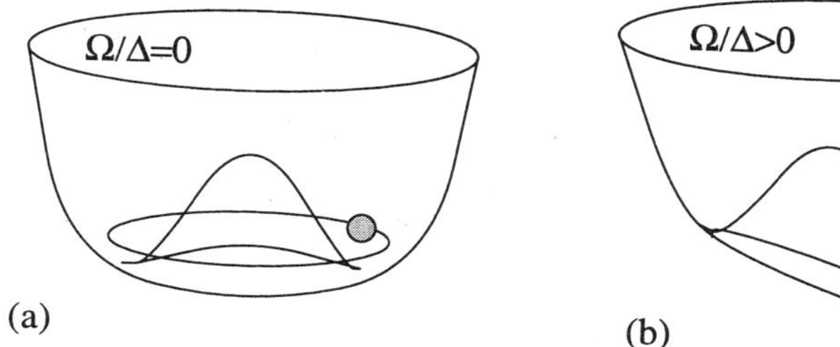

Figure 13. Schematic diagram of the structure of the saddle-point manifold (a) in the absence, and (b) in the presence of the symmetry breaking source Ω. For $\Omega/\Delta = 0$, fluctuations of the supermatrix Q separate into massive longitudinal modes, and massless transverse Goldstone modes. The latter become massive in the presence of symmetry breaking source. An expansion in the vicinity of the saddle-point σ_3^{ar} recovers diagrammatic perturbation theory.

Thus, in the limit $\Delta\tau \gg 1$, longitudinal fluctuations generate only a small contribution to the effective action. (In the Gaussian approximation, their contribution vanishes due to supersymmetry.)

Transverse fluctuations, on the other hand, provide a non-vanishing contribution to the effective action only at second order of the gradient expansion. Using the identity

$$\int d\mathbf{r}\, f_d^2(|\mathbf{r}|) r_\alpha r_\beta = \frac{4\tau^2 D}{\pi\nu\hbar}\delta_{\alpha\beta},$$

where $D = v^2\tau/d$, we obtain the effective action

$$S^{(2)} = \frac{\pi\nu}{8}\int \hbar D \mathrm{str}\,(\partial Q)^2,$$

where $Q(\mathbf{r}) = T^{-1}(\mathbf{r})\sigma_3^{\mathrm{ar}}T(\mathbf{r})$. Finally, taking $\Omega\tau \ll 1$, an expansion of the effective action to leading order in the symmetry breaking sources Ω and J yields

$$\langle Z \rangle = \left\langle e^{-S_J[Q]} \right\rangle, \qquad \langle \cdots \rangle_Q = \int DQ \cdots e^{-S[Q]}, \tag{49}$$

where

$$S[Q] = \frac{\pi\nu}{8}\int \mathrm{str}\left[\hbar D\,(\partial Q)^2 + 2i\Omega_+\sigma_3^{\mathrm{ar}}Q\right]. \tag{50}$$

In the same approximation, the leading contribution to the source term takes the form

$$S_J[Q] = \frac{\pi\nu}{2}\int d\mathbf{R}\, d\mathbf{r}\, f_d(|\mathbf{r}|)\, \bar{J}(\mathbf{R}+\mathbf{r}/2)Q(\mathbf{R})J(\mathbf{R}-\mathbf{r}/2). \tag{51}$$

We have thus succeeded in expressing the response of weakly disordered metallic grains in the form of a functional supersymmetric non-linear σ-model (50). Spontaneous symmetry breaking induced by the external source Ω leads to the existence of Goldstone modes associated with the diffusion modes. The action (50) can be compared to the Landau-Ginzburg free energy of a superconductor or a ferromagnet.

Magnetic Field

An extension of the present theory to include an external magnetic field introduces a magnetic vector potential into the action.

$$\mathcal{L} = \frac{i}{2}\int \bar{\Psi}\left[E - \frac{\Omega_+}{2}\sigma_3^{\mathrm{ar}} - \frac{1}{2m}\left(\hat{p} - \frac{e}{c}\sigma_3^{\mathrm{tr}}\mathbf{A}\right)^2 - V\right]\Psi.$$

The magnetic field breaks time-reversal symmetry and couples to the symmetry breaking matrix σ_3^{tr}. Treating the magnetic field as a weak perturbation, and proceeding as before, it is straightforward to show that the effective action takes a diffusive form involving a covariant derivative,

$$S[Q] = \frac{\pi\nu}{8} \int \mathrm{str}\left[\hbar D\left(\tilde{\partial}Q\right)^2 + 2i\Omega_+\sigma_3^{\mathrm{ar}}Q\right], \qquad \tilde{\partial} = \partial - \frac{ie}{\hbar c}\mathbf{A}\left[\sigma_3^{\mathrm{tr}}, \right]. \qquad (52)$$

From this result one can identify the degrees of freedom of Q corresponding to the diffuson and Cooperon modes. Those degrees of freedom which commute with σ_3^{tr} remain unperturbed by the weak magnetic field and represent the diffuson modes. The remaining degrees of freedom, which correspond to the Cooperon modes, acquire a mass and are frozen out by the magnetic field. This structure compares to the phenomenological analysis made in the previous section.

Correlators

Another virtue of the supersymmetry approach is that, within this representation, statistical correlators of wavefunctions as well as spectra can be obtained with the same efficiency. As an example, employing Eqs. (39) and (51), the two-point correlator of DoS fluctuations is given by

$$R_2(\Omega) = \frac{1}{64}\left\langle\left[\int \frac{d\mathbf{r}}{L^d}\mathrm{str}\left(\sigma_3^{\mathrm{ar}}\otimes\sigma_3^{\mathrm{bf}}Q(\mathbf{r})\right)\right]^2\right\rangle_Q - 1, \qquad (53)$$

while the density response function is given by

$$G^-(0,\mathbf{r})G^+(\mathbf{r},0) = \frac{\pi^2\nu^2}{8}\left\langle\mathrm{str}\left[\sigma_3^{\mathrm{bf}}(\mathbf{I}^{\mathrm{ar}} - \sigma_3^{\mathrm{ar}})Q(0)\sigma_3^{\mathrm{bf}}(\mathbf{I}^{\mathrm{ar}} + \sigma_3^{\mathrm{ar}})Q(\mathbf{r})\right]\right\rangle_Q$$

Perturbation Theory

To help digest the form of the effective action we will focus on several applications identified in the previous section. Specifically, we will examine the interpolation of level statistics from the perturbative to the non-perturbative limit, and examine the role of weak localization and scaling. We begin by describing a perturbative expansion of the action in the vicinity of the global saddle-point σ_3^{ar}. For this purpose a number of convenient parametrizations in terms of the generators of the coset exist. Those most commonly used include

$$Q = \begin{cases} e^{-W}\sigma_3^{\mathrm{ar}}e^W & \text{exponential,} \\ W + \sigma_3^{\mathrm{ar}}(\mathbf{I}^{\mathrm{ar}} - W^2)^{1/2}, & \text{square root,} \\ (\mathbf{I}^{\mathrm{ar}} + W)^{-1}\sigma_3^{\mathrm{ar}}(\mathbf{I}^{\mathrm{ar}} + W) & \text{rational,} \end{cases} \qquad (54)$$

where $[W, \sigma_3^{\mathrm{ar}}]_+ = 0$.¶

Level Statistics. To apply the perturbation theory, we turn first to the question of level statistics. Employing the exponential parametrization, to leading order in W, the effective action takes the form

$$S_0[W] = \frac{1}{8L^d}\int \mathrm{str}W\hat{\Pi}^{-1}W, \qquad \hat{\Pi}^{-1} = -\left(\frac{\pi}{\Delta}\right)\left[\hbar D\partial^2 + i\Omega_+\right], \qquad (55)$$

¶While, order by order, the perturbative expansion is independent of the parametrization, a formal identification with diagrammatic perturbation theory is achieved most easily in the square root parametrization. In this case, each diagram finds its counterpart in the perturbative expansion.

while an expansion of the vertices (53), gives

$$R_2(\Omega) = \frac{1}{256} \left\langle \left[\int \text{str} \left(\sigma_3^{\text{bf}} W^2 \right) \right]^2 \right\rangle_W, \qquad \langle \cdots \rangle_W = \int DW \cdots e^{-S_0[W]}.$$

The average over fields W can be determined by applying Wick's theorem and employing the corresponding contraction rules. Focusing on unitary and orthogonal symmetry, for general P and R, the contraction rules take the form

$$\langle \text{str} \left[W(\mathbf{r}) P W(\mathbf{r}') R \right] \rangle_W$$
$$= \Pi(\mathbf{r}, \mathbf{r}') \left[\text{str} P \text{str} R - \text{str} P \sigma_3^{\text{ar}} \text{str} R \sigma_3^{\text{ar}} + (2 - \beta) \text{str} \left(P \sigma_3^{\text{ar}} \tilde{R} \sigma_3^{\text{ar}} - P \tilde{R} \right) \right],$$
$$\langle \text{str} \left[W(\mathbf{r}) P \right] \text{str} \left[W(\mathbf{r}') R \right] \rangle_W$$
$$= \Pi(\mathbf{r}, \mathbf{r}') \text{str} \left[PR - P \sigma_3^{\text{ar}} R \sigma_3^{\text{ar}} + (2 - \beta) \left(P \sigma_3^{\text{ar}} \tilde{R} \sigma_3^{\text{ar}} - P \tilde{R} \right) \right],$$

where, as usual, $\beta = 1, 2$ for orthogonal and unitary symmetry respectively, and $\tilde{R} = C^T R^T C$ reflects the operation of charge conjugation. Applying these identities, we recover the Altshuler-Shklovskii correction (24)

$$R_2(\Omega) = \frac{1}{\beta} \text{tr} \hat{\Pi}^2, \tag{56}$$

previously found within diagrammatic perturbation theory. This contribution is associated with the exchange of two diffuson or Cooperon ladders between two closed loops (see Fig. 14a). Performing the trace we find that, for $\Omega \gg E_c = \hbar D/L^2$, $R_2(\Omega) \sim g^{-d/2} (\Delta/\Omega)^{2-d/2}$. In the zero-dimensional limit this result recovers the perturbative random matrix limit. Interestingly, in two-dimensions, the numerical prefactor identically vanishes, and a non-trivial dependence of R_2 on Ω emerges from higher order weak localization corrections[55].

Scaling and Renormalization Group. As a second application, we consider scaling and the renormalization of the diffusion constant. To do so, we employ a perturbative renormalization group separating slow from fast ($|\mathbf{q}| \gg \Lambda$) degrees of freedom. Employing the parametrization $Q = T_<^{-1} Q_> T_<$, where $T_<$ includes slow degrees of freedom and $Q_> = T_>^1 \sigma_3^{\text{ar}} T_>$ involves the fast, we obtain

$$S[Q] = \frac{\pi \nu}{8} \int \text{str} \left[\hbar D \left((\partial Q_>)^2 + 2[Q_>, \partial Q_>] \cdot \mathbf{\Phi}_< + [Q_>, \mathbf{\Phi}_<]^2 \right) + 2i\Omega_+ U_< Q_> \right],$$

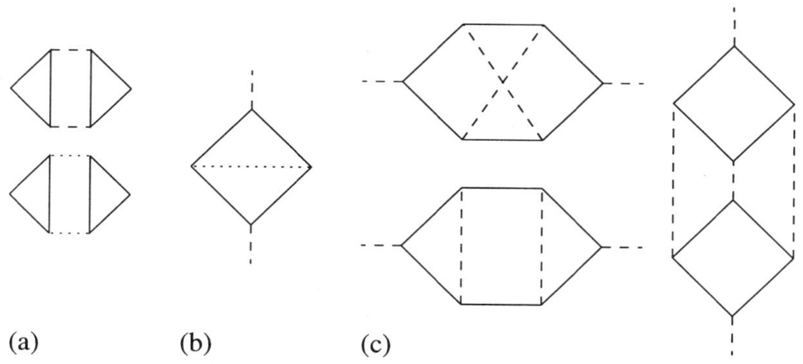

Figure 14. Diagrams contributing to (a) $R_2(\Omega)$ and (b) the weak localization correction to the diffusion constant at one loop. Diagrams contributing to weak localization corrections involving just diffusons are shown in (c) at two loop order.

where $\Phi_< = T_<^{-1}\partial T_<$ and $U_< = T_<\sigma_3^{\mathrm{ar}}T_<^{\mathrm{ar}}$.

A perturbative expansion of $Q_>$ in the generators of the coset yields the series $S = S_< + S_>^{(0)} + S_>^{(1)} + \cdots$ where

$$S_< = \frac{\pi\nu}{8}\int \mathrm{str}\left[\hbar D\,(\partial Q_<)^2 + 2i\Omega_+ Q_<\right],$$

$$S_>^{(0)} = -\frac{\pi\nu}{8}\int \hbar D\mathrm{str}\,(\partial W_>)^2,$$

$$S_>^{(1)} = \frac{\pi\nu}{8}\int \hbar D\mathrm{str}\left[8[W_>,\partial W_>]\Phi_< + 2\left(\Phi_<\sigma_3^{\mathrm{ar}}W_>\right)^2 + 4\left(\Phi_<\sigma_3^{\mathrm{ar}}\right)^2 W_>^2\right]$$
$$+\frac{\pi\nu}{8}\int 2i\Omega_+\mathrm{str}\left[U_<\sigma_3^{\mathrm{ar}}(W_> + W_>^2/2)\right].$$

At one loop, the renormalised action takes the simple form

$$S[Q] = S_< + \left\langle S_>^{(1)}\right\rangle_{W_>}.$$

Applying the contraction rules only one term survives and we obtain

$$\left\langle S_>^{(1)}\right\rangle_{W_>} = \frac{\pi\nu}{8}\int \hbar D \times 2(2-\beta)\mathrm{str}\left[\Phi_<\sigma_3^{\mathrm{ar}}C^T\Phi^T C\sigma_3^{\mathrm{ar}} - \Phi_< C^T\Phi^T C\right].$$

Using the fact that $\Phi_<$ is off-diagonal in ar-space, $C^T(\Phi_<^{\mathrm{ar}})^T C = \Phi_<^{\mathrm{ra}}$, and the identity $\mathrm{str}[\Phi_<,\sigma_3^{\mathrm{ar}}]^2 = \mathrm{str}(\partial Q_<)^2$, we recover the usual diffusive non-linear σ-model with

$$D \to D\left[1 - (2-\beta)\Pi(0,0)\right] = D\left[1 - (2-\beta)\frac{\Delta}{\pi}\int_\Lambda \frac{d\mathbf{q}}{(2\pi)^d}\frac{1}{\hbar D\mathbf{q}^2 + i\Omega_+}\right].$$

This result recovers the standard weak localization correction to the diffusion constant at one loop (see Fig. 14b). It relies on the existence of the Cooperonic mode and is therefore destroyed by the application of a magnetic field ($\beta = 2$). Weak localization corrections arising from the interaction of diffuson modes do exist but only at two loop order (Fig. 14c) and higher.

This completes our brief survey of perturbation theory. We remark that, as well as recovering known results from diagrammatic perturbation theory, the non-linear σ-model affords the possibility of seeking other non-perturbative stationary points of the effective action[56]. Of particular interest are frequency scales comparable to the average level spacing. A stationary phase saddle-point of the σ-model can be identified at $Q = \sigma_3^{\mathrm{ar}} \otimes \sigma_3^{\mathrm{bf}}$. A surprising connection between the perturbation theory in the vicinity of this saddle-point and the conventional saddle-point was recently stressed by Andreev and Altshuler[56].

Zero-Mode

Having examined the short-time perturbative dynamics within the framework of diagrammatic perturbation theory, we now switch attention to consider the long-time non-perturbative limit. In particular, taking $\Omega \ll E_c = \hbar D/L^2$, contributions to the effective action involving spatial fluctuations of Q are strongly suppressed. In this limit, the contribution at leading order in $1/g$ arises from the zero-mode. In parentheses we note that this argument has to be applied with some care: Indeed, each individual non-zero mode configuration is small in $1/g$. However, it may still happen – and does so in systems with localization – that the *cumulative* contribution of all these modes still overpowers the zero-mode. In the simplest case, the zero-mode corresponds to $Q = Q_0$ independent of coordinates, and the effective action takes the universal form

$$S[Q_0] = i\frac{\pi\Omega_+}{4\Delta}\mathrm{str}\left(\sigma_3^{\mathrm{ar}}Q_0\right), \tag{57}$$

independent of dimensionality, diffusion constant, or geometry. In fact, as will be shown below, Eq. (57) is an exact representation of statistical correlators of random matrix ensembles.

Reducing the functional integral to a definite integral allows statistical correlators to be evaluated directly. However, zero-mode integrations require an explicit parametrization of the saddle-point manifold. For this purpose several parametrizations exist in the literature. Each has its own merit and the optimum choice of parametrization depends on the precise nature of the effective action and source. As an application of the zero-mode integration, we will apply the effective action (57) to the two-point correlator of DoS fluctuations for unitary ensembles (i.e. in the absence of time-reversal symmetry). To do so, we will employ the parametrization introduced by Efetov in Ref. 5.

Employing the result of Eq. (53), in the zero-dimensional limit, the two-point correlator of DoS takes the form

$$R_2(\Omega) = \frac{1}{64} \text{Re} \left\langle \left[\text{str} \left(\sigma_3^{\text{ar}} \otimes \sigma_3^{\text{bf}} Q_0 \right) \right]^2 \right\rangle_{Q_0} - 1.$$

Separating the components into eigenvalues and rotations, the zero-dimensional Q-matrix takes the form

$$Q_0 = UH\bar{U},$$

where matrices H and U have the block structure

$$H = \begin{pmatrix} \cos\hat{\theta} & i\sin\hat{\theta} \\ -i\sin\hat{\theta} & -\cos\hat{\theta} \end{pmatrix}_{\text{ar}}, \qquad U = \begin{pmatrix} u_1 u_2 & 0 \\ 0 & v \end{pmatrix}_{\text{ar}}.$$

These matrices are further separated into components

$$\hat{\theta} = \begin{pmatrix} i\theta_1 & 0 \\ 0 & \theta \end{pmatrix}_{\text{bf}} \otimes \mathbf{I}^{\text{tr}}, \qquad \theta_1 > 0, \qquad 0 < \theta < \pi,$$

$$u_1 = \exp \begin{bmatrix} 0 & -2\bar{\eta} \\ 2\eta & 0 \end{bmatrix}_{\text{bf}}, \qquad \eta = \begin{pmatrix} \eta_1 & 0 \\ 0 & -\eta_1^* \end{pmatrix}_{\text{tr}}, \qquad u_2 = \begin{pmatrix} e^{i\chi\sigma_3^{\text{tr}}} & 0 \\ 0 & e^{i\phi\sigma_3^{\text{tr}}} \end{pmatrix}_{\text{bf}},$$

$$v = \exp \begin{bmatrix} 0 & -2i\bar{\kappa} \\ 2i\kappa & 0 \end{bmatrix}_{\text{bf}}, \qquad \kappa = \begin{pmatrix} \kappa_1 & 0 \\ 0 & -\kappa_1^* \end{pmatrix}_{\text{tr}},$$

where $\bar{u}_i \equiv u_i^\dagger$, $\bar{v} \equiv \sigma_3^{\text{bf}} v^\dagger \sigma_3^{\text{bf}}$. Here the coordinates θ_1 and θ represent the eigenvalues of the non-compact bosonic and compact fermionic sector respectively.

Employing this parametrization we find

$$\text{str} \left[\sigma_3^{\text{ar}} Q_0 \right] = 4(\lambda - \lambda_1),$$
$$\text{str} \left[\sigma_3^{\text{fb}} \otimes \sigma_3^{\text{ar}} Q \right] = 4(\lambda_1 + \lambda) - 16 \left(\lambda_1 - \lambda \right) \left[\eta_1^* \eta_1 - \kappa_1^* \kappa_1 \right],$$

where $\lambda = \cos\theta$, $\lambda_1 = \cosh\theta_1$, $\mu = (1 - \lambda^2)^{1/2}$, and $\mu_1 = (\lambda_1^2 - 1)^{1/2}$. Finally, taking into account the invariant measure,

$$dQ_0 = \frac{1}{128\pi^2} \frac{\mu\mu_1}{(\lambda_1 - \lambda)^2} d\theta d\theta_1 d\phi d\chi d\eta_1^* d\eta_1 d\kappa_1^* d\kappa_1,$$

we obtain

$$R_2(\Omega) = \frac{1}{2} \text{Re} \int_1^\infty d\lambda_1 \int_{-1}^1 d\lambda \exp \left[i\Omega_+(\lambda_1 - \lambda) \right]. \tag{58}$$

Explicit integration over the eigenvalues λ, λ_1 recovers the random matrix result (7).

In conclusion, we have demonstrated that, by exploiting the ensemble average, an effective field theory can be developed which is capable of describing the response of disordered conductors from time scales short as compared to the typical transport or diffusion time, to times in excess of the Heisenberg time where properties become universal. As an appendix to this section, we will employ the same supersymmetry approach to show that the zero-dimensional σ-model is equivalent to the supermatrix representation of correlators of random matrix ensembles.

Random Matrix Theory

As indicated above, the supersymmetry approach provides a convenient way of representing statistical correlators of random matrix ensembles. In the following, we will employ this approach to obtain an expression for the generating function of two-point correlators, $\langle \mathcal{Z} \rangle$, in the form of a zero dimensional integral over Q-matrices.

Expressed as a definite integral over $8 \times N$-component supervectors, Ψ, the generating function of the matrix Green function

$$G^{\pm} = \frac{1}{E - H \pm i0}, \qquad H = H_s + i\alpha H_a,$$

of an arbitrary complex Hermitian matrix takes the form

$$\mathcal{Z} = \int d[\Psi] e^{-\mathcal{L}[\Psi] - \mathcal{L}_J[\Psi]},$$

where

$$\mathcal{L}[\Psi] = \frac{i}{2} \bar{\Psi} \left[E - \frac{\Omega_+}{2} \sigma_3^{\mathrm{ar}} - \hat{H}_s - i\alpha\sigma_3^{\mathrm{tr}} \hat{H}_a \right] \Psi, \qquad \mathcal{L}_J = i \left(\bar{J}\Psi + \bar{\Psi}J \right)$$

The complex $N \times N$ Hamiltonian is comprised of symmetric (H_s) and antisymmetric (H_a) matrix elements drawn at random from a Gaussian ensemble (2). For generality, a parameter α allows a smooth interpolation from the Gaussian orthogonal $(\alpha = 0)$ to the unitary $(\alpha = 1)$ ensemble.

Taking our notation from Ref. 57, the ensemble average over the Gaussian distribution generates an interaction of the supervectors,

$$\langle Z \rangle = \int d[\Psi] e^{-\mathcal{L}_0[\Psi] - \mathcal{L}_I[\Psi] - \mathcal{L}_J},$$

$$L_0[\Psi] = \frac{i}{2} \bar{\Psi} \left[E - \frac{\Omega_+}{2} \sigma_3^{\mathrm{ar}} \right] \Psi, \qquad L_I[\Psi] = \frac{\lambda^2}{4N} \mathrm{str} \left(A^2 + \alpha^2 (A\sigma_3^{\mathrm{tr}})^2 \right),$$

where λ denotes the variance of the random distribution, and the 8×8 supermatrix takes the form $A = i\Psi \otimes \bar{\Psi}$. Note that, without the introduction of the composite vector Ψ, the interaction of the fields generated by the ensemble average could not be expressed in terms of the single dyadic product A.

In order to decouple the interaction by means of a Hubbard-Stratonovich transformation, it is convenient to introduce the matrix

$$\tilde{A} = \frac{a_+ + a_-}{2} A + \frac{a_+ - a_-}{2} \sigma_3^{\mathrm{tr}} A \sigma_3^{\mathrm{tr}}, \qquad a_\pm = (1 \pm \alpha^2)^{1/2},$$

such that $\mathcal{L}_I = (\lambda^2/4N)\mathrm{str}\tilde{A}^2$. As a result we obtain

$$\langle Z \rangle = \int d[\Psi] d[Q] \exp \left[-\frac{N}{4} \mathrm{str} Q^2 - \frac{\lambda}{2} \bar{\Psi}\tilde{Q}\Psi - \mathcal{L}_0[\Psi] - \mathcal{L}_J[\Psi] \right],$$

where

$$\tilde{Q} = \frac{a_+ + a_-}{2} Q + \frac{a_+ - a_-}{2} \sigma_3^{\mathrm{tr}} Q \sigma_3^{\mathrm{tr}}.$$

Performing the Gaussian integral over Ψ we obtain

$$\langle Z \rangle = \int d[Q] \exp\left[-\frac{N}{4}\mathrm{str}\left(Q^2 - 2\ln\mathcal{G}^{-1}\right) + \frac{i}{2}\bar{J}\mathcal{G}J \right],$$

where the supermatrix Green function takes the form

$$\mathcal{G}^{-1} = E - \frac{\Omega_+}{2}\sigma_3^{\mathrm{ar}} - i\lambda\tilde{Q}.$$

To proceed, we first establish the mean-field, or saddle-point of the action to leading order in N. Taking the frequency Ω/Δ, and symmetry breaking source α to be parametrically smaller than N, the saddle-point equation takes the form

$$Q_{\mathrm{sp}} = -i\lambda\frac{1}{E - i\lambda Q_{\mathrm{sp}}}.$$

Analyticity then requires the solution

$$Q_{\mathrm{sp}} = -\frac{i}{2\lambda}E + \left(1 - \frac{E^2}{4\lambda^2}\right)^{1/2}\sigma_3^{\mathrm{ar}}.$$

From this result we recover the semi-circular average DoS (5),

$$\langle \nu(E) \rangle = \frac{1}{8}\frac{N}{\pi\lambda}\mathrm{Re}\left\langle \mathrm{str}\left(\sigma_3^{\mathrm{bf}} \otimes \sigma^{\mathrm{ar}} Q_{\mathrm{sp}}\right)\right\rangle_{Q_{\mathrm{sp}}} = \frac{N}{\pi\lambda}\left(1 - \frac{E^2}{4\lambda^2}\right)^{1/2}.$$

Taking the energy at the center of the band, $E = 0$, and expanding to leading order in N, we obtain the effective zero-dimensional non-linear σ-model

$$\langle Z \rangle = \int d[\Psi]d[Q]e^{-S_0[Q] - S_J[Q]}, \tag{59}$$

with the effective action

$$S_0[Q] = i\frac{\pi\Omega_+}{4\Delta}\mathrm{str}\left(\sigma_3^{\mathrm{ar}} Q\right) - \frac{N\alpha^2}{8}\mathrm{str}\left[Q, \sigma_3^{\mathrm{tr}}\right]^2. \tag{60}$$

In particular, for $\alpha = 0$, we recover the zero-dimensional action (57) obtained previously. Time-reversal symmetry breaking implied by the application of H_a leads to a freezing out of the Cooperon degrees of freedom for $\alpha \sim \sqrt{N}$.

We have seen how the application of ensemble averaging allowed quantum statistical properties of weakly disordered metals to be cast in the form of a functional field integral involving the low-lying modes of density relaxation. Does an analogous theory exist for systems which are irregular but not random? The key to the construction of an effective field theory of clean chaotic structures relies on the recognition that the spectrum itself provides a statistical ensemble. By averaging over an interval of energy it is possible to establish a "ballistic" non-linear σ-model within the same framework. It is to the derivation of this effective action that we now turn.

QUANTUM CHAOS: THE BALLISTIC σ-MODEL

The derivation of the diffusive non-linear σ-model relied explicitly on the impurity average over realizations of the random potential. Can an analogous approach be developed for systems which, while not belonging to a statistical ensemble, can be considered as quasi-classical? The first suggestion that such a description was possible came in an insightful work by Muzykantskii and Khmel'nitskii[6]. Guided by the quasi-classical Boltzmann description of density relaxation in the ballistic regime of weakly disordered conductors, they proposed a generalization of the diffusive action. Although still phenomenological, this "ballistic action" established a crucial bridge between the semi-classical description of level statistics offered by the Trace formula, and the universal random matrix correlations.

In a subsequent development by Andreev et $al.$[7] a formal justification of the phenomenological approach was obtained. The crucial step involved the realization that a statistical ensemble is provided by the spectrum itself. By averaging over a range of energy levels, the ballistic σ-model, proposed by Muzykantskii and Khmel'nitskii, was constructed explicitly. Whether, when properly regularised, correlations converge to well defined distributions is a matter of ongoing research in quantum chaos (and not just within the restricted subset of field theoretical approaches). The question behind is whether energy averaging in a individual system does actually suffice to define smooth distributions or whether extra averaging over external parameters needs to be introduced. To the best of our knowledge a conclusive answer is still outstanding.

In its first application, the statistical average over energy was applied directly to the Green function $\hat{G} = (E - \hat{H})^{-1}$. However, as a matter of convenience, to obtain the ballistic model here we will exploit a more recent development by Zirnbauer[58]. This approach requires the introduction of the generalized Green function

$$\hat{G}^+(\varphi) = \frac{1}{1 - e^{i\varphi + }\hat{U}^\dagger}, \qquad \hat{G}^- = (\hat{G}^+)^\dagger,$$

where $\varphi^+ = \varphi + i0$, and the Hamiltonian is expressed through the unitary time evolution operator $\hat{U} = \exp[i\hat{H}t/\hbar]$. With this definition, the DoS

$$\nu(\varphi) = \frac{1}{\pi}\mathrm{tr}\left[\mathrm{Re}\ \hat{G}^-(\varphi) - \frac{1}{2}\right], \tag{61}$$

is periodic on the interval $[0, 2\pi]$. Since, when projected onto the unitary evolution operator $\hat{U}$, the semi-infinite spectrum of the Hamiltonian generates a multiple covering of the unit circle, caution is required in deducing spectral statistics from eigenphase statistics. However, if the number of levels inside one wind of the phase is large, and the statistical correlations decay quickly, the statistics of each interval become separable.

Hamiltonian

To keep our discussion concrete, we will focus on the particular case of a Hamiltonian describing a particle confined to a two-dimensional closed system of size L and subject either to a slowly varying inhomogeneous magnetic field $\mathbf{B}(\mathbf{r}) = \partial_\mathbf{r} \times \mathbf{A}(\mathbf{r})$,

$$\hat{H} = \frac{1}{2m}\left(\hat{p} - \frac{e}{c}\mathbf{A}\right)^2, \tag{62}$$

or, to make contact with our previous analysis, to a slowly varying random scalar potential (8). The correlation length ξ of the inhomogeneous magnetic field is chosen to be large such that $\xi \gg v/\max(\omega_0, \omega_{\rm rms})$ where ω_0 and $\omega_{\rm rms}$ represent respectively the average and standard deviation of the random cyclotron frequency,

$$\omega_c(\mathbf{r}) = \frac{eB(\mathbf{r})}{mv},$$

and v denotes the velocity at the Fermi energy E. Finally, in both cases, the strength of the random potential is adjusted such that the dynamics can be treated quasi-classically, (i.e. $\lambda \ll \ell$ where ℓ denotes the scattering mean free path of the potential). To keep our discussion simple, in contrast to the previous analysis, we will limit our discussion to the case in which time-reversal symmetry is fully broken. In practice, this confines attention to time scales $t \gg L_c/v$, where $L_c = (eB_{\rm rms}/hc)^{1/2}$ represents the typical magnetic length.

Although, in the present case, it is possible to define a statistical ensemble involving the random scalar or vector potential, we will suppose a single realization of each and implement the procedure of phase averaging described above.

Field Integral

Taking our notation from above, statistical two-point correlators of advanced and retarded Green functions can be obtained from the generating function

$$\mathcal{Z}(J) = \int D\psi\, e^{-\mathcal{L}[\psi] - \mathcal{L}_J[\psi]}, \tag{63}$$

$$\mathcal{L}[\psi] = {\rm tr}\left[\bar\psi\left(\mathbf{I} - e^{-i\sigma_3^{\rm ar}(\bar\varphi - \sigma_3^{\rm ar}\varphi^+/2)}\begin{pmatrix}\hat U & 0 \\ 0 & \hat U^\dagger\end{pmatrix}_{\rm ar}\right)\psi\right], \qquad \mathcal{L}_J[\psi] = {\rm tr}\left(\bar\psi J + \bar J \psi\right),$$

where the trace is taken over the Hilbert space. Here the structure and symmetry properties of the supervector fields mirror those introduced previously. In this case ψ only has four components, in ar and bf-blocks. The corresponding Green functions are obtained from the generating function by

$$\hat G^{\mp} = \frac{1}{2}{\rm tr}\left[\frac{\delta}{\delta \bar J_{\mp}} \otimes \frac{\delta}{\delta J_{\mp}}\right]_{\rm bf} \mathcal{Z}\Big|_{J=0}.$$

Phase Averaging

To obtain the statistical average over the center of mass $\bar\varphi$, we will exploit an ingenious and general integral identity introduced by Zirnbauer[58]. For a unitary subgroup $U(N)$ of $GL(N)$, there holds the identity

$$\int_{\Theta(N)} d\Theta \exp\left(i\left[\bar\psi_+^\alpha \Theta \psi_+^\alpha + \bar\psi_-^\alpha \Theta^\dagger \psi_-^\alpha\right]\right) = \int d\mu(z, \bar z) \exp\left[\bar\psi_+^\alpha z^{\alpha\beta}\psi_-^\beta + \bar\psi_-^\alpha \bar z^{\alpha\beta}\psi_+^\beta\right] \tag{64}$$

where $d\Theta$ represents the Haar measure of $U(N)$, and

$$d\mu(z, \bar z) = d(z, \bar z)\,{\rm sdet}\,(\mathbf{I} - \bar z z)^{-N}$$

denotes the integration measure with $d(z, \bar z)$ defining the flat Berezin measure on the symmetric space of supermatrices $U(1, 1/2)/[U(1/1) \times U(1/1)]$. The matrices z and $\bar z = z^\dagger \sigma_3^{\rm bf}$ are precisely those parametrising the non-linear supermatrix manifold $Q = T^{-1}\sigma_3^{\rm ar}T$, where

$$T = \begin{pmatrix}\mathbf{I} & z \\ \bar z & \mathbf{I}\end{pmatrix}, \qquad T^{-1} = \begin{pmatrix}\mathbf{I} & -z \\ -\bar z & \mathbf{I}\end{pmatrix}\begin{pmatrix}(\mathbf{I} - z\bar z)^{-1} & 0 \\ 0 & (\mathbf{I} - \bar z z)^{-1}\end{pmatrix}. \tag{65}$$

Eq. (64) can be thought of as a Hubbard-Stratonovich transformation decoupling the interaction generated amongst the fields by the matrices Θ. Remarkably, in contrast to Gaussian averaging, the Hubbard-Stratonovich fields already belong to the non-linear manifold, apparently circumventing the need for separation of massive from massless degrees of freedom.

In the present case this result may be applied directly: $\alpha = (a, \sigma)$, where $a = 1, \ldots, n$ denotes the Hilbert space indices, $\sigma = \mathrm{b/f}$, and the average over the $U(1)$ phase $\bar{\varphi}$ corresponds to $N = 1$. Averaging the generating function $\mathcal{Z}$, and applying Eq. (64), we obtain

$$\langle \mathcal{Z} \rangle_{\bar{\varphi}} \equiv \frac{1}{2\pi} \int_0^{2\pi} d\bar{\varphi}\, \mathcal{Z} = \int d\mu(z, \bar{z}) \int D\psi\, e^{-\mathcal{L}' - \mathcal{L}_J},$$

$$\mathcal{L}'[\psi] = \mathrm{tr}\bar{\psi}\mathcal{G}^{-1}\psi, \qquad \mathcal{G}^{-1} = \begin{bmatrix} \mathbf{I} & -e^{i\varphi_+/2}\hat{U}z \\ -e^{i\varphi_+/2}\hat{U}^\dagger \bar{z} & \mathbf{I} \end{bmatrix}_{\mathrm{ar}}.$$

Notice that z and $\bar{z}$, and hence Q, are non-local in the Hilbert space indices. Integrating over the superfields ψ we then obtain

$$\langle \mathcal{Z} \rangle = \int d\mu(z, \bar{z})\mathrm{sdet}\mathcal{G}^{-1}e^{-S_J}, \qquad S_J = -\mathrm{tr}\bar{J}\mathcal{G}J. \tag{66}$$

By making use of the parametrization in Eq. (65), from which $dQ = d(z, \bar{z})$, as well as the identity

$$\sigma_3^{\mathrm{ar}}T + T\sigma_3^{\mathrm{ar}} = 2\sigma_3^{\mathrm{ar}}, \qquad T^{-1} = \frac{1}{2}(Q\sigma_3^{\mathrm{ar}} + \mathbf{I}),$$

Eq. (66) can be expressed in the coordinate free or invariant form,

$$\langle \mathcal{Z} \rangle = \int dQ e^{-S-S_J} \tag{67}$$

$$S = -\mathrm{str}\ln\left[\frac{1}{2}\left(\begin{array}{cc} e^{-i\varphi_+/2}\hat{U}^\dagger - \mathbf{I} & 0 \\ 0 & e^{-i\varphi_+/2}\hat{U} - \mathbf{I} \end{array}\right)(Q\sigma_3^{\mathrm{ar}} + \mathbf{I}) + \mathbf{I}\right]$$

$$= -\mathrm{str}\ln\left[\mathbf{I} - \tanh\left(\frac{i}{4}\varphi_+\sigma_3^{\mathrm{ar}} + \frac{1}{2}\ln\hat{U}\right)Q\right]. \tag{68}$$

This result, which is *exact*, is also quite general. The statistical average properties of a general unitary operator $\hat{U}$ are expressed as an integral over a non-linear manifold of supermatrices, Q. In particular, the same procedure can be equally applied to a quantum chaotic map, $\hat{U}$[59]. Applied directly to the unitary (orthogonal or symplectic) matrix ensemble, Eq. (68) correctly recovers the statistical correlations of Dyson's circular ensemble[58]. In the following section we will apply this action to study the quantum statistical properties of the Hamiltonian (62). By implementing a semi-classical approximation we will identify the low-lying modes of density relaxation and thereby obtain a ballistic analogue of the diffusive σ-model.

Semi-classics

Applying Eq. (68) to the Hamiltonian (62), and taking $\varphi_+ = \Omega_+ t/\hbar$, we obtain the effective action

$$S = -\mathrm{str}\ln\left(\mathbf{I} - i\tan\left[\left(\frac{\Omega_+}{2}\sigma_3^{\mathrm{ar}} + \hat{H}\right)\frac{t}{2\hbar}\right]Q\right). \tag{69}$$

Until now the calculation has involved only formal manipulations of the average generating function. To continue we have to apply approximations, i.e. we need to identify contributions to the field integral whose action is small. The physical significance of these 'soft modes' will become clear below. At any rate, however, a word of caution is due at this point: It has turned out that the σ-model (69) as it stands is *ill-defined* [60]. The point is that, besides the soft modes we are going to discuss below, a host of unphysical zero-mode configurations to the field integral exist. The presence of these modes is intimately related to the above question whether or not energy (or phase) averaging actually suffices to give smooth correlation functions. The question if and how these extra modes can be eliminated via suitably designed averaging procedures is a matter of current research. We repeat that this problem is not exclusive to the field theoretical approach taken here but does affect statistical approaches to chaos in general. Here we take a naïve approach, i.e. we will assume that the unphysical modes have been regularized away such that only the below discussed contributions – which are insensitive to extra averaging procedures[60] – survive. At present it can not be safely excluded that the picture discussed here will have to undergo qualitative changes.

Our starting point will be to assume the legitimacy of the semi-classical expansion $\delta\hat{H} = \hat{H} - H(\mathbf{x})$, where $H \equiv W(\hat{H})$ denotes the Wigner transform of the quantum Hamiltonian:

$$W(\hat{H}) \equiv H(\mathbf{x}) = \int d\mathbf{r}\, e^{-i\mathbf{r}\cdot\mathbf{P}/\hbar} H(\mathbf{R} + \mathbf{r}/2, \mathbf{R} - \mathbf{r}/2),$$

$$= \int \frac{d\mathbf{p}}{h^d} e^{i\mathbf{R}\cdot\mathbf{p}/\hbar} H(\mathbf{P} + \mathbf{p}/2, \mathbf{P} - \mathbf{p}/2), \tag{70}$$

$$H(\mathbf{r}_1, \mathbf{r}_2) = \left\langle \mathbf{r}_1 | \hat{H} | \mathbf{r}_2 \right\rangle, \qquad H(\mathbf{p}_1, \mathbf{p}_2) = \left\langle \mathbf{p}_1 | \hat{H} | \mathbf{p}_2 \right\rangle. \tag{71}$$

The phase space coordinate $\mathbf{x} \equiv (\mathbf{R}, \mathbf{P})$ at which we take the Wigner transform will be fixed later such that $\delta\hat{H}$ is small. Translational invariance of the generating function in energy ensures that an expansion in $\delta\hat{H}$ does not involve factors of the absolute energy $H(\mathbf{x})$. In this notation, an expansion leads to the approximation $S \simeq S_0 + S_R$, where

$$S_0 = i\frac{t}{2\hbar}\text{str}\left[\frac{\Omega_+}{2}\sigma_3^{\text{ar}}Q + \delta\hat{H}Q\right], \qquad S_R = -\frac{t^2}{8\hbar^2}\text{str}\left[\delta\hat{H}Q\right]^2 \tag{72}$$

$$S_J = -\frac{1}{2}\int d\mathbf{x}\left[\bar{J}Y^{-1}\left(\sigma_3^{\text{ar}}Q + \mathbf{I}\right)YJ\right],$$

where $Y = \text{diag}(e^{-iH(\mathbf{x})t/2\hbar}, e^{+iH(\mathbf{x})t/2\hbar})$

In the following, those contributions to the Hamiltonian which vary smoothly on a scale comparable to the mean free path, ℓ, will be included in S_0 and treated within a semi-classical approximation. For reasons which will become apparent, quantum corrections to the classical expansion are included through contributions to S_R.

To facilitate the semi-classical expansion it is convenient to switch to the Wigner representation. Treating each term separately, let us consider the first order term in t. Since, in the Wigner representation, the trace of a product of operators involves the product of the Wigner transforms, we find

$$\text{str}\left(\delta\hat{H}Q\right) = \text{str}\left(T^{-1}\sigma_3^{\text{ar}}[T, \delta\hat{H}]\right) = \int d\mathbf{x}\,\text{str}\left(W(T^{-1})\sigma_3^{\text{ar}}W([T, \delta\hat{H}])\right)$$

$$= -i\hbar \int d\mathbf{x}\,\text{str}\left(T^{-1}(\mathbf{x})\sigma_3^{\text{ar}}\{H(\mathbf{x}), T(\mathbf{x})\}\right),$$

where $\{H, \ \}$ represents the classical Poisson bracket

$$\{O_1, O_2\} = \partial_{\mathbf{q}} \cdot O_1 \partial_{\mathbf{p}} O_2 - \partial_{\mathbf{p}} O_1 \cdot \partial_{\mathbf{q}} O_2,$$

and $d\mathbf{x} \equiv d\mathbf{r}d\mathbf{p}/h^d$. Here, for simplicity, we assume that H commutes with Q in the superindices. As a result, we obtain

$$S_0 = i\frac{t}{4\hbar} \int d\mathbf{x}\,\mathrm{str}\left[\Omega_+\sigma_3^{\mathrm{ar}}Q - 2i\hbar T^{-1}\sigma_3^{\mathrm{ar}}\{H,T\}\right]. \tag{73}$$

In the semi-classical approximation ($\hbar \to 0$), the contribution from this leading order term dominates the short-time dynamics.

Mode Locking

In its present form the effective action (73) is ill-defined. The propagator generates the classical flow of phase space density, and thereby admits arbitrary fluctuations transverse to the constant energy shell. This circumstance admits to the existence of a continuum of zero modes. Anticipating (from our previous analysis) the relevance of a single zero mode, it is necessary to seek a mechanism of mode-locking.

The mechanism by which fluctuations of Q transverse to the constant energy shell become frozen out remains an important and outstanding question. Recent, and as yet preliminary, investigations by Zirnbauer[60] suggest a resolution in which the ambiguity in the quantization of an individual classical Hamiltonian can be used to identify a statistical ensemble (cf. the above remarks on 'extra-averaging'). In the semi-classical approximation, ensemble averaging generates a mass which strongly suppresses transverse fluctuations of Q.

To proceed, we will adopt a more circumspect approach and apply the condition of mode-locking as an ansatz to be checked self-consistently. We therefore take the degrees of freedom transverse to the constant energy shell to be locked, which fixes the normalization,

$$\int d\mathbf{x} = \frac{h}{t\Delta}, \tag{74}$$

where the integration runs over the energy shell. Taken together, Eqs. (73) and (74) determine the ballistic σ-model

$$\langle \mathcal{Z} \rangle = \int DQ e^{-S_0-S_J} \tag{75}$$

with the in-shell action

$$S_0 = i\frac{\pi}{2\Delta} \int \mathrm{str}\left[\Omega_+\sigma_3^{\mathrm{ar}}Q - 2i\hbar T^{-1}\sigma_3^{\mathrm{ar}}\{H,T\}\right], \tag{76}$$

where the supermatrix fields $Q = T^{-1}\sigma_3^{\mathrm{ar}}T$ depend on the $2d-1$ phase space coordinates parametrising the constant energy shell, $\mathbf{x}_\| = (\mathbf{r},\mathbf{p})_{2d-1}$. With this definition the integration measure is normalized such that $\int \equiv \int d\mathbf{x}_\| = 1$.

Perturbation Theory

To make sense of the functional integral (76) we must identify the low lying modes of the action. To do so, we adopt a perturbative expansion of the effective action around the zero mode. Employing the exponential parametrization (54), we obtain

$$S[W] = -\frac{\pi}{\Delta} \int \mathrm{str}\left[W\hat{\Pi}^{-1}W\right] + O(W^4), \qquad \hat{\Pi}^{-1} = \hbar\sigma_3^{\mathrm{ar}}\{H,\ \} - i\Omega_+. \tag{77}$$

It is, therefore, tempting to associate the low energy degrees of freedom with eigenmodes of the unitary (*reversible*) evolution operator $e^{-\{H,\ \}t}$. However this identification is incorrect.

48

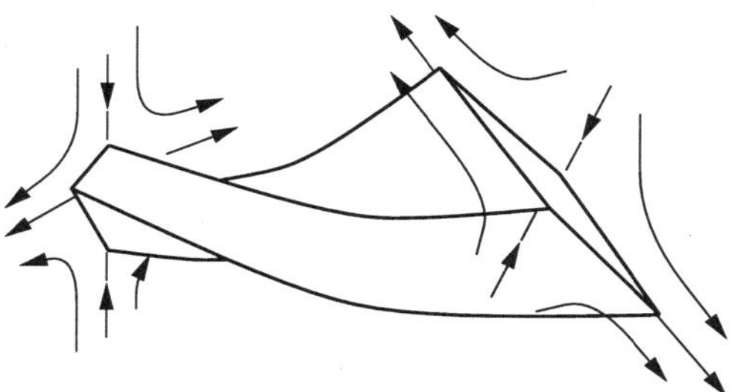

Figure 15. Schematic diagram showing the tendency for a hyperbolic system to stretch exponentially rapidly along the unstable direction and contract along the stable direction.

As with any functional integral there is a need to define an appropriate regularization. For example the functional integral (75) may be understood as the limit $a \to 0$ of a product of definite integrations over a discretized space, where a denotes the discretization cell size. This admits to functions $T(\vec{x}_\parallel)$ which are *smooth* and square integrable. In seeking such a basis, the eigenfunctions of the classical evolution operator seem to be the natural choice. However, the intricate nature of chaotic classical evolution cause these eigenfunctions to lie outside the Hilbert space: The chaotic dynamics of probability densities involves contraction along stable manifolds, together with stretching along unstable ones (see Fig. 15). Thus, in the course of time, an initially non-uniform distribution turns into a singular function on the unstable manifold, which in turn covers the whole energy shell densely. Therefore the eigenfunctions of $\hat{\Pi}$ are not square integrable and their contribution to the functional integral cannot be directly recovered by the discretization procedure involved in evaluating the functional integral.

Thus, in choosing a convenient basis one has to take account of the regularization. Its primary effect is to truncate the contraction along the stable manifold and thereby render the classical evolution *irreversible*. It is the eigenfunctions of this regularized classical evolution operator that serve as a suitable basis for the quantum mechanical correlator. Remarkably, as the strength of the regulator is taken to zero, the spectrum, $\{\gamma_n\}$, of the resulting operator, known in the literature as the Perron-Frobenius operator, reflects intrinsic irreversible properties of the *purely classical dynamics*[61, 62, 63, 64, 65].

In ergodic systems, the leading eigenvalue $\gamma_0 = 0$ is non-degenerate, and manifests the conservation of probability density. Thus any initial density relaxes, in the course of time, to the state associated with γ_0. If, in addition, this relaxation is exponential

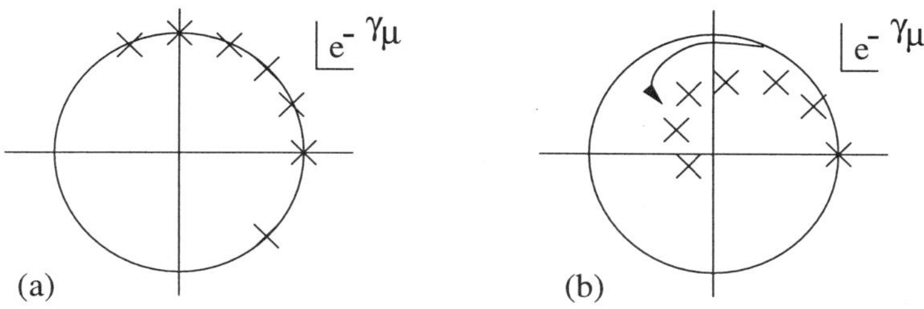

Figure 16. (a) Formal representation of the eigenvalues of the reversible evolution operator which lie on the unit circle. In non-integrable systems, the eigenvalues form an infinitely degenerate and dense set. (b) Poles or resonances of the coarse-grained, regularised Perron-Frobenius operator. In systems which are uniformly hyperbolic, all resonances lie within the unit circle (apart from the zero mode), and correspond to modes of density relaxation.

in time, then the Perron-Frobenius spectrum has a gap associated with the slowest decay rate. Thus, the first non-zero eigenvalue has the property $\gamma_1' \equiv \mathrm{Re}(\gamma_1) > 0$. This gap sets the ergodic time scale, $t_c = 1/\gamma_1'$ over which the classical dynamics relaxes to equilibrium. In the case of disordered metallic grains it coincides with the transport time, t_D, while in ballistic systems or quantum billiards it is of order of the flight time across the system.

In the limit $\Omega \gg \hbar\gamma_1'$ the perturbative expansion (77) is justified, and R_2 takes the form of Eq. (56) but where the diffusion operator is replaced by the Perron-Frobenius operator. This result coincides with that obtained within the "diagonal approximation" of periodic orbit theory[47] thereby establishing a direct correspondence of the field theoretic and semi-classical approaches. In the limit $\Omega \ll \hbar\gamma_1'$ the dominant contribution to the semi-classical action (76) arises from the ergodic classical distribution, the zero-mode $\{H, T\} = 0$. Taking this contribution alone, the action coincides with the *universal* action (57) from which one can deduce that correlations coincide with those of random matrix ensembles.

Although much attention has been paid to the spectral properties of the Perron-Frobenius operator, little is known about the resonance spectrum in dynamical systems. Uniformly hyperbolic or axiom A systems, such as billiards with constant negative curvature, are characterized by exponential decay of classical correlation functions and the resonance spectrum of the Perron-Frobenius operator has a gap. Irregular boundary scattering in two-dimensional billiards on flat surfaces is characterized by isolated resonances, associated with macroscopic modes, together with gapless resonances (of low spectral weight) associated with a degeneracy of weakly unstable periodic orbits.

Although the resonance spectra of the Perron-Frobenius operator are well defined, states associated with the resonances are singular. This has important implications on the role of quantum weak localization corrections to the perturbative result.

Weak Localization

Having identified the degrees of freedom associated with modes of classical phase space density relaxation we now turn our attention to weak localization corrections. To do so, we digress from our consideration of a system in a magnetic field and temporarily restore T-invariance (for the significance of T-invariance, cf. the above discussion on (one-loop) weak localization). As with the disordered conductor, such a generalization implies an action involving Q-matrices of twice the rank,

$$S[Q] = i\frac{\pi}{4\Delta} \int \mathrm{str}\left[\Omega_+\sigma_3^{\mathrm{ar}}Q - 2i\hbar T^{-1}\sigma_3^{\mathrm{ar}}\{H, T\}\right], \tag{78}$$

where, under the Wigner transform, the symmetry properties of the generators under charge conjugation induce the constraint

$$T^\dagger(\mathbf{x}) = CT^T(\bar{\mathbf{x}})C^T,$$

where $\bar{\mathbf{x}} = (-\mathbf{p}, \mathbf{q})$.

Previously, we have seen that the principle mechanism of weak localization arises from the enhancement of return probability due to the interference of Feynman paths with their time-reversed counterpart. At a perturbative level, this mechanism is described by a one-loop contribution to the perturbation theory. Applying the perturbative expansion described previously, to one loop order, the effective action takes the form

$$S[W] = -\frac{\pi}{2\Delta} \int \mathrm{str}\left[W\left(1 - W^2\right)\hat{\Pi}^{-1}W\right] + O(W^6). \tag{79}$$

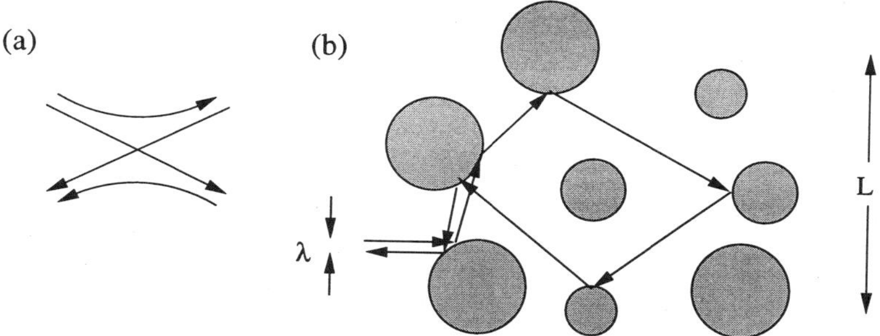

Figure 17. Schematic diagram showing (a) the junction corresponding to the Hikami box in ballistic structures, and (b) the insertion of the Hikami box into a configuration of trajectories in the Lorentz gas.

Following Aleiner and Larkin[66], from a perturbative expansion, we obtain the weak localization correction to the (retarded component of the) propagator

$$2\pi\nu\delta\Pi(1,2) = \Pi(1,\bar{2})\Pi(\bar{2},2) + \Pi(1,\bar{1})\Pi(\bar{1},2)$$
$$+ \int d3\,\Pi(1,3)\Pi(\bar{3},2)(2i\Omega - \hbar\{H, \})_3\Pi(3,\bar{3}), \qquad (80)$$

where we have employed the short-hand notation $1 = (\mathbf{p}_1, \mathbf{q}_1)_{\parallel}$ etc.[||] Here the first two terms represent vertex corrections while the third term accounts for the standard weak localization correction.

The mechanism of weak localization is reminiscent of that encountered in disordered systems. In particular, for the weak localization correction (80) to operate a particle must, at some point, return with opposite momentum (see Fig. (17)). At the level of classical trajectories such an event is ruled out (save for normal reflection from a boundary). We might, however, hope that weak localization corrections are reinstated by the regularization scheme described previously. However, in contrast to the perturbative contribution to the two-point correlator of the DoS, the quantum correction (80) relies explicitly on the eigenfunctions of the Perron-Frobenius operator. These eigenfunctions are strongly sensitive to the regularization scheme, and become singular as the regulator is taken to zero.

We are, therefore, faced with a dilemma. The singular contribution to the quantum correction is ill-defined. To circumvent these difficulties, Aleiner and Larkin[66] introduced a coarse graining of the space to stabilize the theory. Operationally, this involved the introduction of a weak but finite *quantum* impurity potential. This regularising scheme, which serves to mimic the quantum resolution, introduced a new quantum time scale into the action. Taking the resolution to be set by the Fermi wavelength λ, the time scale on which weak localization corrections can operate is set by the time scale for two trajectories which are initially at a separation λ to become separated by some macroscopic scale, L. With a typical Lyapunov exponent of the system, Γ, this time scale, known as the "Ehrenfest time", is given by

$$t_E = \Gamma^{-1}\ln\left(\frac{L}{\lambda}\right). \qquad (81)$$

[||]Notice that, as a consequence of particle conservation, Eq. (80) generates a correction only to the non-zero modes of the Perron-Frobenius operator (i.e. $\int d1\,\delta\Pi(1,2) = 0$).

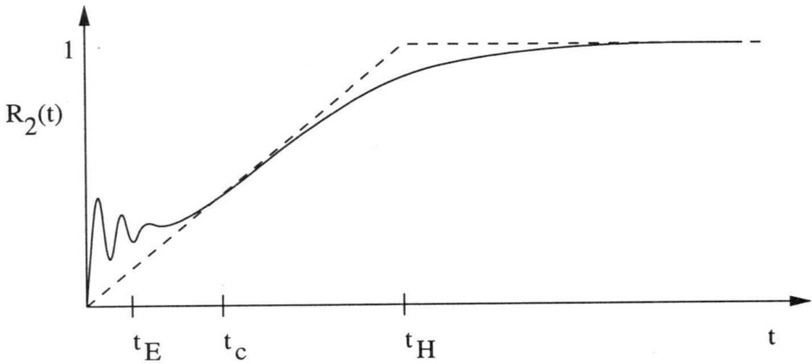

Figure 18. Schematic diagram showing the dynamic structure factor of the two-point correlator of DoS fluctuations.

To assess the implications that these conclusions have on level statistics we can consider the dynamic structure of the two-point correlator of DoS fluctuations (see Fig. 18). According to the analysis above it is possible to identify three relevant time scales, the Ehrenfest time t_E, the lowest non-zero eigenvalue of the Perron-Frobenius operator $t_c = 1/\gamma'_1$, and the Heisenberg time $t_H = \hbar/\Delta$. On time scales shorter than t_E, the results above show that weak localization corrections can not operate and the "diagonal approximation" (together with the random matrix zero mode contribution) is exact. On time scales in excess of t_E, weak localization corrections emerge. However, whether or not the dominant contributions to these corrections arise solely from properties of the classical relaxation modes, or are sensitive to the artificial coarse-graining of the system, depends on the relative magnitude of the quantum (t_E) and classical (t_c) times scales. Indeed, the $\hbar$ dependence of t_E suggests that, in the limit $\hbar \to 0$, all weak localization corrections outside the ergodic regime vanish.

Faced with the scenarios outlined above, the following conclusion can be drawn. If the weak localization correction is controlled by the classical dynamics of the Perron-Frobenius operator, the particular mechanism of coarse-graining is irrelevant. It serves merely to regularise singularities in the classical phase space density and thereby open the channel to weak localization. If, however, the weak localization corrections depend explicitly on the coarse-graining, consistency requires a careful consideration of the leading quantum processes contributing corrections to the classical dynamics. Taking the second point of view, in the next section we examine the leading quantum corrections to the classical action for the Hamiltonian with a smoothly varying inhomogeneous scalar or vector potential.

Quantum Irreversibility

The manner in which classical irreversibility enters the action depends sensitively on the dominant mechanism of quantum scattering or "diffraction". Operationally, this requires the identification and exploitation of slow and fast degrees of relaxation in the system. To illustrate the general approach, let us consider the case of the inhomogeneous magnetic and scalar fields.

As indicated above, irreversibility enters through a mechanism of quantum scattering and can appear only at second order (or higher) in the Hamiltonian. We therefore focus on the second order term in the expansion, S_R (72). Expressed in terms of the

Wigner transform,

$$S_R = -\frac{t^2}{8\hbar^2} \int d\mathbf{X} d\mathbf{X}' O(\mathbf{X}, \mathbf{X}') \mathrm{str}\left[Q(\mathbf{X})Q(\mathbf{X}')\right], \tag{82}$$

where

$$O(\mathbf{X}, \mathbf{X}') = \int d\mathbf{r} d\mathbf{r}' e^{i(\mathbf{r}\cdot\mathbf{P}+\mathbf{r}'\cdot\mathbf{P}')/\hbar} \left\langle \mathbf{R} - \mathbf{r}/2|\hat{H}|\mathbf{R}' + \mathbf{r}'/2 \right\rangle \left\langle \mathbf{R}' - \mathbf{r}'/2|\hat{H}|\mathbf{R} + \mathbf{r}/2 \right\rangle, \tag{83}$$

$$= \int \frac{d\mathbf{p} d\mathbf{p}'}{h^{2d}} e^{-i(\mathbf{p}\cdot\mathbf{R}+\mathbf{p}'\cdot\mathbf{R}')/\hbar} \left\langle \mathbf{P} - \mathbf{p}/2|\hat{H}|\mathbf{P}' + \mathbf{p}'/2 \right\rangle \left\langle \mathbf{P}' - \mathbf{p}'/2|\hat{H}|\mathbf{P} + \mathbf{p}/2 \right\rangle.$$

To employ this representation, we will apply Eq. (83) to specific examples.

Scalar Potential. Applied to the scattering from a scalar potential, Eq. (83) takes the form

$$O(\mathbf{X}, \mathbf{X}') = L^d \delta^d(\mathbf{R} - \mathbf{R}') \int \frac{d\mathbf{r}}{L^d} e^{i\mathbf{r}\cdot(\mathbf{P}-\mathbf{P}')/\hbar} V(\mathbf{R} - \mathbf{r}/2)V(\mathbf{R} + \mathbf{r}/2). \tag{84}$$

This expression is further simplified by applying the ansatz that the low-lying modes of density relaxation are governed by field configurations of Q that vary slowly in space. The validity of this approximation can be checked self-consistently. In this case, the kernel above can be replaced by the coarse grained average $\langle V(\mathbf{R} - \mathbf{r}/2)V(\mathbf{R} + \mathbf{r}/2)\rangle$. With this assumption S_R takes the form of a collision integral with

$$O(\mathbf{X}, \mathbf{X}') \simeq L^d \delta^d(\mathbf{R} - \mathbf{R}')|\tilde{V}(\mathbf{P} - \mathbf{P}')|^2, \tag{85}$$

where the Fourier transform $\tilde{V}$ represents the Born scattering amplitude.

Taking the matrices Q to be locked over a scale $\hbar/t$ perpendicular to the constant energy shell, and assuming that, in this range, the matrix element $\tilde{V}$ depends only on the angle of scattering, we obtain

$$S_R = -\frac{\pi^2}{2\Delta^2} \int \frac{d\mathbf{R}}{L^d} \frac{d\mathbf{n} d\mathbf{n}'}{S_d^2} |\tilde{V}(\mathbf{n} - \mathbf{n}')|^2 \mathrm{str}\left[Q(\mathbf{R}, \mathbf{n})Q(\mathbf{R}, \mathbf{n}')\right]. \tag{86}$$

As the simplest example, taking the potential to be δ-correlated, $|\tilde{V}(\mathbf{n} - \mathbf{n}')|^2 = \hbar\Delta/2\pi\tau$, the physical regulator takes the form

$$S_R = -\frac{\pi\hbar}{4\Delta\tau} \int \frac{d\mathbf{R}}{L^d} \frac{d\mathbf{n} d\mathbf{n}'}{S_d^2} \mathrm{str}\left[Q(\mathbf{R}, \mathbf{n})Q(\mathbf{R}, \mathbf{n}')\right], \tag{87}$$

where $\mathbf{n} = \mathbf{p}/p_F$. If, on the other hand, the potential is taken to be a smooth, slowly varying function, the effective action takes the form

$$S_R = \frac{\pi\hbar}{4d\Delta\tau_{\mathrm{tr}}} \int \frac{d\mathbf{R}}{L^d} \frac{d\mathbf{n}}{S_d} \mathrm{str}\left(\partial_\mathbf{n} Q(\mathbf{R}, \mathbf{n})\right)^2, \tag{88}$$

where

$$\int \frac{d\mathbf{n}}{S_d} |\tilde{V}(\mathbf{n} - \mathbf{n}')|^2 (\mathbf{n} - \mathbf{n}')_\alpha (\mathbf{n} - \mathbf{n}')_\beta = \frac{\delta_{\alpha\beta}}{\pi d} \frac{\hbar\Delta}{\tau_{\mathrm{tr}}} \tag{89}$$

defines the transport mean free time. Interpreting the kinetic component of the action as the left-hand side of a Boltzmann operator, Eqs. (87) and (88) serve as a collision integral introducing irreversibility into the effective action**. Taking $\hbar/\Delta\tau \gg 1$, a moment expansion of the fields can be performed. As a result, the effective action takes the form of the diffusive action. Before showing explicitly how the diffusive action is recovered, we turn to the inhomogeneous magnetic field.

**A construction of similar type has recently been performed by Khmel'nitskii and Muzykantskii[67].

Inhomogeneous Magnetic Field. For a smoothly varying inhomogeneous vector potential, we can employ the same approximation as above. Assuming the low-lying modes to be those associated with field configurations of Q that vary slowly in space, we find

$$O(\mathbf{X}, \mathbf{X}') \simeq 2^d L^d \delta^d (\mathbf{R} - \mathbf{R}') \int \frac{d\mathbf{p}}{L^d h^d} e^{-2i\mathbf{p} \cdot \mathbf{R}/\hbar}$$
$$\times \left\langle \mathbf{P} - \mathbf{p}/2 |\hat{H}| \mathbf{P}' + \mathbf{p}/2 \right\rangle \left\langle \mathbf{P}' - \mathbf{p}/2 |\hat{H}| \mathbf{P} + \mathbf{p}/2 \right\rangle$$

Further integration over $\mathbf{R}$ generates the approximation

$$O(\mathbf{X}, \mathbf{X}') \simeq L^d \delta^d (\mathbf{R} - \mathbf{R}') |\widetilde{H}(\mathbf{P} - \mathbf{P}')|^2, \tag{90}$$

where $\widetilde{H}(\mathbf{P} - \mathbf{P}') = \left\langle \mathbf{P} |\hat{H}| \mathbf{P}' \right\rangle$ again represents the general Born scattering amplitude. Again, taking the Born amplitude to depend only on the angle of scattering, S_R takes the form of Eq. (86).

Applying this approximation to the inhomogeneous magnetic field in the two-dimensional geometry, and taking the Born scattering amplitude as short-ranged, we obtain the effective action

$$S_0[Q] = i \frac{\pi}{2\Delta} \int \frac{d\mathbf{R}}{L^2} \frac{d\theta}{2\pi} \mathrm{str} \left[\Omega_+ \sigma_3^{\mathrm{ar}} Q + 2i\hbar T^{-1} \sigma_3^{\mathrm{ar}} \left(v\hat{\mathbf{n}} \cdot \partial_\mathbf{R} T - \frac{\omega_c v}{2c} \partial_\theta T \right) \right],$$
$$S_R[Q] = \frac{\pi \hbar}{8\Delta \tau_{\mathrm{tr}}} \int \frac{d\mathbf{R}}{L^2} \frac{d\theta}{2\pi} \mathrm{str} \left(\partial_\theta Q \right)^2,$$

where $\hat{\mathbf{n}} = (\cos\theta, \sin\theta)$ and

$$\frac{\hbar \Delta}{2\pi \tau_{\mathrm{tr}}} \sim \omega_{\mathrm{rms}}^2 \int \frac{d\theta}{2\pi} (\theta/2)^2 \cot^2 (\theta/2).$$

identifies the transport mean free time. As with the scalar potential, taking $\hbar/\Delta \tau_{\mathrm{tr}} \gg 1$, a moment expansion of the fields is justified. Employing the parametrization,

$$T(\mathbf{X}_\|) = \exp \left[i\hat{\mathbf{n}} \cdot \mathbf{J}(\mathbf{R}) \right] \widetilde{T}(\mathbf{R}),$$

where $[\mathbf{J}, \sigma_3^{\mathrm{ar}}]_+ = 0$, and expanding the effective action to quadratic order in the generators $\mathbf{J}$ we obtain

$$S_0[\widetilde{Q}] = i \frac{\pi}{2\Delta} \int \frac{d\mathbf{R}}{L^2} \mathrm{str} \left[\Omega_+ \sigma_3^{\mathrm{ar}} \widetilde{Q} - \hbar \left(v \left[(\partial_\mathbf{R} \widetilde{T}) \widetilde{T}^{-1}, \sigma_3^{\mathrm{ar}} \right] \cdot \mathbf{J} + i\omega_c \sigma_3^{\mathrm{ar}} \epsilon^{ij} J_i J_j \right) \right],$$
$$S_R[\widetilde{Q}] = \frac{\pi \hbar}{2\Delta \tau_{\mathrm{tr}}} \int \frac{d\mathbf{R}}{L^2} \mathrm{str}\, \hat{\mathbf{J}}^2$$

where $\widetilde{Q} = \widetilde{T}^{-1} \sigma_3^{\mathrm{ar}} \widetilde{T}$. Integrating over the degrees of freedom $\mathbf{J}$ we finally obtain the effective action

$$S_0[\widetilde{Q}] = \frac{\pi}{4\Delta} \int \frac{d\mathbf{R}}{L^2} \mathrm{str} \left[\hbar \widetilde{D}_{xx} \partial_\mathbf{R} \widetilde{Q} \cdot \begin{pmatrix} \mathbf{I} & -\beta \widetilde{Q} \\ \beta \widetilde{Q} & \mathbf{I} \end{pmatrix} \cdot \partial_\mathbf{R} \widetilde{Q} + 2i\Omega_+ \sigma_3^{\mathrm{ar}} \widetilde{Q} \right],$$
$$= \frac{\pi}{4\Delta} \int \frac{d\mathbf{R}}{L^2} \mathrm{str} \left[\hbar \widetilde{D}_{xx} (\partial_\mathbf{R} \widetilde{Q})^2 + \hbar \widetilde{D}_{xy} \widetilde{Q} \epsilon^{ij} \partial_{R_i} \widetilde{Q} \partial_{R_j} \widetilde{Q} + 2i\Omega_+ \sigma_3^{\mathrm{ar}} \widetilde{Q} \right], \tag{91}$$

where

$$\beta = \omega_c \tau_{\mathrm{tr}}, \qquad \widetilde{D}_{xx} = \frac{1}{1+\beta^2} \frac{v^2 \tau_{\mathrm{tr}}}{2}, \qquad \widetilde{D}_{xy} = \beta \widetilde{D}_{xx}. \tag{92}$$

Taking the magnetic field to be homogeneous, the effective action recovers the diffusive non-linear σ-model (91) with an additional contribution involving the "topological term" introduced by Pruisken[68]. For an inhomogeneous field configuration, modes of density relaxation are described by the diffusion operator

$$\hat{\Pi} = - \left(\frac{\pi}{\Delta} \right) \left[\hbar \partial_{\mathbf{R}} \cdot (\widetilde{D}_{xx} \partial_{\mathbf{R}}) + \hbar \sigma_3^{\text{ar}} \epsilon^{ij} \partial_{R_i} \left(\widetilde{D}_{xy} \partial_{R_j} \right) + i\Omega_+ \right]. \tag{93}$$

This operator, which has a structure similar to the passive scalar operator, describes *advected* diffusion[69], and its long-time properties coincide with the conventional diffusion operator[70, 71]. The short-time properties of the non-hermitian operator depend explicitly on the nature of the random field distribution.

This concludes our preliminary investigation of the properties of the quantum ballistic action. While, from a pedagogical point of view, the ballistic action (76) gives considerable insight into generic properties of the quantum chaotic systems, operationally, the generalized action is complex. It seems likely that, from a practical point of view, this approach will be of use only when there is some separation of fast and slow degrees of freedom already on the level of the action. For the inhomogeneous scalar or vector potential, the fast relaxation of momentum degrees of freedom allowed the ballistic action to be reduced to an effective diffusive action involving only spatial degrees of freedom. On can envisage other scenarios such as the rough billiard or quantum kicked rotor where the spatial degrees of freedom relax faster than the momentum. In such cases, one can expect a quantitative analysis of the quantum dynamics.

DISCUSSION

To conclude, we have discussed mechanisms of quantum coherence in disordered and chaotic structures. For this general class of systems, a quasi-classical approximation allows quantum statistical properties to be expressed in the form of a functional integral involving a supersymmetric non-linear σ-model. The merit of the field theoretic approach lies in its ability to account within a single framework for the shortest time scales where the behavior is well described by the Trace formula, to the longest times where the quantum properties become universal. The formal connection between the field theoretic approach and that developed to study random statistical ensembles reveals a "dictionary" in which coherence phenomena associated with the physics of disordered conductors, such as weak localization, find a counterpart in the more general class of quantum chaotic systems.

Moreover, the field theoretic approach provides a solid platform on which other mechanisms of quantum coherence can be addressed. In particular, our analysis has focussed on aspects of *non-interacting* quantum systems. Manifestations of strong electron interaction play a crucial role in defining properties of disordered electronic structures[72]. The interplay of mesoscopic quantum interference phenomena with phase coherence effects such as the proximity effect in superconducting-normal metal structures are only just beginning to be explored.

REFERENCES

1.   C. E. Porter, ed. *Statistical Theories of Spectral Fluctuations*, Academic Press, New York, 1965.
2.   D. Delande, in Proceedings of Les-Houches Summer School, Session LII, 1989, eds. M. -J. Giannoni, A. Voros and J. Zinn-Justin, North-Holland, Amsterdam 1991.

3.   For a review see L. P. Kouwenhoven *et al.*, *Electron Transport in Quantum Dots*, proceedings of the Summer School on Mesoscopic Electron Transport (Kluwer 1997); C. M. Marcus *et al.*, *Quantum Chaos in Open versus Closed Quantum Dots: Signatures of Interacting Particles*, to appear in Chaos, Solitons and Fractals (July 1997), cond-mat/9703038.

4.   For a review, see M. C. Gutzwiller, "Chaos in Classical and Quantum Mechanics", (Springer, NY, 1990).

5.   K. B. Efetov, Adv. Phys. **32**, 53 (1983); K. B. Efetov, *Supersymmetry in Disorder and Chaos*, Cambridge University Press, New York (1997).

6.   B. A. Muzykantskii, D. E. Khmel'nitskii, JETP Lett. **62**, 76 (1995)

7.   A. V. Andreev, O. Agam, B. D. Simons and B. L. Altshuler, Phys. Rev. Lett. **76** 3947 (1996).

8.   L. P. Gor'kov and G. M. Eliashberg, Sov. Phys. JETP **21**, 940 (1965).

9.   F. Haake, Signatures of Quantum Chaos (Springer, Berlin, 1991).

10.   E. P. Wegner, Ann. Math. **53**, 36 (1953).

11.   O. Bohigas, M. J. Giannoni and C. Schmidt, Phys. Rev. Lett. **52**, 1 (1984); J. Physique Lett. **45**, L1615 (1984).

12.   F. J. Wegner, Z. Phys. B **35**, 207 (1979).

13.   L. Schäfer and F. J. Wegner, Z. Phys. B **38**, 113 (1980).

14.   K. B. Efetov, A. I. Larkin, and D. E. Khmelnitskii, Sov. Phys. JETP **52**, 568 (1980).

15.   A. Houghton, A. Jevicky, R. D. Kenway, and A. M. M. Pruisken, Phys. Rev. Lett. **45**, 394 (1980).

16.   K. Junglich and R. Oppermann, Z. Phys. B **38**, 93 (1980).

17.   S. Hikami, Phys. Rev. B **24**, 2671 (1981).

18.   A. J. MaKane and M. Stone, Ann. Phys. (NY) **131**, 36 (1981).

19.   M. L. Mehta, Random Matrices (Academic Press, New York, 1991).

20.   T. Guhr, A. Mueller-Groeling, and H. A. Weidenmueller, preprint cond-mat/9707301

21.   F. J. Dyson, J. Math. Phys. **3**, 140, 157, 166 (1962); F. J. Dyson and M. L. Mehta, J. Math. Phys. **4**, 701 (1963).

22.   M. R. Zirnbauer, J. Math. Phys. **37**, 4986 (1996).

23.   R. A. Webb, S. Washburn, C. P. Umbach, and R. B. Laibowitz, Phys. Rev. Lett **54**, 2696 (1985); R. A. Webb et al., in: Physics and Technology of Submicron Structures, eds., H. Heinrich, G. Bauer, and F. Kuchar (Springer, Berlin, 1988), Vol. 83, p. 98.

24.   D. E. Khmelnitskii, Physica B+C **126**, 235 (1984).

25.   G. Bergmann, Phys. Rep. **107**, 1 (1984).

26.   P. A. Lee and T. V. Ramakrishnan, Rev. Mod. Phys. **57**, 287 (1985).

27.   S. Chakravarty and A. Schmid, Phys. Rep. **140**, 193 (1986).

28.   A. G. Aronov and Y. V. Sharvin, Rev. Mod. Phys. **59**, 755 (1987).

29.   B. L. Altshuler, A. G. Aronov, M. E. Gershenson, and Y. V. Sharvin, in: Physics Reviews, ed., I. M. Khalatnikov (Harwood Academic Publishers, Switzerland, 1987), p. 225.

30.   L. P. Gor'kov, A. I. Larkin, and D. E. Khmelnitskii, JETP Lett. **30**, 228 (1979).

31.   B. L. Altshuler, A. G. Aronov, and B. Z. Spivak, JETP Lett. **33**, 94 (1981).

32.   D. Y. Sharvin and Y. V. Sharvin, JETP Lett. **34**, 272 (1981).

33.   D. J. Thouless, in: Ill Condensed Matter, eds., R. Balian, R. Maynard, and G. Toulous, Les Houches, Session XXXI (North-Holland, Amsterdam, 1978), p. 1.

34.   E. Abrahams, P. W. Anderson, D. C. Licciardello, and T. V. Ramakrishnan, Phys. Rev. Lett. **42**, 673 (1979).

35.   B. L. Altshuler and B. I. Shklovskii, JETP **64**, 127 (1986).

36.   Y. Imry, Europhys. Lett. **1**, 249 (1986).

37.   P. A. Lee and A. D. Stone, Phys. Rev. Lett. **55**, 1622 (1985).

38.   B. L. Altshuler, JETP Lett. **41**, 648 (1985).

39.   K. B. Efetov, Sov. Phys. JETP **82**, 872 (1982); *ibid* **83**, 833 (1982).

40.   V. N. Prigodin, B. L. Altshuler, K. B. Efetov, and S. Iida, Phys. Rev. Lett. **72**, 546 (1994).

41.   M. F. Crommie, C. P. Lutz and D. M. Eigler, Nature **363**, 524-527 (1993); M. F. Crommie, C. P. Lutz, D. M. Eigler, and E. J. Heller, Surface Rev. and Lett. **2** (1), 127-137 (1995).

42.   M. C. Gutzwiller, J. Math. Phys. **8**, 1979 (1967); **10**, 1004 (1969); **11**, 1791 (1970); **12**, 343 (1971).

43.   M. V. Berry, Proc. Roy. Soc. London A **400**, 229 (1985).

44.   M. V. Berry, in: Chaos and Quantum Physics, eds., M. -J. Gianonni, A. Voros, and J. Zinn-Justin, Les Houches, Session LII 1989 (North-Holland, Amsterdam, 1991), p. 251.

45.   J. H. Hannay and A. M. O. de Almeida, J. Phys. A **17**, 3429 (1984).

46.   P. Cvitanović and B. Eckhardt, J. Phys. A **24**, L237 (1991).

47. O. Agam, B. L. Altshuler and A. V. Andreev, Phys. Rev. Lett. **75**, 4389 (1995).
48. E. Bogomolny and J. P. Keating, Phys. Rev. Lett. **77**, 1472 (1996).
49. S. F. Edwards and P. W. Anderson, J. Phys. F **5**, 965 (1975).
50. V. J. Emery, Phys. Rev. B **11**, 239 (1975).
51. J. J. M. Verbaarschot and M. R. Zirnbauer, J. Phys. A **17**, 1093 (1985).
52. F. A. Berezin, *Introduction to Superanalysis* (Reidel, Dodrecht, 1987).
53. J. J. M. Verbaarschot, H. A. Weidenmüller and M. R. Zirnbauer, Phys. Rep. **129**, 367 (1985).
54. B. D. Simons, O. Agam and A. V. Andreev, J. Math. Phys. **38**, 1982 (1997).
55. V. E. Kravtsov and I. V. Lerner, Phys. Rev. Lett. **74**, 2563 (1995).
56. A. V. Andreev and B. L. Altshuler, Phys. Rev. Lett. **75**, 902 (1995); A. V. Andreev, B. D. Simons and B. L. Altshuler, J. Math. Phys. **37**, 4968 (1996).
57. A. Altland, S. Iida, and K. B. Efetov, J. Phys. A **26**, 3545 (1993).
58. M. R. Zirnbauer, J. Phys. A **29**, 7113 (1996); in this volume
59. O. Agam, A. Altland and M. R. Zirnbauer, to be published.
60. M. R. Zirnbauer, in this volume; A. Altland and M. R. Zirnbauer, to be published
61. P. Gaspard, G. Nicolis, A. Provata, and S. Tasaki, Phys. Rev. E. **51**, 74 (1995).
62. D. Ruelle, Phys. Rev. Lett **56**, 405 (1986).
63. M. Pollicot, Ann. Math. **131**, 331 (1990).
64. G. Nicolis and C. Nicolis, Phys. Rev. A. **38**, 427 (1988).
65. H. H. Hasegawa and D. J. Driebe, Phys. Rev. E. **50**, 1781 (1994).
66. I. L. Aleiner and A. I. Larkin, Chaos, Solitons and Fractals **8**, 1179 (1997).
67. D. E. Khmel'nitskii, private communication
68. H. Levine, S. Libby, and A. A. M. Pruisken, Nucl. Phys. **B240**, 30, 49, 71 (1985).
69. P. Hedegard and A. Smith, Phys. Rev. B **51**, 10869 (1995).
70. M. B. Isichenko, Rev. Mod. Phys. **64**, 961 (1992).
71. L. Zielinski et al., Preprint cond-mat/9704058
72. B. L. Altshuler and A. G. Aronov, in: Electron-Electron Interactions in Disordered Systems, eds., A. L. Efros and M. Pollak (North-Holland, Amsterdam, 1985), p. 1.

SUPERSYMMETRIC GENERALIZATION OF DYSON'S BROWNIAN MOTION (DIFFUSION)

Thomas Guhr

Max Planck Institut für Kernphysik
Postfach 103980
69029 Heidelberg, Germany

INTRODUCTION

The spectral fluctuation properties of an overwhelmingly rich variety of quantum systems can be modeled with a simple, phenomenological approach, the Theory of Random Matrices [1,2]. The key assumption is that the matrix elements of the Hamilton operator $\mathcal{H}$ in Schrödinger's equation are just random numbers. This idea is due to Wigner. It indeed leads to a very satisfactory description of the fluctuation properties in a wide class of many–body systems, ranging from nuclei to molecules, in disordered systems and also in systems with few degrees of freedom which are classically chaotic. More recently, it has been shown that this statistical concept can also be extended to classical wave phenomena, such as elastomechanics, and to quantum systems described by the Dirac equation, such as Quantum Chromodynamics. The list is still incomplete. A detailed review was recently given in Ref. 3.

This concept in its basic version, although enormously successful in a huge number of cases, is bound to fail if we have to deal with systems which undergo a crossover transition between different statistics, or different universality classes. The physically most interesting situations are

- transitions from regular to chaotic fluctuation properties

- breaking of time–reversal invariance

- breaking of symmetries

We stress the difference between the breaking of time–reversal invariance and symmetry breaking. In quantum mechanics, the time–reversal operator is anti–unitary while an operator corresponding to a symmetry of the Hamiltonian is unitary. If a symmetry such as angular momentum, parity etc. holds, the Hamiltonian can be reduced and can be written in block–diagonal form. Each block belongs to a given quantum number, i.e. a representation of the symmetry such as positive or negative parity. On the other hand, the time–reversal operator cannot be used to reduce the Hamiltonian. We refer to the discussion of the applications in various physical situations in Ref. 3.

Supersymmetry and Trace Formulae: Chaos and Disorder
Edited by Lerner *et al.*, Kluwer Academic / Plenum Publishers, New York, 1999

This article is meant to be a pedagogical presentation of the extension of earlier work [4], see also Ref. 3. It is organized as follows: after outlining, for the convenience of the reader, the basic concepts of Random Matrix Theory which are needed to address crossover transitions, we investigate a simplified case study, the diffusion in vector spaces. We then discuss the diffusion in matrix spaces and, eventually, give a summary and a conclusion.

RANDOM MATRIX THEORY AND CROSSOVER TRANSITIONS

For readers with little background in Random Matrix Theory, we try to give an elementary, self–contained introduction into the field. After sketching the basic concepts of the statistical model by addressing the classical scenario of rotation invariant ensembles, we turn to crossover transition which require the treatment of broken rotation invariance. The reader who is familiar with this background might want to skip this section.

Rotation Invariant Ensembles

A random matrix model is formulated in terms of $N \times N$ matrices H whose entries are the matrix elements of $\mathcal{H}$ in some basis of Hilbert space. According to the physical symmetry constraints, H is Hermitean if time–reversal invariance is fully broken. In the case of conserved time–reversal invariance, H is either real symmetric or quaternion real, i.e. self–dual Hermitean, if the system does or does not show Kramers degeneracies, respectively. We mention in passing that it is advantageous to write $2N$ for the dimension of self–dual Hermitean matrices. In any case, N is the cut–off dimension in the Hilbert space and is assumed to be large. Eventually, to compensate the cut–off, we wish to take the limit $N \to \infty$. Statistical observables, particularly correlation functions, are obtained by averaging over the ensemble of all these random matrices H. In a measurement, however, one works out the correlations by a proper average over the spectrum. It should be emphasized that the ensemble average and the spectral average yield, for large N, identical results. This is referred to as ergodicity, see also the discussion in Ref. 3.

A crucial assumption is the equivalence of all bases in Hilbert space, the so–called rotation invariance of the ensemble of matrices H. Thus, the statistical distribution of the matrices H must be invariant under a change of basis. The additional assumption of statistical independence of the independent elements of H leads to the Gaussian distributions

$$P^{(1)}_{N\beta}(H, v) = \sqrt{\frac{\beta}{4\pi v^2}}^{\,-N+N(N-1)\beta/2} \exp\left(-\frac{\beta}{4v^2}\operatorname{tr} H^2\right). \tag{1}$$

with variances $2v^2/\beta$. This defines the Gaussian Orthogonal (GOE), Unitary (GUE) and Symplectic (GSE) Ensembles. To motivate these designations, we notice that a real symmetric H is diagonalized by an orthogonal matrix, a Hermitean H by a unitary matrix and a self–dual Hermitean H by a symplectic matrix. The parameter β labels these cases and takes values $\beta = 1, 2, 4$ for the GOE, GUE and GSE, respectively. In the following, in keeping with the terminology in most of the litaerature, we want to refer to the fluctuations modeled by the Gaussian ensembles as chaotic.

The rotation invariance of the distribution (1) is reflected in the trace in the exponent, implying that the distribution depends only on the eigenvalues of the matrices H. We diagonalize these matrices $H = U^{-1}XU$ where X is diagonal and contains the distinct eigenvalues $x_1, \ldots, x_N$. However, when using the distribution (1) we have to take into account that it is a probability density. The expression $P_{N\beta}^{(1)}(H, v)d[H]$ is a differential probability where $d[H]$ is the volume element in matrix space. In eigenvalue angle coordinates it reads

$$d[H] \;=\; |\Delta_N(X)|^\beta d[X]d\mu(U) \qquad \text{where} \qquad \Delta_N(X) \;=\; \prod_{n>m}(x_n - x_m) \qquad (2)$$

is the Vandermonde determinant and where $d\mu(U)$ is the invariant measure of the diagonalizing group. Hence, the probability density function in the curved space of the eigenvalues is given by

$$P_{N\beta}^{(E)}(x_1, \ldots, x_N) = C_{N\beta}(v) \exp\left(-\frac{\beta}{4v^2} \sum_{n=1}^{N} x_n^2\right) |\Delta_N(X)|^\beta \qquad (3)$$

with $C_{N\beta}(v)$ a normalization constant. As the probability density is rotation invariant, the group integral over U is trivial and yields the group volume. This constant is absorbed in $C_{N\beta}(v)$.

The correlation functions of k energies are obtained by integrating out $N - k$ eigenvalues,

$$R_{\beta k}(x_1, \ldots, x_k) = \frac{N!}{(N-k)!} \int_{-\infty}^{+\infty} dx_{k+1} \cdots \int_{-\infty}^{+\infty} dx_N \, P_{N\beta}^{(E)}(x_1, \ldots, x_N) \, . \qquad (4)$$

This gives the probability of finding a level around each of the positions $x_1, \ldots, x_k$, regardless of labeling. The reader immediately realizes the importance of the Vandermonde determinant $\Delta_N(X)$ in the probability density (3). Whenever two levels degenerate, the integrand in Eq. (4) is anihilated and the correlation vanishes at these points. This is how Random Matrix Theory models the famous Wigner–von Neumann level repulsion. It is this feature which has the strongest impact on the correlation properties. In fact, it turns out that the specific Gaussian form of the distribution (1) is not important, if we are only interested in fluctuations on the scale of the local mean level spacing. A wide class of distributions gives identical fluctuations, provided they do not have structure on the scale of the mean level spacing. This remarkable universality, already inherent in the early formulations of Random Matrix Theory, recently received renewed interest, see the review in Ref. 3.

Crossover Transitions and Crossover Ensembles

As already stressed in the introduction, the concepts just outlined are not sufficient if we have to deal with crossover transitions. In many situations, such transitions are governed by one physical parameter and we can set up a statistical model by decomposing the matrix H into two parts. We consider the random Hamiltonian

$$H = H^{(0)} + \alpha H^{(1)} \qquad (5)$$

as a function of the transition parameter α. Each of the three matrices belongs to the same symmetry class. The elements of $H^{(1)}$ are Gaussian distributed according to $P_{N\beta}^{(1)}(H^{(1)}, v)$ in Eq. (1). The distribution $P_N^{(0)}(H^{(0)})$ of the elements of $H^{(0)}$, however, is completely arbitrary. If, for example, $P_N^{(0)}(H^{(0)})$ sets all off–diagonal matrix elements to zero, the ensemble of matrices $H^{(0)}$ is free of the Wigner–von Neumann level repulsion and does therefore model a class of regular, non–chaotic systems. Thus, the ensemble of matrices (5) interpolates between regular and chaotic statistics. Another situation is obtained if $P_N^{(0)}(H^{(0)})$ restricts the symmetry of $H^{(0)}$ to a sub–symmetry of $H^{(1)}$. In the case of a Hermitean $H^{(1)}$, for example, $P_N^{(0)}(H^{(0)})$ can put all imaginary parts of the elements of $H^{(0)}$ to zero, thereby effectively restricting $H^{(0)}$ to a real symmetric matrix. This choice makes the ensemble of matrices (5) a model for the transition from conserved to broken time–reversal invariance. For the time being, we want to exclude symmetry breaking from the discussion because this would require to take into account a block structure of the matrices.

A precise interpretation of the parameter α is not always easy. The reader might want to think of α as a measure of the ratio between the chaotic and the non–chaotic or regular portions in classical phase space if the regularity chaos transition is addressed. In many case which deal with the breaking of time–reversal invariance, α can be associated with a magnetic field.

The eigenvalues of the matrices H defined in Eq. (5) form the spectrum of our crossover model. How can we calculate the correlation functions for this ensemble? We have to perform an average over the probability densities $P_{N\beta}^{(1)}(H^{(1)}, v)$ and $P_N^{(0)}(H^{(0)})$. It is unpleasant that the eigenvalues of H depend on the transition parameter α. Fortunately, this dependence can be removed by a simple change of variables, i.e. by replacing $\alpha H^{(1)}$ with $H^{(1)}$. Thus, the ensemble is now formed of matrices $H = H^{(0)} + H^{(1)}$ whose eigenvalues are no functions of α any more. Consequently, the transition parameter appears in the variance of the Gaussian distribution and we have to use $P_{N\beta}^{(1)}(H^{(1)}, \alpha v)$ instead of $P_{N\beta}^{(1)}(H^{(1)}, v)$. It is convenient to fix the energy scale by setting $\sqrt{2}v = 1$. Moreover, it is useful to introduce the new transition parameter

$$ t = \alpha^2/2 , \tag{6} $$

which, in the sequel, will be interpreted as a fictitious time. The Gaussian distribution now reads

$$ P_{N\beta}^{(1)}(H^{(1)}, t) = \sqrt{\frac{\beta}{4\pi t}}^{\,-N+N(N-1)\beta/2} \exp\left(-\frac{\beta}{4t}\mathrm{tr}\,(H^{(1)})^2\right) . \tag{7} $$

It still depends only on the eigenvalues $x_n^{(1)}$, $n = 1, \ldots, N$ of $H^{(1)}$. However, this does not solve our problem because the eigenvalues x_n, $n = 1, \ldots, N$ of H are complicated functions of the $x_n^{(1)}$, of the diagonalizing matrices and of $H^{(0)}$. This is referred to as breaking of rotation invariance, according to the definition in the previous section. It is highly non–trivial to obtain the distribution $P_{N\beta}^{(E)}(X, t) = P_{N\beta}^{(E)}(x_1, \ldots, x_N, t)$ of the eigenvalue matrices X of $H = H^{(0)} + H^{(1)}$. Once we have this distribution, the correlation functions are given by

$$ R_{\beta k}(x_1, \ldots, x_k, t) = \frac{N!}{(N-k)!} \int_{-\infty}^{+\infty} dx_{k+1} \cdots \int_{-\infty}^{+\infty} dx_N\, P_{N\beta}^{(E)}(x_1, \ldots, x_N, t) , \tag{8} $$

which extends the definition (8). The evolution of the distribution $P_{N\beta}^{(E)}(X, t)$ from the distribution of the matrices $H^{(0)}$ for $t \to 0$ to the distributions (3) for large t can be expressed as a diffusion in the curved space of the eigenvalues, governed by the fictitious time t. This diffusion will be discussed in the following. It can be viewed as a reformulation of Dyson's Brownian motion.

Remarkably, there is a second diffusion which, conveniently, takes place in a much smaller space. It is the supersymmetric generalization [4] of the diffusion just mentioned. To prepare the ground, we write the correlation functions as an average over the Green functions,

$$ \widehat{R}_{\beta k}(x_1, \ldots, x_k, t) = \frac{1}{\pi^k} \int d[H^{(1)}] P_{N\beta}^{(1)}(H^{(1)}, t) $$

$$\int d[H^{(0)}]P_N^{(0)}(H^{(0)}) \prod_{p=1}^{k} \text{tr}\, \frac{1}{x_p^- - H^{(0)} - H^{(1)}} \ . \tag{9}$$

Here, we set $x_p^- = x_p - i\varepsilon$ and the limit $\varepsilon \to 0$ has to be taken. The functions $\widehat{R}_{\beta k}(x_1,\ldots,x_k,t)$ always allow reconstruction of the functions $R_{\beta k}(x_1,\ldots,x_k,t)$. We simply have to restrict every trace to its imaginary or δ function part, which is simply the spectrum, and to drop the real or principal value part. This operation is symbolically denoted by $\Im$. We disregard some trivial contributions, for details see Refs. 3, 4. Advantageously, these correlation functions can be written as the derivatives

$$\widehat{R}_{\beta k}(x_1,\ldots,x_k,t) \;=\; \frac{1}{(2\pi)^k}\, \frac{\partial^k}{\prod_{p=1}^{k}\partial J_p}\, Z_{\beta k}(x+J,t)\bigg|_{J=0} \tag{10}$$

of the normalized generating function

$$Z_{\beta k}(x+J,t) = \frac{1}{\pi^k}\int d[H^{(1)}]P_{N\beta}^{(1)}(H^{(1)},t)$$

$$\int d[H^{(0)}]P_N^{(0)}(H^{(0)}) \prod_{p=1}^{k} \frac{\det\left(x_p^- + J_p - H^{(0)} - H^{(1)}\right)}{\det\left(x_p^- - J_p - H^{(0)} - H^{(1)}\right)} \ . \tag{11}$$

Here, the diagonal matrices x and J contain the k energies and the k source variables, $x = \text{diag}(x_1, x_1, \ldots, x_k, x_k,) \otimes 1_{\zeta_\beta}$, $J = \text{diag}(-J_1, +J_1, \ldots, -J_k, +J_k,) \otimes 1_{\zeta_\beta}$. We have introduced the numbers $\zeta_1 = \zeta_4 = 2$ and $\zeta_2 = 1$ and the $\zeta_\beta \times \zeta_\beta$ unit matrices 1_{ζ_β}.

The supersymmetric diffusion describes the time evolution of the generating functions $Z_{\beta k}(x+J,t)$. It takes place in the $2k$–dimensional space of the eigenvalues x and the source variables J for $\beta = 2$ and in a closely related, slightly larger space for $\beta = 1$ and $\beta = 4$. These spaces will be identified as the curved spaces of the eigenvalues of certain supermatrices.

For later purposes we emphasize that, in almost all physics applications, we wish to consider these functions in the limit $N \to \infty$. To remove the dependence on the level density from the true correlations, the functions $R_{\beta k}(x_1,\ldots,x_k,t)$ have to be unfolded. The mean level spacing D is given by $D = 1/R_1(0)$. We note that it depends on N. We then define the dimensionless energies $\xi_p = x_p/D$ on the scale of the mean level spacing. The new energy variables ξ_p are held fixed while the limit is taken. We also have to measure the transition parameter on the scale of the mean level spacing, i.e. the ratio α/D also has to be held fixed while the limit $N \to \infty$ is taken. We introduce the parameter

$$\tau \;=\; t/D^2 \tag{12}$$

as the fictitious time on the unfolded scale. The unfolded correlation functions can then be written as

$$X_{\beta k}(\xi_1,\ldots,\xi_k,\tau) = \lim_{N\to\infty} D^k R_{\beta k}(D\xi_1,\ldots,D\xi_k,D^2\tau) \ . \tag{13}$$

In particular we have $X_{\beta 1}(\xi_1) = 1$ by construction.

DIFFUSION IN VECTOR SPACES AS A CASE STUDY

Before discussing the diffusion in the space of matrices, we turn to a case study in the, much simpler, spaces of vectors. Since the aim of this section is to give a pedagocical introduction into the concepts using the simplest possible scenario, we restrict ourselves to real vector spaces. We start with the trivial case of two dimensions before we generalize this discussion to arbitrarily many dimensions.

Two Dimensions

We consider a distribution $P(\vec{r}, t)$ of real two–component vectors $\vec{r} = (x, y)$. We assume that the time evolution of this distribution is diffusive and given by

$$\frac{1}{4}\Delta P(\vec{r}, t) = \frac{\partial}{\partial t}P(\vec{r}, t) \qquad \text{with} \qquad \lim_{t \to 0} P(\vec{r}, t) = P^{(0)}(\vec{r}) , \tag{14}$$

where $\Delta = \partial^2/\partial \vec{r}^2$ is the usual Laplacean in two dimensions. The distribution $P^{(0)}(\vec{r})$ is the initial condition at time $t = 0$. To solve the diffusion problem (14), it is useful to construct the diffusion kernel $P^{(1)}(\vec{r}, t)$ which has to satisfy

$$\frac{1}{4}\Delta P^{(1)}(\vec{r}, t) = \frac{\partial}{\partial t}P^{(1)}(\vec{r}, t) \qquad \text{with} \qquad \lim_{t \to 0} P^{(1)}(\vec{r}, t) = \delta(\vec{r}) . \tag{15}$$

Since the vector space under consideration is a simple Cartesean one, the kernel is just the Gaussian

$$P^{(1)}(\vec{r}, t) = \frac{1}{\pi t} \exp\left(-\frac{\vec{r}^2}{t}\right) . \tag{16}$$

It can be verified in a straightforward calculation that the solution $P(\vec{r}, t)$ of the diffusion problem (14) is expressible as the convolution

$$P(\vec{r}, t) = \int P^{(1)}(\vec{r} - \vec{r}^{(0)}, t)\, P^{(0)}(\vec{r}^{(0)})\, d^2 r^{(0)} \tag{17}$$

of the kernel with the initial condition.

We now assume that the initial condition depends only on the length $r^{(0)} = |\vec{r}^{(0)}|$ of the vector $\vec{r}^{(0)}$. This has some remarkable consequences. Obviously, it is useful to introduce polar coordinates. We rotate the (x, y) coordinate system such that the x axis is parallel to $\vec{r}$. With $\vec{r}^{(0)} = r^{(0)}(\cos\varphi^{(0)}, \sin\varphi^{(0)})$ and $d^2 r^{(0)} = r^{(0)} dr^{(0)} d\varphi^{(0)}$ we find

$$P(\vec{r}, t) = \int_0^\infty \Gamma(r, r^{(0)}, t)\, P^{(0)}(r^{(0)})\, r^{(0)} dr^{(0)} \tag{18}$$

where

$$\Gamma(r, r^{(0)}, t) = \int_0^{2\pi} P^{(1)}(\vec{r} - \vec{r}^{(0)}, t)\, d\varphi^{(0)} . \tag{19}$$

Due to the angular average, this function depends only on the radial coordinate $r^{(0)}$. Moreover, by construction, it also depends only on the length $r = |\vec{r}|$ of the second vector, but not on its orientation. This is most easily seen in the special coordinate system we have chosen. We have $\vec{r} \cdot \vec{r}^{(0)} = r r^{(0)} \cos\varphi^{(0)}$ and thus

$$\begin{aligned}
\Gamma(r, r^{(0)}, t) &= \frac{1}{\pi t} \exp\left(-\frac{r^2 + (r^{(0)})^2}{t}\right) \int_0^{2\pi} \exp\left(-\frac{2 r r^{(0)} \cos\varphi^{(0)}}{t}\right) d\varphi^{(0)} \\
&= \frac{2}{t} \exp\left(-\frac{r^2 + (r^{(0)})^2}{t}\right) I_0\left(\frac{2 r r^{(0)}}{t}\right)
\end{aligned} \tag{20}$$

where I_ν is the modified Bessel function of order ν. This in turn implies that the solution (18) of the diffusion problem is also a function of the lenght r only, i.e. we have $P(\vec{r}, t) = P(r, t)$.

We conclude that initial conditions which depend only on the radial coordinate are propagated entirely in the radial space and the solution of the diffusion problem is, at every time t, only a function of the radial coordinate. To find the diffusion equation in this radial space, we simply have to perform an angular average over Eqs. (14) and (15). We arrive at the diffusion problem

$$\frac{1}{4}\Delta_r P(r,t) = \frac{\partial}{\partial t}P(r,t) \qquad \text{with} \qquad \lim_{t\to 0} P(r,t) = P^{(0)}(r) , \qquad (21)$$

where the radial Laplacean is given by

$$\Delta_r = \frac{1}{r}\frac{\partial}{\partial r}r\frac{\partial}{\partial r} = \frac{\partial^2}{\partial r^2} + \frac{1}{r}\frac{\partial}{\partial r} . \qquad (22)$$

The function $\Gamma(r,r^{(0)},t)$ is the kernel of this diffusion problem and satisfies

$$\frac{1}{4}\Delta_r\Gamma(r,r^{(0)},t) = \frac{\partial}{\partial t}\Gamma(r,r^{(0)},t) \qquad \text{with} \qquad \lim_{t\to 0}\Gamma(r,r^{(0)},t) = \frac{\delta(r-r^{(0)})}{\sqrt{rr^{(0)}}} . \qquad (23)$$

We notice that the proper δ function on the right hand side of the limit relation involves the square roots of the Jacobians r and $r^{(0)}$. It is easily checked that $\Gamma(r,r^{(0)},t)$ obeys this limit relation. For small t, the argument of the modified Bessel function in Eq. (20) becomes large and we may use the approximation $I_\nu(z) \simeq \exp(z)/\sqrt{2\pi z}$ for large $|z|$. The limit $t \to 0$ then yields the right hand side of the limit relation in Eq. (23).

We identify the formula (18) as the convolution in the curved, radial space that solves the diffusion problem (21).

Arbitrarily Many Dimensions

In the case of two dimensions, the reformulation of the Cartesean diffusion problem in terms of a diffusion in the curved radial space can be helpful. However, we have not gained much because we have removed just one variable, the angular one, at the price of having to deal with Bessel functions. The situation is very different in higher dimensional spaces. In d dimensions, the reduction to the curved radial space removes $d - 1$ angular variables and the diffusion takes place in only one dimension. This advantage is well worth to be paid for by having to handle Bessel functions. We consider the space of real d–component vectors $\vec{r} = (x_1,\ldots,x_d)$. The diffusion of the distribution $P(\vec{r},t)$ is still described by the Eqs. (14), (15), provided we interpret $\vec{r}$ as the d–component vector just defined. Again, the kernel is a Gaussian

$$P^{(1)}(\vec{r},t) = \frac{1}{(\pi t)^{d/2}} \exp\left(-\frac{\vec{r}^2}{t}\right) . \qquad (24)$$

and the solution of the diffusion problem is the convolution

$$P(\vec{r},t) = \int P^{(1)}(\vec{r}-\vec{r}^{(0)},t)\, P^{(0)}(\vec{r}^{(0)})\, d^d r^{(0)} \qquad (25)$$

of the kernel with the initial condition. As before we assume that the initial condition $P^{(0)}(r^{(0)})$ depends only on the length $r^{(0)} = |\vec{r}^{(0)}|$ of the vector $\vec{r}^{(0)}$. Thus we introduce spherical coordinates in d dimensions. The volume element reads $d^d r^{(0)} = (r^{(0)})^{d-1}dr^{(0)}d\Omega_d^{(0)}$ where $d\Omega_d^{(0)}$ is the solid angle element in d dimensions. We find

$$P(r,t) = \int_0^\infty \Gamma(r,r^{(0)},t)\, P^{(0)}(r^{(0)})\, (r^{(0)})^{d-1}dr^{(0)} \qquad (26)$$

where

$$\Gamma(r,r^{(0)},t) = \int d\Omega_d^{(0)} P^{(1)}(\vec{r}-\vec{r}^{(0)},t) . \qquad (27)$$

Anticipating the result of the calculation to follow, we wrote the left hand side of Eq. (26) as a function of the length $r = |\vec{r}|$ of the vector $\vec{r}$ only. Extending the procedure in the two–dimensional case, we choose a coordinate system in which the $x_d^{(0)}$ axis is parallel to $\vec{r}$. We then have $\vec{r} \cdot \vec{r}^{(0)} = rr^{(0)} \cos \vartheta^{(0)}$ and a factor of $\sin^{d-2} \vartheta^{(0)}$ appears in $d\Omega_d^{(0)}$. Collecting everything we arrive at

$$\Gamma(r, r^{(0)}, t) = \frac{1}{(\pi t)^{d/2}} \exp\left(-\frac{r^2 + (r^{(0)})^2}{t}\right)$$

$$\cdot \; \frac{2\pi^{(d-1)/2}}{\Gamma((d-1)/2)} \int_0^\pi \exp\left(-\frac{2rr^{(0)} \cos \vartheta^{(0)}}{t}\right) \sin^{d-2} \vartheta^{(0)} d\vartheta^{(0)}$$

$$= \left(\frac{2}{t}\right)^{d/2} \exp\left(-\frac{r^2 + (r^{(0)})^2}{t}\right) \frac{I_{(d-2)/2}(2rr^{(0)}/t)}{(2rr^{(0)}/t)^{(d-2)/2}} \; , \tag{28}$$

where we used a standard formula in the theory of Bessel functions. Functions of this kind which are a result of such an angular average are often referred to as zonal spherical functions. For $d = 2$ we obtain the previous formula. For $r \to 0$ and $r^{(0)} \to 0$ the function $\Gamma(r, r^{(0)}, t)$ has a finite limit,

$$\lim_{r, r^{(0)} \to 0} \Gamma(r, r^{(0)}, t) = \left(\frac{2}{t}\right)^{d/2} \frac{1}{2^{(d-2)/2} \Gamma(d/2)} \; . \tag{29}$$

As before, we obtain the diffusion in the curved space of the radial coordinate by performing the angular avergage over the diffusion equations in the d–dimensional Cartesean space. The result reads

$$\frac{1}{4} \Delta_r P(r, t) = \frac{\partial}{\partial t} P(r, t) \qquad \text{with} \qquad \lim_{t \to 0} P(r, t) = P^{(0)}(r) \; , \tag{30}$$

where the radial Laplacean in d dimensions is given by

$$\Delta_r = \frac{1}{r^{d-1}} \frac{\partial}{\partial r} r^{d-1} \frac{\partial}{\partial r} = \frac{\partial^2}{\partial r^2} + \frac{d-1}{r} \frac{\partial}{\partial r} \; . \tag{31}$$

We identify the function $\Gamma(r, r^{(0)}, t)$ as the kernel of this diffusion problem,

$$\frac{1}{4} \Delta_r \Gamma(r, r^{(0)}, t) = \frac{\partial}{\partial t} \Gamma(r, r^{(0)}, t) \qquad \text{with} \qquad \lim_{t \to 0} \Gamma(r, r^{(0)}, t) = \frac{\delta(r - r^{(0)})}{(rr^{(0)})^{(d-1)/2}} \; . \tag{32}$$

Once more, the proper δ function in the radial space contains the square root of the Jacobians, here r^{d-1} and $(r^{(0)})^{d-1}$. Formula (28) indeed satisfies this limit relation for $t \to 0$, as is easily seen by using $I_\nu(z) \simeq \exp(z)/\sqrt{2\pi z}$ for large $|z|$.

The representation (26) for the solution of the diffusion problem as a single radial integral is, together with the explicit result (28) for the kernel, a considerable simplification of the original d–dimensional problem.

The kernel $\Gamma(r, r^{(0)}, t)$ acquires a remarkable feature when the number of dimensions d is odd. We consider a three–dimensional space, $d = 3$. The Bessel function in Eq. (28) can now be expressed in elementary functions, $I_{1/2}(z) = \sqrt{2} \sinh(z)/\sqrt{\pi z}$, and we obtain the simple result

$$\Gamma(r, r^{(0)}, t) = \frac{2}{\sqrt{\pi t}} \exp\left(-\frac{r^2 + (r^{(0)})^2}{t}\right) \frac{\sinh(2rr^{(0)}/t)}{rr^{(0)}} \; . \tag{33}$$

In any odd dimension d, the kernel $\Gamma(r, r^{(0)}, t)$ can be written in elementary functions. This is an immediate consequence of the angular integral in Eq. (28).

DIFFUSION IN MATRIX SPACES

We now generalize the approach outlined in the previous to the case of matrices. As already stressed in the introduction, this will lead to two diffusion processes. First, we discuss transitions in terms of a diffusion for the joint probability density in matrix space. Thereby, we summarize some of the results of Dyson's Brownian motion model. Second, we show that this approach can be generalized to a diffusion in the space of supermatrices. Here, it is the generating functions of the correlators whose evolution is described by a diffusion. The supersymmetric diffusion takes place in a much smaller space of supermatrices.

Diffusion in the space of ordinary matrices

Dyson [5, 6] was the first to realize that crossover transitions can be formulated as Brownian motion of the eigenvalues. He constructed the Smoluchowski equation for the joint probability density. Recent reviews can be found in the books by Haake [2] and Mehta [1], see also the discussion by Lenz and Haake [7]. There is also a connection to the Pechukas gas and the Calogero–Sutherland–Moser model, see Refs. 3, 2. Here, we want to take a slightly different route and use explicitly the above mentioned universality of the unfolded correlations. It allows us to restrict ourselves to the Gaussian probability density (7).

The representation (9) of the correlation functions suggests to change integration variables from $H^{(1)}$ to $H = H^{(0)} + H^{(1)}$. This will remove the breaking of the rotation invariance from the traces. We then have to deal with the distribution function

$$P_{N\beta}(H, t) = \int P_{N\beta}^{(1)}(H - H^{(0)}, t) P_N^{(0)}(H^{(0)}) d[H^{(0)}] , \qquad (34)$$

which is easily identified as the formal solution of a diffusion problem. We introduce the matrix gradient $\partial/\partial H$ and the Laplacean $\Delta = \mathrm{tr}\partial^2/\partial H^2$. The evolution of the total probability density is given by the diffusion

$$\frac{\beta}{4}\Delta P_{N\beta}(H, t) = \frac{\partial}{\partial t}P_{N\beta}(H, t) \qquad \text{with} \qquad \lim_{t\to 0} P_{N\beta}(H, t) = P_N^{(0)}(H) . \qquad (35)$$

The kernel of this diffusion problem has to satisfy

$$\frac{\beta}{4}\Delta P_{N\beta}^{(1)}(H, t) = \frac{\partial}{\partial t}P_{N\beta}^{(1)}(H, t) \qquad \text{with} \qquad \lim_{t\to 0} P_{N\beta}^{(1)}(H, t) = \delta(H) , \qquad (36)$$

and is therefore precisely the Gaussian distribution (7). The formal solution of the diffusion problem (35) is just the convolution (34).

For initial conditions which depend only on the eigenvalues $X^{(0)}$ of $H^{(0)}$, the diffusion process is restricted to the curved space of the eigenvalues. For the joint probability density, one obtains

$$\frac{\beta}{4}\Delta_X P_{N\beta}(X, t) = \frac{\partial}{\partial t}P_{N\beta}(X, t) \qquad \text{and} \qquad \lim_{t\to 0} P_{N\beta}(X, t) = P_N^{(0)}(X) . \qquad (37)$$

The "radial part" of the Laplacean is given by

$$\Delta_X = \frac{1}{|\Delta_N(X)|^\beta} \sum_{n=1}^{N} \frac{\partial}{\partial x_n}|\Delta_N(X)|^\beta \frac{\partial}{\partial x_n} . \qquad (38)$$

The notation $\Delta_N(X)$ for the Vandermonde determinant should not be confused with the symbol Δ_X standing for the radial part of the Laplacian. This diffusion equation (37) is solved by the convolution

$$P_{N\beta}(X,t) = \int \Gamma_{N\beta}(X,X^{(0)},t) P_N^{(0)}(X^{(0)}) |\Delta_N(X^{(0)})|^\beta d[X^{(0)}] \qquad (39)$$

and the kernel $\Gamma_{N\beta}(X,X^{(0)},t)$ is given by the group integral

$$\Gamma_{N\beta}(X,X^{(0)},t) = \int P_{N\beta}^{(1)}(U^{-1}XU - X^{(0)},t) d\mu(U) , \qquad (40)$$

which can be viewed as a Bessel function in the space of matrices. Unfortunately, this integral is not known explicitly for $\beta = 1$ and $\beta = 4$. Similar to the case of vectorspaces with odd dimension, however, there is a dramatic simplification in the unitary case $\beta = 2$ which allows one to calculate the group integral. This is the famous Harish–Chandra–Itzykson–Zuber integral [8]. It was first used in this context by Mehta and Pandey [9]. For the joint probability density of the eigenvalues, it yields

$$P_{N2}^{(E)}(X,t) = P_{N2}(X,t)\Delta_N^2(X) = \frac{1}{\sqrt{2\pi t}^N} \int d[X^{(0)}] \Delta_N(X^{(0)})$$

$$\exp\left(-\frac{1}{2t}\mathrm{tr}(X - X^{(0)})^2\right) P_N^{(0)}(X^{(0)})\Delta_N(X) . \qquad (41)$$

For example, to study the transition from GOE to GUE [9], one chooses $P_N^{(0)}(X^{(0)}) \propto \exp(-\mathrm{tr}(X^{(0)})^2/2)/|\Delta_N(X^{(0)})|$. The integrals over $X^{(0)}$ can be worked out and give the joint probability density of the interpolating ensemble.

According to Eq. (8), we can use Eq. (41) to compute the k–level correlation functions for crossover transitions towards the GUE. An alternative procedure was recently given by Pandey [10]. He starts from an exact formal solution of the diffusion equation in terms of representation functions of the unitary group $U(N)$. His method is equivalent to calculating the Harish–Chandra–Itzykson–Zuber integral.

An important property of the correlation functions was obtained by French et $al.$ [11]. Starting from the Fokker–Planck or Smoluchowski equation closely related to the diffusion equation discussed above, these authors derived hierarchic relations among the correlation functions (13) on the unfolded scale. These hierarchic equations involve the fictitious time τ and acquire the form

$$\frac{\partial}{\partial\tau}X_{\beta k}(\xi_1,\ldots,\xi_k,\tau) = \sum_{p=1}^{k} \frac{1}{|\Delta_k(\xi)|^\beta} \frac{\partial}{\partial\xi_p} |\Delta_k(\xi)|^\beta \frac{\partial}{\partial\xi_p} X_{\beta k}(\xi_1,\ldots,\xi_k,\tau)$$

$$-\beta \sum_{p=1}^{k} \frac{\partial}{\partial\xi_p} \int_{-\infty}^{+\infty} \frac{X_{\beta(k+1)}(\xi_1,\ldots,\xi_k,\xi_{k+1},\tau)}{\xi_p - \xi_{k+1}} d\xi_{k+1} \qquad (42)$$

where $\Delta_k(\xi) = \prod_{p<q}(\xi_p - \xi_q)$. For $k = 1$ one obtains a closed equation which coincides with the famous Pastur equation [12].

Diffusion in the space of supermatrices

We now turn to the supersymmetric generalization of the diffusion just discussed. We follow the presentation in Ref. 4, see also Ref. 3. The supersymmetry method was developed by Efetov [13] in the context of disordered systems and first applied to Random Matrix Theory by Verbaarschot et $al.$ [14]. The average over the Gaussian ensembles of matrices $H^{(1)}$ is replaced, via a Hubbard–Stratonovitch transform, with an average over supermatrices of the form

$$\sigma = \begin{bmatrix} \sigma^{(11)} & \sigma^{(12)} \\ \sigma^{(21)} & \sigma^{(22)} \end{bmatrix} . \tag{43}$$

Here, the diagonal blocks $\sigma^{(11)}$ and $\sigma^{(22)}$ are ordinary matrices with commuting entries. The entries of the off–diagonal blocks $\sigma^{(12)}$ and $\sigma^{(21)}$, however, are anticommuting, i.e. Grassmann, variables. The overall symmetries of these matrices σ reflect the symmetries of the matrices $H^{(1)}$. Whereas the dimension of the matrices $H^{(1)}$ is N, the dimension of the matrices σ is, importantly, given by $\zeta_\beta 2k$, i.e. proportional to the index k of the k–point correlation functions. Applying the standard steps of the super-symmetry method [13, 14], the generating function (11) is expressed as the integral

$$Z_{\beta k}(x + J, \alpha) = \int d[H^{(0)}] P_N^{(0)}(H^{(0)}) \int d[\sigma] Q_{\beta k}(\sigma, t)$$

$$\mathrm{detg}^{-1}\left((x^\pm + J - \sigma) \otimes 1_N - 1_{\zeta_\beta 2k} \otimes H^{(0)} \right) . \tag{44}$$

over the supermatrices σ. Determinant and trace in superspace are denoted by detg and trg, respectively. Here, 1_N and $1_{\zeta_\beta 2k}$ are unit matrices of dimension N and $\zeta_\beta 2k$, respectively. With $c_{\beta k}$ a normalization constant, the graded probability density is given by

$$Q_{\beta k}(\sigma, t) = c_{\beta k} \exp\left(-\frac{\beta}{4t} \mathrm{trg}\sigma^2 \right) . \tag{45}$$

In the limit $t \to 0$ it reduces to the superspace δ function $\delta(\sigma)$. For vanishing transition parameter t, $Z_{\beta k}$ therefore becomes the generating function

$$Z_k^{(0)}(x + J) = \int d[H^{(0)}] P_N^{(0)}(H^{(0)})$$

$$\mathrm{detg}^{-1}\left((x^\pm + J) \otimes 1_N - 1_{\zeta_\beta 2k} \otimes H^{(0)} \right) \tag{46}$$

for the correlations of the ensemble of matrices $H^{(0)}$.

Because of the Gaussian form of the graded probability density (45) we can formulate a diffusion process in superspace. We shift $\sigma \to \sigma - x - J$ and diagonalize the supermatrix $\sigma = u^{-1} s u$ where s is the diagonal matrix of the eigenvalues of σ. There are $2k$ eigenvalues for the GUE and $3k$ eigenvalues for the GOE and the GSE. We notice that these eigenvalues are commuting variables. We now replace the matrix $x + J$ by an diagonal matrix r having the same form as s. These steps yield

$$Z_{\beta k}(r, t) = \int Q_{\beta k}(\sigma - r, t) Z_k^{(0)}(s) d[\sigma] . \tag{47}$$

There are only $2k$ energies and source variables in $x + J$, whereas the matrix r contains $3k$ variables for $\beta = 1$ and $\beta = 4$. The convolution integral (47) is related to a diffusion equation. Since $Z_k^{(0)}(s)$ depends only on s, the diffusion takes place in the curved space of the eigenvalues of the supermatrix. In analogy to the previous section, the diffusion equation has the form

$$\frac{\beta}{4} \Delta_r Z_{\beta k}(r, t) = \frac{\partial}{\partial t} Z_{\beta k}(r, t) \qquad \text{with} \qquad \lim_{t \to 0} Z_{\beta k}(r, t) = Z_k^{(0)}(r) \tag{48}$$

and holds for all three symmetry classes and for arbitrary initial generating functions $Z_k^{(0)}(r)$. The Laplacean Δ_r is the "radial part"

$$\Delta_r = \frac{1}{B_{\beta k}(r)} \sum_{j,p} \frac{\partial}{\partial r_{pj}} B_{\beta k}(r) \frac{\partial}{\partial r_{pj}} . \tag{49}$$

of the full Laplacean $\Delta = \mathrm{trg}\partial^2/\partial\sigma^2$, defined in terms of the matrix gradient $\partial/\partial\sigma$. As usual, the radial part involves the Jacobians or Berezinians $B_{\beta k}(r)$ of the transformation to radial coordinates. Explicit expressions for these quantities are not given here and may be found in Ref. 4. We notice that Eq. (37) describes a diffusion of the probability density of the eigenvalues in the space of ordinary matrices whereas Eq. (48) describes a diffusion of the generating function for the correlations in the space of supermatrices.

The corresponding diffusion kernel is given by the angular average

$$\gamma_{\beta k}(s, r, t) \;=\; \int Q_{\beta k}(u^{-1}su - r, t)\, d\mu(u) \tag{50}$$

where $d\mu(u)$ is the Haar measure. It also satisfies the diffusion Eq. (48), and an initial condition at $t = 0$ not given here. This implies that we can write the solution of Eq. (48) in the form

$$Z_{\beta k}(r, t) = \int \gamma_{\beta k}(s, r, t) Z_k^{(0)}(s) B_{\beta k}(s) d[s] \ . \tag{51}$$

The diffusion process cannot be formulated for the k variables x but only for the $2k$ or $3k$ variables r, respectively. Therefore, the role of the joint probability density in ordinary space is here played by the generating function rather than the correlation functions.

In the unitary case $\beta = 2$, the kernel $\gamma_{\beta k}$ is explicitly known: It is the supersymmetric Harish–Chandra–Itzykson–Zuber integral [15]. The solution of the diffusion equation reads

$$Z_{2,k}(r, t) = \frac{1}{B_k(r)} \int G_k(r - s, t) Z_k^{(0)}(s) B_k(s) d[s] \tag{52}$$

where the relevant part of the kernel is the Gaussian

$$G_k(s - r, t) = \frac{1}{\sqrt{2\pi t}^{2k}} \exp\left(-\frac{1}{2t}\mathrm{trg}(s - r)^2\right) \ . \tag{53}$$

The Jacobian or Berezinian is written as $B_{2,k}(s) = B_k^2(s)$ where $B_k(s) = \det[1/(s_{p1} - is_{q2})]_{p,q=1,\ldots,k}$ reflects the determinant structure [1] of the GUE correlation functions. The correlation functions can now be calculated straightforwardly [4, 16]. They are given by the $2k$–dimensional integral representation

$$R_{2,k}(x_1, \ldots, x_k, t) = \frac{(-1)^k}{\pi^k} \int G_k(s - x, t) \Im Z_k^{(0)}(s) B_k(s) d[s] \tag{54}$$

which is exact for all initial conditions and also exact for any finite level number N. The determinant structure of the pure GUE correlations is only preserved if $\Im Z_k^{(0)}(s)$ is a product of functions, each depending on of the eigenvalues.

It can easily be seen [4] that the transition on the unfolded scale is a diffusive process as well. With $\rho = r/D$, $z_k^{(0)}(\rho) = \lim_{N\to\infty} Z_k^{(0)}(r)$, and $z_{\beta k}(\rho, \tau) = \lim_{N\to\infty} Z_{\beta k}(r, t)$ one finds

$$\frac{\beta}{4}\Delta_\rho z_{\beta k}(\rho, \tau) = \frac{\partial}{\partial\tau} z_{\beta k}(\rho, \tau) \qquad \text{with} \qquad \lim_{\tau\to 0} z_{\beta k}(\rho, \tau) = z_k^{(0)}(\rho) \ . \tag{55}$$

The diffusion processes on the original and the unfolded scale are the same, only the initial conditions are different, see Ref. 4. Therefore, the integral representation (51) must hold on the unfolded scale as well,

$$z_{\beta k}(\rho, \tau) \;=\; \int \gamma_{\beta k}(s, \rho, \tau)\, z_k^{(0)}(s)\, B_{\beta k}(s) d[s] \ . \tag{56}$$

In particular, in the unitary case $\beta = 2$ we find

$$z_{2,k}(\rho, \tau) = \frac{1}{B_k(\rho)} \int G_k(\rho - s, t) z_k^{(0)}(s) B_k(s) d[s] \tag{57}$$

for the generating function. From that, the $2k$–dimensional integral representation

$$X_{2,k}(\xi_1, \ldots, \xi_k, \tau) = \frac{(-1)^k}{\pi^k} \int G_k(s - \xi, \tau) \Im z_k^{(0)}(s) B_k(s) d[s] \tag{58}$$

for the correlation functions on the unfolded scale is obtained. In the case $k = 2$, a further simplification arises because the two–level correlation function depends only on $r = \xi_1 - \xi_2$. The remaining four integrals can be trivially reduced to two integrals and one has [4, 16]

$$X_{2,2}(r, \tau) = \frac{4}{\pi^3 \tau} \int\limits_{-\infty}^{+\infty} \int\limits_{-\infty}^{+\infty} \exp\left(-\frac{1}{4\tau}(t_1^2 + t_2^2)\right)$$
$$\frac{t_1 t_2}{(t_1^2 + t_2^2)^2} \sinh \frac{rt_1}{2\tau} \sin \frac{rt_2}{2\tau} \Im z_2^{(0)}(t_1, t_2) dt_1 dt_2 \ . \tag{59}$$

The trigonometric and the hyperbolic functions can be directly traced back to the eigenvalues in the Boson–Boson and the Fermion–Fermion block.

We compare the hierarchic equations (42) and the diffusion equation (55). The former couple the correlation functions with index k to all those with index up to $k + 1$ and thus cannot be viewed as describing a diffusion process. The diffusion equation (55) in superspace, on the other hand, does not couple different values of k. In this sense, the diffusion equation (55) diagonalizes the hierarchic equations (42).

SUMMARY AND CONCLUSION

Crossover transition between various fluctuation properties are a common feature of many physical systems, see the discussion in Ref. 3. Within Random Matrix Theory, it is easily possible to construct ensembles which interpolate between different statistics. The price one has to pay is the loss of rotation invariance in these ensembles. Diffusion in matrix spaces, a variant of Dyson's Brownian motion, is the framework that allows one to study the evolution of the fluctuation properties with a fictitious time, provided by the squared transition parameter. We tried to convince the reader that this diffusion is a straightforward extension of the diffusion in vector spaces. The kernels can be viewed as generalized Bessel function in these matrix spaces. Unfortunately, the kernel is only explicitly known in the unitary case.

Moreover, the diffusion in the space of ordinary matrices has a very natural extension in the space of supermatrices. Here, however, it is the generating function of the correlators that is subjected to a diffusion, whereas it is the probability density function in the space of ordinary matrices. Thus, the correlations are accessible in a much more direct way since no further intergrations are neccessary once the time evolution of the generating functions is obtained. This considerable advantage is reflected in the much smaller dimension of the space in which the diffusion takes place. In both diffusions, it is the curved space of eigenvalues. In the case of ordinary matrices, this dimension is the level number N, and we have to take the limit $N \to \infty$ in the end. In the case of supermatrices, the dimension is $2k$ or $3k$, respectively, when k–point functions are studied. This dimension is therefore independent of N.

The difference in the spaces and the dimensions of the two diffusions is intertwined with the most important advantage of the diffusion in the space of supermatrices, the invariance under scaling. The diffusion problem is the same on all scales, in particular on the original and on the unfolded scale. The difference in the correlations stems from the initial condition alone, the time evolution is always the same. This remarkable feature has no analog in the space of ordinary matrices. It leads to closely related integral representations for all correlations on the different scales. Unfortunately and similar to the case of the diffusion in the space of ordinary space, explicit results are only known in the unitary case. Explicit results for the orthogonal and symplectic case are highly desirable. The diffusion discussed here does not apply, in the strict sense, to the case of symmetry breaking. However, the results of Ref. 17 where this problem was solved suggest that even the diffusion in the space of supermatrices is embedded into a slightly more genral structure.

The convenient features of the diffusion in the space of supermatrices makes it well suited for applications. In Refs. 4, 16, $2k$–dimensional integral representations were obtained for the correlation functions, on the original and the unfolded scale, in the transition regime from Poisson regularity to the GUE. With the help of these convenient integral representations, a detailed discussion of various scales emerging in this particular transition could be identified and studied [18, 19].

Finally, we mention that the concept of the diffusion is not restricted to classical Random Matrix Theory. In recent years, Chiral Random Matrix Theory [20] received much interest. In involves random matrices of a different structure which reflect the chiral symmetry of the Dirac operator. In Refs. 21, 22 it was shown that there are similar diffusions in Chiral Random Matrix Theory, in the space of ordinary and super-matrices. The kernel for both diffusions was calculated in Ref. 23 in the unitary case. It requires integrals over two unitary matrices. In contrast to classical Random Matrix Theory, Chiral Random Matrix Theory exhibits a second scale, the microscopic scale, in the center of the spectrum, in addition to the scale of the local mean level spacing in the bulk of the spectrum. Thus, the convenient scaling invariance of the diffusion in the space of supermatrices turns out to be particularly useful in Chiral Random Matrix Theory.

REFERENCES

1. M.L. Mehta, "Random Matrices", 2nd ed., Academic, New York (1991).
2. F. Haake, "Quantum Signatures of Chaos", Springer, Berlin (1991).
3. T. Guhr, A. Müller–Groeling and H.A. Weidenmüller, Random Matrix Theories in Quantum Physics: Common Concepts, *Physics Reports, Phys. Rep.* 299:189 (1998).
4. T. Guhr, *Ann. Phys. (NY)* 250:145 (1996).
5. F.J. Dyson, *J. Math. Phys.* 3:1191 (1962).
6. F.J. Dyson, *J. Math. Phys.* 13:90 (1972).
7. G. Lenz and F. Haake, *Phys. Rev. Lett.* 65:2325 (1990).
8. C. Itzykson and J.B. Zuber, *J. Math. Phys.* 21:411 (1980).
9. M. L. Mehta and A. Pandey, *J. Phys. A* 16:2655,L601 (1983).
10. A. Pandey, *Chaos, Solitons and Fractals* 5:1275 (1995).
11. J.B. French, V.K.B. Kota, A. Pandey and S. Tomsovic, *Ann. Phys. (NY)* 181:198 (1988).
12. L.A. Pastur, *Teor. Mat. Fis.* 10:102 (1972).
13. K.B. Efetov, *Adv. in Phys.* 32:53 (1983).
14. J.J.M. Verbaarschot, H.A. Weidenmüller and M.R. Zirnbauer, *Phys. Rep.* 129:367 (1985).
15. T. Guhr, *J. Math. Phys.* 32:336 (1991).
16. T. Guhr, *Phys. Rev. Lett.* 76:2258 (1996).
17. T. Guhr and H.A. Weidenmüller, *Ann. Phys. (NY)* 199:412 (1990).
18. T. Guhr and A. Müller–Groeling, *J. Math. Phys.* 38:1870 (1997).

19.   K. Frahm, T. Guhr and A. Müller–Groeling, Between Poisson and GUE Statistics: Role of the Breit–Wigner Width, Annals of Physics (NY), in press, `cond-mat/9801298`.

20.   E. Shuryak and J.J.M. Verbaarschot, *Nucl. Phys. A* 560:306 (1993); J.J.M. Verbaarschot, *Phys. Rev. Lett.* 72:2531 (1994).

21.   T. Guhr and T. Wettig, Universal Spectral Correlations of the Dirac Operator at Finite Temperatures, *Nuclear Physics B*, 506:589 (1997).

22.   A.D. Jackson, M.K. Şener and J.J.M. Verbaarschot, *Nuclear Physics B*, 506:612 (1997).

23.   T. Guhr and T. Wettig, *J. Math. Phys.* 37:6395 (1996).

WHAT HAPPENS TO THE INTEGER QUANTUM HALL EFFECT IN THREE DIMENSIONS?

J. T. Chalker

Theoretical Physics, University of Oxford,
1, Keble Road, Oxford OX1 3NP, United Kingdom.

INTRODUCTION

The integer quantum Hall effect is often advertised as one of the most striking phenomena to occur in the two-dimensional electron gas. It is nevertheless natural and interesting to ask whether it can also take place in a three-dimensional system, and that question is the subject of this article. More specifically, two remarkable aspects of the quantum Hall effect in two dimensions are that the Hall conductance shows plateaus, in which it is constant for a range of Landau level filling factors, and that, when the Hall conductance lies on a plateau, current flows through the sample via edge states. Here we ask, for three-dimensional systems, whether the Hall conductance can also show plateaus, and whether there then exist surface states, analogous to the edge states of a two-dimensional system. Of course, for there to be much to write about, the answers to these questions must in some circumstances be 'yes', and our focus will partly be on the conditions necessary to favour the quantum Hall effect in three dimensions.

The area has a substantial and growing literature. Early work has been reviewed by Halperin [1]. The new ideas described in the present article are the result of a collaboration with A. Dohmen [2], and many were developed independently by Balents and Fisher [3]. Recent theoretical work includes that in Refs. [4, 5, 6, 7, 8, 9]. Experimental work [10, 11, 12, 13], which began over ten years ago, is summarised in the concluding section.

It is useful to start by analogy with the two-dimensional integer quantum Hall effect, which one explains [14] in terms of the properties of single-particle states in disorder-broadened Landau levels. Most of these states, including those in Landau level tails, are Anderson-localised, and electrons occupying them do not participate in transport at low temperatures. Near the centre of each Landau level, however, the localisation length diverges and there exist extended, current-carrying states. Crucially, as a function of energy, there is an alternating sequence of localised and extended states. What should one expect for a *three*-dimensional, disordered system of non-interacting electrons in a strong magnetic field? There seem to be two possibilities. The first, illustrated in Fig. 1, is that there exist localised states only over one energy range, centred on the Lifshitz tail at the band edge. In this event, there is no three-dimensional quantum Hall

Supersymmetry and Trace Formulae: Chaos and Disorder
Edited by Lerner *et al.*, Kluwer Academic / Plenum Publishers, New York, 1999

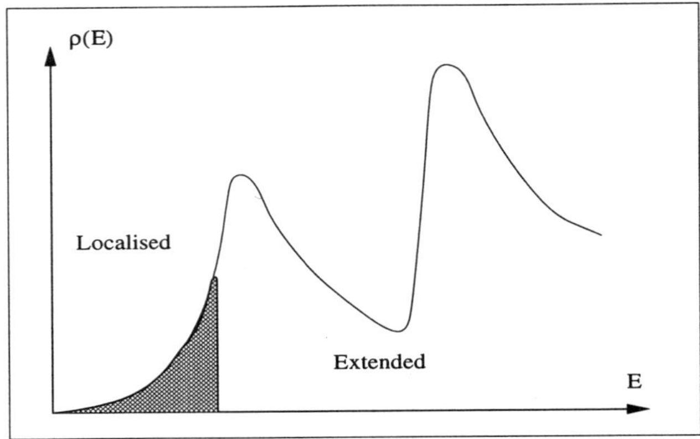

Figure 1. Possible dependence of localisation properties and density of states, $\rho(E)$, on energy, E, for the lowest Landau levels of a three-dimensional conductor in a strong magnetic field. In this case, there is only one band of localised states.

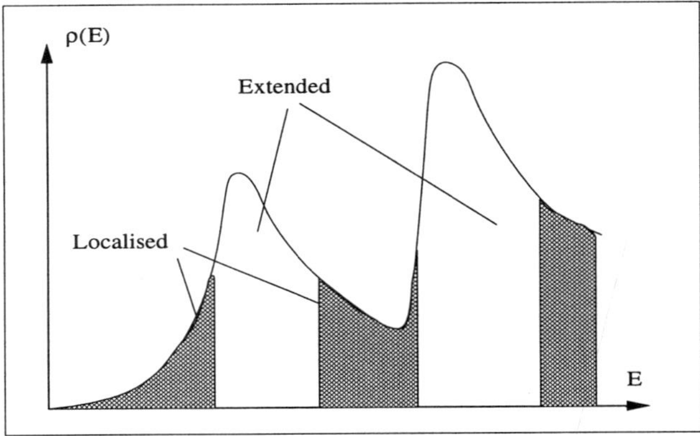

Figure 2. Alternative behaviour for the lowest Landau levels of a three-dimensional conductor in a strong magnetic field, with more than one band of localised states.

effect. The alternative, more interesting possibility is that there is a sequence of bands of localised states, occuring where the density of states has minima, and separated by bands of extended states, as illustrated in Fig. 2. In this event, if the Fermi energy is increased, starting from the band edge, the system will pass through a sequence of phases: insulator - metal - quantum Hall conductor - metal ... and so on.

A central question is clear: how does one ensure the behaviour shown in Fig. 2 over that of Fig. 1? The answer is also reasonably clear. One should pick a *layered* system, consisting of a stack of two-dimensional electron gases, each chosen so that it would, in isolation, display the quantum Hall effect. We shall argue that the quantum Hall effect in such a stack persists after the introduction of inter-layer coupling, provided the coupling is not too strong. Indeed, a plausible phase diagram, in the plane of Fermi energy and inter-layer coupling strength, is that illustrated in Fig. 3. Briefly, the grounds for expecting this phase diagram are as follows. In the tails of Landau levels, the localisation length in an isolated layer is short, so that localised states are stable against weak interlayer coupling, and a quantum Hall conductor is anticipated, while near the centre of each Landau level the localisation length in an isolated layer is large, so that weak interlayer coupling should be sufficient to mix states from different layers,

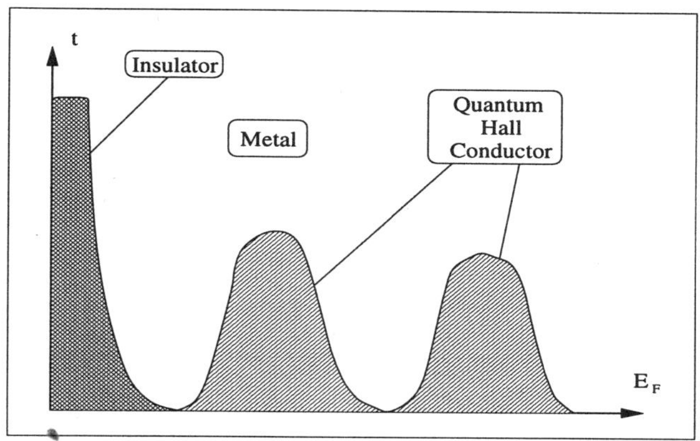

Figure 3. Conjectured phase diagram for a layered conductor in a strong magnetic field, as a function of inter-layer coupling, t, and Fermi energy, E_F.

forming a metallic phase. If the interlayer coupling is stronger, more states are likely to be extended, and for sufficiently strong coupling it is likely that the behaviour switches from that of Fig. 2 to that of Fig. 1. We substantiate these ideas in the following two sections.

PHASE DIAGRAM OF A LAYERED CONDUCTOR IN A STRONG MAGNETIC FIELD

One can give arguments of several kinds in favour of the phase diagram suggested in Fig. 3. In this section, we outline some of these.

First, recall the scaling flow diagram for the two-dimensional integer quantum Hall effect [15]. In this, one considers the dependence of the conductivity tensor on the length-scale at which it is measured, under the assumption that knowledge of its value at one scale is sufficient to determine its evolution at larger scales. From the scaling flow suggested by experiment, one concludes, amongst other things, that the localisation length in a two-dimensional quantum Hall system is divergent at those combinations of Fermi energy and magnetic field strength for which the Hall conductance takes the values $(n + 1/2)e^2/h$, with n integer. Now suppose that it is legitimate to apply this result to a sample consisting of N coupled layers, at least for N finite. If the Fermi energy is swept through one Landau level of the layers, the Hall conductance of the N-layer sample as a whole will increase by Ne^2/h, passing through N delocalisation points in the process. The localisation length of an N-layer sample should therefore diverge at N discrete energies, a fact previously established for $N = 2$ in the context of spin-degenerate Landau levels [16, 17]. For large N, the delocalisation points crowd together, and the minimum value of the localisation length between each divergence gets larger, resulting in a metallic phase which occupies a window of energy, in the thermodynamic limit, $N \to \infty$, as in Fig. 3.

Second, consider the shape of the boundary to the metallic phase at weak inter-layer coupling, defined by the dependence of the critical energy, $E_c(t)$ on the coupling, t, for small t. We can determine this shape as follows. In an isolated layer, the localisation length, $\xi(E)$, is known to vary with energy, E, according to $\xi(E) \sim |E - E_c(0)|^{-\nu_{2d}}$, where $\nu_{2d} \approx 2.3$ is the two-dimensional localisation length exponent. We ask how strong the interlayer coupling must be, in order to mix two localised states from adjacent layers

77

and form a metal. We expect, at the critical value, that the matrix element of this coupling between the two states will be of the same order as their energy separation. Since the states have a normalised probability density spread over an area ξ^2, with a phase correlation length of order the magnetic length, l_B, this matrix element is tl_B/ξ, while the energy splitting is $[\rho_{2d}\xi^2]^{-1}$, where ρ_{2d} is the density of states of a single layer. Equating these, we obtain $t \sim \xi^{-1} \sim |E_c(t) - E_c(0)|^{\nu_{2d}}$, and hence the form of the metallic phase boundary shown in Fig. 3.

Third, one can try to make use of the representation of quantum Hall systems in terms of quantum spin chains [18]. Since the relevant, supersymmetric spin chains have not been solved, it is necessary to proceed by analogy with the behaviour of conventional, Heisenberg spin chains. The analogy relates the plateau transition in an isolated layer (a system with two space dimensions) to a phase transition as parameters are varied in a spin-half Heisenberg antiferromagnetic chain (with one space and one time dimension). In general, the required spin chain has exchange interactions of alternating strength, $J \pm \Delta J$. It is then in a gapped spin-Peierls phase, equivalent, under the analogy, to a plateau phase of a quantum Hall system. Varying ΔJ through zero takes the system from one phase to another, with a divergent correlation length only at the critical point, $\Delta J = 0$. Thus the phase diagrams of the two systems are in correspondence, though not, of course, their critical properties. An equivalent analogy [4, 8] relates a three-dimensional conductor in a magnetic field to a spin system in two space and one time dimensions, constructed by coupling parallel copies of the Heisenberg chains described above. The critical point in an isolated chain, at $\Delta J = 0$, expands for coupled chains [19] into a Néel-ordered phase, stable over a finite range of ΔJ. This phase is the analogue of the metallic phase in the disordered conductor. It separates spin-Peierls phases, entered at large $|\Delta J|$, which are the analogues of quantum Hall plateau phases.

Further, as shown recently by Wang [9], the phase diagram of Fig. 3 can be derived using a renormalisation group analysis of the appropriate, layered non-linear sigma model.

NUMERICAL CALCULATIONS OF THE PHASE DIAGRAM

The phase diagram suggested in Fig. 3 can also be obtained from numerical calculations on a suitable model. We choose to study a three-dimensional version of the network model [20], which in two dimensions is known to provide a convenient example of a system in the universality class of the integer quantum Hall plateau transition. The model is intended to represent electron motion in a strong magnetic field and smooth disordered potential, in terms of phase-coherent guiding-centre drift. The model consists of set of directed links, joined at nodes. Electrons propagate along links in the direction of the local guiding-centre drift, and tunnel between links at nodes. Disorder is introduced via the phase shifts that electrons aquire in traversing a link: an independent, fixed, random phase is associated with each link. The model is illustrated in Fig. 4; further details can be found in Refs. [20] and [2]. It contains two parameters, which determine the tunneling probabilities at the nodes. Physically, the in-plane tunneling strength can be interpreted as the Fermi energy, while the interplane tunneling corresponds simply to the interlayer coupling, denoted by t in Fig. 3.

We study this model numerically, using standard transfer matrix methods [21] to calculate the localisation length, $\xi(M)$, in quasi one-dimensional samples of cross-section $M \times M$ and length L, for $M \leq 12$ and $L \leq 8 \cdot 10^4$. To obtain information on bulk properties, we apply periodic boundary conditions in the directions transverse to L,

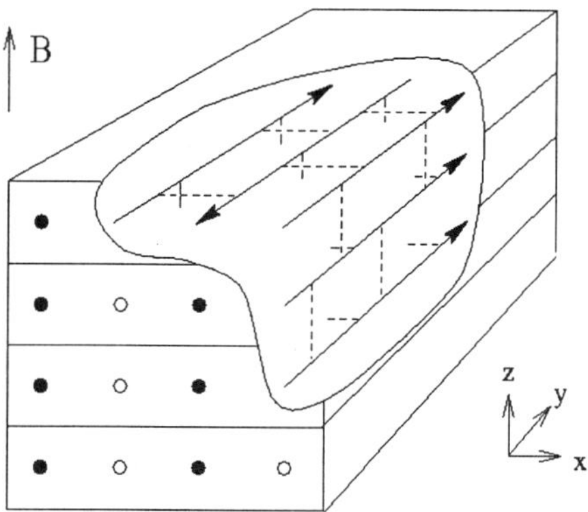

Figure 4. Three-dimensional network model.

and examine the localisation length as a function of cross-section, M. In the insulator and quantum Hall conductor, $\xi(M)$ extrapolates in the limit $M \to \infty$ to a finite value, which is the bulk localisation length for a sample that is macroscopic in all three dimensions. By contrast, in the metallic phase, $\xi(M)$ diverges in the limit $M \to \infty$. The phase diagram we obtain in this way is displayed in Fig. 5.

This phase diagram for the three-dimensional network model has the form anticipated in the preceeding section and in Fig. 3. In particular, the shape of the phase boundary at weak interlayer coupling is roughly quadratic, as it should be if $t \sim |E_c(t) - E_c(0)|^{\nu_{2d}}$ and $\nu_{2d} \approx 2.3$. The most important limitation of the model is that only one Landau level is retained for each layer. As a consequence, only one 'period' of the phase diagram of Fig. 3 is reproduced in Fig. 5. More seriously, for the same reason, the model cannot capture the likely disappearance of the quantum Hall conductor at strong inter-layer coupling, because this involves mixing between different Landau levels of each layer.

SURFACE STATES

The bulk properties of the insulator and quantum Hall conductor are identical, since states at the Fermi energy are localised in both phases, within the bulk of the sample. Indeed, for the three-dimensional network model that we study numerically, this symmetry is built in exactly. In order to distinguish the two phases, it is necessary to examine the properties of their surface states. Surface states at the Fermi energy are localised in the insulator, but extended – and responsible for carrying the Hall current – in the quantum Hall conductor.

To display this difference using numerical calculations, we examine quasi one-dimensional samples with hard-wall boundary conditions at the edges of the layers, and with the long axis of the samples parallel to the layers. In a quantum Hall conductor, we expect one surface channel per layer of the sample, arising from the edge states that would exist in each layer in the absence of inter-layer coupling. The presence of these surface channels is revealed in the spectrum of Lyapunov exponents of the quasi one-dimensional sample. The Lyapunov exponents give, essentially, the decay rates

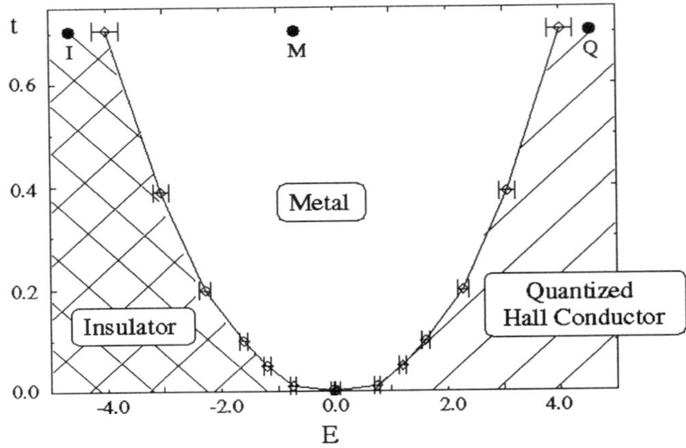

Figure 5. Numerically determined phase diagram for a layered conductor in a strong magnetic field, as a function of interlayer coupling, t, and Fermi energy, E_F.

(or inverse localisation lengths) for waves travelling along the sample in each mode. In the insulator, all of the Lyapunov exponents remain non-zero, even in the limit of large sample cross-section, the smallest of them converging to the inverse of the bulk localisation length. By contrast, in the quantum Hall conductor with hard-wall boundary conditions, M of the Lyapunov exponents in an M layer sample are small, as shown in Fig. 6. The fact that these highly transmitting modes are associated with surface states is made clear by examining the corresponding eigenvectors of the transfer matrix, shown in the inset to Fig. 6: these eigenvectors have their density concentrated overwhelmingly near the sample surface.

We next consider the properties of these surface states in more detail. We argue that they constitute a novel type of conductor, two-dimensional and chiral, which is critical in the sense that the only length scales are set by the sample dimensions. Deep in a quantum Hall plateau, the localisation length in the bulk is short and the bulk states decouple from the surface. To focus on the chiral conductor, it is convenient to discard the bulk altogether and to examine a two-dimensional model [2, 22]. A suitable way to define such a model is again as a variant of the network model, this time in the form illustrated in Fig. 7. In this figure, each full line corresponds to the edge chanel of one layer of the three-dimensional sample. Guiding centre drift is in the same direction in each layer, as indicated by the arrows in the figure, and so the model for surface states differs from that for the two-dimensional integer quantum Hall effect in being *directed*. Scattering between the edge states of adjacent layers occurs in the model at nodes, indicated by dashed lines in the figure. The amplitude, t, for this scattering is the only parameter of the model. Randomness enters, as before, via random scattering phases associated with links, chosen for simplicity uniformly from $(0, 2\pi)$. Remembering that the model describes the surface of a sample, we should apply periodic boundary conditions in the chiral direction, wrapping the system of Fig. 7 around a cylinder of circumference C and height L.

Properties of the surface states depend crucially on the value of the aspect ratio, L/C. To see this, consider a sample connected to leads at either end of the cylinder, and imagine the track of an electron injected from one lead. If the cylinder is short, $L \ll C^{1/2}$, the electron will escape via one or other lead before it can circumnavigate

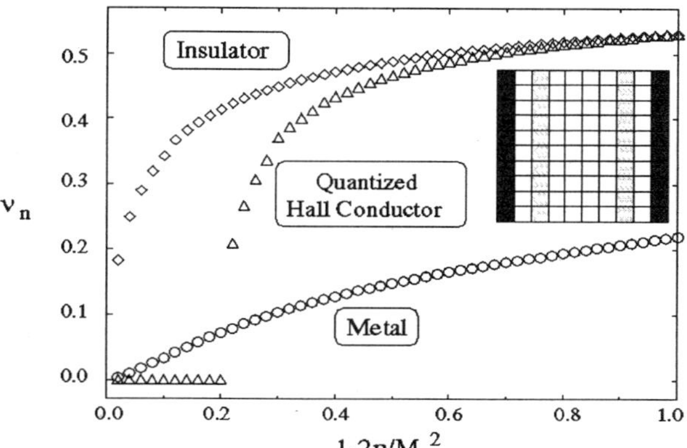

Figure 6. Spectra of positive Lyapunov exponents, ν_n, $n = 1 \ldots M^2/2$, at three points in the phase diagram, marked I, Q, and M in Fig. 5, for sample size $M = 10$. Inset: amplitude distribution for the transfer matrix eigenvector with the smallest positive Lyapunov exponent, in the quantised Hall conductor. Squares represent links of the model and are shaded according to the mean probability flux, p, carried by that link: black, $p > 0.04$; grey, $0.04 \geq p > 2 \cdot 10^{-4}$; white, $2 \cdot 10^{-4} > p$.

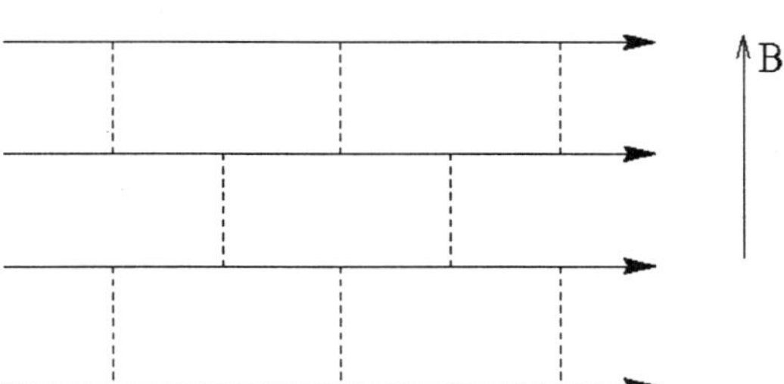

Figure 7. The directed network model for the two-dimensional chiral metal.

the cylinder. In consequence, if we consider the amplitude for propagation from one point to another, in terms of a sum of contributions from different Feynman paths, each distinct path will have an independent phase. Because of this, the non-trivial interference effects that result, for example, in localisation are absent, and motion is diffusive in the direction perpendicular to the layers of the sample. In fact, the disorder-averaged conductance, $\langle G \rangle$, of the sample can be calculated exactly [2], giving $\langle G \rangle = \sigma C/L$ with $\sigma = (e^2/h)t^2/(1 - t^2)$. This Ohm's-law dependence of $\langle G \rangle$ on C and L actually holds good (with the same σ) not only for the chiral metal, $L \ll C^{1/2}$, but also for the quasi one-dimensional metallic regime, $C^{1/2} < L < C$. Behaviour in the opposite limit of a long cylinder, $L \gg C$, is however very different. In this limit, typical particle trajectories wrap many times around the cylinder before leaving via one of the contacts. Correspondingly, the different paths joining a pair of points can be grouped into classes: members of a given class traverse the same set of links but in different orders, and all have related phases. Interference between Feynman paths related in this way results in localisation in the direction parallel to the cylinder axis, and the conductance decreases exponentially with L, behaviour which is natural since the sample is quasi one-dimensional. The fact that the chiral conductor is critical reveals itself in this regime through the proportionality of the localisation length, ξ, to the sample circumference, C, which hence sets the length scale for the system. The localisation length can be obtained as the sample size at which $\langle G \rangle = e^2/h$, yielding [2] $\xi = A \cdot C$ with $A = 2t^2/(1 - t^2)$, a relation that has been established more formally by Gruzberg, Read and Sachdev [7].

Whilst the dependence of the mean conductance on sample geometry is Ohmic in both the chiral metal and the quasi one-dimensional metal, these regimes are distinguished in the behaviour of conductance fluctuations [5, 6, 7]. The chiral metal presents of order C/L^2 independent, parallel regions for conductance along the cylinder axis, and each of these contributes additively to fluctuations, so that [5, 6, 7] $\langle (\delta G)^2 \rangle \sim (e^2/h)^2 C/L^2$; by contrast, $\langle (\delta G)^2 \rangle \sim (e^2/h)^2$ in the quasi one-dimensional metal.

Finally, we note that, just as models for the quantum Hall effect in two dimensions can be mapped onto models for antiferromagnetic spin chains, using supersymmetry or replicas, so also can chiral surface states be represented as spin chains [4, 6, 7], but with ferromagnetic coupling arising from the directed nature of the underlying problem.

EXPERIMENTS

Quantisation of the Hall conductance has been observed in three-dimensional, layered samples of two kinds: semiconductor multi-quantum wells [10], and spin-density wave phases of Bechgaard salts [11]. In both cases, plateaus in σ_{xy} are accompanied by very small values of the in-plane, dissipative conductivity, σ_{xx}. Moreover, specific evidence for the existence of the chiral metal at the surface of a quantum Hall conductor has been obtained recently, indirectly in organic conductors [12], and directly in multi-quantum well samples, by Gwinn and collaborators [13]. In these latter experiments, the dependence on sample dimensions is studied for the conductance, G_{zz}, in the direction perpendicular to the layers. In the metallic phase, G_{zz} is expected, and found, to be proportional to sample cross-section. The same is true in the quantum Hall conductor at high temperatures, when contributions to transport from the bulk presumably short-circuit the surface states. In the same phase at low temperatures, however, when the bulk conductivity is small, G_{zz} is found to be proportional to sample circumference, rather than cross-section [13], in direct and striking confirmation of theoretical ideas.

ACKNOWLEDGEMENTS

I am grateful to Axel Dohmen for his collaboration, and to M. Janssen, D. E. Khmelnitskii, and B. Shapiro for valuable discussions. The work was supported in part by EPSRC Grant GR/J8327.

REFERENCES

1. B.I. Halperin, Jpn. J. Appl. Phys. **26** Suppl.26-3, 1913 (1987).
2. J.T. Chalker and A. Dohmen, Phys. Rev. Lett. **75**, 4496 (1995).
3. L. Balents and M.P.A. Fisher, Phys. Rev. Lett. **76**, 2153 (1996).
4. Y.B. Kim, Phys. Rev. B **53**, 16420 (1996).
5. H. Mathur, Phys. Rev. Lett. **78**, 2429 (1997); Y.-K. Yu, cond-mat/9611137.
6. L. Balents, M.P.A. Fisher, and M.R. Zirnbauer, Nucl. Phys. B **483**, 601 (1997); S. Cho, L. Balents, and M.P.A. Fisher, cond-mat/9708038.
7. I.A. Gruzberg, N. Read, and S. Sachdev, Phys. Rev. B **55**, 10593 (1997); *ibid* **56** 13218 (1997).
8. Z.Q. Wang, Phys. Rev. Lett. **78**, 126 (1997).
9. Z.Q. Wang, Phys. Rev. Lett **79**, 4002 (1997).
10. H.L. Störmer, J.P. Eisenstein, A.C. Gossard, W. Wiegmann, and K. Baldwin, Phys. Rev. Lett. **56** 85 (1986).
11. D. Poilblanc, G. Montambaux, M. Heritier, and P. Lederer, Phys. Rev. Lett. **58** 270 (1986).
12. S. Hill, P.S. Sandhu, J.S. Qualls, J.S. Brooks, M. Tokumoto, N. Kinoshita, T. Kinoshita, and Y. Tanaka, Phys. Rev. B **55**, R4891 (1997).
13. D.P. Druist, P.J. Turley, E.G. Gwinn, K. Maranowski, and A.C. Gossard, UCSB preprint.
14. For reviews, see *The Quantum Hall Effect*, edited by R.E. Prange and S.M. Girvin (Springer, Berlin, 1990); and M. Jansen, O. Viehweger, U. Fastenrath and J. Hajdu *Introduction to the Theory of the Integer Quantum Hall Effect* (VCH, Weinheim, 1994).
15. D.E. Khmel'nitskii, Pis'ma Zh. Eksp. Teor. Fiz. **38**, 454 (1983) [JETP Lett. **38**, 552].
16. D.K.K. Lee and J.T. Chalker, Phys.Rev.Lett. **72**, 1510(1994).
17. Z.Q. Wang, D.-H. Lee, and X.G. Wen, Phys. Rev. Lett. **72**, 2454 (1994).
18. N. Read, unpublished; D.-H. Lee, Phys. Rev. B **50**, 10788 (1994); M. R. Zirnbauer, Annalen der Physik **3**, 513 (1994).
19. M.A. Martindelgado, R. Shankar, and G. Sierra, Phys. Rev. Lett. **77**, 3443 (1996).
20. J.T. Chalker and P.D. Coddington, J. Phys. C **21**, 2665 (1988).
21. A. MacKinnon and B. Kramer, Phys. Rev. Lett. **47**, 1546 (1981); Z. Phys. B **53**, 1 (1983).
22. The directed network model was first introduced, in a different context, in: L. Saul, M. Kardar, and N. Read, Phys. Rev. A **45**, 8859 (1992).

TRACE FORMULAS IN CLASSICAL DYNAMICAL SYSTEMS

Predrag Cvitanović

Department of Physics & Astronomy
Northwestern University
2145 Sheridan Road
Evanston, IL 60208-3112

INTRODUCTION

Dynamics is posed in terms of local equations, but the ergodic averages require global information. How can we use a local description of a flow to learn something about the global behavior? We shall now relate global averages to the eigenvalues of appropriate evolution operators. The eigenvalues are given by the zeros of corresponding determinants, and one way to evaluate determinants is to expand them in terms of traces, using the matrix identity $\log \det = \operatorname{tr} \log$. Traces of evolution operators can be evaluated as integrals over Dirac delta functions, and in this way the spectra of evolution operators become related to periodic orbits. If there is one idea that one should learn about chaotic dynamics, it is this: there is a fundamental local $\leftrightarrow$ global duality which says that

the spectrum of eigenvalues is dual to the spectrum of periodic orbits

For dynamics on the circle, this is called Fourier analysis; for dynamics on well-tiled manifolds, Selberg traces and zetas; and for generic nonlinear dynamical systems the duality is embodied in the formulas that we will now introduce: trace formulas, zeta functions and spectral determinants. These objects are to dynamics what partition functions are to statistical mechanics.

We start by discussing the necessity of studying the averages of observables in chaotic dynamics, and then cast the formulas for averages in a multiplicative form that motivates the introduction of evolution operators and the further formal developments to come.

Dynamical averaging

In chaotic dynamics detailed prediction is impossible, as any finitely specified initial condition, no matter how precise, will fill out the entire accessible phase space. Hence for chaotic dynamics one cannot follow individual trajectories for a long time; what is attainable is a description of the geometry of the set of possible outcomes, and

Supersymmetry and Trace Formulae: Chaos and Disorder
Edited by Lerner *et al.*, Kluwer Academic / Plenum Publishers, New York, 1999

evaluation of the long time averages. Examples of such averages are transport coefficients for chaotic dynamical flows, such as escape rate, mean drift and diffusion rate; power spectra; and a host of mathematical constructs such as generalized dimensions, entropies and Lyapunov exponents. Here we outline how such averages are evaluated within the framework of the periodic orbit theory. The key idea is to replace the expectation values of observables by the expectation values of generating functionals. This associates an evolution operator with a given observable, and relates the expectation value of the observable to the leading eigenvalue of the evolution operator.

Time averaging

Let a be any "observable", a function that associates to each point in phase space a number, a vector, or a tensor. The observable reports on a property of the dynamical system. The velocity field $a_i(x) = v_i(x)$ is an example of a vector observable; the length of this vector, or perhaps a temperature measured in an experiment at instant τ are examples of scalar observables. We define the *integrated observable A^t* as the time integral of the observable a evaluated along the trajectory of the initial point x_0,

$$A^t(x_0) = \int_0^t d\tau\, a(f^\tau(x_0))\,. \tag{1}$$

If the dynamics is given by an iterated mapping and the time is discrete, the integrated observable is given by

$$A^t(x_0) = \sum_{k=0}^{t-1} a(f^k(x_0)) \tag{2}$$

(we suppress possible vectorial indices for the time being). For example, if the observable is the velocity, $a_i(x) = v_i(x)$, its time integral $A_i^t(x_0)$ is the trajectory $A_i^t(x_0) = x_i(t)$. Another familiar example, for Hamiltonian flows, is the action associated with a trajectory passing through a phase space point x_0 (this function is the key to the quasiclassical quantization):

$$A^t(x_0) = \int_0^t d\tau\, \dot{\mathbf{q}}(\tau) \cdot \mathbf{p}(\tau)\,, \qquad x = [p, q]\,, \quad x_0 = [p(0), q(0)]\,.$$

The *time average* of the observable along the trajectory is defined by

$$\overline{a(x_0)} = \lim_{t \to \infty} \frac{1}{t} A^t(x_0)\,. \tag{3}$$

If a does not behave too wildly as a function of time — for example, $a_i(x) = x_i$ is bounded for bounded dynamical systems — $A^t(x_0)$ is expected to grow not faster than t, and the limit (3) might exist.

The time average depends on the trajectory, but not on the initial point. Starting at a later phase space point $f^T(x_0)$

$$\overline{a(f^T(x_0))} = \lim_{t \to \infty} \frac{1}{t} \int_T^t d\tau\, a(f^\tau(x_0))$$

$$= \lim_{t \to \infty} \frac{1}{t} \left(\int_t^T d\tau\, a(f^\tau(x_0)) + \int_0^t d\tau\, a(f^\tau(x_0)) \right) = \overline{a(x_0)}$$

one obtains a finite contribution to $A^t(x_0)$ which vanishes in the limit $t \to \infty$.

The integrated observable $A^t(x_0)$ and the time average $\overline{a(x_0)}$ take a particularly simple form when evaluated on a periodic orbit. Define

$$A_p = \int_0^{T_p} d\tau\, a(f^\tau(x_0))\,, \qquad x_0 \in p\,, \tag{4}$$

where p is a prime cycle, and T_p is its period. A_p is a loop integral along the cycle, so it is an intrinsic property of the cycle, independent of the starting point $x_0 \in p$. (If the observable a is not a scalar but a vector or matrix we might have to be more precise in defining an average which is independent of the starting point around the cycle). If the trajectory retraces itself r times, we just obtain A_p repeated r times. Evaluation of the asymptotic time average (3) requires therefore only a single traversal of the cycle:

$$\overline{a_p} = \frac{1}{T_p} A_p\,. \tag{5}$$

However, $\overline{a(x_0)}$ is in general certainly a very wild function of x_0; for a nice hyperbolic system it takes the same value $\langle a \rangle$ for almost all initial x_0, but a different value (5) on any periodic orbit, that is, on a dense set of points. For example, for an open system such as the Sinai gas (an infinite two-dimensional periodic array of scattering disks) the phase space is dense with initial points that correspond to periodic runaway trajectories. The mean distance squared traversed by any such trajectory grows as $x(t)^2 \sim t^2$, and its contribution to the diffusion rate $D \approx x(t)^2/t$, (3) evaluated with $a(x) = x(t)^2$, diverges. Seemingly there is a paradox; even though intuition says the typical motion should be diffusive, we have an infinity of ballistic trajectories.

For chaotic dynamical systems, this paradox is resolved by robust averaging, that is, averaging also over the initial x, and worrying about the measure of the "pathological" trajectories.

Space averaging

The *space average* of a quantity a that may depend on the point x of phase space $\mathcal{M}$ and on the time t is given by the $(d+1)$-dimensional integral over the $(d+1)$ coordinates of the dynamical system:

$$\langle a(t) \rangle = \frac{1}{|\mathcal{M}|} \int_{\mathcal{M}} dx\, a(x(t))$$

$$|\mathcal{M}| = \int_{\mathcal{M}} dx = \text{volume of } \mathcal{M}\,. \tag{6}$$

The space $\mathcal{M}$ is assumed to have finite dimension and volume. For open systems like the 3-disk pinball we chose $\mathcal{M}$ such that it encloses the non–wandering set, and we discard all runaway trajectories. We define the *expectation value* $\langle a \rangle$ of an observable a to be the asymptotic time and space average over the "phase space" $\mathcal{M}$

$$\langle a \rangle = \lim_{t \to \infty} \frac{1}{|\mathcal{M}|} \int_{\mathcal{M}} dx\, \frac{1}{t} \int_0^t d\tau\, a(f^\tau(x))\,. \tag{7}$$

We use the same $\langle \cdots \rangle$ notation as for the space average (6), and distinguish the two by the presence of the time variable in the argument: if the quantity $\langle a(t) \rangle$ being averaged depends on time, then it is a space average, if it does not, it is the expectation value $\langle a \rangle$.

The expectation value is a space average of time averages. Each $x \in \mathcal{M}$ is used as a starting point of a time average. The averaging over space has the great advantage that it smears over the starting points which were problematic for the time average (like the periodic points). While easy to define, the expectation value $\langle a \rangle$ turns out not to be particularly tractable in practice. Such averages are more conveniently studied by introducing an auxiliary variable β, and investigating instead of $\langle a \rangle$ the space averages of form

$$\left\langle e^{\beta \cdot A^t} \right\rangle = \frac{1}{|\mathcal{M}|} \int_{\mathcal{M}} dx \, e^{\beta \cdot A^t(x)}. \tag{8}$$

In the present context β is an auxiliary variable of no particular physical meaning. In most applications β is a scalar, but if the observable is a $(d+1)$-dimensional vector $a_i(x) \in \mathbb{R}^{d+1}$, then β is a conjugate vector $\beta \in \mathbb{R}^{d+1}$; if the observable is a $(d+1) \times (d+1)$ tensor, β is also a rank-2 tensor, and so on. We will here mostly limit ourselves to scalar values of β.

If the limit $\overline{a(x_0)}$ for the time average (3) exists for "almost all" initial x_0 and the system is ergodic and mixing, we expect the time average along almost all trajectories to tend to the same value $\bar{a}$, and the integrated observable A^t to tend to $t\bar{a}$. The space average (8) is an integral over exponentials, which therefore also grows exponentially with time. So as $t \to \infty$ we would expect the space average of $\langle \exp(\beta \cdot A^t) \rangle$ itself to grow exponentially with time

$$\left\langle e^{\beta \cdot A^t} \right\rangle \sim e^{ts(\beta)} .$$

Hence we expect the rate of growth of the space average to go to a (parameter β dependent) limit

$$s(\beta) = \lim_{t \to \infty} \frac{1}{t} \ln \left\langle e^{\beta \cdot A^t} \right\rangle . \tag{9}$$

Now we have the first argument for why it is smarter to compute $\langle \exp(\beta \cdot A^t) \rangle$ rather than $\langle a \rangle$: the expectation value of the observable (7) and the moments of the integrated observable (1) can be computed by evaluating the derivatives of $s(\beta)$

$$\begin{aligned}
\left. \frac{\partial s}{\partial \beta} \right|_{\beta=0} &= \lim_{t \to \infty} \frac{1}{t} \left\langle A^t \right\rangle = \langle a \rangle , \\
\left. \frac{\partial^2 s}{\partial \beta^2} \right|_{\beta=0} &= \lim_{t \to \infty} \frac{1}{t} \left(\left\langle A^t A^t \right\rangle - \left\langle A^t \right\rangle \left\langle A^t \right\rangle \right) \\
&= \lim_{t \to \infty} \frac{1}{t} \left\langle (A^t - t \langle a \rangle)^2 \right\rangle ,
\end{aligned} \tag{10}$$

and so forth. If we can compute the function $s(\beta)$, we have the desired expectation value without having to estimate any infinite time limits from finite time data.

Suppose we could evaluate $s(\beta)$ and its derivatives. What are such formulas good for? A typical application is to the problem of describing a particle scattering elastically off a two-dimensional triangular array of disks. If the disks are sufficiently large to block any infinite length free flights, the particle will diffuse chaotically, and the transport coefficient of interest is the diffusion constant given by $x(t)^2 \approx 4Dt$. In contrast to D estimated numerically from trajectories $x(t)$ for finite but large t, the above formulas yield the asymptotic D without any extrapolations to the $t \to \infty$ limit. For example, for $a_i = v_i$ and zero mean drift $\langle v_i \rangle = 0$, the diffusion constant is given by the curvature of $s(\beta)$ at $\beta = 0$,

$$D = \lim_{t \to \infty} \frac{1}{2dt} \left\langle x(t)^2 \right\rangle = \frac{1}{2d} \sum_{i=1}^{d} \left. \frac{\partial^2 s}{\partial \beta_i^2} \right|_{\beta=0} , \tag{11}$$

so if we can evaluate derivatives of $s(\beta)$, we can compute transport coefficients that characterize deterministic diffusion. The periodic orbit theory yields an explicit closed form expression for D.

Averaging in open systems

If the $\mathcal{M}$ is a compact region or set of regions to which the dynamics is confined for all times, (7) is a sensible definition of the expectation value. However, if we have a repeller or an attractor for which trajectories can exit the region without ever returning we might be in trouble. For example, a pinball in our 3-disk pinball might fly off to Mars, contributing to the expectation value in a not very useful way.

For an attractor whose basin is not included in $\mathcal{M}$, and for sufficiently large t a trajectory originating in $\mathcal{M}$ might never renter it again,

$$\int_{\mathcal{M}} dy\, \delta(y - f^t(x_0)) = 0 \qquad \text{for } t > t_{exit}.$$

For a repeller $f^t(x_0)$ will eventually leave the region of integration unless we choose x_0 to be on the repeller, so the identity

$$\int_{\mathcal{M}} dy\, \delta(y - f^t(x_0)) = 1 \qquad \text{iff } x_0 \in \text{ non–wandering set} \tag{12}$$

might be correct only on a fractal set $\mathcal{M}$ of zero Lebesgue measure. Clearly, we need to modify the definition of the expectation value so it refers only to the dynamics on the set of trajectories which are confined for all times, that is the non–wandering set.

Designate by $\mathcal{M}$ a phase space region that encloses all interesting initial points, say the 3-disk Poincaré section constructed from the disk boundaries and all possible incidence angles. The volume of the phase space containing all trajectories which start out within the phase space region $\mathcal{M}$ and recur within that region at the time t

$$|\mathcal{M}(t)| = \int_{\mathcal{M}} dx\, dy\, \delta(y - f^t(x)) \sim |\mathcal{M}| e^{-\gamma t}. \tag{13}$$

is expected to decay exponentially, with the *escape rate* γ. The integral over x takes care of all possible initial pinballs; the integral over y checks whether they are still within $\mathcal{M}$ by the time t. If the dynamics is bounded, and $\mathcal{M}$ envelops the entire accessible phase space, $|\mathcal{M}(t)| = $ const. for all t. However, if trajectories exit $\mathcal{M}$ the recurrence volume $|\mathcal{M}(t)|$ decreases with time. For example, any trajectory that falls off a pinball table is gone for good.

For open systems we replace the space average (8) by

$$\left\langle e^{\beta \cdot A^t} \right\rangle = \frac{1}{|\mathcal{M}|} \int_{\mathcal{M}} dx\, e^{\beta \cdot A^t(x) + \gamma t}. \tag{14}$$

The interpretation is that by compensating for the loss of trajectories with factor $e^{\gamma t}$ we have a correctly normalized $\beta = 0$ average $\langle 1 \rangle = 1$.

We now turn to the problem of evaluating $\left\langle e^{\beta \cdot A^t} \right\rangle$.

Evolution operators

The above simple shift of focus, from studying $\langle a \rangle$ to studying $\langle \exp\left(\beta \cdot A^t\right) \rangle$ is the key to all what follows. Make the dependence on the flow explicit by rewriting this quantity as

$$\left\langle e^{\beta \cdot A^t} \right\rangle = \frac{1}{|\mathcal{M}|} \int_{\mathcal{M}} dx \int_{\mathcal{M}} dy\, \delta(y - f^t(x))\, e^{\beta \cdot A^t(x)}. \tag{15}$$

Here $\delta(y - f^t(x))$ is the Dirac delta function: for a deterministic flow an initial point x maps into a unique point y at time t. Formally, all we have done above is to insert the identity

$$1 = \int_{\mathcal{M}} dy \, \delta(y - f^t(x)) \,, \tag{16}$$

that is, we are averaging over trajectories that remain in $\mathcal{M}$ for all times. However, having made this substitution we have replaced the study of individual trajectories $f^t(x)$ by the study of the evolution of the totality of initial conditions. Instead of temporal averages (a long trajectory which explores the phase space ergodically) we shall always work with finite time, topologically partitioned space averages. We now introduce the *evolution operator* that acts on bounded scalar functions $h(x)$ of the phase space as

$$\mathcal{L}^t \circ h(y) = \int_{\mathcal{M}} dx \, \delta(y - f^t(x)) \, e^{\beta \cdot A^t(x)} h(x) \,, \tag{17}$$

and rewrite the expectation value of the generating function (15) as

$$\left\langle e^{\beta \cdot A^t} \right\rangle = \left\langle \mathcal{L}^t \circ \iota \right\rangle \,,$$

where $\iota(x)$ is the constant function that always returns 1. For chaotic dynamics any smooth intial distribution $\rho(x)$ would be just as good.

We shall refer to the kernel of the evolution operator in (17) as $\mathcal{L}^t(y, x)$ where

$$\mathcal{L}^t(y, x) = \delta(y - f^t(x)) \, e^{\beta \cdot A^t(x)} \,. \tag{18}$$

The evolution operator is different for different observables, as its definition depends on the choice of the integrated observable A in the exponential. Its role, as $t \to \infty$, is to generate the expectation value of a, but before showing that we verify some of its basic properties.

By its definition, the integral over the observable a is additive along the trajectory

$$
\begin{aligned}
A^{t_1 + t_2}(x_0) &= \int_0^{t_1} d\tau \, a(x(\tau)) &+& \int_{t_1}^{t_1 + t_2} d\tau \, a(x(\tau)) \\
&= A^{t_1}(x_0) &+& A^{t_2}(f^{t_1}(x_0)) \,.
\end{aligned}
$$

If $A^t(x)$ is additive along the trajectory, the evolution operator generates a semigroup; $\mathcal{L}^{t_2} \circ \mathcal{L}^{t_1} = \mathcal{L}^{t_1 + t_2}$. The order of the composition of operators can be a little confusing. What is meant is that

$$\mathcal{L}^{t_2} \circ \mathcal{L}^{t_1} \circ h(x) = \int_{\mathcal{M}} dy \, \delta(x - f^{t_2}(y)) e^{\beta \cdot A^{t_2}(y)} (\mathcal{L}^{t_1} \circ h)(y) = \mathcal{L}^{t_1 + t_2} h(x) \,. \tag{19}$$

This semigroup property is the main reason why (15) is preferable to (7) as a starting point for evaluation of dynamical averages; their value in the asymptotic $t \to \infty$ limit can be recovered by means of evolution operators.

The topological entropy, the asymptotic rate of growth of the number of topologically distinct trajectories, is given by the leading eigenvalue of the transition matrix. It is our goal now to generalize this observation and relate general asymptotic averages over chaotic dynamics to eigenvalues of evolution operators. As we shall now show, in the case of piecewise-linear approximations to dynamics these operators reduce to finite matrices, but for generic smooth flows, they are infinite-dimensional linear operators, and finding smart ways of computing their eigenvalues requires some thought.

Transfer operators

We start by showing that for a piecewise-linear map with a finite Markov partition, the evolution operator is a finite dimensional transfer operator, a straightforward generalization of the topological transition matrices.

Consider as an example of expanding 1-d map $f(x)$, with $|f'_s(x)| > 1$ and monotone on two non-overlapping intervals, a piecewise-linear 2–branch repeller with slopes Λ_0 and Λ_1:

$$f(x) = \begin{cases} \Lambda_0 x & \text{if } x \in \mathcal{M}_0 = [0, 1/\Lambda_0] \\ \Lambda_1(1-x) & \text{if } x \in \mathcal{M}_1 = [1 - 1/\Lambda_1, 1] \end{cases} . \tag{20}$$

As both $f(\mathcal{M}_0)$ and $f(\mathcal{M}_1)$ map onto the entire unit interval $\mathcal{M}$, the partition into $\{\mathcal{M}_0, \mathcal{M}_1\}$ is a finite Markov partition with the unrestricted 2-letter alphabet $s_n \in \{0, 1\}$, with the transition matrix

$$T = \begin{pmatrix} 1 & 1 \\ 1 & 1 \end{pmatrix} .$$

The transition matrix T counts the allowed transitions, so $T_{dm} = 0$ or 1; however, it is just as easy to keep track of matrix elements T_{dm} that take other ranges of values. Due to the piecewise linearity, the simplest of evolution operators, the Perron-Frobenius operator acts on a piecewise constant function $\rho(x) = (\rho_0, \rho_1)$ as a $[2 \times 2]$ transfer matrix with matrix elements

$$\begin{pmatrix} \rho_0 \\ \rho_1 \end{pmatrix} \to \mathcal{L} \circ \rho = \begin{pmatrix} \frac{1}{|\Lambda_0|} & \frac{1}{|\Lambda_1|} \\ \frac{1}{|\Lambda_0|} & \frac{1}{|\Lambda_1|} \end{pmatrix} \begin{pmatrix} \rho_0 \\ \rho_1 \end{pmatrix} . \tag{21}$$

More generally, T_{dm} is the "penalty" paid for choosing a particular transition, and T is a finite memory *Markov* or *transfer* matrix. Piecewise-linear approximations to dynamical systems yield finite-dimensional transfer operators, provided that the symbolic dynamics is a subshift of finite type.

If we increase the number of steps remembered, the transition matrix grows big quickly, as N-ary dynamics with n-step memory requires an $[N^n \times N^n]$ matrix. As the matrix is very sparse, we shall find it convenient to use the Markov graph representations for T.

Trace formulas

Our extraction of the spectrum of $\mathcal{L}$ commences with the evaluation of the trace. To compute an expectation value we have to integrate over all the values of the kernel $\mathcal{L}^t(x, y)$. If $\mathcal{L}^t$ were a matrix we would be computing a weighted sum of its eigenvalues which is dominated by the leading eigenvalue as $t \to \infty$. As the trace of $\mathcal{L}^t$ is also dominated by the leading eigenvalue as $t \to \infty$, we might just as well look at the trace

$$\operatorname{tr} \mathcal{L}^t = \int dx \, \mathcal{L}^t(x, x) = \int dx \, \delta\left(x - f^t(x)\right) e^{\beta \cdot A^t(x)} . \tag{22}$$

Assume that $\mathcal{L}$ has a spectrum of discrete eigenvalues $s_0, s_1, s_2, \cdots$ ordered so that $\operatorname{Re} s_\alpha \leq \operatorname{Re} s_{\alpha+1}$. We ignore for the time being the question of what function space the eigenfunctions belong to, as we shall compute the eigenvalues directly, without constructing the corresponding eigenfunctions.

By definition, the trace is the sum over eigenvalues

$$\operatorname{tr}\mathcal{L}^t = \sum_{\alpha=0}^{\infty} e^{s_\alpha t}\,, \tag{23}$$

so if we can evaluate the above delta-function integral, we have a formula for eigenvalue sums (for the time being we choose not to worry about convergence of such sums). Integrating Dirac delta functions is easy: $\int_{\mathcal{M}} dx\, \delta(x) = 1$ if $0 \in \mathcal{M}$, zero otherwise. Integral over a one-dimensional Dirac delta function picks up the Jacobian of its argument evaluated at all of its zeros:

$$\int dx\, \delta(h(x)) = \sum_{x \in \mathrm{Zero}\,[h]} \frac{1}{|h(x)'|}\,, \tag{24}$$

and in d dimensions the denominator is replaced by

$$\int dx\, \delta(h(x)) = \sum_{x \in \mathrm{Zero}\,[h]} \frac{1}{\left|\det \frac{\partial h(x)}{\partial x}\right|}\,. \tag{25}$$

An evolution operator propagates a density of inital conditions $h(x)$ to time t:

$$\mathcal{L}^t \circ h(x) = \int dy\, \delta(x - f^t(y)) h(y) = \sum_{y = f^{-t}(x)} \frac{h(y)}{|f^t(y)'|}\,.$$

As the case of discrete time mappings is somewhat simpler, we first derive the trace formula for maps, and then for flows. The final formula (39) covers both cases.

Hyperbolicity assumption

According to (25) the trace (22) picks up a contribution whenever $x - f^T(x) = 0$, that is whenever x belongs to a periodic orbit. The contribution of a prime cycle p of period T_p for a map f can be evaluated by restricting the integration to an infinitesimal neighborhood $\mathcal{M}_p$ around the cycle, with $x_p \in p$ a cycle point

$$\operatorname{tr}{}_p\mathcal{L}^{T_p} = \int_{\mathcal{M}_p} dx\, \delta\big(x - f^{T_p}(x)\big) = \frac{1}{|\det(\mathbf{1} - \mathbf{J}_p)|} = \prod_{i=1}^{d} \frac{1}{|1 - \Lambda_{p,i}|}\,. \tag{26}$$

(Here we set the observable $e^{A_p} = 1$ for the time being). Periodic orbit Jacobian matrix $\mathbf{J}_p$ is also known as the monodromy matrix, and its eigenvalues $\Lambda_{p,1}, \Lambda_{p,2}, \ldots, \Lambda_{p,d}$ as the Floquet multipliers. The integral can be carried out only if $\mathbf{J}_p$ has no eigenvalue of unit magnitude. Sort the eigenvalues $\Lambda_{p,1}, \Lambda_{p,2}, \ldots, \Lambda_{p,d}$ of the p-cycle $[d \times d]$ transverse Jacobian matrix $\mathbf{J}_p$ into sets $\{e, m, c\}$:

$$\text{expanding eigenvalues:} \quad e = \{\Lambda_{p,i} : |\Lambda_{p,i}| > 1\}$$
$$\text{marginal eigenvalues:} \quad m = \{\Lambda_{p,i} : |\Lambda_{p,i}| = 1\}$$
$$\text{contracting eigenvalues:} \quad c = \{\Lambda_{p,i} : |\Lambda_{p,i}| < 1\}$$

We assume that no eigenvalue is marginal (as we shall show in (37), the longitudinal $\Lambda_{p,d+1} = 1$ eigenvalue for flows can be eliminated by restricting the consideration to the transverse Jacobian matrix $\mathbf{J}_p$), and factorize the trace (26) into a product over the expanding and the contracting eigenvalues

$$|\det(\mathbf{1} - \mathbf{J}_p)|^{-1} = \frac{1}{|\Lambda_p|} \prod_e \frac{1}{1 - 1/\Lambda_{p,e}} \prod_c \frac{1}{1 - \Lambda_{p,c}}\,, \tag{27}$$

where $\Lambda_p = \prod_e \Lambda_{p,e}$ is the product of expanding eigenvalues. In the above both $\Lambda_{p,c}$ and $1/\Lambda_{p,e}$ are smaller than 1 in absolute value, and as long as they are real we are allowed to drop the absolute value brackets $|\cdots|$.

The *hyperbolicity assumption* requires that the stabilities of all cycles included in the trace sums be exponentially bounded away from unity:

$$|\Lambda_{p,i}| > e^{\lambda_e T_p} \qquad \text{for any } p, \text{ any expanding eigenvalue } i \in e$$
$$|\Lambda_{p,i}| < e^{-\lambda_c T_p} \qquad \text{for any } p, \text{ any contracting eigenvalue } i \in c\,, \tag{28}$$

where $\lambda_e, \lambda_c > 0$ are strictly positive bounds on the expanding, contracting cycle Lyapunov exponents. If a dynamical system satisfies the hyperbolicity assumption (for example, the well separated 3-disk system clearly does), the $\mathcal{L}^t$ spectrum will be relatively easy to control. If the expansion/contraction is slower than exponential, let us say $|\Lambda_{p,i}| \sim T_p{}^2$, the system may exhibit "phase transitions", and the analysis is much harder.

A trace formula for maps

If the evolution is given by a discrete time mapping, and all periodic points have stability eigenvalues $|\Lambda_{p,i}| \neq 1$ strictly bounded away from unity, the trace $\mathcal{L}^n$ is given by the sum over all periodic points i of period n:

$$\operatorname{tr} \mathcal{L}^n = \int dx\, \mathcal{L}^n(x,x) = \sum_{x_i \in \operatorname{Fix} f^n} \frac{e^{\beta \cdot A^n(x_i)}}{|\det(1 - \mathbf{J}^n(x_i))|}\,. \tag{29}$$

Here $\operatorname{Fix} f^n = \{x : f^n(x) = x\}$ is the set of all periodic points of period n. The weight follows from the properties of the Dirac delta function (25) by taking the determinant of $\partial_i(x_j - f^n(x)_j)$. If a trajectory retraces itself r times, its Jacobian matrix is $\mathbf{J}_p^r$, where $\mathbf{J}_p$ is the $[d \times d]$ Jacobian matrix evaluated along a single traversal of the prime cycle p. As we saw in (4), the integrated observable A^n also has a similar property: If a prime cycle p trajectory retraces itself r times, $n = rn_p$, we obtain A_p repeated r times, $A^n(x_i) = r A_p$, $x_i \in p$. A prime cycle is a single traversal of the orbit, and its label is a non-repeating symbol string. There is only one prime cycle for each cyclic permutation class. For example, the four cycle points $\overline{0011} = \overline{1001} = \overline{1100} = \overline{0110}$ belong to the same prime cycle $p = 0011$ of length 4. As both the stability of a cycle and the weight A_p are the same everywhere along the orbit, each prime cycle of length n_p contributes n_p terms to the sum, one for each cycle point. Hence (29) can be rewritten as a sum over all prime cycles and their repeats

$$\operatorname{tr} \mathcal{L}^n = \sum_p n_p \sum_{r=1}^{\infty} \frac{e^{r\beta \cdot A_p}}{|\det(1 - \mathbf{J}_p^r)|} \delta_{n, n_p r}\,, \tag{30}$$

with the Kronecker delta $\delta_{n, n_p r}$ projecting out the periodic contributions of total period n. This constraint is awkward, and will be more awkward still for the continuous time flows, where it will yield a series of Dirac delta spikes (38). Such sums are familiar from the density-of-states sums of statistical mechanics, where they are dealt with in the same way as we shall do here: we smoothen this distribution by taking a Laplace transform which rids us of the $\delta_{n, n_p r}$ constraint.

We define the trace formula for maps to be the Laplace transform of $\operatorname{tr} \mathcal{L}^n$ which, for discrete time mappings, is simply the generating function for the trace sums

$$\sum_{n=1}^{\infty} z^n \operatorname{tr} \mathcal{L}^n = \operatorname{tr} \frac{z\mathcal{L}}{1 - z\mathcal{L}} = \sum_p n_p \sum_{r=1}^{\infty} \frac{z^{n_p r} e^{r\beta \cdot A_p}}{|\det(1 - \mathbf{J}_p^r)|}\,. \tag{31}$$

Expressing the trace as (23), the sum of the eigenvalues of $\mathcal{L}$, we obtain the *trace formula for maps*:

$$\sum_{\alpha=0}^{\infty} \frac{ze^{s_\alpha}}{1 - ze^{s_\alpha}} = \sum_{p} n_p \sum_{r=1}^{\infty} \frac{z^{n_p r} \, e^{r\beta \cdot A_p}}{\left| \det \left(1 - \mathbf{J}_p^r\right) \right|} . \tag{32}$$

This is our first example of the duality between the spectrum of eigenvalues and the spectrum of periodic orbits, announced in the introduction to this chapter.

A trace formula for transfer operators

For a piecewise-linear map with a finite Markov partition, we can explicitly evaluate the trace formula. We illustrate this by the piecewise linear repeller (20). By the piecewise linearity and the chain rule $\Lambda_p = \Lambda_0^{n_0} \Lambda_1^{n_1}$, where the cycle p contains n_0 symbols 0 and n_1 symbols 1, the trace (29) reduces to

$$\operatorname{tr} \mathcal{L}^n = \sum_{m=0}^{n} \binom{n}{m} \frac{1}{|1 - \Lambda_0^m \Lambda_1^{n-m}|} = \sum_{k=0}^{\infty} \left(\frac{1}{|\Lambda_0| \Lambda_0^k} + \frac{1}{|\Lambda_1| \Lambda_1^k} \right)^n , \tag{33}$$

as is also clear form the explicit form of the transfer matrix (21).

The Perron-Frobenius operator trace formula for the piecewise-linear map (20) follows from (31):

$$\operatorname{tr} \frac{z\mathcal{L}}{1 - z\mathcal{L}} = \frac{\frac{z}{|\Lambda_0 - 1|} + \frac{z}{|\Lambda_1 - 1|}}{1 - z \left(\frac{1}{|\Lambda_0 - 1|} + \frac{1}{|\Lambda_1 - 1|} \right)} . \tag{34}$$

A trace formula for flows

As any point along a cycle returns to itself exactly at each cycle period, the eigenvalue corresponding to the eigenvector along the flow necessarily equals unity for all periodic orbits. Hence for flows the trace integral $\operatorname{tr} \mathcal{L}^t$ requires a separate treatment for the longitudinal direction. To evaluate the contribution of a prime cycle p of period T_p, restrict the integration to an infinitesimally thin tube $\mathcal{M}_p$ enveloping the cycle and choose a local coordinate system with a longitudinal coordinate $dx_\parallel$ along the direction of the flow, and d transverse coordinates $x_\perp$

$$\operatorname{tr}_p \mathcal{L}^t = \int_{\mathcal{M}_p} dx_\perp dx_\parallel \, \delta\left(x_\perp - f_\perp^t(x)\right) \delta\left(x_\parallel - f_\parallel^t(x)\right) . \tag{35}$$

(Here we again set the observable $\exp(\beta \cdot A^t) = 1$ for the time being). Let $v(x)$ be the magnitude of the velocity at the point x along the flow. $v(x)$ is strictly positive, as otherwise the orbit would stagnate for infinite time at $v(x) = 0$ points, and get nowhere. Therefore we can trade in the longitudinal variable $x_\parallel$ for the flight time τ, with jacobian $dx_\parallel = v \, d\tau$. In terms of the time variable $f_\parallel^t(x)$ contributes for every T_p return to the same point along the trajectory, and the delta function along the flow becomes $\delta(\tau - (t + \tau) \bmod T_p)$, independent of τ. So the integral along the trajectory yields a contribution whenever the time t is a multiple of the cycle period T_p

$$\int_{\mathcal{M}_p} dx_\parallel \, \delta\left(x_\parallel - f_\parallel^t(x)\right) = \sum_{r=1}^{\infty} \delta(t - rT_p) \int_p d\tau$$

$$= T_p \sum_{r=1}^{\infty} \delta(t - rT_p) . \tag{36}$$

Linearization of the periodic flow in a plane perpendicular to the orbit yields

$$\int_{\mathcal{M}_p} dx_\perp \delta\left(x_\perp - f_\perp^{rT_p}(x)\right) = \frac{1}{\left|\det\left(1 - \mathbf{J}_p^r\right)\right|} \, , \tag{37}$$

where $\mathbf{J}_p$ is the p-cycle $[d{\times}d]$ *transverse* Jacobian matrix, and as in (28) we have to assume hyperbolicity, that is that the magnitudes of all transverse eigenvalues are bounded away from unity.

Substituting (36), (37) into (35), we obtain an expression for $\operatorname{tr}\mathcal{L}^t$ as a sum over all prime cycles p and their repetitions

$$\operatorname{tr}\mathcal{L}^t = \sum_p T_p \sum_{r=1}^{\infty} \frac{e^{r\beta \cdot A_p}}{\left|\det\left(1 - \mathbf{J}_p^r\right)\right|} \delta(t - rT_p) \, . \tag{38}$$

A Laplace transform smoothes the above sum over Dirac delta spikes and yields*

$$\int_{0_+}^{\infty} dt\, e^{-st} \operatorname{tr}\mathcal{L}^t = \operatorname{tr}\frac{1}{s - \mathcal{A}} = \sum_{\alpha=0}^{\infty} \frac{1}{s - s_\alpha} = \sum_p T_p \sum_{r=1}^{\infty} \frac{e^{r(\beta \cdot A_p - sT_p)}}{\left|\det\left(1 - \mathbf{J}_p^r\right)\right|} \, .$$

Here $\mathcal{A}$ is the infinitesimal evolution generator, $\mathcal{L}^t = e^{t\mathcal{A}}$, and $(s - \mathcal{A})^{-1}$ is its resolvent. The *classical trace formula* for flows

$$\sum_{\alpha=0}^{\infty} \frac{1}{s - s_\alpha} = \sum_p T_p \sum_{r=1}^{\infty} \frac{e^{r(\beta \cdot A_p - sT_p)}}{\left|\det\left(1 - \mathbf{J}_p^r\right)\right|} \tag{39}$$

is our second example of the duality between the (local) cycles and (global) eigenvalues[†].

If T_p takes integer values, we can replace $e^{-s} \to z$ throughout. We see that the trace formula for maps (32) is a special case of the trace formula for flows. The relation between the continuous and discrete time cases can be summarized as follows:

$$\begin{aligned} T_p &\leftrightarrow n_p \\ e^{-s} &\leftrightarrow z \\ e^{t\mathcal{A}} &\leftrightarrow \mathcal{L}^n \end{aligned} \tag{40}$$

We could now proceed to estimate the location of the leading singularity of $\operatorname{tr}\mathcal{L}(s)$ by extrapolating finite cycle length truncations of (39) by methods such as Padé approximants. However, it pays to first perform a simple resummation which converts this divergence into a *zero* of a related function. We shall accomplish this in (49), after we complete our offering of trace formulas.

---

*We should caution the reader that in taking the Laplace transform we have ignored a possible $t \to 0_+$ volume term, as we do not know how to regularize the delta function kernel in this limit. In the quantum (or heat kernel) case this limit gives rise to the Weyl mean density of states. A more careful treatment should regularize the divergent sum in (39) and might assign to such volume term some interesting role in the theory of classical resonance spectra.

[†]Continuous time flow traces weighted by the cycle periods were introduced by Bowen[1] who treated them as Poincaré section suspensions weighted by the "time ceiling" function. They were used by Perry and Pollicott[2]. The derivation presented here was designed[3] to parallel as closely as possible the derivation of the Gutzwiller semiclassical trace formula.

An asymptotic trace formula

In order to illuminate the above manipulations and relate them to something we already possess intuition about, we now derive a heuristic sum from the exact trace formula (32). The Laplace transforms (32) or (39) are designed to capture the time $\to \infty$ asymptotic behavior of the trace sums. By the hyperbolicity assumption (28) for $t = T_p r$ large the cycle weight approaches

$$\left| \det \left(\mathbf{1} - \mathbf{J}_p^r \right) \right| \to \left| \Lambda_p \right|^r , \tag{41}$$

where Λ_p is the product of the expanding eigenvalues of $\mathbf{J}_p$. Define $\Gamma(z)$ as the approximate trace formula obtained by replacing the cycle weights $\left| \det \left(\mathbf{1} - \mathbf{J}_p^r \right) \right|$ by $\left| \Lambda_p \right|^r$ in (32). Equivalently, think of this as a replacement of the evolution operator (18) by a transfer operator. For concreteness consider a dynamical system whose symbolic dynamics is complete binary, for example the 3-disk system. In this case distinct periodic points that contribute to the nth periodic points sum (30) are labelled by all admissible itineraries composed of sequences of letters $s_i \in \{0, 1\}$:

$$\Gamma(z) = \sum_{n=1}^{\infty} z^n \Gamma_n = \sum_{n=1}^{\infty} z^n \sum_{x_i \in \mathrm{Fix} f^n} \frac{e^{\beta \cdot A^n(x_i)}}{|\Lambda_i|}$$

$$= z \left\{ \frac{e^{\beta \cdot A_0}}{|\Lambda_0|} + \frac{e^{\beta \cdot A_1}}{|\Lambda_1|} \right\} + z^2 \left\{ \frac{e^{2\beta \cdot A_0}}{|\Lambda_0|^2} + \frac{e^{\beta \cdot A_{01}}}{|\Lambda_{01}|} + \frac{e^{\beta \cdot A_{10}}}{|\Lambda_{10}|} + \frac{e^{2\beta \cdot A_1}}{|\Lambda_1|^2} \right\}$$

$$+ z^3 \left\{ \frac{e^{3\beta \cdot A_0}}{|\Lambda_0|^3} + \frac{e^{\beta \cdot A_{001}}}{|\Lambda_{001}|} + \frac{e^{\beta \cdot A_{010}}}{|\Lambda_{010}|} + \frac{e^{\beta \cdot A_{100}}}{|\Lambda_{100}|} + \ldots \right\}$$

Both the cycle averages A_i and the stabilities Λ_i are the same for all points $x_i \in p$ in a cycle p. Summing over repeats of all prime cycles we obtain[‡]

$$\Gamma(z) = \sum_p \frac{n_p t_p}{1 - t_p} , \qquad t_p = z^{n_p} e^{\beta \cdot A_p} / |\Lambda_p| . \tag{42}$$

If the weights $e^{\beta \cdot A^n(x)}$ are multiplicative along the flow, and the flow is hyperbolic, for given β the magnitude of each $|\exp(\beta \cdot A^n(x_i))/\Lambda_i|$ term is bounded by some constant M^n, the total number grows as 2^n (or e^{hn} in general), and the sum is convergent for sufficiently small $|z| < 1/2M$. For large n the n-th level sum (29) tends to the leading $\mathcal{L}^n$ eigenvalue $e^{n s_0}$. Hence s_0 is determined by the smallest $z = e^{-s_0}$ for which (32) diverges,

$$\Gamma(z) \approx \sum_{n=1}^{\infty} \left(z e^{s_0} \right)^n = \frac{z e^{s_0}}{1 - z e^{s_0}} . \tag{43}$$

If one is interested only in the leading eigenvalue of $\mathcal{L}$, it suffices to consider the approximate trace $\Gamma(z)$. We will use this fact below to motivate the introduction of dynamical zeta functions (53).

Spectral determinants

The problem with trace formulas (32), (39) and (42) is that they diverge at $z = e^{-s_0}$, respectively $s = s_0$, that is, precisely where one would like to use them. While

---

[‡]Note that we could not perform such sum over r in the exact trace formula (32) as $\left| \det \left(\mathbf{1} - \mathbf{J}_p^r \right) \right| \neq \left| \det \left(\mathbf{1} - \mathbf{J}_p \right) \right|^r$; the correct way to resum the exact trace formulas is given in (51).

this does not prevent numerical estimation of some "thermodynamic" averages for iterated mappings, in the case of the Gutzwiller trace formula this leads to a perplexing observation that crude estimates of the radius of convergence seem to put the entire physical spectrum out of reach. We shall now partially cure this problem by going from trace formulas to determinants. Determinants tend to have larger analyticity domains because if $\operatorname{tr} \mathcal{L}/(1 - z\mathcal{L}) = \frac{d}{dz} \ln \det(1 - z\mathcal{L})$ diverges at a particular value of z, then $\det(1 - z\mathcal{L})$ might have an isolated zero there, and a zero of a function is easier to determine than its radius of convergence.

We establish the relation between determinants and traces first for maps and then for flows. The inverse of the eigenvalues of a linear operator are given by the zeros of the determinant

$$\det(1 - z\mathcal{L}) = \prod_\alpha (1 - z\lambda_\alpha)\,. \tag{44}$$

For finite matrices this is the characteristic determinant; for operators this is the Hadamard representation of the *spectral determinant*. Consider first the case of maps, for which the evolution operator advances the densities by integer steps in time. In this case we can use the formal matrix identity

$$\ln \det(1 - M) = \operatorname{tr} \ln(1 - M) = -\sum_{n=1}^\infty \frac{1}{n} \operatorname{tr} M^n\,, \tag{45}$$

to relate the spectral determinant of an evolution operator for a map to its traces, that is, periodic orbits:

$$\det(1 - z\mathcal{L}) = \exp\left(-\sum_n \frac{z^n}{n} \operatorname{tr} \mathcal{L}^n\right)$$
$$= \exp\left(-\sum_p \sum_{r=1}^\infty \frac{1}{r} \frac{z^{n_p r} e^{r\beta \cdot A_p}}{\left|\det\left(\mathbf{1} - \mathbf{J}_p^r\right)\right|}\right)\,. \tag{46}$$

The inverse relation which recovers the trace formula (32) from the spectral determinant is given by the derivative

$$\operatorname{tr} \frac{z\mathcal{L}}{1 - z\mathcal{L}} = -z\frac{d}{dz} \ln \det(1 - z\mathcal{L})\,. \tag{47}$$

Spectral determinants of transfer operators

For a piecewise-linear map (20) with a finite Markov partition, an explicit formula for the spectral determinant follows by substituting the trace formula (34) into (46):

$$\det(1 - z\mathcal{L}) = \prod_{k=0}^\infty \left(1 - \frac{t_0}{\Lambda_0^k} - \frac{t_1}{\Lambda_1^k}\right)\,, \tag{48}$$

where $t_s = z/|\Lambda_s|$. The eigenvalues are simply

$$e^{s_k} = \frac{1}{|\Lambda_0|\Lambda_0^k} + \frac{1}{|\Lambda_1|\Lambda_1^k}\,.$$

If the map is a symmetric 2-branch repeller, with $\Lambda = \Lambda_1 = -\Lambda_0$, all odd eigenvalues vanish, and the even eigenvalues are given by $e^{s_k} = 2/\Lambda^{k+1}$, k even. If the system is bounded and no trajectories escape, the leading eigenvalue of $\mathcal{L}^t$ is exactly 1, that is

$s(0) = 0$. Asymptotically the spectrum is dominated by the lesser of the two fixed point slopes $\Lambda = \Lambda_0$ (if $|\Lambda_0| < |\Lambda_1|$, otherwise $\Lambda = \Lambda_1$), and the eigenvalues e^{s_k} fall off exponentially as $1/\Lambda^k$, just as in the single fixed-point examples[§]. The exponential spacing of eigenvalues leads to entire spectral determinant. It is this property that will generalize to piecewise smooth flows with finite Markov paritions and single out spectral determinants rather than the trace formulas or dynamical zeta functions as the tool of choice for evaluation of spectra.

Spectral determinant for flows

We write the formula for the spectral determinant for flows by analogy to (46), the determinant for maps

$$\det(s - \mathcal{A}) = \exp\left(-\sum_p \sum_{r=1}^{\infty} \frac{1}{r} \frac{e^{r(\beta \cdot A_p - sT_p)}}{\left|\det\left(\mathbf{1} - \mathbf{J}_p^r\right)\right|}\right), \tag{49}$$

and then check that the trace formula (39) is the logarithmic derivative of the spectral determinant so defined

$$\operatorname{tr} \frac{1}{s - \mathcal{A}} = \frac{d}{ds} \ln \det(s - \mathcal{A}). \tag{50}$$

To recover $\det(s - \mathcal{A})$ integrate both sides $\int_{s_0}^{s}$. With z set to $z = e^{-s}$ in (46), the spectral determinant (49) has the same form for both maps and flows.

We now note that the r sum in (49) is close in form to the expansion of a logarithm. This observation enables us to recast the spectral determinant into an infinite product over periodic orbits as follows:

Let $\mathbf{J}_p$ be the p-cycle $[d \times d]$ transverse Jacobian matrix, with eigenvalues $\Lambda_{p,1}$, $\Lambda_{p,2}$, ..., $\Lambda_{p,d}$. Expanding $1/(1 - 1/\Lambda_{p,e})$, $1/(1 - \Lambda_{p,c})$ in (27) as geometric series, substituting back into (49), and resumming the logarithms, we find that the spectral determinant is formally given by the infinite product

$$\det(s - \mathcal{A}) = \prod_{k_1=0}^{\infty} \cdots \prod_{l_c=0}^{\infty} \frac{1}{\zeta_{k_1 \cdots l_c}(s)}$$

$$1/\zeta_{k_1 \cdots l_c}(s) = \prod_p \left(1 - \frac{e^{\beta \cdot A_p - sT_p}}{|\Lambda_p|} \frac{\Lambda_{p,e+1}^{l_1} \Lambda_{p,e+2}^{l_2} \cdots \Lambda_{p,d}^{l_c}}{\Lambda_{p,1}^{k_1} \Lambda_{p,2}^{k_2} \cdots \Lambda_{p,e}^{k_e}}\right). \tag{51}$$

The observable being averaged contributes through the A_p term, which is the p cycle average of the multiplicative weight $e^{A^t(x)}$. By its definition (1), for maps the weight is a product along the cycle points

$$e^{A_p} = \prod_{j=0}^{n_p-1} e^{a(f^j(x_p))},$$

and for the flows the weight is an exponential of the integral (4) along the cycle

$$e^{A_p} = \exp\left(\int_0^{T_p} d\tau \, a(x(\tau))\right).$$

[§]Here are some situations in which everything is mathematically rigorous, and the transfer operator is well defined: expanding Hölder case, weighted subshifts of finite type, expanding differentiable case, see Bowen[1]: expanding holomorphic case, see Ruelle[4]; piecewise monotone maps of the interval, see Hofbauer and Keller[5] and Baladi and Keller[6].

This formula is correct for scalar weighting functions; more general matrix valued weights (such at the stability matrix) would require a time-ordering prescription.

The infinite product formula (51) is in practice not much more than a shorthand notation for the periodic orbit weights contributing to the spectral determinant; more work will be needed to bring such cycle formulas into a tractable form.

Now we are finally poised to deal with the problem posed at the beginning of this chapter; how do we evaluate the averages of (7)? The eigenvalues of the dynamical averaging evolution operator are given by the values of s for which the spectral determinant

$$\det(s - \mathcal{A}(\beta)) = \exp\left(-\sum_p \sum_{r=1}^{\infty} \frac{1}{r} \frac{e^{(\beta \cdot A_p - sT_p)r}}{|\det(1 - \mathbf{J}_p^r)|}\right). \tag{52}$$

of the evolution operator (18) vanishes. If we can compute the leading eigenvalue $s_0(\beta)$ and its derivatives, we are done.

Dynamical zeta functions

As we saw in the (42), if one is interested only in the leading eigenvalue of $\mathcal{L}^t$, the size of the p cycle neighborhood can be approximated by $1/|\Lambda_p|^r$, the dominant term in the $rT_p = t \to \infty$ limit, where $\Lambda_p = \prod_e \Lambda_{p,e}$ is the product of the expanding eigenvalues of the Jacobian matrix $\mathbf{J}_p$. With this replacement the spectral determinant (51) is approximated by the *dynamical zeta function*

$$1/\zeta(s) = \exp\left(-\sum_p \sum_{r=1}^{\infty} \frac{1}{r} t_p^r\right). \tag{53}$$

Resumming the logarithms using $\sum_r t_p^r / r = -\ln(1 - t_p)$ we obtain the *Euler product representation* of the dynamical zeta function:

$$1/\zeta(s) = \prod_p (1 - t_p), \qquad t_p = \frac{1}{|\Lambda_p|} e^{\beta \cdot A_p - sT_p}. \tag{54}$$

The approximate trace formula (42) plays the same role vis-a-vis the dynamical zeta function

$$\Gamma(s) = \frac{d}{ds} \zeta(s)^{-1} \tag{55}$$

as the exact trace formula (39) plays vis-a-vis the spectral determinant (49), see (50) and (47). The heuristically derived dynamical zeta function now reemerges as the $1/\zeta_{0\cdots0}(z)$ part of the *exact* spectral determinant; other factors in the infinite product (51) affect the non-leading eigenvalues of $\mathcal{L}$. The dynamical zeta function is useful because it also vanishes at e^s equal to e^{s_0}, the leading eigenvalue of $\mathcal{L}^t$, defined implicitly as the largest solution of either of the equations

$$\det(s - \mathcal{A}) = 0, \quad 1/\zeta(s) = 0. \tag{56}$$

Dynamical zeta functions for transfer operators

Ruelle constructed the first dynamical zeta function as a generalization of the topological zeta function

$$\zeta(z) = \exp \sum_{n=1}^{\infty} \frac{z^n}{n} \operatorname{tr} T^n, \qquad \operatorname{tr} T^n = \sum_{x_i \in \operatorname{Fix} f^n} \operatorname{tr} \prod_{j=0}^{n-1} g(f^j(x_i)).$$

Here the sum goes over all periodic points x_i of period n, and $g(x)$ is a (matrix valued) weighting function, with weight evaluated multiplicatively along the trajectory of x_i.

By the chain rule the stability of any n-cycle of a 1-d map factorizes as $\Lambda = \prod_{j=1}^{n} f'(x_i)$, so the cycle stability is the simplest example of a multiplicative cycle weight, and indeed - via the Perron-Frobenius evolution operator - the historical motivation for Ruelle's more abstract construction.

We now show that for a piecewise-linear map with a finite Markov partition, the dynamical zeta function is given by a finite polynomial, a straightforward generalization of the topological transition matrix. For a finite $[N \times N]$ dimensional matrix the determinant is given by

$$\prod_p (1 - z^{n_p} t_p) = \sum_{n=1}^{N} z^n c_n \,,$$

where c_n = sum over all non-self-intersecting closed paths of length n together with products of all non-intersecting closed paths of total length n. We illustrate this by the piecewise linear repeller (20). Due to the piecewise linearity, the stability of any n-cycle factorizes as $\Lambda_{s_1 s_2 \ldots s_n} = \Lambda_0^m \Lambda_1^{n-m}$, so the sum (42) takes a particularly simple form

$$\Gamma_n = \left(\frac{1}{|\Lambda_0|} + \frac{1}{|\Lambda_1|} \right)^n .$$

The dynamical zeta function (53) evaluated from the trace via (55) is indeed the $1/\zeta_{00\ldots0}$ part of the corresponding spectral determinant (48)

$$1/\zeta(z) = 1 - z/|\Lambda_0| - z/|\Lambda_1| \,. \tag{57}$$

More generally, piecewise-linear approximations to dynamical systems yield polynomial or rational polynomial cycle expansions, provided that the symbolic dynamics is a subshift of finite type.

We see that the dreaded exponential proliferation of cycles is a bogus anxiety; all that really matters in this example are the stabilities of the two fixed points. Clearly the information carried by individual cycles is highly redundant.

Topological zeta function for flows

The leading zero of a topological zeta function describes the growth rate of the number of topologically distinct trajectories with time. We can apply the method we used in deriving (39) to the problem of deriving the topological zeta functions for flows. We construct the time-weighted density of prime cycles of period t by analogy to (38):

$$\Gamma(t) = \sum_p \sum_{r=1} T_p \delta(t - rT_p) \tag{58}$$

A Laplace transform smoothes the sum over Dirac delta spikes and yields the *topological trace formula* and *topological zeta function* for the continuous time flows:

$$\sum_p \sum_{r=1} T_p \int_{0_+}^{\infty} dt\, e^{-st} \delta(t - rT_p) = -\frac{\partial}{\partial s} \ln 1/\zeta_{\text{top}}(s)$$

$$1/\zeta_{\text{top}}(s) = \prod_p \left(1 - e^{-sT_p} \right) . \tag{59}$$

Zeta functions, historical remarks

For a function to be deserving of the appellation "zeta function", one expects it to have an Euler product type representation, and perhaps also satisfy a functional equation. Various kinds of zeta functions are reviewed by A. Voros in ref. 7. Historical antecedents of the dynamical zeta function are the fixed-point counting functions introduced by Weil[8], Lefschetz[9], Artin and Mazur[10], and the determinants of transfer operators of statistical mechanics[11]. Ref. 12 contains a concise survey of the many zeta functions of mathematics.

In his review article Smale[13] already intuited, by analogy to the Selberg Zeta function, that the spectral determinant is the right generalization for continuous time flows. Ruelle was motivated by the Artin-Mazur zeta function and the statistical mechanics analogy, so he did not consider the spectral determinant a more natural object. In dynamical systems theory dynamical zeta functions arise naturally only for piecewise linear mappings; for smooth flows the natural object for study of classical and quantal spectra are the spectral determinants introduced above.

The nomenclature has not settled down yet; what we call evolution operators here is called transfer operators[14], Perron-Frobenius operators[15] and/or Ruelle-Araki operators elsewhere. We follow Ruelle in our usage of the term "dynamical zeta function", but elsewhere in the literature function (54) is often called the Ruelle zeta function. Ruelle[16] points out the corresponding transfer operator T was never considered by either Perron or Frobenius; a more appropriate designation would be the Ruelle-Araki operator. Here we have consistently refered to kernels such as (18) as evolution operators. Determinants similar to or identical with our spectral determinants are sometimes called Selberg Zetas, Selberg-Smale zetas[17], functional determinants, Fredholm determinants, or even - to maximize confusion - dynamical zeta functions[18]. A Fredholm determinant is a notion that applies to trace class operators only - as we consider here a somewhat wider class of determinants, we prefer to refer to them losely as "spectral determinants".

Resumé

The description of a chaotic dynamical system in terms of cycles can be visualized as a tessellation of the dynamical system, with a smooth flow approximated by its *periodic orbit skeleton*, each region $\mathcal{M}_i$ centered on a periodic point x_i of the topological length n, and the size of the region determined by the linearization of the flow around the periodic point. The integral over such topologically partitioned phase space yields a trace formula, and via a Laplace transform of this trace formula the spectral problem is recast into a problem of determining zeros of a spectral determinant.

Spectral determinants and dynamical zeta functions arise in classical and quantum mechanics because in both the dynamical evolution can be described by the action of linear evolution operators on infinite-dimensional vector spaces. The critical step in the derivation of spectral determinants and dynamical zeta functions is the hyperbolicity assumption, that is the assumption that all cycle stability eigenvalues are bounded away from unity, $|\Lambda_{p,i}| \neq 1$. By dropping the prefactors in this limit we have given up on any possibility of recovering the precise distribution of starting x (which should anyhow be impossible due to the exponential growth of errors), but in exchange we gain an effective description of the asymptotic behavior of the system. The pleasant surprise is that the infinite time behavior of an unstable system turns out to be as easy to determine as its short time behavior. To learn how this is accomplished the reader should turn to more comprehensive expositions of the subject, refs. 17, 19 in particular.

REFERENCES

1. R. Bowen, Equilibrium states and the ergodic theory of Anosov diffeomorphisms, *Springer Lecture Notes in Math.* **470** (1975).
2. W. Parry and M. Pollicott, *Ann. Math.* **118**, 573 (1983).
3. P. Cvitanović and B. Eckhardt, Periodic orbit expansions for classical smooth flows, *J. Phys.* **A 24**, L237 (1991).
4. D. Ruelle, Zeta functions for expanding maps and Anosov flows, *Inventiones math.* **34**, 231 (1976).
5. F. Hofbauer and G. Keller, Ergodic properties of invariant measures for piecewise monotonic transformations, *Math. Z.* **180**, 119 (1982).
6. V. Baladi and G. Keller, Zeta functions and transfer operators for piecewise monotone transformations, *Comm. Math. Phys.* **127**, 459 (1990).
7. A. Voros, in: Zeta functions in geometry, *in: Advanced Studies in Pure Mathematics* **21**, N. Kurokawa and T. Sunada, eds., Math. Soc. Japan, Kinokuniya, Tokyo (1992), p.327-358.
8. A. Weil, Numbers of solutions of equations in finite fields, *Bull. Am. Math. Soc.* **55**, 497 (1949).
9. D. Fried, Lefschetz formula for flows, *in:* "The Lefschetz centennial conference," *Contemp. Math.* **58**, 19 (1987).
10. E. Artin and B. Mazur, *Annals. Math.* **81**, 82 (1965).
11. Ya.G. Sinai, Gibbs measures in ergodic theory, *Russ. Math. Surveys* **166**, 21 (1972).
12. Kiyosi Itô, ed. , "Encyclopedic Dictionary of Mathematics," MIT Press, Cambridge (1987).
13. S. Smale, Differentiable dynamical systems, *Bull. Am. Math. Soc.* **73**, 747 (1967).
14. D. Ruelle, "Statistical Mechanics, Thermodynamic Formalism," Addison-Wesley, Reading MA (1978).
15. Y. Oono and Y. Takahashi, *Progr. Theor. Phys* **63**, 1804 (1980); S.-J. Chang and J. Wright, *Phys. Rev. A* **23**, 1419 (1981); Y. Takahashi and Y. Oono, *Progr. Theor. Phys* **71**, 851 (1984).
16. D. Ruelle, Functional determinants related to dynamical systems and the thermodynamic formalism, (Lezioni Fermiane, Pisa) preprint IHES/P/95/30 (March 1995).
17. P. Gaspard, "Chaos, Scattering and Statistical Mechanics," Cambridge Univ. Press, Cambridge (1997).
18. M. Sieber and F. Steiner, *Phys. Lett.* **A 148**, 415 (1990).
19. P. Cvitanović, et al, "Classical and Quantum Chaos: A Cyclist Treatise," http://www.nbi.dk/ChaosBook/, Niels Bohr Institute, Copenhagen (1997).

THEORY OF EIGENFUNCTION SCARRING

Lev Kaplan,[1] Eric J. Heller,[2]

[1]Department of Physics and Society of Fellows
[2]Department of Physics and Harvard-Smithsonian Center for Astrophysics
Harvard University
Cambridge, MA 02138

INTRODUCTION

Random matrix theory (RMT) serves as a very useful backdrop for the discussion of classical chaos and its effects on quantum mechanics. Two important suggestions began this fruitful association: the conjecture by Bohigas, Giannoni, and Schmit[1] associated RMT phenomenology with quantum eigenstates of classically chaotic systems. Berry's conjecture[2] stated that quantum eigenstates of classically chaotic systems should locally look like random superpositions of plane waves of the same (local) wavevector magnitude k, producing Gaussian random fluctuations in position space. These two closely related ideas turn out to be the clay out of which the truth is molded. Much research has centered on modifications to the extreme uniformity of RMT, modifications imposed by known constraints on the quantum dynamics. In fact there is no Hamiltonian system whose eigenstates are known to hold to the rigors of true Gaussian randomness.

The paper is an account of recent progress toward understanding one such modification to RMT, namely, the phenomenon of scarring of eigenfunctions[3]. Gutzwiller periodic orbit theory of the energy spectra of classically chaotic systems has enjoyed much success, and it would be strange if periodic orbits had no visible manifestation in the eigenstates. Indeed, exact solutions of the Schrödinger equation for classically chaotic systems sometimes exhibit obvious and apparently nonrandom patterns of concentration on periodic orbits. Other non-coordinate space representations such as projections onto coherent states (Husimi phase space) naturally show the same effect. This is a fully quantum phenomenon, although the explanation of scarring in terms of classical periodic orbits and their stability exponents[3] is based on semiclassical arguments. Classically there is no such accumulation of density near a periodic orbit in the long time limit for a bounded chaotic system. Scars are interesting because they stand out against the monotonous backdrop of wavefunction randomness over most of phase space. It has been shown that explicitly constructed "random" wavefunctions (random superpositions of plane waves with constant wavevector magnitude) show a ridge pattern that can be thought of as a precursor of true scarring[4].

Supersymmetry and Trace Formulae: Chaos and Disorder
Edited by Lerner *et al.*, Kluwer Academic / Plenum Publishers, New York, 1999

Scars have now been seen in many studies, including experimental work in microwave cavities[5, 6], tunnel junctions[7], and the hydrogen atom in a uniform magnetic field[8, 9].

To the casual observer the scarring phenomenon finds its proof in visual evidence (*e.g.* heavily scarred wavefunctions when plotted in coordinate space). However there are quantitative reasons for scarring; the first theory for them[3] was based on wavepacket motion (in the semiclassical limit) along a periodic orbit, and used the *linearized* short time dynamics near the orbit. This was a Husimi phase space theory; subsequent work by Bogomolny[10] and Berry[11] were coordinate space and Wigner phase space theories, respectively. These also were based purely on the linearized dynamics in the vicinity of an unstable periodic orbit. Reference [3] showed that scar strength for the heavily scarred states, as measured by the ratio of actual density to statistically expected density, was expected to be a function of the Lyapunov exponent λ only, tending to c/λ for small λ, where c is a constant. (Note the independence of $\hbar$.) That paper also remarked that scars are sometimes much stronger than this. At the other extreme, Steiner's work with systems of constant negative curvature was claimed to show no scarring when it was supposed to exist[12]. These factors plus confusion over the definition and measures of scar strength have understandably caused much discussion of whether there is indeed a theory of scar strength. The work by Fishman, Agam, and co-workers[13, 14, 15] has provided additional perspectives, and a proposed measure for scars, which we discuss below.

In this paper we point out that very often the simple linear theory, together with a proper account of gaussian fluctuations and symmetry, is sufficient to understand scar strength. We further show that scarring stronger than this may sometimes be understood in terms of identifiable *nonlinear* homoclinic recurrences associated with a given periodic orbit, the effect of which recurrences will turn out to be $\hbar$-dependent.

In the following section, we hope to clarify the concept of scarring. Then we provide a review of the original linear theory, for completeness and to establish the important concept of the spectral envelope, which we use heavily. After presenting the properties of scars as a localization phenomenon, we turn to considerations of discrete symmetries, and of nonlinear fluctuations about the envelope, which together explain many cases of enhanced wavefunction scarring. Connections with semiclassical theory are made here, and emphasis is placed on the constraints which the short-time dynamics places on the stationary properties of the system. Various measures of scarring are described. This is followed by numerical studies which show the expected amount of scarring according to the linear theory supplemented by gaussian random fluctuations, in the first statistical study of a large number of scarred states. Numerical evidence is also presented for the effects on scarring of individual homoclinic orbits, in a situation where the long-time recurrences are not random. We end with a discussion of the state of the theory of wavefunction scarring.

DEFINITION OF SCARRING

We begin with a definition, close to what was already given in the 1989 Les Houches proceedings[16]:

Definition: A quantum eigenstate of a classically chaotic system has a *scar* of a periodic orbit if its density on the classical invariant manifolds near the periodic orbit is enhanced over the statistically expected density.

Alternatively, an unstable periodic orbit is scarred when some eigenstates of the system have greater amplitude, and others less amplitude, along the orbit than would be predicted based on gaussian random fluctuations. Also, a wavepacket launched on or near such an orbit will have a tendency to return to the orbit, having larger overlaps with itself at long times than a wavepacket launched elsewhere in phase space. Scars can appear as strong enhancements in the eigenfunction coordinate space density surrounding periodic orbits, especially near self-conjugate points along the classical orbit, as shown by Bogomolny[10]. Scar strength S, as measured by the projection of scarred eigenstates onto a coherent state centered on the scarring periodic orbit and aligned along the stable and unstable manifolds, is generically a function only of λ, the Lyapunov exponent for one period of the periodic orbit, and is $\hbar$-independent. For small λ, $S \to C/\lambda$, where C is a constant obtained by considering the linear theory at short times combined with random long-time fluctuations. Enhancements of the scarring phenomenon can occur in the presence of strong, isolated long-time recurrences associated with homoclinic orbits. Symmetry factors also must be included if one is to obtain a quantitatively correct picture of scarring.

Weak scars are not always visible to the "naked eye"; there can be scarring nonetheless, according to the definition given above. The amount of scarring associated with specific eigenstates varies significantly from state to state, though we find below that the variance is in accordance with a theoretically predicted ranges. Orbits with larger instability exponents exhibit less scarring on average, and a statistical analysis may be necessary in such cases to determine that scarring is indeed present. It needs to be emphasized that scarring is not in violation of ergodicity of wavefunctions in the sense of Schnirelman, Zelditch, and Colin de Verdiere[17], because the phase space area affected by scarring vanishes in the semiclassical limit, forming a narrower and narrower region around the periodic orbit.

Another important point is that scars are not merely associated with a one dimensional line along a periodic orbit; rather, they are associated also with the stable and unstable manifolds of that orbit. For this reason, a phase space study of scarring may often be more illuminating than a coordinate space projection. Scars do not disappear as $\hbar \to 0$, except in the sense that the total amount of scarring is expected to become distributed over an ever increasing number of eigenstates in that limit, while the region of phase space in which the eigenstates are scarred is simultaneously decreasing.

LINEAR THEORY OF SCARRING

For completeness and context we need to review the linear theory of scarring, first discussed in Reference [3]. Consider an unstable fixed point of a classical map located at the origin, with the stable and unstable manifolds oriented along the p (vertical) and q (horizontal) axes, respectively. Linearizing the map around the fixed point, we obtain to first order

$$q' = e^{\lambda t} q$$
$$p' = e^{-\lambda t} p,$$

(1)

where λ is the Lyapunov exponent for one iteration of the orbit. For simplicity we do not discuss the case of a wavepacket centered near, but not on, a periodic orbit (see Reference [3], and also Section of the present paper). We now take a gaussian wavepacket

$$g_\sigma(q) = (\frac{4\pi\hbar^2}{\sigma^2})^{1/4} e^{-q^2/2\sigma^2},$$

(2)

which corresponds to a classical distribution centered on the origin with width σ in the q direction and width $\sigma_p = \hbar/\sigma$ in the p direction ($\hbar \ll \sigma \ll 1$, e.g. $\sigma \sim \sqrt{\hbar}$). Now for a small enough $\hbar$, the initial wavepacket and its short-time iterates are contained within the linear regime, and we have the time-evolved wavepacket $g_t(q) = U^t g_\sigma(q)$ given by the expression above in Eq. 2 with σ replaced by $\sigma_t = e^{\lambda t}\sigma$ (here U is a quantum operator corresponding locally to the classical dynamics given by Eq. 1). Classically this corresponds to a horizontal stretching and vertical shrinking of the gaussian distribution in phase space.

The overlap

$$A(t) = \langle g_t | g \rangle = \frac{e^{i\theta t}}{\sqrt{\cosh(\lambda t)}} \tag{3}$$

is easily found by Gaussian integration (notice that the autocorrelation function $A(t)$ is independent of σ, the width of the initial wavepacket). Here θ is the quantum phase associated with the fixed point (semiclassically it is given by $S/\hbar$, S being the action for one traversal of the periodic orbit, plus Maslov phases arising from caustics). This time domain behavior can be fourier transformed to obtain an envelope in the (quasi-)energy spectrum, centered at $E = \theta$ and with a width which depends only on λ, scaling linearly with λ for small λ.

We remark here that the situation for a fixed point of period $P > 1$ is similar. In this case the linear autocorrelation function is nonzero only at integer multiples of P, and the corresponding spectrum has P identical bumps, each of a width and height related to the instability of the entire orbit. Additional time scales are present in a continuous-time system, which are not directly relevant to the phenomenon of scarring, but which produce a background spectrum relative to which scarring can manifest itself. Some of these issues are addressed in Section VII of Kaplan and Heller [18].

In the case of exact linearity, or where the evolving wavefunctions are allowed to escape to infinity at long times (as in an inverted harmonic oscillator), the preceding is all that there is to be said about the spectrum of the wavepacket. The width of the spectral bump then corresponds to a decay rate. But in a closed, unitary system, the escaping probability must eventually start returning to the origin. In a classically mixing system, this will begin happening not later than by the mixing time, this being the time required for a classical distribution corresponding to a minimum uncertainty wavepacket to spread through all of phase space on a mesh of size $\hbar$. The mixing time scales as $T_{\text{mix}} \sim \log(N)/\overline{\lambda}$, where $\overline{\lambda}$ is the "typical" exponent for the entire system, and N is the total number of states in the available phase space. The key point is that the Fourier transform of $A(t)$ for small λ localizes the spectrum (local density of states) to a region of width $\sim \lambda$, smaller than the whole quasi-energy interval. In effect the initial state is in a resonance mode which decays more slowly than a random state. A random state should decay in a time of the order of a single time step for a discrete map. (The reason for the single step decay is simple: a random state having a random (*i.e.* RMT) local density of states spectrum has a quasi-energy uncertainty of the whole interval ($\delta\epsilon = \delta\omega/\hbar = 2\pi$). Now $\hbar\delta\epsilon\delta\tau \sim \delta\omega\delta\tau \sim \hbar$ implies $\delta\tau \sim \mathcal{O}(1)$, *i.e.* one time step). The corresponding resolved spectrum for a wavepacket launched on a periodic orbit thus *cannot* be picked from an *a priori* RMT local density of states, as Figure 1 shows. Now, to complete the point, we recall that the spectral line intensities are the squared projections of eigenstates onto the local "test" Gaussian. *The intensities are the "support" of the envelope, and are thus required to be (upon local average) larger in the peak region of the envelope than RMT predicts by a factor of order λ^{-1}.* This enhancement of the overlaps (that is, the enhancement by a factor of order λ^{-1} over the statistical expectation of $1/N$, where N is the dimension of the Hilbert space) means that at a minimum there must be states with a projection onto the test state of order λ^{-1} larger than what is statistically expected. However, if this projection were to be

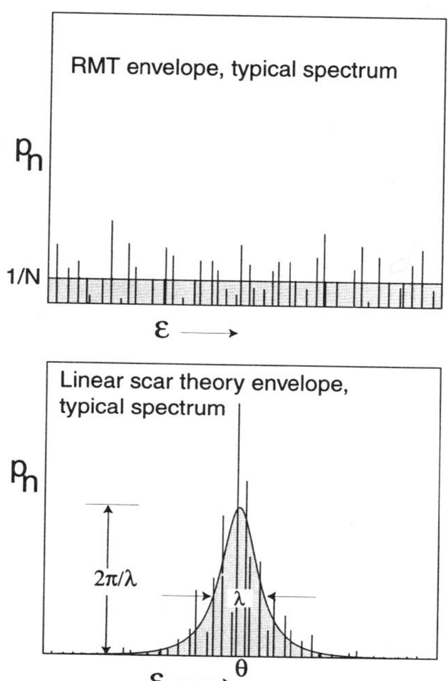

Figure 1. The short time dynamics of the localized wavepacket imposes an envelope in the local density of states which the resolved spectrum (coming from long-time dynamics up to times of order of the Heisenberg time) must obey. The envelope has a peak at quasi-energy $\epsilon = \theta$, a width $\delta\epsilon \sim \mathcal{O}(\lambda)$, and a height $\sim \mathcal{O}(1/\lambda)$.

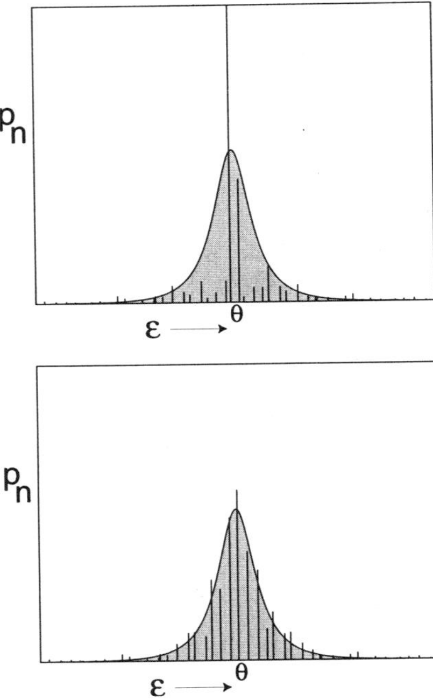

Figure 2. The short time dynamics by itself does not predict whether "totalitarian" (top) or "egalitarian" (bottom) filling of the local density of states envelope occurs. Both spectra have the same short time local density of states envelope.

shared in an egalitarian fashion among all the available states, then most or all of the states in the peak region would be enhanced by this factor. On the other hand, if only a small fraction f of the available states are enhanced, then these states must have larger projections $\mathcal{O}(\lambda^{-1}f^{-1})$ onto the test state in order to support the local density of states envelope. These two extremes are illustrated in Figure 2. The short time dynamics, which depends only on the linear or "tangent" map around the periodic orbit on which the test Gaussian is centered, cannot tell us without further assumptions which extreme (or intermediate) regime is realized; it only tells us that some states must be enhanced. The egalitarian case corresponds to the least striking type of scarring, since each state is enhanced at most by a factor of order λ^{-1}. If λ is not too small, this enhancement is not even competitive with the fluctuations expected from RMT, and we might conclude by cursory inspection that individual states are not scarred at all. Indeed, this would be a justifiable definition, although the *systematic*, statistically significant enhancement of many nearby states in the egalitarian case would still reveal the underlying mechanism of scar localization. In effect this definition was adopted by Steiner and co-workers[12] in their studies of the eigenstates of the hyperbolic billiards, which appear to live close to the egalitarian limit. In the opposite "totalitarian" extreme, enhancements are very large, and scars are obvious in pictures of eigenstates, even for orbits which are very unstable. It is important to note however that for small λ even the linear (short time) theory in the "worst case" egalitarian scenario predicts strong scarring, well above the typical RMT fluctuations, of strength $1/\lambda$.

These considerations extend easily to include the possible dependence of scar strength on $\hbar$ or on the density of states. If the density of states is such that only one or a few states can exist within a quasi-energy width λ, then effectively only the "totalitarian" option exists. This is a strong localization regime, where one or a few states carry the total scar intensity. At the ideal unitary limit of an overlap of 1, one state is entirely localized to the periodic orbit region. (This was the basis for our conclusion that the bouncing ball modes in the stadium billiard persist up to infinite energy[19]). Starting from this extreme, as N increases, the scar strength of individual eigenstates *could* decrease as fast as $1/N$, in the egalitarian limit, although the intensity *enhancement factor* would still remain finite, as the average intensity is also decreasing as $1/N$. (In a billiard system there is a $\sqrt{E}$ increase in the number of affected states with increasing E: the density of states is independent of E but the energy width δE of the scar "resonance" scales as $\sqrt{E}\lambda$, where again λ is the Lyapunov exponent for a complete period of the orbit. The time required to traverse this orbit goes as $1/\sqrt{E}$, thus $\delta E \sim \sqrt{E}\lambda$).

We remark that scarring can become no weaker than the egalitarian limit defined above for any given periodic orbit, even as $\hbar \to 0$. Suppose that the egalitarian limit is the usual circumstance as $\hbar \to 0$. Then scars become less dramatic but do not disappear as $\hbar \to 0$ as measured by the test states whose area in a surface of section is h. However this area (projected onto coordinate space, say) amounts to a diminishing portion (going as $\sqrt{\hbar}$ for a phase space Gaussian with an aspect ratio of order unity) of the total coordinate space volume. These subtleties have caused much confusion over whether scars "disappear" as $\hbar \to 0$.

Husimi projections and phase space tubes

The projection of Gaussian wavepackets, or other distributions localized around periodic orbits, onto the eigenstates as a test of their localization properties was introduced earlier [3]. Subsequently, the idea of detecting and quantifying scars by integrating over tubes in phase space surrounding the periodic orbit has been discussed[14]. The tube should be of diameter $\sqrt{\hbar}$ normal to the direction of the orbit. The diameter originally used was $\hbar$ but $\sqrt{\hbar}$ is more appropriate and is used in more recent work (S. Fishman, private communication). Of course, the structure of the linear local dynamics around the periodic orbit must also be considered here. In particular, for certain alignments of the stable and unstable manifolds with respect to the p and q directions, a tube of width $\hbar$ in position space and width 1 in momentum would be equally optimal. This is consistent with the findings of Li [20], where certain orbits in a stadium billiard show optimal scarring in coordinate space with a tube size scaling as the wavelength (instead of as the square root of the wavelength).

The phase space tube approach is closely related to the Gaussian wavepacket projection, as we shall now show. In two dimensions, suppose we average the Gaussian projections over the whole length of the periodic orbit (instead of taking an overlap with a Gaussian centered at just one periodic point). The mean wavepacket momentum points along the orbit. Then we have, for an orbit pointing along the x-axis, the average projection S given by

$$S = \frac{1}{L_x} \int dx_0 |\langle \alpha(x_0, p_{x0}, y_0, 0)|\psi_E\rangle|^2 = \mathrm{Tr}(\rho_L|\psi_E\rangle\langle\psi_E|), \tag{4}$$

where

$$\rho_L(x\;,\;y,x',y') = \frac{1}{L_x} \int dx_0 \exp[-(x'-x_0)^2/2\sigma_x^2\hbar - (x-x_0)^2/2\sigma_x^2\hbar$$
$$-(y'-y_0)^2/2\sigma_y^2\hbar - (y-y_0)^2/2\sigma_y^2\hbar + i(x-x')p_{0x}/\hbar]$$
$$\sim \exp[-(x'-x)^2/4\sigma_x^2\hbar - (y'-y_0)^2/2\sigma_y^2\hbar - (y-y_0)^2/2\sigma_y^2\hbar + i(x-x')p_{0x}/\hbar](5)$$

Now we Wigner transform this density matrix:

$$\rho_L^W(\mathbf{q},\mathbf{p}) = 2^d \int_{-\infty}^{\infty} e^{-2i\mathbf{p}\cdot\mathbf{s}/\hbar} \rho_L(\mathbf{q}+\mathbf{s},\mathbf{q}-\mathbf{s})\,d\mathbf{s}$$
$$= \rho_L^W(x,p_x;y,p_y)$$
$$\sim \exp\left[-(y-y_0)^2/\sigma_y^2\hbar - \sigma_y^2(p_y-p_{y_0})^2/\hbar - \sigma_x^2(p_x-p_{x_0})^2/\hbar\right]. \qquad (6)$$

We see *this is a phase space tube surrounding the classical trajectory, of diameter* $\propto \sqrt{\hbar}$, if σ_y^2, the aspect ratio in the y–p_y subspace, is chosen to be of order unity. Hence, the relation between the Fishman *et al.* phase space tube and the Husimi projection is that the tube is the Husimi projection averaged over the entire length of the orbit. The advantage of this smearing is the same as its disadvantage: it is insensitive to the local direction of the classical invariant manifolds. In this sense the tube is a somewhat duller probe for scarring, while at the same time it has the advantage of providing a more universal description of scars.

Limitations of the linear theory

The linear theory does not say how the localization, predicted from the short time dynamics, actually manifests itself in the long time behavior of the autocorrelation function $A(t)$. The linear theory can have no information about whether the totalitarian or egalitarian limit of scar intensity distribution over the available eigenstates is approached. This information comes from longer time dynamics which necessarily involves more than the linearized tangent map of the periodic orbit. No systematic study of scar strength over a large enough ensemble to unambiguously test the predictions of the linear theory has until now been undertaken.

SCARRING AS A LOCALIZATION PHENOMENON

Consider a compact classical phase space of area A, with chaotic dynamics given by a discrete-time evolution (area-preserving map of A onto itself), and no conserved quantities. A two-dimensional billiard can be reduced to such a discrete time one-dimensional map by the surface of section technique. (Although we restrict ourselves to one spatial dimension for specificity, the concepts are completely generalizable.) Our results can be extended to a situation in which conserved quantities (such as energy) are present, by considering flow between phase space-localized states (usually coherent states), and taking account of the energy spread contained in such states. This problem is treated in elsewhere [21], and is also mentioned briefly in Section VII of Kaplan and Heller [18]. We will assume in the present discussion that all of phase space is classically accessible from any smooth starting distribution.

If the area is an integer multiple of Planck's constant h, $A = Nh$, the system can be quantized (with a choice of quantization conditions), to obtain an N-dimensional Hilbert space. Because the underlying classical dynamics is completely ergodic, one might expect the eigenstates to appear random in any natural basis, such as that of position, momentum, or Gaussian states. Thus, let $|a\rangle$ be such a physically-motivated basis and $|n\rangle$ be the basis of eigenstates. Then we expect the overlaps $f_{an} = \langle a|n\rangle$ to be Gaussian variables with the normalization condition $< |f_{an}|^2 > = 1/N$. This does *not* mean that all energy eigenstates have equal overlaps with all the trial basis states, *i.e.* $|f_{an}|^2 \neq 1/N$ for all a, n. In fact, such a Gaussian distribution (predicted by random matrix theory, which is based on the absence of a preferred basis for analyzing the dynamics), leads to $< |f_{an}|^4 > = F/N^2$, where $F = 3$ if both $|a\rangle$ and $|n\rangle$ are real (convenient if, for example, the dynamics is time-reversal invariant), and $F = 2$ otherwise. This is a quantum fluctuation result and is already a deviation from the classical expectation of $F = 1$. Localization, however, is taken to mean an additional deviation of the f_{an} distribution, away from Gaussian form, towards a distribution with a longer tail. In particular, the inverse participation value (IPR) $< N^2 |f_{an}|^4 >$ (where the average can be taken over trial states, energy eigenstates, or both, and also over an ensemble of systems) is in the presence of such localization enhanced from its ergodic value of F. Higher moments and the behavior of the tail can also be investigated.

Let us now consider the connection with the time domain. We define the auto-correlation function $A(t) = \langle a|U^t|a\rangle$, where U is the discrete time evolution operator (a completely analogous notation can be written down for continuous time). The fourier transform of $A(t)$ is the weighted spectrum (local density of states) $S(E) = \sum_n \delta(E - E_n)|f_{an}|^2$. The squared autocorrelation function

$$|A(t)|^2 = \left| \sum_n |f_{an}|^2 e^{-iE_n t} \right|^2 \tag{7}$$

can be thought of as a wavepacket-specific form factor, similar to the usual spectral form factor $F(t) = \sum_{mn} e^{-i(E_n - E_m)t}$, but weighting each term by the heights of the corresponding lines in the spectrum $S(E)$ above.

For long times, in the absence of degeneracies, one easily obtains the relation

$$< |A(t)|^2 >_t \equiv \lim_{T_{\max} \to \infty} \frac{1}{T_{\max}} \sum_{t=0}^{T_{\max}-1} |A(t)|^2$$
$$= \sum_n |f_{an}|^4, \tag{8}$$

where on the left hand side a time average must be taken over times long compared to the Heisenberg time T_H (generically $T_H \sim N$). Within RMT, both sides of Eq. 8 are predicted to approach F/N, in the semiclassical limit $N \to \infty$. Localization is associated with an enhancement in the long-time return probability $< |A(t)|^2 >_t$. We will see in Section how this is possible in the case of scarring. We will also see there that short-time unstable orbits induce nontrivial correlations $< A^*(t + \Delta)A(t) >$ at long times t. These will be seen to correspond to eigenvalue-eigenstate correlations (through the formation of an envelope in the spectrum) in the energy domain.

BEYOND THE LINEAR THEORY

Homoclinic orbits

In keeping with our treatment of the linear theory of scarring, we will now discuss long time recurrences in the autocorrelation function from a semiclassical point of view. Let us consider a homoclinic orbit which begins near the fixed point $(0, 0)$ along the unstable manifold at large negative times and again approaches the fixed point along the stable manifold at large positive times. Specifically, let the orbit $\mathcal{HC}$ be given by $\{(q_t, p_t)\}_{t=-\infty\ldots\infty}$, such that $(q_t, p_t) = (ae^{\lambda t}, 0)$ for $t \to -\infty$ and $(q_{t'}, p_{t'}) = (0, be^{-\lambda t'})$ for $t' \to +\infty$. Then we claim that a thin vertical strip cut out of the initial Gaussian near $q = ae^{\lambda t}$ at time t will intersect the same Gaussian as a long horizontal strip at $p' = be^{-\lambda t'}$ at a much later time t'. Note that because the wavepacket is contained well inside the linearizable region around the fixed point, the dynamics from time t to time t' can be divided into three parts. First, the tall, narrow distribution shrinks vertically and stretches horizontally as its center moves out horizontally at an exponential rate along the unstable manifold (for, let us say, τ_1 steps). This is followed by complicated nonlinear dynamics which eventually brings the center of the distribution back into the linearizable region, this time along the stable manifold of the fixed point. Now the part of the distribution which is in the linear region again begins to stretch horizontally and shrink vertically, becoming a narrow horizontal strip moving in towards the original wavepacket. We will denote by τ_3 the time spent in this last stage of the evolution. The first and third parts of the dynamics allow the breadth of the initial distribution centered on $(ae^{\lambda t}, 0)$ and the height of the final distribution centered on $(0, be^{-\lambda t'})$ both to be small compared to the size of the Gaussian wavepacket. All of this is illustrated in Figure 3.

The overlaps of the Gaussian with the vertical and horizontal strips, as well as the effects of the linear dynamics in stages one and three are easy to write down analytically. There is also an amplitude factor Q which measures the stretching of the distribution in the nonlinear stage of the dynamics, from time $t + \tau_1$ until time $t' - \tau_3$. Finally, there is a phase ϕ_{nonlin} associated with this nonlinear excursion. The total contribution to the wavepacket autocorrelation function at time $t' - t$ coming from this homoclinic excursion is given by a product of five factors:

$$A_{\mathcal{HC}} = e^{-q_t^2/2\sigma^2} \cdot e^{i\tau_1\theta}e^{-\lambda\tau_1/2} \cdot Q(t + \tau_1, t' - \tau_3)e^{i\phi_{\text{nonlin}}} \cdot e^{i\tau_3\theta}e^{-\lambda\tau_3/2} \cdot e^{-p_{t'}^2/2\sigma_p^2} . \tag{9}$$

The factors $e^{-\lambda\tau_1/2}$ and $e^{-\lambda\tau_3/2}$ are instability factors associated with the linearized motion of the wavepacket, while $e^{i\tau_1\theta}$ and $e^{i\tau_3\theta}$ are the corresponding phases. The suppression factors $e^{-q_t^2/2\sigma^2}$ and $e^{-p_{t'}^2/2\sigma_p^2}$ are associated with the fact that the initial and final points of the excursion are both off-center relative to the gaussian wavepacket. The total correlation function $A(T)$ is given semiclassically by a sum of terms of the form given above over all homoclinic excursions of length $t' - t = T$.

$$A_{\text{SC}}(T) = \sum_{\mathcal{HC}} \delta_{t'_{\mathcal{HC}} - t_{\mathcal{HC}} - T} A_{\mathcal{HC}} . \tag{10}$$

Effect of short-time dynamics

One might think naively that the various contributions to the sum in Eq. 10 are all independent and uncorrelated at long times, but this is not the case. In fact, correlations are present both among the different contributions to the autocorrelation function at a fixed time, and also among homoclinic contributions of different excursion lengths T. Let us consider the homoclinic orbit of the previous subsection, with the mapping taking us from $(ae^{\lambda t}, 0)$ to $(0, be^{-\lambda t'})$ in the time interval $T = t' - t$. Now of course the same homoclinic orbit takes the Δ_1-step iterate of the original point, $(ae^{\lambda(t+\Delta_1)}, 0)$ to the Δ_3-step iterate of the final point, $(0, be^{-\lambda(t'+\Delta_3)})$, in a time $T + \Delta_3 - \Delta_1$. In particular, taking $\Delta_1 = \Delta_3$, we have a family of excursions, all of the same length, associated with one homoclinic orbit. An important thing to notice is that all these contributions come

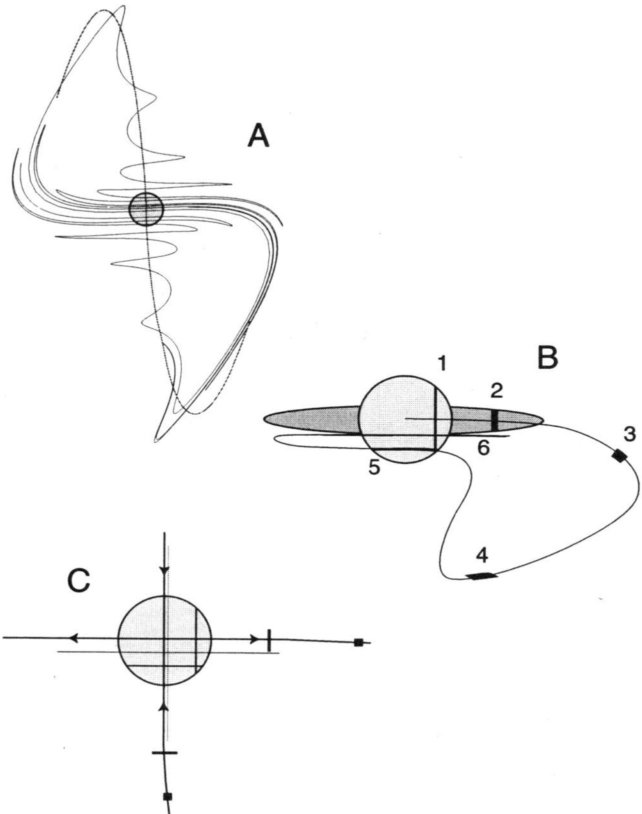

Figure 3. Nonlinear recurrences resulting from following a homoclinic classical orbit. Linear wavepacket spreading is followed by a nonlinear excursion, and finally by an approach back to the fixed point along the stable manifold. Three levels of abstraction are shown. Diagram A shows a portion of the stable manifold, and a longer portion of the unstable manifold. B is more schematic, showing how a small rectangle in the initial disk (representing the initial state in phase space) is compressed and stretched, finally returning along the stable manifold. Note that this particular rectangle is special in that it returns soon, guided by a primary homoclinic trajectory. In C, we see a regularized (normal form) version of B. Two types of correlation are seen in this Figure. In B, for example, "1" and "5" have exactly the same phase relation as "2" and "6". Also "5" and "6" are related by a phase associated with one iteration of the periodic orbit.

with the same phase, the extra phase in the final Δ_3 steps approaching the fixed point being exactly canceled by the missing Δ_1 steps at the beginning of the trip (there is only one phase associated with the fixed point, and it is the same in stages one and three). ("1" and "5" in Figure 3 have exactly the same phase relation as "2" and "6".) The phase ϕ_{nonlin} coming from the nonlinear dynamics in stage two is, of course, independent of $\Delta_{1,3}$. Naturally, the exponential prefactor

$$
\begin{aligned}
&e^{-q_{t+\Delta_1}^2/2\sigma^2}\, e^{-p_{t'+\Delta_3}^2/2\sigma_p^2}\, e^{-\lambda(\tau_1-\Delta_1)/2}\, e^{-\lambda(\tau_3+\Delta_3)/2}\\
&= e^{-(q_t e^{\Delta_1\lambda})^2/2\sigma^2}\, e^{-(p_{t'}e^{-\Delta_3\lambda})^2/2\sigma_p^2}\, e^{-\lambda(\tau_1-\Delta_1)/2}\, e^{-\lambda(\tau_3+\Delta_3)/2}
\end{aligned}
\tag{11}
$$

will depend on $\Delta_{1,3}$, so only a finite number of the infinite family of contributions for $\Delta_1 = \Delta_3$ will be of significant size (this number, the number of iterations for which one tends to stay near the periodic orbit, scales as $1/\lambda$.) However, all of these will add exactly in phase. This is an important difference between the classical and semiclassical long-time dynamics of the system. The presence of this coherence is the underlying reason for the fact that in quantum mechanics the probability for coming back to an unstable periodic orbit at long times is enhanced over the classical value of $1/N$, which fact leads to scarring in the eigenstate domain, as seen in Section .

We also notice that if we take $\Delta_1 \neq \Delta_3$, so the new excursion length is different from the original one, the difference in phase is given by $e^{i\theta(\Delta_3-\Delta_1)}$, and each excursion of length T also contributes to the autocorrelation function at $T + \Delta \equiv T + \Delta_3 - \Delta_1$, with this extra phase and a somewhat different amplitude prefactor. Thus, as will be explained in more detail in the following section, nontrivial correlations will be present in the autocorrelation function at nearby times, $< A^*(T + \Delta)A(T) >$, for small Δ. This is an effect that is qualitatively easy to understand even in a classical picture in terms of a "reloading" of the original wavepacket. Because the original wavepacket is centered on a fixed point of the map, any significant recurrence at time T is expected to be accompanied by a recurrence for all times $T + \Delta$, where Δ is within the decay time associated with the fixed point. In effect, any new long-time recurrences get convoluted with the (linear) short-time dynamics of the system around the fixed point. In the energy domain, this corresponds (by fourier transform) to a *multiplication* of the original linear envelope by an oscillating function associated with the long-time recurrence.

RANDOM RECURRENCE MODEL

Correlations among homoclinic excursions

In a chaotic system, the number of homoclinic orbits increases exponentially with the excursion length at long times. Let us for the moment assume that the nonlinear phases associated with the terms in Eq. 10 are uncorrelated at long times. We also assume a uniform distribution of homoclinic points q_t and $p_{t'}$ along the unstable and stable manifolds. We then obtain for the average square of the long-time semiclassical autocorrelation function

$$
< |A_{\text{SC}}|^2 >_{\text{diag}} = \mathcal{N} \int dq_t \int dp_{t'}\, e^{-q_t^2/\sigma^2} e^{-p_{t'}^2/\sigma_p^2},
\tag{12}
$$

where the "diag" subscript indicates that we are working in a diagonal approximation (no correlations between the homoclinic orbits). The mean squared amplitude factor associated with the excursions as well as the densities of homoclinic points along the two manifolds have been incorporated into the normalization factor $\mathcal{N}$. This normalization factor can easily be fixed by noticing that classically, in the absence of any coherence effects, the return probability must approach $1/N$ at long times.

We now correct the assumption made in the previous paragraph and include the fact that, as discussed in the preceding section, the contributions from homoclinic excursions associated with a single homoclinic orbit all add in phase. We then have

$$< |A_{\rm SC}| >^2 = \mathcal{N}' \int dq_t \int dp_{t'} \left| \sum_{\Delta_1} e^{-(q_t e^{\Delta_1 \lambda})^2/2\sigma^2} e^{-(p_{t'} e^{-\Delta_1 \lambda})^2/2\sigma^2} \right|^2 , \qquad (13)$$

where the normalization factor $\mathcal{N}'$ is given by

$$< |A_{\rm SC}| >^2_{\rm diag} = \mathcal{N}' \int dq_t \int dp_{t'} \sum_{\Delta_1} \left| e^{-(q_t e^{\Delta_1 \lambda})^2/2\sigma^2} e^{-(p_{t'} e^{-\Delta_1 \lambda})^2/2\sigma^2} \right|^2 = 1/N . \qquad (14)$$

We now see that the enhancement factor $< |A_{\rm SC}| >^2 \, / \, < |A_{\rm SC}| >^2_{\rm diag}$ is given by

$$\frac{\sum_{\Delta_1,\tilde{\Delta}_1} \int dq_t \int dp_{t'} g_{\tilde{\Delta}_1}(q_t, p_{t'}) g_{\Delta_1}(q_t, p_{t'})}{\sum_{\tilde{\Delta}_1} \int dq_t \int dp_{t'} g_{\tilde{\Delta}_1}(q_t, p_{t'}) g_{\tilde{\Delta}_1}(q_t, p_{t'})}$$
$$= \frac{\sum_{\Delta_1} \int dq_t \int dp_{t'} g_{\tilde{\Delta}_1}(q_t, p_{t'}) g_0(q_t, p_{t'})}{\int dq_t \int dp_{t'} g_0(q_t, p_{t'}) g_0(q_t, p_{t'})} , \qquad (15)$$

where g_0 is a Gaussian distribution of width σ in position and σ_p in momentum, and g_{Δ_1} is the same distribution stretched by a factor of $e^{\Delta_1 \lambda}$ horizontally and compressed vertically by the same factor. In Eq. 15 we have used the relations

$$\int dq_t \int dp_{t'} g_{\tilde{\Delta}_1}(q_t, p_{t'}) g_{\tilde{\Delta}_1}(q_t, p_{t'}) = \int dq_t \int dp_{t'} g_0(q_t, p_{t'}) g_0(q_t, p_{t'}) \qquad (16)$$

and

$$\int dq_t \int dp_{t'} g_{\tilde{\Delta}_1}(q_t, p_{t'}) g_{\Delta_1}(q_t, p_{t'}) = \int dq_t \int dp_{t'} g_{\tilde{\Delta}_1 - \Delta_1}(q_t, p_{t'}) g_0(q_t, p_{t'}) . \qquad (17)$$

Returning to the discussion following Eq. 2 we notice that the overlap of the original Gaussian wavepacket $a(q)$ with the same wavepacket stretched by a factor $e^{\Delta_1 \lambda}$ is just the linearized autocorrelation function $A_{\rm lin}(\Delta_1)$. The overlap of the corresponding classical distributions is given by $|A_{\rm lin}(\Delta_1)|^2$, and finally we have

$$< |A_{\rm SC}| >^2_{\rm coherent} = \frac{1}{N} \sum_{\Delta_1} |A_{\rm lin}(\Delta_1)|^2 . \qquad (18)$$

A similar analysis can be performed for the case $\Delta \equiv \Delta_3 - \Delta_1 \neq 0$, and there we find

$$< A^*(T + \Delta) A(T) >_{\rm coherent} = \frac{1}{N} \sum_{\Delta_1} A^*_{\rm lin}(\Delta + \Delta_1) A_{\rm lin}(\Delta_1) . \qquad (19)$$

This shows the unmistakable effect of the short-time correlations at long times. The expressions obtained here quantify the connection between short-time and long-time behavior which was already suggested in the previous section.

Effect on spectra

We now claim that the time domain correlations given in Eq. 19 are just the right correlations to produce spectral fluctuations which multiply the original linear envelope. One might have thought that the oscillations in the spectrum caused by long-time recurrences simply get added to the original linear envelope, instead of multiplying it, which would produce completely different spectral behavior on small energy scales. Indeed suppose the spectrum has the form

$$S(E) = S_{\text{lin}}(E) S_{\text{fluct}}(E) \,, \tag{20}$$

where $S_{\text{lin}}(E) = 2\pi \sum_T e^{iET} A_{\text{lin}}(T)$ is the spectrum given by the linear dynamics, and $S_{\text{fluct}}(E)$ is the fluctuating part, with the property that $K \equiv\, < |S_{\text{fluct}}(E)|^2 >$ is an $E-$independent constant (an ensemble average is implied in the definition of K). Now we have

$$< A^*(t + \Delta)A(t) > =< \int dE' S_{\text{lin}}^*(E') S_{\text{fluct}}^*(E') e^{iE'(t+\Delta)} \int dE S_{\text{lin}}(E) S_{\text{fluct}}(E) e^{-iEt} > \,. \tag{21}$$

Averaging over t, and inserting the expression for $S_{\text{lin}}(E)$ from above,

$$< A^*(t + \Delta)A(t) > \; = \; < 2\pi \int dE |S_{\text{lin}}(E)|^2 |S_{\text{fluct}}(E)|^2 e^{iE\Delta} >$$

$$= \; < \frac{1}{2\pi} \int dE |S_{\text{fluct}}(E)|^2 e^{iE\Delta} \sum_{TT'} e^{iE(T'-T)} A_{\text{lin}}^*(T') A_{\text{lin}}(T) > \tag{22}$$

Finally, inserting the fluctuation intensity K and performing the energy integral, we obtain

$$< A^*(t + \Delta)A(t) > = K \sum_T A_{\text{lin}}^*(T + \Delta) A_{\text{lin}}(T) \,, \tag{23}$$

which agrees with the result in Eq. 19 above, obtained by taking into account the phase coherence of all the homoclinic excursions associated with a single homoclinic orbit.

Heisenberg time dynamics and symmetries

Up until now we have been assuming that the nonlinear recurrences associated with the homoclinic orbits are sums of random contributions, under the constraint of correlations due to the short-time linear behavior near the periodic orbit. The picture here is that random "new" recurrences get smeared out by a function associated with the reloading effect. This leads to Gaussian random fluctuations in the spectrum at all scales short compared to the inverse of the mixing time, all of these uncorrelated fluctuations multiplying the initial linear envelope. But now we have to include an additional constraint coming from Heisenberg time dynamics, namely unitarity and discreteness of the spectrum. In the absence of scarring, these long-time correlations cause the mean return probability $|A_{\text{QM}}(t)|^2$ to converge to a value of $2/N$ at long times, $3/N$ if the eigenstates and the initial wavepacket are both purely real. We may reasonably suppose that this long-time constraint, associated with the statistics of very long excursions away from the periodic orbit, is independent of the short-time behavior near the periodic orbit. Thus, if we suppose that what we in the previous section called stage two of the dynamics does not know about the trajectory's approach to the periodic orbit at times $|t| \rightarrow \infty$, then the same long-time constraints should be present in the case of scarring. So we finally obtain a discrete spectrum, with intensities given by a χ^2 variable of 2 degrees of freedom (1 degree of freedom for purely real f_{an}), except that in the case of scarring this RMT line spectrum multiplies the original short-time envelope. The line height associated with energy E_n is given by

$$|f_{an}|^2 = \frac{1}{N} S_{\text{lin}}(E_n)|r_n^2|\,, \tag{24}$$

where r_n is a Gaussian variable, real or complex, with variance one. In particular, the first two moments of this distribution are given by

$$< |f_{an}|^2 > = \frac{1}{N} S_{\text{lin}}(E_n) \quad < |f_{an}|^4 > = \frac{F}{N^2} S_{\text{lin}}^2(E_n)\,, \tag{25}$$

where $F = 2$ or 3 as explained above. We may note here that the linear spectrum $S_{\text{lin}}(E)$ is purely real because of the unitarity of the time evolution, $A(-t) = A^*(t)$. This holds also for the full spectrum $S(E)$, and for any smoothed spectrum obtained by fourier transforming $A(t)$ for all times $|t| < T_{\text{max}}$.

The inverse participation ratio for the wavepacket is given by

$$
\begin{aligned}
N < \sum_n |f_{an}|^4 > &= F \int dE\ S_{\text{lin}}^2(E) \\
&= F \sum_T |A_{\text{lin}}(T)|^2 \\
&= F \sum_T [\cosh(\lambda T)]^{-1} \\
&\to cF/\lambda \quad \text{for small } \lambda\,.
\end{aligned}
\tag{26}
$$

Note that the IPR enhancement factor is a number that depends only on the Lyapunov exponent of the periodic orbit. Higher moments of the f_{an} distribution can be computed easily in terms of $A_{\text{lin}}(T)$, always taking into account the proper quantum fluctuation factors.

If the spectrum $S(E)$ is ensemble-averaged while preserving the periodic orbit (*i.e.* if we average over the nonlinear dynamics only), we recover the original linear envelope spectrum, the same spectrum which is present in the case of an open system.

Finally we need to mention here an issue that has been the cause of some misunderstanding in the literature, namely the issue of spatial symmetries. In symmetric systems like the stadium billiard, many periodic orbits either remain invariant under a reflection operation, or they get shifted by some fraction of the total period, or else they get mapped to their time-reversed counterparts. In all these cases, one expects extra correlations in the contributions coming from the homoclinic excursions which are related by such a symmetry operation. Thus, consider the simplest case of a parity symmetry, where the initial wavepacket centered on the periodic orbit remains unchanged under the symmetry. Of course we know from quantum mechanics that the wavepacket can overlap only with states in the even part of the Hilbert space, so the IPR is increased by a factor of two from the value that would otherwise be expected. We can also see this effect semiclassically, because each homoclinic excursion away from the invariant periodic orbit has a counterpart related by parity, and the contributions from the two have the same amplitude and phase, so they add constructively. More complicated situations can be treated similarly, and we will not go into the details here. In the example described in the following Section, we have intentionally desymmetrized the system of interest to eliminate the enhanced scarring effects.

MODEL SYSTEM: GENERALIZED BAKER'S MAPS

As a testing ground for our predictions about nonlinear scarring, we will use the generalized baker's maps, a class of bernoulli systems which are a paradigm of hard chaotic behavior. In addition to having no stable regions of phase space, these systems have the property that the long-time semiclassical dynamics and the semiclassical eigenstates can be computed efficiently (*i.e.* in a time that scales as a power law, rather than exponentially in $1/\hbar)^{22}$. This is useful for studying a phenomenon such as scarring, where predictions about the system are made based on our expectations about the statistical properties of the semiclassical behavior. These predictions, obtained in the previous sections, can then be independently compared with the exact semiclassical and the full quantum results, allowing us to distinguish errors in the statistical arguments from errors inherent in the semiclassical approximation itself. As we will see in Section , semiclassically computed measures of scarring agree only roughly with the results of a full quantum computation for any particular periodic orbit in a given realization of a chaotic quantum system. The fluctuations of the semiclassically computed scarring strengths (when considering an ensemble of orbits with a fixed Lyapunov exponent) are however virtually identical (in mean and variance) to those of the quantum scarring strengths. Any anomalous scarring for a given orbit must be attributed to correlations in new long-time recurrences near the Heisenberg time, such correlations being present in both the semiclassical and full quantum computations.

We will discuss briefly the definition and properties of the classical, semiclassical, and quantum generalized baker's map, referring the reader to the literature on the subject for more details [23, 24]. Classically, the map is defined as a map of the unit square onto itself, where the square is initially cut up into M vertical strips with widths w_m ($\sum_{m=0}^{M-1} w_m = 1$) and height 1. (In the original baker's map, $M = 2$ and the two strips each have width $1/2$. This leads to a constant Lyapunov exponent of $\log(2)$ everywhere in phase space. In the generalized version of these maps, this restriction is lifted, allowing different periodic orbits of the same length to have different instability factors.) Each strip is stretched horizontally and compressed vertically, in an area-preserving way, so that the width becomes 1 and the height becomes w_m. The deformed horizontal strips are finally stacked on top of each other, in some order (conventionally, the left-to-right order of the initial arrangement corresponds to the bottom-to-top order of the final one). If we define $s_m = \sum_{j<m} w_j$ to be the left edge of the m-th strip, we have

$$x' = (x - s_m)/w_m$$
$$p' = w_m p + s_m \tag{27}$$

for x in the m-th strip, $s_m \le x < s_{m+1}$. Trajectories and periodic orbits can be labeled symbolically by strings of digits, each between 0 and $M-1$, the T-th digit indicating the horizontal strip in which the particle is found at time T. A simple nontrivial example is the $\ldots 11111 \ldots$ orbit for the case $M = 3$. The instability exponent of this orbit is given by $|\log(w_1)|$, and homoclinic orbits of various instabilities and phases can be constructed (*e.g.* $\ldots 111112001202011111 \ldots$). This is in fact the periodic orbit that we will use for the numerical data on scarring in Section .

A quantum mechanical version of this system can be obtained readily for any value of Planck's constant given by $h = 1/N$. N is the dimension of the resulting Hilbert space. The position basis consists of states $|j\rangle$, $j = 0 \ldots N - 1$, corresponding semiclassically to vertical strips located at $x_j = (j + \epsilon_1)/N$. Similarly, the momentum basis is formed by $|\tilde{k}\rangle$, $\tilde{k} = 0 \ldots N - 1$, with the states living at $p_{\tilde{k}} = (\tilde{k} + \epsilon_2)/N$. The two bases are related by a discrete fourier transform. The numbers $\epsilon_{1,2} \in [0,1)$ are constants which provide the quantization conditions for the system (they define the phases associated with going around the torus in the vertical and horizontal directions, respectively).

To define the baker's map dynamics we write N as a sum of integers $N = \sum_{m=0}^{M-1} N_m$, approximating the classical division of phase space into strips ($|N_m - w_m N| <$ 1). Then the leftmost N_0 position states are mapped into the bottom N_0 momentum states by a discrete fourier transform, and similarly for each of the other strips. Finally, we transform back to the position basis. So the one-step evolution operator in the position basis has the form

$$U = \left[F_N^{-1} \right] \begin{bmatrix} e^{i\theta_0} F_{N_0} & 0 & \cdots & 0 \\ 0 & e^{i\theta_1} F_{N_1} & \cdots & 0 \\ \vdots & \vdots & \ddots & \vdots \\ 0 & 0 & \cdots & e^{i\theta_{M-1}} F_{N_{M-1}} \end{bmatrix}, \tag{28}$$

where F_N is a discrete fourier transform matrix on N sites, and the θ_m are arbitrary angles. Different choices of θ_m correspond to different semiclassical theories, all having the same classical limit. The statistical properties of the quantum mechanical system should be independent of the choice of quantization parameters $\epsilon_{1,2}$ and θ_m, so these can be randomly chosen in the context of an ensemble averaging.

The semiclassical one-step propagator can be easily written down in the position basis in terms of the stretching factors w_m and phases θ_m, making use of the symbolic dynamics governing this system.

$$U_{\rm SC}(x', x) = \sum_m \sqrt{w_m} e^{i s_m x'/\hbar + i\theta_m} \delta(x - (s_m + w_m x')). \tag{29}$$

The exact semiclassical $T-$step propagator is obtained by iterating the one-step formula T times (this is permitted as long as we do not impose the quantization condition by forcing x at intermediate times to have one of the N quantum mechanically allowed values). Long-time overlaps of Gaussian wavepackets, as well as semiclassical spectra and eigenstates can be computed in an efficient manner (for example, by fourier transforming to momentum variables and cutting off the high-frequency modes)[22]. These calculations generally show good agreement with the quantum results, although the convergence is not uniform over phase space [25]. In particular, the semiclassical approximation does not see the "diffractive" effects associated with the boundaries between classical strips.

To illustrate the effect of isolated returning orbits as discussed in Section VII of Kaplan and Heller [18], we construct a modified version of the system described above. A rectangle is divided into $2L+1$ strips numbered $-L \ldots +L$. The strips $-L \ldots -1$ all have the same width, and similarly for strips numbered $+1 \ldots +L$. Now the three vertical strips $-L$, 0, and $+L$ get mapped into the horizontal region spanned by strips -1, 0, and $+1$, via the usual generalized baker's map dynamics for $M = 3$ described in the above paragraphs. Simultaneously strip i for $1 < i < L-1$ gets shifted to the right into strip $i+1$, and similarly $-i$ is shifted to the left into $-(i+1)$. In effect, we have two long corridors to the left and right of the hyperbolic region, each of which must be traversed in its entirety before one returns to the center, which is where all of the mixing occurs. This slows down the total mixing in phase space because each excursion away from strip 0 takes at least L steps. An effective symbolic dynamics using the three symbols '0', '$-$', and '$+$' can be used to represent the trajectories, where '$+$'$= +1 + 2 \ldots + L$, and similarly for '$-$'. Then for the periodic orbit $\ldots 0000 \ldots$, the shortest homoclinic excursions involve L steps, and are symbolically represented as $\ldots 000 + 000 \ldots$ and $\ldots 000 - 000 \ldots$. On the other hand, the Lyapunov exponent of the original orbit (and therefore the decay time of the linear dynamics) can be arranged to be of order unity, creating a clean separation of scales between the two effects. Numerical results for this system will be presented towards the end of Section .

NUMERICAL RESULTS

We now proceed to examine the numerical evidence for scarring in the three-strip generalized baker's map described in the previous section. We concentrate on the period one orbit $\ldots 11111 \ldots$, for which the particle stays always in the middle strip. The location of the fixed point is given by $x_{\text{FP}} = p_{\text{FP}} = w_0/(w_0+w_2)$, and the Lyapunov exponent is $\lambda = |\log w_1|$. An ensemble average can be performed over different strip widths w_0, w_1, and w_2, and over the value of Planck's constant $\hbar = 1/2\pi N$. The integers N_m, describing the dimensions of the subspaces corresponding to the three strips, are chosen to be prime to eliminate possible sources of anomalous behavior. (In the original baker's map, such anomalous behavior is associated with values of N which are divisible by powers of 2; this was discussed by us earlier [22].) Also, we demand $N_0 \neq N_2$ to eliminate the parity symmetry. This parity symmetry, also present in the original baker's map and given by $P : x \to 1 - x$, $p \to 1 - p$, would otherwise produce a factor of 2 enhancement in the inverse participation ratio for the $\ldots 11111 \ldots$ orbit, as explained in the concluding paragraph of Section .

For each of 97 realizations of this system, the inverse participation ratio was computed for a circular wavepacket centered on the periodic point. The Lyapunov exponent for the orbit ranged from a low value of 0.28 (corresponding to a linear IPR enhancement factor of 10.4 according to Eq. 26, and not including the quantum fluctuation factor F), to a high value of 1.94, corresponding to a linear prediction of 1.66 for the IPR. The values of N in this ensemble lie between 129 and 419. In Figure 4, the actual quantum value of the IPR is plotted (using squares) *versus* the linear prediction on the horizontal axis. The data is compared to a line of slope 2.2, corresponding to the quantum fluctuation factor F appropriate to this system. F is obtained here by measuring the mean IPR for a wavepacket placed randomly in the phase space instead of on a periodic point. We see large fluctuations in this Figure around the predicted behavior, but the overall linear trend is certainly correct. Fluctuations in the amount of scarring at a given value of the Lyapunov exponent are of the order of twenty percent around the mean. This level of variation in scarring strength from orbit to orbit is quite reasonable taking into account the fact that the effective number of spectral lines under the linear envelope is well under 100 in most cases studied. (This effective dimension of the space in which the wavepacket lives is given by the total dimension N divided by the IPR of the linear envelope.)

For a subset of 46 out of this set of 97 systems, the semiclassical IPR was also computed (this is defined by the fluctuation of the *right* eigenstates of the time evolution). These values are shown on the same plot in Figure 4 using the '+' symbol. The semiclassical IPR is correlated with but does not exactly follow the full quantum value. Deviations from the predictions of the random statistical theory are present in the semiclassical as well as in the quantum calculation. This suggests that cases of excessive (or deficient) scarring must be associated with non-randomness in the new long-time recurrences, and not simply with a breakdown of the semiclassical approximation. Given the small number of parameters defining the system, and the finite dimension N of the Hilbert space, such occasional anomalous behavior is not very surprising.

On the same Figure, we also plot (using triangles) the results for a modified version of the generic baker's map, where random matrices have been substituted for the matrices F_{N_0} and F_{N_2} describing the dynamics of the left and right strips (while leaving the behavior of the middle strip unchanged). This corresponds to explicitly randomizing the long-time recurrences, while preserving the constraint imposed by the short-time dynamics in the region of the fixed point. The full IPR in this case fol-

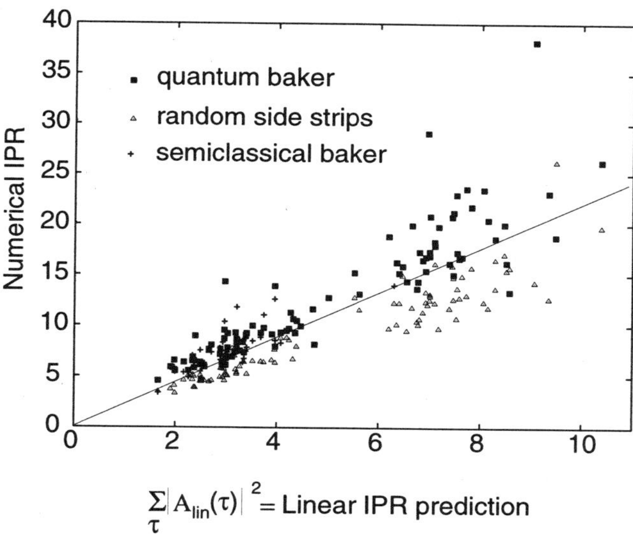

Figure 4. A plot of the actual value of the inverse participation ratio (IPR) for a wavepacket centered on a periodic orbit (squares), *versus* the value predicted by the linear theory. IPR's in the semiclassical approximation are plotted using the '+' symbol, and IPR's in a baker's map with random matrix theory nonlinear behavior are plotted with triangles.

lows the same linear dependence on the short-time envelope prediction, but with a clearly smaller numerical coefficient F. This is consistent with the finding that the generic wavepacket-averaged (non-scarred) value of the IPR for this system is smaller (2.0 compared with 2.2 for the true baker's map).

In Figure 5, the ratio between the actual IPR and the linearly predicted value is plotted (squares) as a function of the effective system size $N_{\mathrm{eff}} = N/\mathrm{IPR}_{\mathrm{lin}}$. We find significant fluctuations around the mean value of 2.5, with no evident trend as N_{eff} increases. In a random matrix theory model, fluctuations around the mean value of the IPR scale as $1/\sqrt{N_{\mathrm{eff}}}$, and this behavior is not inconsistent with the data, although the range of N used is not sufficient to see the convergence. On the same Figure, the corresponding ratio is plotted (using triangles) for the "randomized baker's map" described in the previous paragraph. Here the fluctuations around the mean are only slightly smaller than for the real baker's map (their magnitude being about 15 rather than 20 percent). The mean value of the full-to-linear-IPR ratio is also smaller, as explained in the previous paragraph. For the explicitly randomized system, the only correlations in the new long-time recurrences (after having taken into account the short-time constraint) are due to the finiteness of the Hilbert space, and are expected to become insignificant in the $\hbar \to \infty$ limit.

So far we have been focusing on the second moment of the spectral intensity distribution, for various realizations of the generalized baker's map system. We now turn to a specific system, for which we will be able to look at the entire spectrum and see explicitly the pattern of fluctuations around the linear envelope. We choose the first entry in our data file, with $N = 223$ and three strips of widths $N_0 = 79$, $N_1 = 101$, and $N_2 = 43$. The Lyapunov exponent for the $\ldots 11111 \ldots$ orbit is 0.79, producing by Eq. 26 a linear envelope with an IPR of 3.97. The actual quantum value of the IPR is

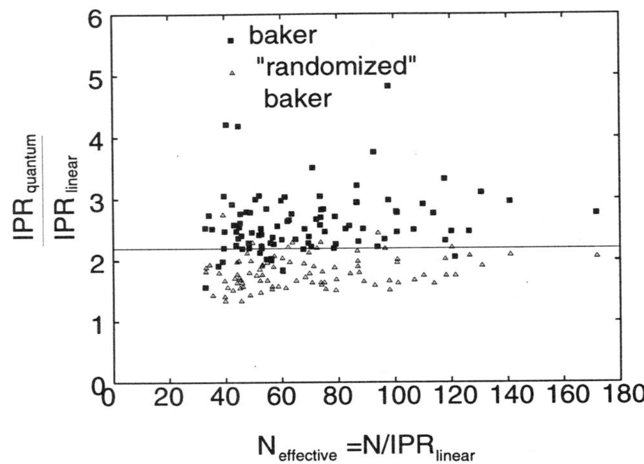

Figure 5. Here the ratio of the full IPR to the value predicted by the linear theory is plotted against the effective number of states that the wavepacket can overlap with (according to the linear theory). Using triangles, the same quantity is plotted for a randomized baker's map.

7.93, so there is a Heisenberg-time fluctuation factor of 2.0 (compared to the expected value 2.2). The semiclassically computed value for the IPR is somewhat larger, at 8.46.

The intensity spectrum for a circular wavepacket centered on the fixed point is plotted in Figure 6a, along with the linear envelope $S_{\text{lin}}(E)$ and an intermediate envelope obtained by fourier transforming $A(T)$ for $|T| < 30$. The intermediate envelope corresponds to a time scale large compared to the linear decay time and mixing time of the system, but still small compared with the Heisenberg time. A semiclassical version of this intermediate envelope is also plotted and is seen to be very similar to the full quantum computation. In Figure 6, a portion of the line spectrum is compared with the same spectrum computed semiclassically. (Again, the semiclassical spectrum is defined by taking the overlaps between the wavepacket and the right eigenstates of the semiclassical evolution. The imaginary parts of the semiclassical energies are ignored.) In Figure 6c, the quantum spectrum of Figure 6a is now plotted after dividing out by the linear envelope. We see that the underlying oscillations around the envelope are in fact energy independent, and are equally strong near the peak and valley of the envelope.

In Figure 7, a histogram of these scaled intensities is plotted, after averaging over several realizations of the system. This is compared to a similar histogram of the unscaled intensities. It is seen that the former comes much closer to obeying an exponential law as predicted by Porter-Thomas.

In Figure 8, the spectral correlation function $< S(\epsilon + E)S(\epsilon) > / < S(\epsilon) >^2$ is plotted in the form of a histogram (plusses), and compared to the correlation function for a scaled spectrum (diamonds). An ensemble average and an average over the energy ϵ has been performed in each case. We can see that the correlation function for the scaled spectrum is uniform with small random fluctuations (we ignore the large correlations for E of the order of a mean level spacing, $E = O(1/N)$). On the other hand, correlations in the unscaled spectrum are very striking, and sharply peaked near $E = 0$.

In Figure 9, a spectrum with a linear and a second-order envelope is plotted for

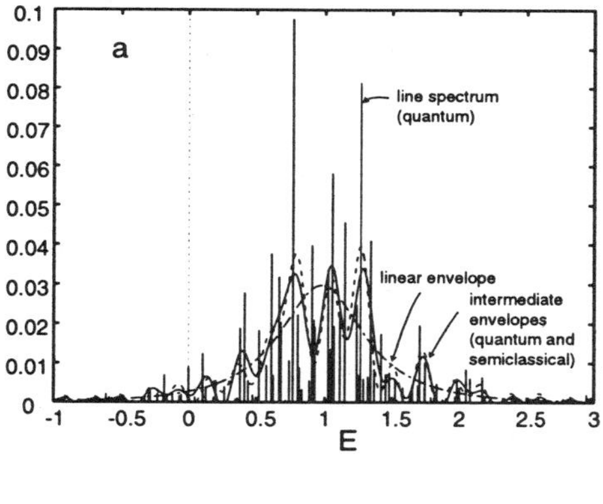

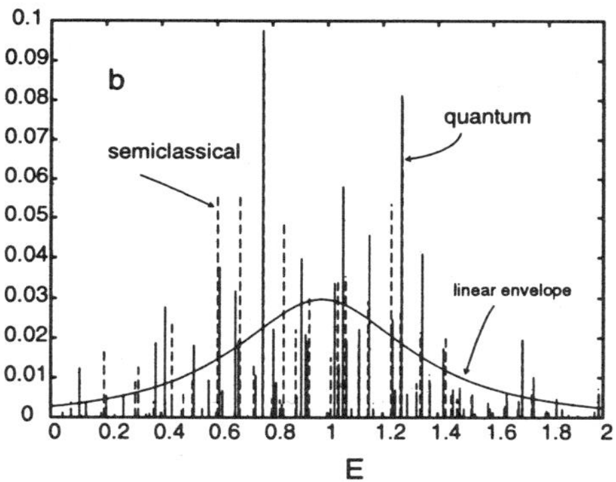

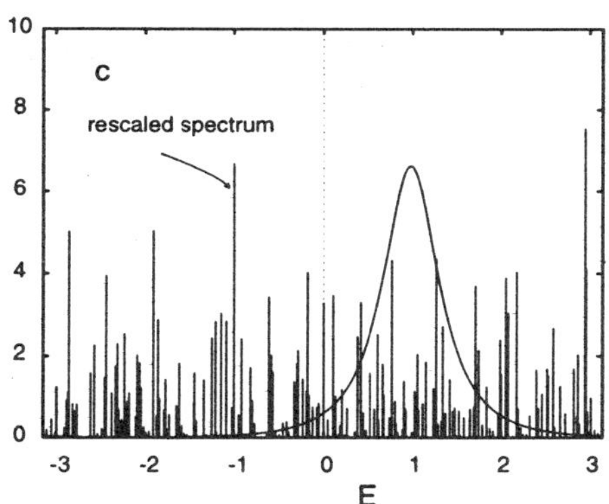

Figure 6. In (a), the full spectrum is plotted along with the linear envelope (dotted line), an intermediate envelope corresponding to $|T| < 30$ (solid line), and a semiclassical intermediate envelope (dashed line). In (b), a portion of the spectrum (solid) is compared to the spectrum obtained using semiclassical eigenstates (dashed). In (c), the quantum spectrum has been divided out by the linear envelope of (a).

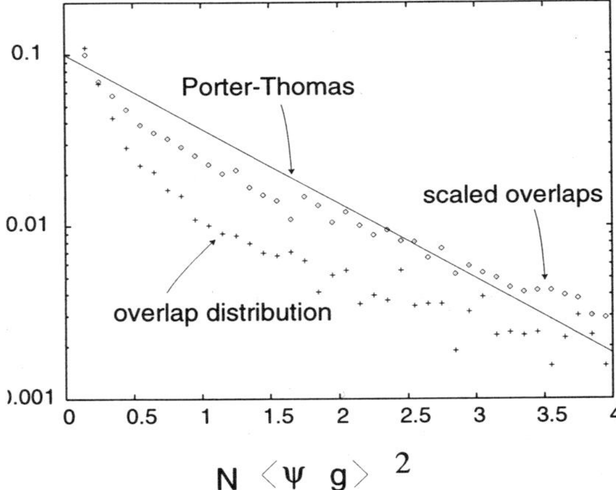

Figure 7. A histogram of scaled spectral intensities (diamonds) after having divided out by the linear envelope, compared to a histogram of raw (unscaled) intensities (plusses). The Porter-Thomas exponential law is plotted as a solid line. All distributions are defined to have a mean value of one.

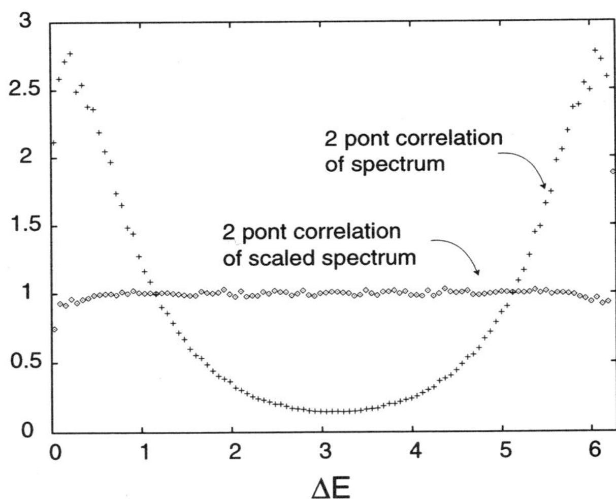

Figure 8. The two-point spectral correlation function of the scaled spectrum (diamonds) is compared to the correlation function of the raw spectrum (plusses), after ensemble and energy averaging.

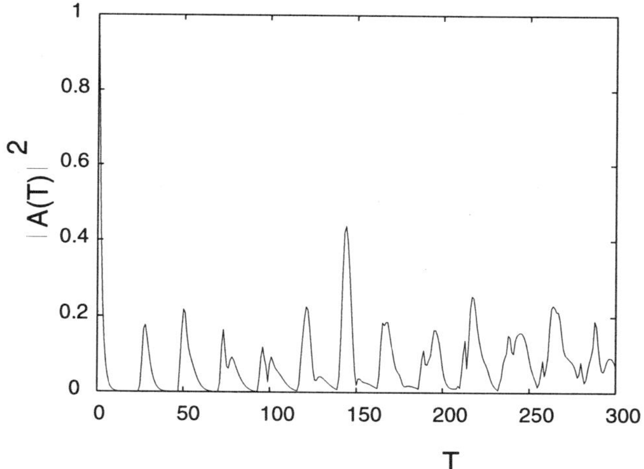

Figure 9. The quantum return probability as a function of time for the modified baker's map with a minimum excursion length of $L = 23$.

a system with isolated homoclinic recurrences, associated with excursions away from the mixing region as discussed at the end Section . We choose a minimum excursion length $L = 23$, with strip 0 having width $N_0 = 17$ and all the others being of width $N_{\pm 1} = \ldots = N_{\pm 23} = 19$. A gaussian wavepacket is launched at the center of the middle strip. The initial linear decay occurs within $1 - 2$ time steps (the Lyapunov exponent at the fixed point being 1.17), but the initial recurrence, corresponding to the homoclinic orbit $\ldots 0000 + 0000 \ldots$ and its parity counterpart $\ldots 0000 - 0000 \ldots$ does not peak until $T = 28$. The next set of recurrences results from the classical orbits $\ldots 0000 + 0^n + 0000 \ldots$ and $\ldots 0000 + 0^n - 0000 \ldots$, and their parity counterparts (n being a nonnegative integer). In all, 12 identifiable peaks with spacing ~ 23 can be identified, the structure not breaking down completely until $T \sim 300$, at which time random mixing begins to take over (classically, homoclinic excursions with more and more insertions of the '0' symbol are becoming important here for entropic reasons). The squared autocorrelation function for the wavepacket is plotted in Figure 9.

In Figure 10, the corresponding energy spectrum is plotted along with a linear envelope (dotted line) and an envelope resulting from the first two sets of recurrences (up to two insertions of '+' or '−'), dashed line. The solid curve includes the effects of the first six sets of homoclinic returns.

WAVEFUNCTION INTENSITY STATISTICS

The nonlinear scarring theory as presented above can be used to study the tail of the wavefunction intensity distribution in chaotic systems[26]. Predictions, validated by numerical experiments, have been obtained for the distribution of eigenstate intensities in a single region of phase space, for phase space-averaged distributions, and also for ensembles which include systems with orbits of different lengths and instability exponents. Power-law tails are naturally obtained in the process of ensemble averaging, under certain assumptions, in sharp contrast with the exponential behavior predicted by RMT.

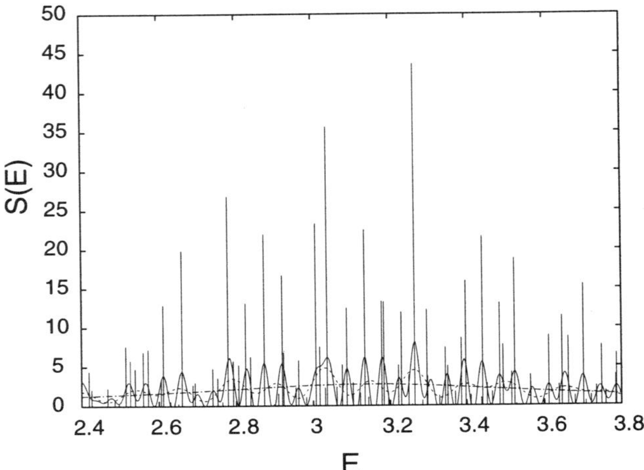

Figure 10. The spectrum corresponding to the wavepacket whose return probability is plotted in the previous Figure, along with a linear envelope (dotted line), an intermediate envelope corresponding to the first two sets of homoclinic recurrences (dashed line), and higher resolution envelope corresponding to the first six sets of recurrences (solid line).

The gaussian wavepacket of Eq. 2 is generalized to be centered at an arbitrary point (q_0, p_0) in phase space, relative to a fixed point of the dynamics, which is still located at the origin. The form is then given by

$$\Psi(q) = \left(4\pi\sigma_p^2\right)^{1/4} \exp\left[-(q - q_0)^2/2\sigma^2 + ip_0(q - q_0)/\hbar\right] \tag{30}$$

We allow this wavepacket to evolve under the linearized dynamics, stretching each time step by a factor e^λ in the q−direction and shrinking by the same factor in the p−direction. The linearized quantum autocorrelation function after time t is given by

$$A_{\rm lin}(q, p, \sigma, \lambda, t) = \frac{\exp i\theta_0 t}{\cosh \lambda t}$$
$$\times \exp\left(-\frac{\cosh \lambda t - 1}{2\cosh \lambda t}\left(\frac{q_0^2}{\sigma^2} + \frac{p_0^2}{\sigma_p^2}\right) - \frac{iqp}{\hbar}\tanh \lambda t\right). \tag{31}$$

Here θ_0 is the phase associated with one iteration of the periodic orbit, given by the classical action in units of $\hbar$, plus any Maslov indices associated with caustics in the classical dynamics. θ_0 determines the location of the peak in the spectral envelope $S_{\rm lin}(E)$, defined to be the fourier transform of $A_{\rm lin}(t)$. Since we will be interested in performing an energy average, we freely set $\theta_0 = 0$.

Now we assume individual spectral lines show chi-squared (Porter-Thomas) fluctuations around the linear envelope. In the case of complex eigenstates, the chi-squared distribution has two degrees of freedom, and in the absence of scarring the probability of having a spectral line height greater than x is given by $P(x) = \exp(-x)$. Here x is normalized to have a mean value of unity, i.e. $x_n = N|\langle n|\Psi\rangle|^2$, where N is the total number of states. Now in the presence of scarring this is modified to

$$P(q, p, \sigma, \lambda, E, x) = \exp\left(-x/S_{\rm lin}(q, p, \sigma, \lambda, E)\right) \tag{32}$$

for a eigenstate with energy E. Now we perform an energy average, remembering that for a map the quasi-energy is defined to lie between 0 and 2π only. We also notice that the tail of the distribution will be dominated by the peak of the spectral envelope at $E = 0$, and we therefore can use a saddle point approximation, obtaining

$$P(q, p, \sigma, \lambda, x) = \frac{1}{2\pi} \int dE\, P(q, p, \sigma, \lambda, E, x)$$
$$\approx \frac{1}{\sqrt{2\pi}} \frac{1}{\sqrt{\frac{-x}{S_{\text{lin}}(0)^2} \frac{\partial^2 S_{\text{lin}}}{\partial E^2}}}\,, \tag{33}$$

where the expression obtained is an asymptotic form valid for large x. For small λ, the sum over time steps can be replaced by an integral, and we have at $q_0 = p_0 = 0$

$$S_{\text{lin}}(E) = \int dt \frac{e^{-iEt}}{\sqrt{\cosh \lambda t}}\,. \tag{34}$$

Now by dimensional analysis, $S_{\text{lin}}(0) = Q/\lambda$ and $\frac{\partial^2 S_{\text{lin}}}{\partial E^2}(0) = -W/\lambda^3$, where Q and W are numerical constants. We thus obtain the first result of this paper, the tail of the intensity distribution for a wavepacket centered on a periodic orbit,

$$P(q_0 = 0, p_0 = 0, \sigma, \lambda, x) = \frac{1}{\sqrt{2\pi}} \frac{Q}{\sqrt{W}} \lambda(x\lambda)^{-1/2} e^{-x\lambda/Q}\,. \tag{35}$$

Notice that the exponential tail has been effectively stretched by a factor of Q/λ, corresponding to the increased height of the peak of the linear envelope at small λ. There is also a linear suppression factor of λ, corresponding to the width of the peak in $S_{\text{lin}}(E)$, and indicating that only a fraction scaling as λ of all the eigenstates are effectively scarred. The remainder of the eigenstates are typically "antiscarred", having on average a lower intensity at the periodic orbit than would be expected based on RMT. This distribution will have a nontrivial inverse participation ratio (IPR), scaling as the inverse of the width of the linear envelope, *i.e.* as $1/\lambda$.

The region of validity of Eq. 35 is

$$1 \ll \lambda^{-1} \ll x \ll N\,. \tag{36}$$

The first inequality ensures that many iterations of the periodic orbit contribute (so the sum over iterations can be replaced by an integral) and the scarring is strong. In fact, however, because of the large value of the numerical constant Q, the formula works well even for exponents as small as $\log 2$, as will be seen in the numerical study below. The second inequality says that we are in the tail of the distribution and the events are all coming from the peak of the linear envelope. The third inequality is a unitarity constraint – obviously our assumption of random fluctuations breaks down for intensities of order unity, when the entire wavefunction would be concentrated in a phase space area of order $\hbar$.

Now we go on to perform a similar analysis integrating over the phase space variables q_0 and p_0. As before, we take the exponential $\exp\left(-x/S_{\text{lin}}(q_0, p_0, \ldots)\right)$ and expand to second order in q_0, p_0 around the maximum $q_0 = p_0 = 0$. Then upon integration by stationary phase we pick up a determinant factor of

$$\frac{2\pi S_{\text{lin}}(0)^2}{x} \frac{1}{\sqrt{\frac{\partial^2 S_{\text{lin}}}{\partial q_0^2} \frac{\partial^2 S_{\text{lin}}}{\partial p_0^2}}}\,. \tag{37}$$

Here we have taken the classical phase space volume to be unity for simplicity. Now for small λ,

$$\frac{\partial^2 S_{\text{lin}}}{\partial q_0^2}(0) = \frac{2}{\sigma^2}\frac{-Z}{\lambda},\tag{38}$$

where Z is yet another numerical constant, and similarly for p_0, with σ replaced by $\sigma_p = \hbar/\sigma$. So the total factor resulting from the phase space integration is $\frac{\pi\hbar}{(x\lambda)}\frac{Q^2}{Z}$, again independent of σ. Combining this with the expression in Eq. 35 above, we obtain the second result, for the distribution of overlap intensities after energy and phase space averaging,

$$P(\lambda, x) = \sqrt{\frac{\pi}{2}}\frac{Q^3}{Z\sqrt{W}}\hbar\lambda(x\lambda)^{-3/2}\exp\left(-x\lambda/Q\right).\tag{39}$$

Here we have picked up a factor of $\hbar$ from the factor of σ in Eq. 38 and the corresponding factor of $\sigma_p = \hbar/\sigma$ associated with the falloff in S_{lin} in the momentum direction. This indicates that the tail is coming entirely from the region near the periodic orbit, specifically from wavepackets that have large classical probability density right on the orbit. (Thus, a measure like the IPR for a generically placed wavepacket will not see the effect of scarring by an individual periodic orbit, when the semiclassical limit has been taken.) The result in Eq. 39 is valid in the regime

$$\max(\log N, \lambda^{-1}) \ll x \ll N.\tag{40}$$

Here $\log N$ is the value of x near which the RMT exponential decay law reaches values of order $\hbar = 1/2\pi N$. In this region, a crossover occurs between the head of the distribution, which is dominated by non-scarred region of phase space and approaches the Porter-Thomas (RMT) prediction, and the tail, dominated by scarring, given by the expression above. The expression Eq. 39 holds also for an ensemble of systems, all having an orbit with instability λ. In principle, we should of course do a sum over all periodic orbits, however the tail will clearly always be dominated by the orbit with smallest λ.

Finally, we now consider an ensemble of systems where the value of the smallest exponent varies from system to system, with distribution $\mathcal{P}(\lambda) = C\lambda^\alpha$ for small λ. Then using Eq. 39 and integrating over λ we obtain

$$P(x) = C\sqrt{\frac{\pi}{2}}\frac{Q^3}{Z\sqrt{W}}Q^{\alpha+1/2}\Gamma(\alpha + 1/2)\hbar x^{-(2+\alpha)}.\tag{41}$$

Note that this is an uncontrolled approximation because we have integrated over λ after having assumed $x\lambda$ was large. However, if we had included higher-order corrections in $(x\lambda)^{-1}$ in Eq. 39, the scaling of $P(x)$ would remain unchanged, i.e.

$$P(x) = Cf(\alpha)\hbar x^{-(2+\alpha)},\tag{42}$$

with the dimensionless function $f(\alpha)$ somewhat different from that given in Eq. 41. An important point is that the tail displays power-law behavior in the intensity x, a strong deviation from the exponential prediction of RMT. As with Eq. 39, this asymptotic form is valid for values of x large compared to $\log N$ and small compared to N. For small x we again expect a crossover to the Porter-Thomas form. For large x we expect a downward correction away from the $x^{-(2+\alpha)}$ form, with a breakdown of the approximation occurring at some fraction of N, depending on α.

Now, we proceed to test numerically these predictions of the nonlinear scarring theory. The system we use for this purpose is the generalized three-strip baker's

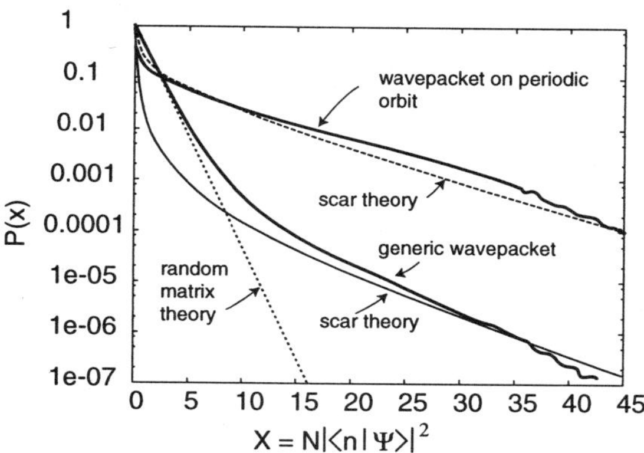

Figure 11. Cumulative wavefunction intensity distribution (a) as measured by a test state centered on a periodic orbit with instability $\lambda = \log 2$, plotted as the upper thick curve with scarring theory prediction given by dashed curve, and (b) averaged over the entire phase space of size $200h$, plotted as lower thick curve with theory given by solid curve. The dotted line is the Porter-Thomas law.

map, described in some detail in Section above. The three widths are normalized to $\sum_{i=0}^{2} w_i = 1$. There is a fixed point of the classical dynamics associated with the middle strip, and the instability associated with this orbit is given by $\lambda = |\log w_1|$. So we choose $w_1 = 1/2$, set $N = 200$, and find numerically the wavefunction intensity distribution at the fixed point after ensemble averaging over the widths $w_{0,2}$. (The predictions are expected to hold for individual systems as well, at sufficiently large values of N. However, for the matrices which we can efficiently diagonalize, N is not large enough to obtain good statistics in the tail without ensemble averaging.) A circular wavepacket of width $\sigma = \sqrt{\hbar}$ is used. The results are compared in Fig. 11 (upper thick curve) with the prediction of Eq. 35, plotted as a dashed curve. The RMT prediction is shown as a dotted line.

Next, we perform a phase space average for the systems described above, collecting statistics for wavepackets uniformly distributed over the entire phase space. The resulting statistics are also plotted in Fig. 11 (lower thick curve), where the theoretical prediction for the tail, given by Eq. 39, is shown as a solid curve. Again, the Porter-Thomas distribution appears as a dotted line. We see a crossover between the two regimes at a value of x of order $\log N$.

Finally, we want to construct an ensemble which will contain systems with orbits of different instability exponents λ. For this purpose, we take a uniform distribution of strip widths w_0 and w_2, each in the range $[0, 1/4]$. The fixed point in the middle strip, with exponent $\lambda = |\log(1 - w_0 - w_2)|$, is always the least unstable periodic orbit. This ensemble has power $\alpha = 1$ and $C = 4^2 = 16$ in the notation of Eq. 41. Averaging over 100 systems, we obtain the statistics plotted in Fig. 12. The power-law prediction given by Eq. 41 is plotted as a solid line on the log-log scale, with the RMT prediction as a dotted curve. Once again, we see a crossover between the two regimes for x of order $\log N \approx 5.3$. We also see a gradual breakdown of the approximation as x approaches values comparable to $N = 200$. Note that the quantitative agreement is in spite of

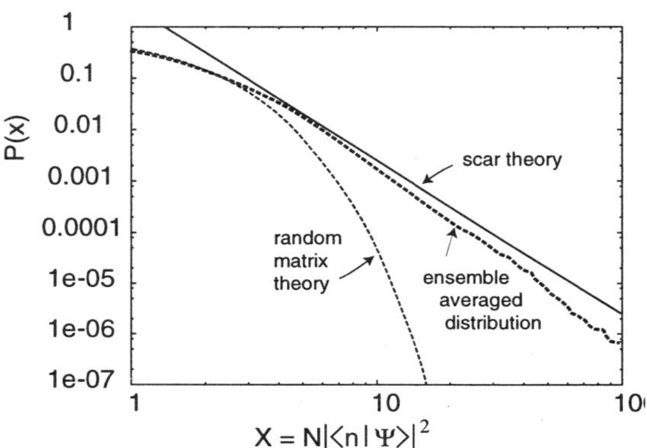

Figure 12. Cumulative wavefunction intensity distribution after ensemble averaging over systems with classical orbits of different instability exponents. Here again $N = 200$, and the dotted curve is the RMT prediction.

the fact that an uncontrolled stationary phase approximation was used in obtaining the overall constant in front of Eq. 41, as explained above. The important thing to notice here is the power-law behavior of the tail, in agreement with the theory, and the dramatic deviation from the predictions of RMT. By $x = 100$, where the approximation $x \ll N$ is clearly beginning to break down, the measured probability is still within a factor of 4 of our prediction and is enhanced by 10^{37} over the Porter-Thomas value.

We have also checked the linear $\hbar$-dependance of the phase-space averaged results Eqs. 39,41 by repeating the preceding numerical analysis with larger matrices ($N = 500, 1000$). In addition, we have constructed an $\alpha = 0$ ensemble by imposing the restriction $w_0 = w_2$ and have observed a x^{-2} power-law behavior in accordance with Eq. 41.

CONCLUSION

By focusing on the importance of nonlinear recurrences, long-time fluctuations, symmetry factors, and the local classical structure around a periodic orbit, we have attempted to clear up some of the long-standing mysteries in the literature on scarring. We have seen that the linear theory, even in a worst-case "egalitarian" scenario, makes a lower bound prediction on scarring strength that is a function of the instability of the orbit only, and independent of energy and $\hbar$. Generically, we expect random long-time fluctuations to be present, associated with nonlinear excursions away from the periodic orbit. When these gaussian random fluctuations are included in the theory, quantitative agreement between theory and numerics is obtained using measures such as the inverse participation ratio, wavefunction intensity distribution, and correlations in the local density of states. Scarring stronger than that predicted by the random nonlinear theory can be obtained in the presence of identifiable homoclinic recurrences.

The formalism developed here lends itself naturally to the investigation of other effects of short-time dynamics on the properties of quantum eigenstates. Such short-time behavior may involve classical structures other than periodic orbits, diffraction, and "quiet time" behavior. (For an explanation of this last concept, see Section VII of Kaplan and Heller [18].) Intensity correlations among eigenstates as well as phase space correlations in the structure of individual eigenstates can be studied. Interesting questions arise in the design of "optimal" measures for observing this class of deviations from random matrix theory behavior (see Section VIII of Kaplan and Heller [18]).

ACKNOWLEDGMENTS

This research was supported by the National Science Foundation under grant number CHE-9321260. We wish to thank the Institute for Theoretical Physics at UCSB, where this research was initiated, for its hospitality. We also wish to thank the Isaac Newton Institute for the Mathematical Sciences in Cambridge, where this work was completed. Important early conversations with S. Tomsovic and further discussions with S. Zelditch are gratefully acknowledged.

REFERENCES

1.   O. Bohigas, M.-J. Giannoni, and C. Schmit, *J. Physique Lett.* **45**, L-1015 (1984).
2.   M. V. Berry, in *Chaotic Behaviour of Deterministic Systems*, ed. by G. Iooss, R. Helleman, and R. Stora (North-Holland 1983) p. 171.
3.   E. J. Heller, *Phys. Rev. Lett.* **53**, 1515 (1984).
4.   P. O'Connor, J.N. Gehlen, and E.J. Heller, *Phys. Rev. Lett.* **58**, 1296 (1987).
5.   S. Sridhar, *Phys. Rev. Lett* **67**, 785 (1991).
6.   J. Stein and H.-J. Stöckman, *Phys. Rev. Lett.* **68**, 2867 (1992).
7.   T. M. Fromhold, P. B. Wilkinson, F. W. Sheard, L. Eaves, J. Miao, and G. Edwards, *Phys. Rev. Lett.* **75**, 1142 (1995); P. B. Wilkinson, T. M. Fromhold, L. Eaves, F. W. Sheard, N. Miura, and T. Takamasu, *Nature* **380**, 608 (1996).
8.   D. Wintgen and A. Honig, *Phys. Rev. Lett.* **63**, 1467 (1989).
9.   K. Muller and D. Wintgen, *J. Phys.* **B 27**, 2693 (1994).
10. E. B. Bogomolny, *Physica* **D 31**, 169 (1988).
11. M. V. Berry, Les Houches Lecture Notes, Summer School on Chaos and Quantum Physics, M-J. Giannoni, A. Voros, and J. Zinn-Justin, eds., Elsevier Science Publishers B.V. (1991); M.V. Berry, *Proc. Roy. Soc.* **A 243**, 219 (1989).
12. R. Aurich and F. Steiner, *Chaos, Solitons and Fractals* **5**, 229 (1995).
13. O. Agam and N. Brenner, *J. Phys.* **A 28**, 1345 (1995).
14. O. Agam and S. Fishman, *Phys. Rev. Lett.* **73**, 806 (1994); O. Agam and S. Fishman, *J. Phys.* **A 26**, 2113 (1993).
15. S. Fishman, B. Georgeot, and R. E. Prange, *J. Phys.* **A 29**, 919 (1996).
16. E. J. Heller, Wavepacket Dynamics and Quantum Chaology in *Chaos and Quantum Physics*, Eds. M. J. Giannoni, A. Voros, and J. Zinn-Justin, Elsevier Science Publishers, Amsterdam (1990).
17. A. I. Schnirelman, *Usp. Mat. Nauk.* **29**, 181 (1974); Y. Colin de Verdiere, *Commun. Math. Phys.* **102**, 497 (1985); S. Zelditch, *Duke Math. J.* **55**, 919 (1987).
18. L. Kaplan and E. J. Heller, Linear and Nonlinear Theory of Eigenfunction Scars, *Ann. Phys. (N. Y.)*, in press.
19. P. W. O'Connor and E. J. Heller, *Phys. Rev. Lett.* **61**, 2288 (1988).
20. B. Li, *Phys. Rev.* **E 55**, 5376 (1997).
21. E. J. Heller, *J. Chem. Phys.* **72**, 1337 (1980); E. J. Heller and M. J. Davis, *J. Phys. Chem.* **86**, 2118 (1982); E. B. Stechel and E. J. Heller, *Ann. Rev. Phys. Chem.* **35**, 563 (1984); E. J. Heller, *Phys. Rev.* **A35**, 1360 (1987).
22. L. Kaplan and E. J. Heller, *Phys. Rev. Lett.* **76**, 1453 (1996); F.-M. Dittes, E. Doron, and U. Smilansky, *Phys. Rev.* **E 49**, R963 (1994).
23. N. L. Balazs and A. Voros, *Europhys. Lett.* **4**, 1089 (1987); N. L. Balazs and A. Voros, *Ann. Phys. (N. Y.)* **190**, 1 (1989); M. Saraceno, *Ann. Phys. (N. Y.)* **199**, 37 (1990); M. Saraceno and A. Voros, *Physica* **D 79**, 206 (1994).

24. P. W. O'Connor, S. Tomsovic and E. J. Heller, *Physica D* **55**, 340 (1992); P. W. O'Connor and S. Tomsovic, *Ann. Phys. (N. Y.)* **207**, 218 (1991).
25. L. Kaplan and E. J. Heller, unpublished.
26. L. Kaplan, unpublished.

NONEQUILIBRIUM EFFECTS IN THE TUNNELING CONDUCTANCE SPECTRA OF SMALL METALLIC PARTICLES

Oded Agam

The Racah Institute of Physics
The Hebrew University
Jerusalem, 91904
Israel

INTRODUCTION

Trace formulae and the non-linear supersymmetric σ-model are basic analytical tools used successfully in the fields of quantum chaos and disordered systems. Both are designed to treat systems with a small number of degrees of freedom. Hence they are limited in their possibility of analyzing many-body systems where interparticle interactions play an important role, and the number of degrees of freedom is large.

On the other hand, many experimental studies of quantum chaos use systems which consist in a large number of interacting particles, for example quantum dots or disordered metallic particles. Having an elaborate single-particle description of these systems, it is of prominent importance to understand the role of interactions, the range of applicability of a single-particle picture, and the interplay between chaos and inter-particle interactions.

In this respect, an important observation is that strong chaotic dynamics, on the level of non-interacting single-particle description, provides us with the possibility of analyzing interacting many-body systems by a systematic perturbative approach. The small parameter of this perturbation theory is $1/g$, where $g = t_H/t_c$ is the dimensionless conductance, i.e. the ratio of the Heisenberg time, t_H (the inverse mean level spacing), to the classical relaxation time, t_c.

The general form of the interaction Hamiltonian in which particles interact via a two-body potential $U(\mathbf{r}, \mathbf{r}')$ is

$$H_{\mathrm{int}} = \frac{1}{2} \sum_{ijkl} \sum_{\sigma\sigma'} U_{ijkl} c_{i\sigma}^\dagger c_{j\sigma'}^\dagger c_{k\sigma'} c_{l\sigma}, \tag{1}$$

where $c_{i\sigma}^\dagger$ and $c_{i\sigma}$ are the creation and annihilation operators for a particle in state ψ_i and spin σ, while

$$U_{ijkl} = \int d\mathbf{r} d\mathbf{r}' U(\mathbf{r}, \mathbf{r}') \psi_i^*(\mathbf{r}) \psi_j^*(\mathbf{r}') \psi_l(\mathbf{r}) \psi_k(\mathbf{r}')$$

Supersymmetry and Trace Formulae: Chaos and Disorder
Edited by Lerner *et al.*, Kluwer Academic / Plenum Publishers, New York, 1999

are the matrix elements of the interaction potential. These matrix elements can be divided into two groups according to their typical magnitude. One contains diagonal matrix elements, namely those U_{ijkl} in which two pairs of indices are identical. All the other matrix elements, which we call off-diagonal, are included in the second group. In appendix A it is shown that the typical magnitude of off-diagonal matrix elements is as small as d/g where d is the single-particle mean level spacing and g is the dimensionless conductance. The same smallness restricts also the fluctuations in the diagonal matrix elements. Therefore, the interaction Hamiltonian of electrons in a quantum dot takes the form

$$H_{\text{int}} = \frac{e^2}{2C} \left(\sum_{i\sigma} c_{i\sigma}^\dagger c_{i\sigma} - N_0 \right)^2 - \lambda \sum_{i,j} c_{i\uparrow}^\dagger c_{i\downarrow}^\dagger c_{j\uparrow} c_{j\downarrow} + \frac{a}{2} \sum_{ij,\sigma\sigma'} c_{i\sigma}^\dagger c_{j\sigma'}^\dagger c_{i\sigma'} c_{j\sigma} + O(d/g). \quad (2)$$

The first term of this formula is the orthodox model [1,2] representing the charging energy of the dot: C is the total capacitance of the dot, $N = \sum_{i\sigma} c_{i\sigma}^\dagger c_{i\sigma}$ is the total number of electrons in the dot, and N_0 is a continuous parameter controlled by the gate voltage. The second term, when $\lambda > 0$, represents an attractive interaction which drives the grain into a superconducting state at sufficiently low temperatures and weak magnetic field. The last term represents the electron-electron interaction in the spin channel, and the coefficient a is of order the single-particle mean level spacing, d.

Strong chaotic dynamics of the noninteracting particles implies that $g \gg 1$. Consequently, all off-diagonal matrix elements U_{ijkl} are proportional to $1/g$, and the mean field approximation (2) for the interacting Hamiltonian (1) is justified. Indeed, many phenomena of normal and superconducting metallic grains are described by the approximation (2). The most prominent one is the Coulomb blockade, which is essentially the quantization of the number of electrons in the grain away from the charge degeneracy point. Because of this quantization, the zero-bias conductance of the system vanishes, while the current I as the function of the source-drain voltage V shows a threshold behavior. The fine structure of the current–voltage curve is associated with the single-electron levels of the system[3,4].

Nevertheless, there are interesting phenomena emerging from the fluctuations of the interaction matrix elements, i.e. with the $O(d/g)$ corrections to (2). In this review we analyze two experiments of Ralph, Black and Tinkham[5,6] and show that fluctuations of the interaction energy, although small as d/g, clearly manifest themselves in the differential conductance spectra of ultrasmall metallic grains. The small effects of fluctuations in the charging energy are especially pronounced due to the fact that the system is driven out of equilibrium, and is able to explore several high excited states at relatively low source-drain voltage.

The experimental system consists of a single aluminum particle connected to external leads via high resistance ($1 - 5$ MΩ) tunnel junctions formed by oxidizing the surface of the particle. The device (see illustration in Fig. 1) is fabricated using electron beam lithography and reactive ion etching to form a bowl-shaped hole in an insulating Si_3N_4 membrane. The opening at the lower edge of the membrane is 3-10 nm in diameter. Al is evaporated onto the bowl-shaped side of the membrane, and subsequently the Al surface is oxidized. The oxide layer forms a tunnel barrier in the vicinity of the small hole in the membrane. The membrane is then flipped up side down and a small amount of Al is deposited. Because of surface tension, the Al forms a layer of electrically isolated particles, a few nanometers in size. Following a second oxidation,

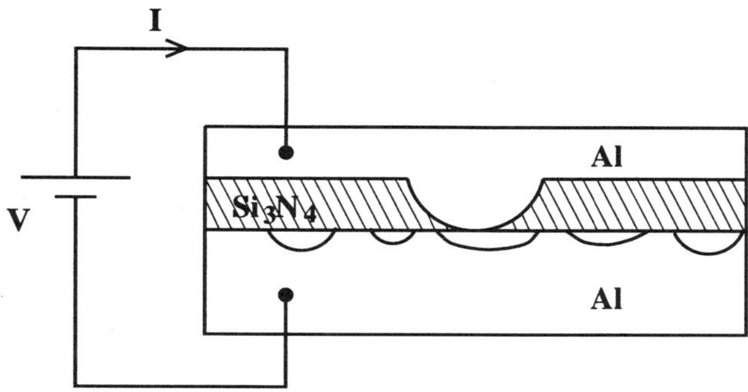

Figure 1. A schematic illustration of the device used in the experiments of Ralph, Black and Tinkham for measuring the differential conductance spectra of ultrasmall aluminum grains.

a thick layer of Al is deposited on top of the particles.* In approximately 25% of the devices one Al particle covers the hole in the nitride membrane, so that the electrons passing between the leads tunnel trough the metal particle.

The capacitances and the resistances of the tunnel junctions are estimated by fitting the large scale $I - V$ curves, $eV \sim e^2/C$, to the Coulomb blockade staircase pattern. From the capacitances one can determine the area of the tunnel junctions and the volume of the grain which is used in turn to estimate the single-particle mean level spacing[5].

In this review we focus our attention on scales of the $I - V$ curves which are much smaller than that of the Coulomb blockade, scales over which the single-particle mean level spacing, d, and the fluctuation in the charging energy, d/g, are resolved. Fig. 2 displays the differential conductance, dI/dV, of two different *normal* metallic particles (of sizes roughly 2.5 and 4.5 nm) as a function of the source–drain bias energy eV. The spectra display three clear features:

1. The low resonances of the differential conductance are grouped in clusters. The distance between nearby clusters is of order the mean level spacing d of the noninteracting electrons in the dot.

2. The first cluster contains only a single resonance.

3. Higher clusters consist of several resonances spaced much more closely than d.

In section 2 it will be shown that these features are manifestations of the interplay between electron-electron interactions and nonequilibrium effects[7]. Each cluster of resonances is identified with one excited single-electron state, and each resonance in turn is associated with a different occupancy configuration of the metal particle's other single-electron states. The appearance of multiple resonances reflects the strongly nonequilibrium state of the particle.

In another experiment[6], Ralph, Black and Tinkham measured the tunneling resonance spectra of ultrasmall *superconducting* grains. The number of electrons in the system was controlled by a gate voltage. The results of this experiment, depicted in Fig. 3, show that:

---

*In other configurations of these devices a gate electrode of ring shape is deposited after flipping the membrane, and the same procedure follows the oxidation of the gate.

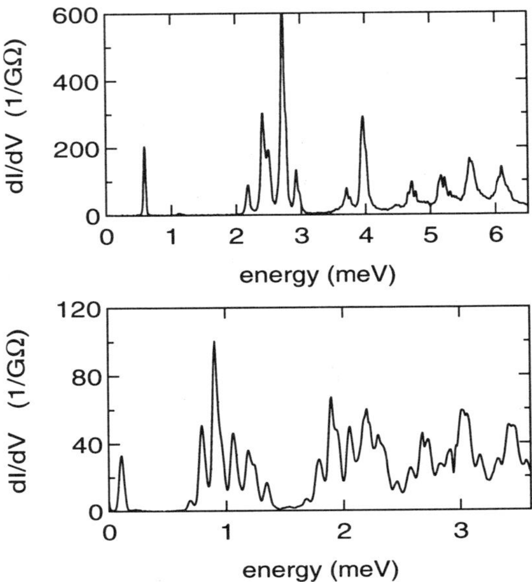

Figure 2. The low temperature (30 mK) differential conductance dI/dV versus bias energy of ultrasmall Al particles with volumes ≈ 40 nm^3 (upper panel) ≈ 100 nm^3 (lower panel). The first resonance is isolated while subsequent resonances are clustered in groups. The distance between nearby groups of resonances is approximately the single-particle mean level spacing d. (From Ref. [5]).

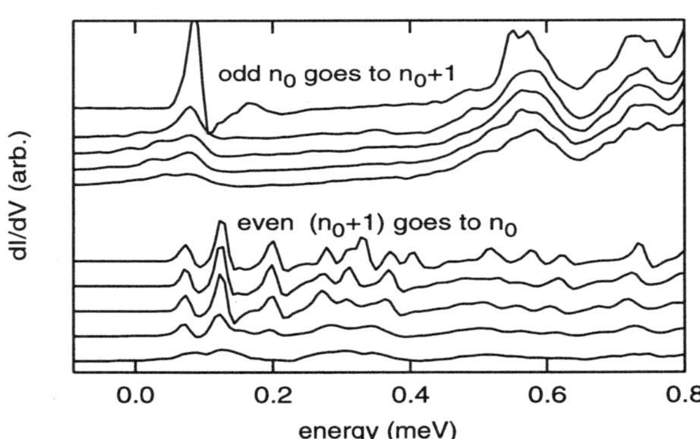

Figure 3. The tunneling resonances of superconducting grains in the odd (upper scans) and the even (lower scans) charging states. Different scans correspond to different value of the gate voltage, and are artificially shifted in energy to align peaks due to the same eigenstate. In contrast with Fig. 2 the first resonance, in the odd charging state, develops a substructure when shifted by the gate voltage. (From Ref. [6]).

1. For the ground state of the grain with an even number of electrons, the first peak of the differential conductance is merely shifted by the gate voltage V_g. The shape of this peak does not change over a large interval of V_g. Contrarily, if the grain contains an odd number of electrons, the height of the first peak rapidly reduces with a change of the gate voltage, and a structure of subresonances develops on the low-voltage shoulder of this peak.

2. The characteristic energy scale between subresonances of the first peak is of the order of the mean level spacing d.

These observations contrast the results for the normal case in which the first peak did not split. Nevertheless, it was suggested in Ref. [6] that the substructure of the first peak is still associated with nonequilibrium steady states of the grain. In section 3, the origin of these nonequilibrium states and the mechanism which generates them will be clarified[8].

The explanations for both experiments discussed here rely on the assumption that the systems are stimulated into steady states which are far from equilibrium, namely that relaxation processes are too slow to maintain the system in equilibrium. In section 4 we summarize the various relaxation processes and estimate the inelastic time, τ_{in}, for electrons in the dot. The results will be summarized in section 5.

NORMAL GRAINS

Our model for the experimental system is given by the Hamiltonian: $H = H_0 + H_T + H_{\text{int}}$. Here H_0 describes the noninteracting electrons in the left (L) and right (R) leads and in the metallic grain[†],

$$H_0 = \sum_{\alpha=L,R} \sum_q \xi_{\alpha q} d^\dagger_{\alpha q} d_{\alpha q} + \sum_l \xi_l c^\dagger_l c_l. \tag{3}$$

Tunneling across the barriers is described by

$$H_T = \sum_{\alpha=L,R} \sum_{q,l} T^{(\alpha)}_{ql} d^\dagger_{\alpha q} c_l + \text{ H.c.}, \tag{4}$$

where $T^{(\alpha)}_{ql}$ are the tunneling matrix elements. Interaction effects given by (1) are taken into account only for the electrons in the grain, but including screening by image charges in the leads.

For the ultrasmall aluminum grains considered here, one can neglect superconducting pairing since the single-particle mean level spacing, ≈ 1 meV, is larger than the BCS superconducting gap which is 0.18 meV[9]. Under this condition, the interaction term of the electrons is generally approximated by the orthodox model, $H_{\text{int}} \approx (e \sum_l c^\dagger_l c_l)^2 / C$, where C is the effective capacitance of the grain[1]. Within this approximation the charging energy depends only on the total number of electrons in the dot, but not on their particular occupancy configuration. The orthodox model is able to account for the Coulomb blockade[1], and the Coulomb staircase behavior of the current as the number of extra tunneling electrons in the dot increases. It can also be generalized to describe features on the scale of the single-particle level spacing[10]. However, the orthodox model cannot account for the clusters of resonances in Fig. 2, since these result from fluctuations, δU, in the interaction energy between pairs of electrons.

[†]Unless explicitly written, from now on single-particle and spin states will be denoted by a single subscript.

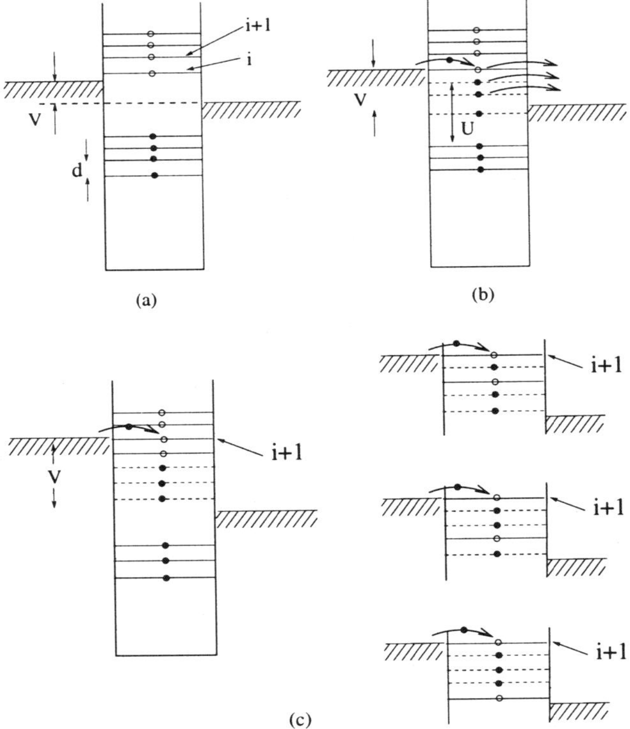

Figure 4. An illustration of transport through the metal particle at various values of the source–drain voltage V. Filled single-particle levels are indicated by full circles and empty ones by open circles. U is the charging energy, and d is the single-particle mean level spacing. (a) The system at small voltage bias within the Coulomb blockade regime; (b) V corresponding to the first resonance in Fig. 2. The thin dashed lines indicate the energy of a level after an electron has tunneled into the dot; (c) V near the first cluster of resonances in Fig. 2. The splitting within the first cluster originates from the sensitivity of level $i + 1$ to the different possible occupation configurations as shown.

We focus our attention on the (experimental) voltage regime where there is no more than one extra tunneling electron in the dot. At small voltage bias, V, within the Coulomb-blockade regime [Fig. 4(a)], current does not flow through the system. Current first starts to flow when one state i inside the grain becomes available for tunneling through the left barrier, say, as illustrated in Fig. 4(b). As the system becomes charged with an additional electron, the potential energy of the other electrons in the dot increases by $U \simeq e^2/C$, and some of the lower energy occupied electronic states are raised above the right lead chemical potential [in Fig. 4(b) these "ghost" states are shown as dashed lines]. Electrons can tunnel out from these states into the right lead leaving the particle in an excited state. There is, however, only one configuration of the electrons which allows an electron to tunnel into level i from the left lead, namely all lower energy levels occupied. This implies that only a single resonance peak appears in the differential conductance at the onset of the current flow through the system (broken spin degeneracy would cause splitting of this peak).

The situation changes when V increases such that electrons can tunnel from the left lead into the next higher available state $i + 1$, as shown in Fig.4 (c). In this case, there are several possible occupancy configurations, on which the exact energy of level $i+1$ depends. The several possible energies of level $i+1$ lead to a cluster of resonances in the differential conductance of the grain. The scenario described above holds provided

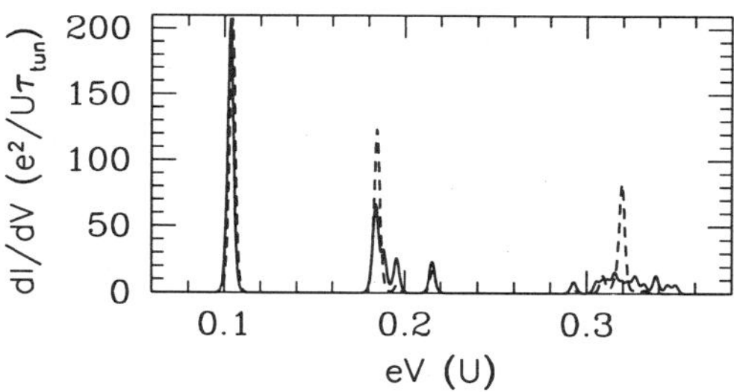

Figure 5. Model differential conductance obtained from nonequilibrium detailed-balance equations: solid line – in the absence of inelastic processes, $1/\tau_{in} = 0$; dashed line – with inelastic relaxation rate larger than the tunneling rate, $1/\tau_{in} = 5/\tau_{tun}$.

that inelastic processes are too slow to maintain equilibrium in the particle.

To explicitly demonstrate the splitting of resonances induced by fixed fluctuations in the interaction energy δU, model detailed-balance equations[10] were solved numerically and the corresponding differential conductance plotted in Fig. 5 by the solid line. The model system consists of 7 equally spaced levels, occupied alternately by 4 or 5 electrons, in a current-carrying steady state. For simplicity, the tunneling rate into each level, $1/\tau_{tun}$ ($\Gamma_{L(R)}(\epsilon_l)$ in the notation of Ref. [10]), is chosen to be uniform, and the voltage is applied by increasing the left chemical potential. The temperature is 1% of the mean level spacing d, and the variance of the fluctuations δU in the interaction energy is $d/5$. In the absence of fluctuations ($\delta U = 0$), dI/dV consists of single resonances spaced by d.

To estimate the fluctuations in the interaction energy consider the Hartree term of the interaction energy, U_H. We wish to calculate the interaction energy difference associated with different occupation configurations of low energy states. Suppose that, as illustrated in Fig. 4(c), these differ by a single occupation number, namely, in one configuration the state j is empty and j' is full while in the other j' is empty and j is full. Then

$$\delta U_H = \int dr_1 dr_2 |\psi_i(\mathbf{r_1})|^2 U(\mathbf{r_1}, \mathbf{r_2}) \left[|\psi_{j'}(\mathbf{r_2})|^2 - |\psi_j(\mathbf{r_2})|^2 \right] ,$$

where the index i labels an electron state other than j or j', $U(\mathbf{r_1}, \mathbf{r_2})$ is the interaction potential. Since wave functions of chaotic systems associated with different energies are statistically independent, $\langle \delta U_H \rangle = 0$ where $\langle \cdots \rangle$ denotes ensemble or energy averaging. We are therefore interested in fluctuations of δU_H which emerge from the non-uniform probability distributions of the single-particle eigenstates in real space. The calculation of $\langle \delta U_H^2 \rangle$ is similar to that presented in appendix A. The result for diffusive systems is

$$\langle \delta U_H^2 \rangle = \left(c \frac{d}{g} \right)^2 ,$$

where $c = \sqrt{2} \alpha \sum_{\mathbf{n}} |\mathbf{n}|^{-4} / \pi$ is a constant of order unity, and α equals two for system with time reversal symmetry and unity for systems without time reversal symmetry[‡].

The above estimate for the fluctuations in the charging energy also applies for general chaotic systems, with $g = \hbar\gamma_1/d$ where γ_1 is the first non-vanishing Perron–Frobenius eigenvalue[12], see appendix A.

The increase of the fluctuations in the interaction energy as g decreases is related to the fact that g is a measure for the uniformity of the single-particle wave functions. The bigger g the more uniform are the wave functions and the less are the fluctuations in the interaction energy. Experimentally we find $g \approx 5$. Unfortunately, an analytical estimate of g requires precise knowledge of the shape and disorder of the particle which we lack. A naive estimate of g in ballistic systems is $\hbar/\tau d$, where τ is the time for an electron at the Fermi energy to cross the system. The metallic grains of the experiment, however, have a roughly pancake shape. Assuming diffusive dynamics one can show that $g\tau d/\hbar \propto (z/r)^2$ where z is the pancake thickness and r is its radius. g is therefore much smaller than $\hbar/\tau d$.

When M available states below the highest accessible energy level (including spin), are occupied by $M' < M$ electrons, there are $\binom{M}{M'}$ different occupancy configurations. The typical width of a cluster of resonances in this case is $W^{1/2}cd/g$ where $W = \min(M - M', M')$. The width of a cluster of resonances therefore *increases* with the source–drain voltage. The distance between nearby peaks of the cluster, on the other hand, *decreases* as $W^{1/2}/\binom{M}{M'}$. This behavior can be seen in Fig. 5.

SUPERCONDUCTING GRAINS

As illustrated by Fig. 4, the splitting of tunneling resonance peaks in normal metallic grains comes from the possibility of forming different occupation configurations of single-particle states at sufficiently high source-drain voltage. These configurations are reached by resonant tunneling provided relaxation processes are sufficiently slow. This picture also explains the observation that the first cluster in Fig. 2 contains only a single peak. However, the data of Fig. 3 shows that the first peak in the tunneling resonance spectra also splits into several subresonances (see illustration in Fig. 6). This behavior appears when the superconducting grain contains an odd number of electrons and the gate voltage is such that the dot is far from the charge degeneracy point.

In this section we show that the development of a substructure in the first peak of the tunneling resonance spectra is also associated with the generation of nonequilibrium steady state. However in contrast with the resonant tunneling mechanism used in the previous section for the high resonance peaks, here the nonequilibrium steady state is reached by inelastic cotunneling processes.

The principal difference between odd and even grains is that all excitations of the latter are of energy larger than the superconducting gap 2Δ. Therefore, a source-drain voltage in the range $V < \Delta/e$ can not induce nonequilibrium states. Odd grains, on the other hand, contain one unpaired electron, which may be shifted to various single-electron levels with characteristic energy scale smaller than the mean level spacing d. For this reason even a small source-drain voltage $d < eV < \Delta$ is sufficient to excite the grain. The mechanism of excitation is inelastic cotunneling[13]. Tunneling into the excited grain requires less energetic electrons, and lead in turn to the substructure on the low-voltage shoulder of the of the first resonance, see Figs. 3 and 6. A closely related problem was considered by Averin and Nazarov[14], however, their theory assumed that relaxation processes prevent the formation of nonequilibrium states. As will be argued

---

‡This estimation does not take into account a change in the potential due to the insertion of an additional electron. It was argued that the latter effect may lead to an even stronger effect, i.e. $\delta U \sim d/\sqrt{g}$.[11]

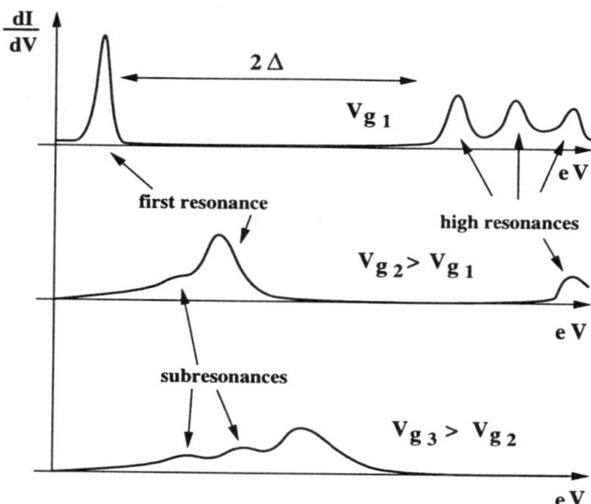

Figure 6. A schematic illustration of the differential conductance of an "odd" superconducting grain as function of the source-drain voltage V, at various gate voltages V_g. Higher resonances are separated by the superconducting gap from the first one, and subresonances are developed as the first resonance is shifted by the gate voltage.

in the next section, relaxation processes in ultrasmall metallic grains are very slow, and therefore will be neglected in our theory.

To describe the effect quantitatively, we construct the master equations governing the time evolution of probabilities of different electronic configurations of supercon- ducting grains allowing for second order cotunneling processes. The solution of these equations for two limiting cases (one in which two levels participate in the transport, and the other when a large number of levels contribute) explains the substructure of the first peak of the differential conductance illustrated in Fig. 6.

As in the previous section the model Hamiltonian consists of three terms: $H = H_0 + H_T + H_{\text{int}}$. H_0, given by (3), describes the noninteracting electrons in the leads and in the dot; H_T, given by (4), is the tunneling Hamiltonian, and the interaction Hamiltonian will be approximated by

$$H_{\text{int}} = \frac{e^2}{2C}(N - N_0)^2 - \lambda \sum_{i,j} c_{i\uparrow}^\dagger c_{i\downarrow}^\dagger c_{j\uparrow} c_{j\downarrow}. \tag{5}$$

where $N = \sum_{j,\sigma} c_{j\sigma}^\dagger c_{j\sigma}$ is the number of electrons in the dot, and N_0 is a continuous parameter controlled by the gate voltage. N_0 determines the finite charging energy required to insert, U_+, or to remove, U_-, one electron,

$$U_\pm = \frac{e^2}{C}\left[\frac{1}{2} \pm (N - N_0)\right], \quad |N - N_0| \le \frac{1}{2}. \tag{6}$$

Consider the experimentally relevant case, $e^2/C \gg \Delta$, so that the grain has well defined number of electrons. If this number is even $N = 2m$, the ground state energy (we will omit charging part of the energy and restore it later), $E_{2m} = F_{2m} + 2\mu m$, can be calculated in the mean field approximation[15, 9] by minimizing thermodynamic potential $F_{2m} = \sum_k(\xi_k - \epsilon_k) + \Delta^2/\lambda$ where $\epsilon_k = (\xi_j^2 + \Delta^2)^{1/2}$, with respect to Δ, and by fixing the chemical potential according to the number of electrons in the grain. All the excited states of even dots are separated from the ground state by a large energy, 2Δ. Considering now the energy spectrum of an odd grain, $N = 2m - 1$, we notice that the second term in Eq. (5) operates only within spin singlet states. Therefore, to calculate the low-lying excited states in this case, we fill the single-electron state j with one electron, and then find the ground state of the remaining $2m$ electrons with state j excluded from the Hilbert space. In the mean field approximation it corresponds to the minimization of the thermodynamic potential $F_{2m-1}^{(j)} = \sum_{k \neq j}(\xi_k - \epsilon_k) + \Delta^2/\lambda + \xi_j$. The excited states with energies smaller than Δ are characterized by a single index, j and will be denoted by $E_{2m-1}^{(j)}$. In what follows we will need the energy cost of introducing an additional electron into the odd state: $U_+ + \varepsilon_j$, where $\varepsilon_j = E_{2m} - E_{2m-1}^{(j)}$. In appendix B it is shown that in the limit $\Delta \gg d$ the result is:

$$\varepsilon_j = \mu_{2m} - \frac{3d}{2} + \frac{\xi_j d}{2\Delta} - \sqrt{\xi_j^2 + \Delta^2}. \tag{7}$$

We turn now to the kinetics of a superconducting grain. Consider the regime where $U_+ = U \leq \Delta$, $U_- \approx \frac{e^2}{2C} \gg U$, and $\frac{e^2}{2C} \gg \Delta \gg d$. We also assume the conductance of the tunnel barriers to be much smaller than e^2/h, and that the source-drain voltage is small $eV < \Delta$. The simplicity brought to the problem in this regime of parameters stems from the fact that there is only one available state with an even number of electrons (because $U_- \gg U_+$ one can only add an electron to grain but not subtract one), and whenever the grain contains an even number of electrons it is in its ground state. This imply that even grains cannot be driven out of equilibrium state, while for odd grains tunneling (and cotunneling) takes place via unique state.

Henceforth, we concentrate on grains with an odd ground state. Let us denote by P_e the probability of finding the grain with an even number of electrons, and by P_j the probability to find the grain in the odd state j. Since these states are spin degenerate in the absence of magnetic field, P_j will denote the sum $P_{j,\uparrow} + P_{j,\downarrow}$. The master equations for the probabilities P_e, P_j have the form

$$\frac{dP_e}{dt} = \sum_j \left[\Gamma_{o \to e}^{(j)} P_j - 2\Gamma_{e \to o}^{(j)} P_e \right], \tag{8}$$

$$\frac{dP_j}{dt} = 2\sum_{i \neq j}[\Gamma_{i \to j} P_i - \Gamma_{j \to i} P_j] + 2\Gamma_{e \to o}^{(j)} P_e - \Gamma_{o \to e}^{(j)} P_j,$$

where $\Gamma_{o \to e}^{(j)}$ and $\Gamma_{e \to o}^{(j)}$ are the transitions rates from the odd j-th state to the even and from the even to odd respectively, while $\Gamma_{i \to j}$ is the rate of transition from the i-th to the j-th odd states. Equations (8) are not independent, so they have to be supplied with the normalization condition $P_e + \sum_j P_j = 1$. Current in the steady state equals to the electron flow through, say, the left barrier, and for positive V it is given by

$$I = e \sum_j \left(\Gamma_{o \to e}^{(j)} + \Gamma_{j \to j} \right) P_j + 2e \sum_{j \neq i} \Gamma_{j \to i}^{(j)} P_j. \tag{9}$$

Transition from the j-th odd state into the even state occurs when $\mu_L > U + \varepsilon_j$. The amplitude of this transition is calculated by first order perturbation theory in the tunneling Hamiltonian (4). Fermi's golden rule yields

$$\Gamma_{o \to e}^{(j)} = g_L \frac{u_j^2 \rho_{Lj} d}{2\pi\hbar} \theta(\mu_L - \varepsilon_j - U), \tag{10}$$

where g_L is the dimensionless conductance of the left tunnel barrier per one spin, $u_j = (1 + \xi_j/\epsilon_j)/2$ is the coherence factor, $\theta(x)$ is the unit step function, and $\rho_{Lj} = \Omega|\psi_j(r_L)|^2$, where Ω is the volume of the grain and $\psi_j(r_L)$ is the value of j-th single-particle wave function at the left point contact r_L. Energies ε_i are given by Eq. (7) and $U = U_+$ is defined in Eq. (6). Similarly, the rate of transition from even state to i-th odd state, by tunneling of an electron from the dot to the right lead, is given by

$$\Gamma_{e\to o}^{(i)} = g_R \frac{v_i^2 \rho_{Ri} d}{2\pi\hbar} \theta(U + \varepsilon_i - \mu_R), \tag{11}$$

where g_R is the dimensionless conductance of the right tunnel barrier, $v_i = (1 - \xi_i/\epsilon_i)/2$, and $\rho_{Ri} = \Omega|\psi_i(r_R)|^2$, where r_R is the position of the right point contact.

A change in the occupation configuration of the odd states occurs via inelastic cotunneling[13]. This mechanism is a virtual process in which an electron tunnels into j-th available level and another electron tunnels out from the i-th level. Calculating this rate by second order perturbation theory in the tunneling Hamiltonian, one obtains

$$\Gamma_{j\to i} = \frac{g_L g_R d^2 u_j^2 v_i^2 \rho_{Lj} \rho_{Ri}(eV - \varepsilon_j + \varepsilon_i)}{8\pi^3\hbar(U + \varepsilon_j - \mu_L)(U + \varepsilon_i - \mu_R)} \tag{12}$$

for $eV > \varepsilon_j - \varepsilon_i$, $\mu_L < U + \varepsilon_j$, $\mu_R < U + \varepsilon_i$, and zero otherwise. $\Gamma_{j\to i}$ diverges in the limits $\mu_L \to U + \varepsilon_j$ and $\mu_R \to U + \varepsilon_i$. It signals that a real transition takes over the virtual one. The region of applicability of Eq. (12) is, therefore, $U + \varepsilon_j - \mu_L > \gamma$ and $U + \varepsilon_i - \mu_R > \gamma$ where $\gamma \sim gd/4\pi$ is the width of a single-particle level in the dot due to the coupling to the leads, $g = g_L + g_R$. However, the interval of biases where Eq. (12) is not valid is narrow, and to the leading approximation in $\hat{H}_T$ our results will be independent of this broadening.

Let us now apply Eqs. (8) and (9) to describe the appearance of the low-voltage substructure of the first peak. We will consider two situations: (i) small voltage such that only one subresonance can emerge on the shoulder of the leading one, and (ii) large voltage, $d \ll eV < \Delta$, where the substructure of the main resonance consists of a large number of subresonances.

In the first case, the chemical potentials of the left and right leads are such that transport through the grain involves only two levels: ε_0 and $\varepsilon_1 < \varepsilon_0$ corresponding to the ground and the first excited states of the odd grain. We solve Eqs. (8) for probabilities P_0, P_1 and P_e using Eqs. (9–12). There are two distinct regimes of the source-drain voltage: (1) $\mu_L < U + \varepsilon_0$ where transport is dominated by cotunneling, and (2) $\mu_L \geq U + \varepsilon_0$ where state "0" is available for resonant tunneling. The substructure of the first resonance in the differential conductance appears in the first regime. Below we show that as μ_L passes through $U + \varepsilon_1$, see Fig. 7, there is a discontinuity in the current-voltage curve. In the first regime, the total current to the leading approximation in g_L, g_R is a sum of two contributions, $I \simeq I_{eq} + I_{ne}$. The first, $I_{eq} = e\Gamma_{0\to 0} + 2e\Gamma_{0\to 1}$, is the equilibrium current coming from cotunneling. The second contribution is associated with the nonequilibrium population of state "1" and is given by

$$I_{ne} = 2e\Gamma_{0\to 1} \times \begin{cases} \frac{\Gamma_{1\to 0} - \Gamma_{0\to 1} + 2\Gamma_{1\to 1} - 2\Gamma_{0\to 0}}{\Gamma_{0\to 1} + \Gamma_{1\to 0}} & \mu_L < U + \varepsilon_1 \\[2mm] 2\left(1 + \frac{\Gamma_{e\to o}^{(1)}}{\Gamma_{e\to o}^{(0)}}\right) & \mu_L > U + \varepsilon_1 \end{cases}$$

Assuming that the voltage drop $eV = \mu_L - \mu_R$ is larger than the energy difference $\tilde{d} = \varepsilon_0 - \varepsilon_1$, the jump in the nonequilibrium current is:

$$\delta I_{ne} = c_1 e \frac{g_L g_R d^2}{8\pi^2 h \tilde{d}}\left(1 - \frac{\tilde{d}}{eV}\right) \qquad eV \sim 2\tilde{d} \tag{13}$$

where $c_1 = 4u_0^2 v_1^2 \rho_{L0} \rho_{R1}$ is a constant of order unity. This jump in the nonequilibrium current leads to the peak in the differential conductance spectra. Formula (13) has

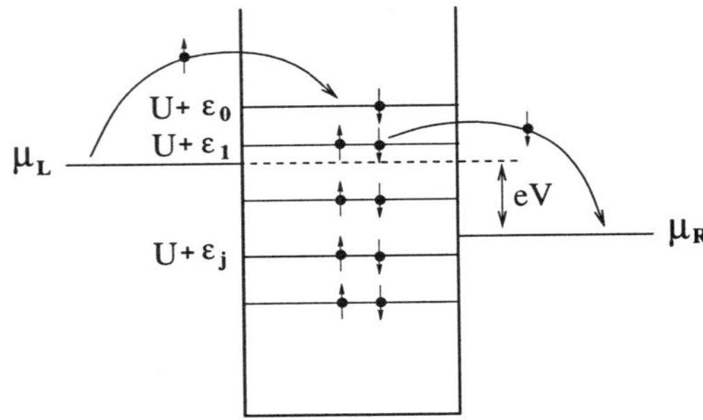

Figure 7. Inelastic cotunneling process can drive an "odd" superconducting grain out of its ground state. In the ground state, the single-particle level indicated by $U + \varepsilon_0$ is occupied by one electron. Excited states are those in which the unpaired electron is shifted to other single-particle levels. In a nonequilibrium steady state, low single-particle levels become available for resonant tunneling, leading to a subresonances structure of the differential conductance shown in Fig. 6. State j shown to be filled with two electrons should be understood as a coherent superposition of double occupied and empty states with weights v_j^2 and u_j^2 respectively.

simple interpretation. Up to numerical prefactors it is a product of two factors: first is the probability of finding the grain with an unpaired electron in state "1". It is proportional to $g_R(d/\tilde{d})(1 - \tilde{d}/eV)$, and increases with the voltage V and as $\tilde{d} = \varepsilon_0 - \varepsilon_1 \to 0$. The second factor is associated with the rate in which the state "1" is filled with an electron, $eg_L d/h$.

The magnitude of the jump (13) should be compared to the jump in the current as μ_L increases above $U + \epsilon_0$, and real transition via the even state become allowed. To the leading order in g_L and g_R, the current in this regime is

$$I = c_2 e \frac{g_R g_L d}{h(g_L + 4g_R)}, \quad \mu_L > U + \varepsilon_0, \tag{14}$$

where c_2 is a constant of order unity having structure similar to c_1. Comparing the current jump, δI_{ne}, with that associated with the resonant tunneling, δI, we find

$$\frac{\delta I_{ne}}{\delta I} \simeq \frac{g_L + 4g_R}{8\pi^2}, \quad eV \sim 2\tilde{d}.$$

Thus nonequilibrium population of the excited level of the odd-grain leads to the appearance of a subresonance at small V, however, its height is much smaller than that of the main resonance.

We turn now to the second regime of the parameters, $d \ll eV < \Delta$, in which many levels contribute to the transport. Again, we focus our attention on the cotunneling regime, $\mu_L < U + \varepsilon_0$. We show that the characteristic amplitude of the subresonances in this regime may become comparable to the amplitude of the main peak.

To the leading order in g_L, g_R, and d/Δ, the steady state solution of the rate equations at $\mu_L = U + \varepsilon_1 + 0$ is $P_0 \simeq 1$, while for the other probabilities we have

$$P_e \simeq \frac{\sum_{i \neq 0} \Gamma_{0 \to i}}{\Gamma_{e \to o}^{(0)}}, \quad P_j \simeq 2\frac{\Gamma_{0 \to j}}{\Gamma_{o \to e}^{(j)}} + 2\frac{\Gamma_{e \to o}^{(j)}}{\Gamma_{o \to e}^{(j)}}P_e. \tag{15}$$

The characteristic number of states contributing to the current (9) is large as $\sqrt{\Delta eV}/d$ so that mesoscopic fluctuations of the tunneling rates and of the inter-level spacings may be neglected. Additional large factor, $\sqrt{\Delta eV}/d$, comes from the summation over the levels in Eq. (15), and we find

$$I \simeq \frac{e^2 g_L g_R}{2\pi^2 \hbar} \frac{V\Delta}{\varepsilon_0 - \varepsilon_j}, \qquad \left\{ \begin{array}{l} d \ll eV \ll \Delta \\ \mu_L = U + \varepsilon_j + 0 \end{array} \right. .$$

Once again, the current jumps each time μ_L passes through $U + \epsilon_j$. This jump for large j (but still such that $U + \varepsilon_j - \mu_L \ll eV$) scales as $1/j^3$, and the ratio of the jump at $j = 1$ to the jump at the resonance level (14) is given by

$$\frac{\delta I_{ne}}{\delta I} \simeq \frac{(g_L + 4g_R)}{8\pi^2} \frac{eV\Delta}{\tilde{d}d}, \quad d \ll eV < \Delta.$$

Noticing that $\tilde{d} = \varepsilon_0 - \varepsilon_1 \simeq d^2/2\Delta$, we see that the first subresonance becomes comparable in height to the main one at voltages as small as $eV \approx 4\pi^2 d^3/\Delta^2(g_L + 4g_R)$.

We conclude by comparing the above results with the experimental data of Ref. [6]. There $\Delta \approx 5d$, the conductances in the normal state are $g_R \approx 10 g_L \approx 1/8$, and the leads are also superconducting. The singularity in the density of states of the leads imply that the effective conductance is increased by factor of $2 - 3$. Neglecting inelastic processes and the Josephson coupling, these parameters imply that when $eV \approx 2d$ the ratio of the subresonances amplitude to that of the first resonance is of order one, while at $eV \sim 2\tilde{d} \approx d/5$ it is of order of 1%. It implies that first subresonance peak associated with tunneling into state "1" cannot be resolved, both because its amplitude and its distance from the main peak are too small. However next subresonance appear already at distance of order d from the main resonance, and for $V > 2d$ have an amplitude comparable with the main resonance.

RELAXATION PROCESSES

Central to our analysis is the assumption that the steady-state occupation configurations of the electrons in the dot are far from equilibrium. This condition holds when the rate $1/\tau_{in}$ of inelastic relaxation processes is smaller than the tunneling rate of an electron into and out of the dot, $1/\tau_{tun}$. In the opposite limit, $1/\tau_{in} > 1/\tau_{tun}$, the system relaxes to equilibrium between tunneling events, and the electrons effectively occupy only one configuration. In this case one expects each resonance cluster to collapse to a single peak. This behavior is illustrated by the dashed line in Fig. 5 where a large inelastic relaxation rate $1/\tau_{in} = 5/\tau_{tun}$ was included in the detailed-balance equations.

The data shown in Figs. 2 and 3 indicate that the metal particle in the experimental system is indeed in a strongly nonequilibrium state. It is useful, however, to consider the various relaxation processes in our system in order to delimit the expected nonequilibrium regime. Relaxation of excited Hartree–Fock states may occur due to: (1) electron-electron interaction in the dot beyond Hartree–Fock; (2) electron-phonon interaction; (3) Auger processes in which an electron in the dot relaxes while another one in the lead is excited; (4) relaxation of an electron in the dot as another electron tunnels out to the lead; (5) thermalization with the leads via tunneling. The last two processes are small corrections since they clearly happen on time scales larger than the tunneling time.

In Ref. [16] it was shown that excited many-body states of closed systems with energy ϵ smaller than $(g/\log g)^{1/2}d$ are merely slightly perturbed Hartree–Fock states. In other words, the overlap between the true many-body state and the corresponding Hartree–Fock approximation is very close to unity. This justifies the use of our model for the low energy resonances since $g \approx 5$ therefore the energy interval $0 < \epsilon < (g/\log g)^{1/2}d$ contains at least the first few excited states. At high source–drain voltage, however, when the dot is excited to energy $g^{1/2}d < \epsilon < gd$, tunneling takes place into quasi-particle states of width $\epsilon^2/(g^2 d)$[17]. This width is larger than the typical separation between nearby resonances but smaller than d. Therefore, electron-electron scattering will obliterate the fine structure of resonances for high energy excitations of the dot.

Consider now the electron–phonon interaction. The temperature, 30 mK, is much smaller than the mean level spacing, therefore, the probability of phonon absorption is negligible, and only emission may take place. The sound velocity in aluminum is $v_s = 6420$ m/sec, therefore the wavelength of a phonon associated with relaxation of energy $\omega \sim d = 1$ meV is approximately 50 Å, the same as the system size. In this regime, we estimate the phonon emission rate to be

$$\frac{1}{\tau_{e-ph}} \sim \left(\frac{2}{3}\epsilon_F\right)^2 \frac{\omega^3 \tau d}{2\rho \hbar^4 v_s^5},$$

where ϵ_F is the Fermi energy (11.7 eV in Al), and ρ is the ion mass density (2.7 g/cm^3 in Al). This rate is that of a clean metal but reduced by a factor of $\tau d/\hbar$ where τ is the elastic mean free time[18]. In ballistic systems, τ is the traversal time across the system of an electron at the Fermi level. Assuming ballistic motion this factor is of order 10^{-3}. The resulting relaxation rate for $\omega = d$ is therefore of order $1/\tau_{e-ph} \approx 10^8$ sec^{-1} which is similar to the tunneling rate $1/\tau_{tun} \approx 6 \cdot 10^8$ sec^{-1} (corresponding to a current of 10^{-10} A through the particle). Thus, by increasing the resistance of the tunnel junctions one should be able to cross over to the near-equilibrium regime shown by the dashed line in Fig. 5.

Relaxation due to Auger process is estimated to be negligible. Two factors reduce this rate considerably: (1) it is exponentially small in w/χ where w is the width of the tunnel junction and χ is the screening length; (2) interaction between electrons on both sides of the tunnel junction can take place only within a very limited volume.

CONCLUSIONS

In this review it was shown that the low-voltage tunneling-resonance spectra of a ultrasmall metallic grains, normal as well as superconducting, reflect nonequilibrium electron configurations. These configurations are reached by resonant tunneling as well as inelastic cotunneling. The first tunneling resonance develops a substructure on energy scales of order of the single-particle mean level spacing, d, while high resonances split due to electron-electron interactions and appear in clusters of width d/g. The latter phenomenon is a result of electron-electron interaction beyond the orthodox model[1]. Relaxation due to electron–phonon interaction, which becomes important for high resistance tunnel barriers, will collapse the clusters. This effect can be used to probe the electron-phonon relaxation rate in nanometer size metal particles.

Appendix A

The purpose of this appendix is to calculate the second moment of off-diagonal matrix elements of the interaction potential $U(\mathbf{r} - \mathbf{r}')$, and show that U_{ijkl} are small as $1/g$, where g is the dimensionless conductance. The subject was discussed in several papers,[19, 20, 21] and is presented here for completeness.

When calculating off-diagonal matrix elements of the interaction potential, it is important take into account screening effects. The relevant two-particle interaction potential is not the bare one, $\tilde{U}(\mathbf{k})$, but rather the statically screened potential: $\tilde{U}_s(\mathbf{k}) = \tilde{U}(\mathbf{k})/[1 + 2\nu\tilde{U}(\mathbf{k}]$ where $\tilde{U}(\mathbf{k})$ is the Fourier transform of the bare two-particle interaction $U(\mathbf{r} - \mathbf{r}')$, and ν is the density of states per unit volume. The contribution to the off-diagonal matrix elements comes only from spatial fluctuations in the electron density (non-zero modes) for which screening is established at very short time, of order of the time it takes for a plasmon to propagate through the system. The latter is much shorter than the relaxation time, t_c, of fluctuations in the electron density. Thus for large ν the screened interaction potential, $U_s(\mathbf{r} - \mathbf{r}')$, is close to a δ-function.

Consider, therefore, the integral

$$U_{ijkl} = \frac{1}{2\nu} \int d\mathbf{r}\psi_i^*(\mathbf{r})\psi_j^*(\mathbf{r})\psi_l(\mathbf{r})\psi_k(\mathbf{r}), \tag{16}$$

where no two indices are the same. Clearly on average $\langle U_{ijkl}\rangle = 0$ since wave functions associated with different eigenvalues are independent and $\langle\psi\rangle = 0$. To estimate the magnitude of the off-diagonal matrix elements we calculate the second moment $\langle|U_{ijkl}|^2\rangle$. The square of the matrix element, $|U_{ijkl}|^2$, contains four pairs of wave functions in the form $\psi_i^*(\mathbf{r})\psi_i(\mathbf{r}')$. Since the correlation between wave functions and eigenenergies are only to order $1/g$, one can approximate these pairs as

$$\psi_i^*(\mathbf{r})\psi_i(\mathbf{r}') \approx \frac{d}{2\pi i}\left[G(\mathbf{r}, \mathbf{r}'; E_i - i\eta) - G(\mathbf{r}, \mathbf{r}'; E_i + i\eta)\right], \tag{17}$$

where d is the single-particle mean level spacing, η is a positive energy which will be taken to zero at the end of the calculation, and $G(\mathbf{r}, \mathbf{r}'; E)$ is the single-particle Green function at energy E. Two basic correlators emerge when calculating the ensemble or the energy average of $|U_{ijkl}|^2$. These correlators, known in disordered diagrammatic nomenclature as the diffuson and Cooperon, are

$$\Pi_\omega(\mathbf{r}, \mathbf{r}') = \langle G(\mathbf{r}, \mathbf{r}'; E + \omega + i\eta)G(\mathbf{r}', \mathbf{r}; E - i\eta)\rangle,$$

and $\langle G(\mathbf{r}, \mathbf{r}'; E + i\eta)G(\mathbf{r}', \mathbf{r}; E + \omega - i\eta)\rangle$. For systems with time reversal symmetry, considered here, these correlators are the same. In the semiclassical limit,

$$\Pi_\omega(\mathbf{r}, \mathbf{r}') = 2\pi\nu \sum_\mu \frac{\bar{\chi}_\mu(\mathbf{r})\chi_\mu(\mathbf{r}')}{-i\omega + \hbar\gamma_\mu}, \tag{18}$$

where the sum is over all classical relaxation modes, i.e. diffusion modes in the case of disordered grains and Perron-Frobenius modes in chaotic systems[§]. γ_μ are the corresponding eigenvalues, and $\bar{\chi}_\mu(\mathbf{r})$ $[\chi_\mu(\mathbf{r}')]$ is the projection of the Perron-Frobenius left [right] eigenfunctions on the real coordinate space at fixed energy E.

With the help of (16), (17) and (18), and assuming all energy differences (such as $E_i - E_j$) to be much smaller than $\hbar\gamma_1$, one obtains

$$\langle|U_{ijkl}|^2\rangle \simeq c'\left(\frac{d}{g}\right)^2$$

where $g = \hbar\gamma_1/d$ is the dimensionless conductance of the system, and c' is a constant of order unity given by

$$c' = \frac{3}{4\pi^2} \sum_{\mu\neq 0, v\neq 0} \int d\mathbf{r}d\mathbf{r}'\mathrm{Re}\frac{\bar{\chi}_\mu(\mathbf{r})\chi_\mu(\mathbf{r}')}{\gamma_\mu/\gamma_1}\mathrm{Re}\frac{\bar{\chi}_v(\mathbf{r}')\chi_v(\mathbf{r})}{\gamma_v/\gamma_1}.$$

[§]For simplicity we consider here chaotic systems in the form of billiards, namely the Hamiltonian contains only a kinetic part, and chaotic dynamics is due to the irregular boundary.

Here we assumed for simplicity that γ_1 is real. In case it contains also an imaginary part, the same formula applies with the substitution $\gamma_1 \rightarrow \mathrm{Re}\{\gamma_1\}$. Notice that there is no zero mode contribution to $\langle |U_{ijkl}|^2\rangle$, since only density fluctuations associated with non zero-modes can induce scattering and contribute to U_{ijkl}. Mathematically this results from the fact that eigenfunctions, $\chi_0(\mathbf{r})$ and $\bar\chi_0(\mathbf{r})$, associated with the zero mode are real, and since Π_ω is always calculated at a finite energy deference, ω, taking its real part excludes the zero-mode contribution.

The rest of this appendix is a semiclassical derivation of formula (18). We begin by writing Green's function in the semiclassical approximation as a sum of two terms[22]:

$$G(\mathbf{r},\mathbf{r}';E \pm i\eta) \simeq G_0(\mathbf{r},\mathbf{r}';E \pm i\eta) + \sqrt{\frac{2\pi}{h^f \hbar}} \sum_l A_{\mathbf{rr}',l} e^{\pm \frac{i}{\hbar} S_{\mathbf{rr}',l}(E)}.$$

Here G_0 is the Weyl contribution associated with "zero length" trajectories. This term is important only at distances $|\mathbf{r} - \mathbf{r}'|$ of order of the particle wavelength and therefore can be neglected. The second term is a sum over all classical trajectories from $\mathbf{r}'$ to $\mathbf{r}$, in which f is the number of degrees of freedom, $S_{\mathbf{rr}',l}(E)$ is the action, and $A_{\mathbf{rr}',l}$ is the corresponding probability amplitude. It is convenient to introduce a local coordinate system in which τ is the time along the trajectory, and $\mathbf{r}_\perp$ are the coordinates perpendicular to the trajectory. In these coordinates[22]

$$|A_{\mathbf{rr}',l}|^2 = \frac{1}{\dot r \dot r'} \mathrm{Det}^{-1}\left(\frac{\partial \mathbf{r}_\perp}{\partial \mathbf{p}'_\perp}\right)_l,$$

where $\dot r$ and $\dot r'$ denote the velocity of the particle at the final and initial points, and $\mathbf{p}'_\perp$ is the conjugate momenta to $\mathbf{r}'_\perp$.

Expanding $S_{\mathbf{rr}',l}(E + \omega)$ to first order in ω, and using the diagonal approximation for the product of the two Green functions, one obtains

$$\Pi_\omega(\mathbf{r},\mathbf{r}') = \frac{2\pi}{h^f \hbar} \sum_l |A_{\mathbf{rr}',l}|^2 e^{i\omega T_l/\hbar} = \frac{2\pi}{h^f \hbar} \int dt P(t) e^{i\omega t/\hbar}, \tag{19}$$

where $T_l = \partial S_{\mathbf{rr}',l}(E)/\partial E$ is the time associated with the l-th trajectory, and

$$P(t) = \sum_l \frac{1}{\dot r \dot r'} \mathrm{Det}^{-1}\left(\frac{\partial \mathbf{r}_\perp}{\partial \mathbf{p}'_\perp}\right)_l \delta(t - T_l). \tag{20}$$

Next we show that $P(t)$ is the projection of the classical propagator in phase space onto configuration space at fixed energy E, namely

$$P(t) = \int d\mathbf{p}' \int d\mathbf{p}\, \delta[E - H(\mathbf{r},\mathbf{p})]\, \langle \mathbf{r},\mathbf{p}|e^{-\mathcal{L}t}|\mathbf{r}',\mathbf{p}'\rangle, \tag{21}$$

where $H(\mathbf{r},\mathbf{p})$ is the classical Hamiltonian of the system and $e^{-\mathcal{L}t}$ is the evolution (Perron-Frobenius) operator for time t. In the coordinate system introduced above, the Hamiltonian function H is the conjugate momentum to the time coordinate τ along the trajectory, therefore

$$P(t) = \int \frac{dH}{\dot r} d\mathbf{p}_\perp \int \frac{dH'}{\dot r'} d\mathbf{p}'_\perp \delta(E - H)\delta(H - H'_t)\delta(\tau - \tau'_t)\delta(\mathbf{r}_\perp - \mathbf{r}'_{\perp t})\delta(\mathbf{p}_\perp - \mathbf{p}'_{\perp t})$$

$$= \frac{1}{\dot r \dot r'} \sum_l \delta(t - T_l) \int_{\Gamma_l} d\mathbf{p}'_\perp \delta(\mathbf{r}_\perp - \mathbf{r}'_{\perp t}),$$

where subscript t denotes the value of the corresponding coordinate after time t starting from the phase space point $(\mathbf{r}', \mathbf{p}')$. Since the energy of the particle is fixed, the integral reduces to a discrete sum over trajectories from $\mathbf{r}'$ to $\mathbf{r}$. The contribution to each trajectory, l, comes from an infinitesimally small region of the coordinate $\mathbf{p}'_\perp$ denoted by Γ_l. Straightforward integration yields the result (20).

Starting now form (21) and using the spectral decomposition of the Perron-Frobenius operator

$$\langle \mathbf{r}, \mathbf{p}|e^{-\mathcal{L}t}|\mathbf{r}', \mathbf{p}'\rangle = f\delta(p^f - p'^f)\sum_\mu e^{-\gamma_\mu t}\bar{\varphi}_\mu(\mathbf{r}, \mathbf{n})\varphi_\mu(\mathbf{r}', \mathbf{n}'),$$

where $\mathbf{n}$ denote the direction of the momentum, while $\bar{\varphi}_\mu$ and φ_μ are the left and right eigenfunctions, we get

$$P(t) = h^f\nu\sum_\mu e^{-\gamma_\mu t}\int\frac{d\mathbf{n}d\mathbf{n}'}{\Omega_f}\bar{\varphi}_\mu(\mathbf{r}, \mathbf{n})\varphi_\mu(\mathbf{r}', \mathbf{n}') = h^f\nu\sum_\mu e^{-\gamma_\mu t}\bar{\chi}_\mu(\mathbf{r})\chi_\mu(\mathbf{r}').$$

Here $\Omega_f = \int d\mathbf{n}$ is the solid angle of a sphere in f dimensions, and $\nu = \int d\mathbf{p}\delta(E - H)/h^f$ defines the density of states per unit volume. Substituting this expression in (19) and performing the time integration yields the required result (18).

Appendix B

In this appendix we derive formula (7) for the cost of introducing an additional electron into an odd state: $\varepsilon_j = E_{2m} - E^{(j)}_{2m-1}$ (ignoring the charging energy). Denote by $F_N = E_N - \mu_N N$ the free energy of the system where E_N is the ground state energy with N electrons, and μ_N is the chemical potential. For the superconducting state with unpaired electron in level j one has

$$F^{(j)}_{2m-1} = \sum_{k\neq j}(\xi_k - \epsilon_k) + \Delta^2/\lambda + \xi_j, \qquad \epsilon_k = \sqrt{\xi_k^2 + \Delta^2}, \qquad (22)$$

where λ is the pairing coupling constant. In the intermediate state where another electron tunneled into the j-th state and pairs with the already existing one: $F_{2m} = \sum_k(\xi_k - \epsilon_k) + \Delta^2/\lambda$. The parameters in these two equations, Δ and μ (which are functions of N), are determined by the relations:

$$\frac{\partial F_N}{\partial \Delta_N} = 0 \qquad\qquad N = -\frac{\partial F_N}{\partial \mu_N}$$

Notice that the single-particle energies ξ_k are measured with respect to the chemical potential, thus for the differentiation with respect to μ_N it it is convenient to introduce $\xi_k = \tilde{\xi}_k - \mu_N$, and $\epsilon_k = \sqrt{(\tilde{\xi}_k - \mu_N)^2 + \Delta_N^2}$, where $\tilde{\xi}_k$ are independent of μ_N.

The above derivatives for the *even case*, $N = 2m$, give

$$\sum_k\frac{1}{2\sqrt{(\tilde{\xi}_k - \mu_{2m})^2 + \Delta_{2m}^2}} = \frac{1}{\lambda}, \qquad 2m = \sum_k\left(1 - \frac{\tilde{\xi}_k - \mu_{2m}}{\sqrt{(\tilde{\xi}_k - \mu_{2m})^2 + \Delta_{2m}^2}}\right), \qquad (23)$$

while for the *odd case*, $N = 2m - 1$,

$$\sum_{k\neq j}\frac{1}{2\sqrt{(\tilde{\xi}_k - \mu^{(j)}_{2m-1})^2 + \Delta_{2m-1}^2}} = \frac{1}{\lambda}, \qquad 2m - 2 = \sum_{k\neq j}\left(1 - \frac{\tilde{\xi}_k - \mu^{(j)}_{2m-1}}{\sqrt{(\tilde{\xi}_k - \mu^{(j)}_{2m-1})^2 + \Delta_{2m-1}^2}}\right).$$

$$(24)$$

We expand $\varepsilon_j = E_{2m} - E_{2m-1}^{(j)} = F_{2m}(\mu_{2m}; \Delta_{2m}) - F_{2m-1}^{(j)}(\mu_{2m-1}; \Delta_{2m-1}) + 2m\mu_{2m} - (2m-1)\mu_{2m-1}$ to linear order in the differences $\Delta_{2m} - \Delta_{2m-1}$, and $\mu_{2m} - \mu_{2m-1}$:

$$\begin{aligned}
\varepsilon_j &\simeq F_{2m}(\mu_{2m}; \Delta_{2m}) - F_{2m-1}(\mu_{2m}; \Delta_{2m}) + \mu_{2m} \\
&= \mu_{2m-1}^{(j)} - \sqrt{(\tilde{\xi}_j - \mu_{2m})^2 + \Delta_{2m}^2}.
\end{aligned} \tag{25}$$

We are left, now, with the problem of finding the dependence of $\mu_{2m-1}^{(j)}$ on the unpaired electron state j. For this purpose we choose μ_{2m} as the reference point and calculate the difference $\mu_{2m-1}^{(j)} - \mu_{2m}$. From Eqs. (23) and (24) we have

$$\begin{aligned}
2m - 2 &= \sum_{k \neq j} \left(1 - \frac{\tilde{\xi}_k - \mu_{2m-1}^{(j)}}{\sqrt{(\tilde{\xi}_k - \mu_{2m-1}^{(j)})^2 + \Delta_{2m-1}^2}} \right) \\
&= \sum_k \left(1 - \frac{\tilde{\xi}_k - \mu_{2m}}{\sqrt{(\tilde{\xi}_k - \mu_{2m})^2 + \Delta_{2m}^2}} \right) - \left(1 - \frac{\tilde{\xi}_j - \mu_{2m}}{\sqrt{(\tilde{\xi}_j - \mu_{2m})^2 + \Delta_{2m}^2}} \right) \\
&\quad - (\mu_{2m-1}^{(j)} - \mu_{2m}) \sum_k \left(\frac{-1}{\sqrt{(\tilde{\xi}_k - \mu_{2m})^2 + \Delta_{2m}^2}} + \frac{(\tilde{\xi}_k - \mu_{2m})^2}{\left((\tilde{\xi}_k - \mu_{2m})^2 + \Delta_{2m}^2 \right)^{3/2}} \right).
\end{aligned}$$

(The correction to Δ vanishes in the limit where the single-particle mean level spacing d is much smaller than the superconducting gap, $d \ll \Delta$). Since the first sum in the second line equals $2m$ we have

$$\begin{aligned}
-2 &= 1 - \frac{\tilde{\xi}_j - \mu_{2m}}{\sqrt{(\tilde{\xi}_j - \mu_{2m})^2 + \Delta_{2m}^2}} + \sum_k \frac{(\mu_{2m-1}^{(j)} - \mu_{2m})\Delta_{2m}^2}{\left((\tilde{\xi}_k - \mu_{2m})^2 + \Delta_{2m}^2 \right)^{3/2}} \\
&\simeq 1 - \frac{\tilde{\xi}_j - \mu_{2m}}{\Delta} + (\mu_{2m-1}^{(j)} - \mu_{2m}) \int \frac{d\xi}{d} \frac{\Delta^2}{(\xi^2 + \Delta^2)^{3/2}}.
\end{aligned}$$

Thus

$$\mu_{2m} \simeq \mu_{2m-1}^{(j)} + \frac{d}{2}\left(3 - \frac{\xi_j}{\Delta} \right),$$

and substituting this result in (25) we obtain (7).

ACKNOWLEDGMENTS

This review summarize results of collaboration with I. L. Aleiner, B. L. Altshuler, D. C. Ralph, M. Tinkham, and N. S. Wingreen, whom it is my pleasure to thank. I would like also to thank N. Brenner for many useful comments on the manuscript.

REFERENCES

1.   D. V. Averin and K. K. Likharev, in *Mesoscopic Phenomena in Solids*, eds. B. L. Altshuler, P. A. Lee, and R. A. Webb (Elsevier, NY, 1991) pp. 173–271.

2.   M. Kastner, Rev. Mod. Phys., **64**, 849 (1992).

3.   D.V. Averin and A.N. Korotkov, Sov. Phys. JETP. **97**, 1161 (1990).

4.   A.T. Johnson *et. al.*, Phys. Rev. Lett. **69**, 1592 (1992); S. Tarucha *et. al.*, *ibid*, **77**, 3613 (1996).

5.   D.C. Ralph, C.T. Black, and M. Tinkham, Physica B **218**, 258 (1996).

6.   D.C. Ralph, C.T. Black and M. Tinkham, Phys. Rev. Lett **78**, 4087 (1997).

7.   O. Agam, N. S. Wingreen, B. L. Altshuler, D. C. Ralph, and M. Tinkham, Phys. Rev. Lett, **78**, 1956 (1997).

8.   O. Agam and I. L. Aleiner, Phys. Rev. B, **56**, R5759 (1997).

9.   J. von Delft, A. D. Zaikin, D. S. Golubev, and W. Tichy Phys. Rev. Lett. **77**, 3189 (1996) ; R. A. Smith and V. Ambegaokar, Phys. Rev. Lett. **77**, 4962 (1996).

10.   D. V. Averin and A. N. Korotkov, Zh. Eksp. Teor. Fiz. **97**, 1661 (1990) [Sov. Phys. JETP **70**, 937 (1990)]. This work neglects fluctuations of the interaction energy ($\delta U = 0$).

11.   Ya. M. Blanter, A. D. Mirlin and B. A. Muzykanskii, Phys. Rev. Lett. **78** 2449 (1997).

12.   O. Agam, B. L. Altshuler, and A. V. Andreev, Phys. Rev. Lett. **75**, 4389 (1995).

13.   D.V. Averin and A.A. Odintsov, Phys. lett. A **140**, 251 (1990); D.V. Averin and Yu.N. Nazarov, Phys. Rev. Lett. **65**, 2446 (1990).

14.   D.V. Averin and Yu.N. Nazarov, Phys. Rev. Lett. **68**, 1993 (1992).

15.   M. Tinkham, *Introduction to superconductivity*, (McGraw–Hill, New York, 1980).

16.   B. L. Altshuler, Y. Gefen, A. Kamenev, and L. S. Levitov, Phys. Rev. Lett. **78**, 2803 (1997).

17.   U. Sivan, Y. Imry, and A. G. Aronov, Europhys. Lett. **28**, 115 (1994).

18.   M. Yu Reizer and A. V. Sergeyev, Zh. Eksp. Teor. Fiz. **90**, 1056 (1986) [Sov. Phys. JETP **63**, 616 (1986)].

19.   Ya. M. Blanter, Phys. Rev. B **54**, 12807 (1996).

20.   Ya. M. Blanter and A. D. Mirlin, Phys. Rev. E **55**, 6514 (1997).

21.   I. L. Aleiner and L. I. Glazman, cond-mat/9710195 (1997).

22.   M. C. Gutzwiller, *Chaos in Classica and Quantum Mechanics* (Springer, New York, 1990).

PAIR CORRELATIONS OF QUANTUM CHAOTIC MAPS FROM SUPERSYMMETRY

M R Zirnbauer

Institut f. Theor. Physik, Universität Köln, Germany
email: *zirn@thp.Uni-Koeln.DE*

A conjecture due to Bohigas, Giannoni and Schmit (BGS), stating that the energy level correlations of quantum chaotic systems generically obey the laws of random matrix theory, is given a precise formulation for quantized symplectic maps. No statement is made about any individual quantum map. Rather, a few–parameter ensemble of maps is considered, such that the deterministic map is composed with a diffusion operator on average. The ensemble is a "quantum" one, which is to say that the diffusion operator contracts to the identity in the classical limit. It is argued that the BGS conjecture is true on average over such an ensemble, provided that the classical map is mixing. The method used is closely related to the supersymmetric formalism of Andreev et al for chaotic Hamiltonian systems.

1. INTRODUCTION

One of the goals of semiclassical analysis is to characterize the energy level correlations of quantum mechanical systems in the limit $\hbar \to 0$. To gain an understanding of these correlations, physicists have traditionally used semiclassical trace formulas, relating the density of energy levels to the periodic orbits of the classical dynamics (see e.g. the contributions by Bogomolny, Keating, and Smilansky to this volume). In brief, the qualitative picture that has emerged is that integrable systems have Poisson statistics, whereas for a generic chaotic system one expects random matrix (or Wigner–Dyson) statistics. This picture is supported by a large body of numerical evidence, and applies equally to Hamiltonian systems and symplectic maps. Bohigas, Giannoni, and Schmit (BGS) [1] are credited for having put forth the random matrix conjecture for quantum chaotic systems. The principal development toward an analytical theory is due to Berry [2] who augmented Gutzwiller's trace formula with sum rule arguments to determine the form factor for times much shorter and much longer than the Heisenberg time. Berry's work has recently been refined by Bogomolny and Keating [3].

In these lectures, I will describe a supersymmetric formalism that offers an alternative to the trace formula approach and promises rigorous results for chaotic symplectic maps. The formalism bears much similarity to that of Andreev et al[4] for chaotic Hamiltonian systems, reviewed at this workshop by Ben Simons. Only the simplest case of maps without unitary or anti–unitary symmetries will be treated here. In particular,

Supersymmetry and Trace Formulae: Chaos and Disorder
Edited by Lerner *et al.*, Kluwer Academic / Plenum Publishers, New York, 1999

time reversal invariance will be assumed to be broken. The starting point of our theory is an expression for the two-level correlation function which involves two ratios of quantum spectral determinants (Sect. 3). The determinants are written as traces over a Fock space of fermions, and the inverse determinants as traces over a Fock space of bosons (Sect. 4). By the introduction of generalized coherent states, the two-level correlation function can then be expressed as a Berezin integral over the Riemannian symmetric superspace of type $AIII|AIII$ (Sect. 6). A quick review of the basic mathematics underlying the Berezin integral is given in Sect. 5.

The BGS conjecture is not a precise statement, as the meaning of the word "generic" is left undefined. (The qualification "generic" serves to exclude some prominent counterexamples, such as arithmetic billiards and the cat maps, which are paradigms of classical chaos but fail to obey random matrix statistics when quantized in the canonical way.) It turns out that in our formalism it is impossible to prove the conjecture in the sense originally intended, i.e. as a statement about an individual quantum system. The technical obstacle that prevents us from making progress for individual systems is the nonexistence of the semiclassical limit (Sect. 7) for the Berezin integral representation of the two-level correlation function, or any other correlation function. To enforce such a limit one needs to impose some sort of coarse graining, or ultraviolet regulator, on the Berezin integral. The way to implement regularization is to compose the map with a stochastic Hamiltonian flow and average (Sect. 8).

From experience with the supersymmetry formalism applied to disordered electron systems [5] it is known that the so–called zero mode approximation to the Berezin integral gives the random matrix answer. The strategy therefore is to establish sufficient conditions for this approximation, which is of the saddle point type, to be justified. The key issue now becomes the stability of the manifold of saddle points. Without ensemble averaging, the Hessian of the saddle point manifold is not positive, and the saddle point approximation is false. Averaging over the stochastic Hamiltonian flow with "time" parameter ϵ stabilizes the saddle point manifold and leads to a good asymptotic expansion (Sect. 9). If the classical map is mixing we argue from power counting that stability can achieved with a time parameter ϵ that vanishes with $\hbar$ as $\hbar^\alpha$ where $0 < \alpha < 1$. Thus we propose that the BGS conjecture holds on average over an ensemble of quantum chaotic maps *all of which have the same classical limit.*

My original intention was to finish with a discussion of the finite $\hbar$ corrections to the universal random matrix result. These are a very timely issue, as the ballistic nonlinear σ model calculation of Andreev et al disagrees with the result of Bogomolny and Keating derived from the diagonal approximation. (The difference lies in the weights given to the contributions from periodic orbits with repetition number $r \geq 2$.) There exist strong indications that the result of Bogomolny and Keating is the correct one. I believe that the resolution of the discrepancy will teach us something profound about the ballistic nonlinear σ model, but this definitely exceeds the scope of the present lectures and must be left for a future publication.

2. PRELIMINARIES

We begin with some basic definitions.

Let M, called the phase space, be a $2d$–dimensional compact manifold with a symplectic structure ω, i.e. an antisymmetric second rank tensor that is expressed in local coordinates q_k, p_k by $\omega = \sum_{k=1}^{d} \mathrm{d}p_k \wedge \mathrm{d}q_k$. Via ω, every function $f : M \to \mathbb{R}$ is associated with a Hamiltonian vector field $\Xi_f = \sum_{k=1}^{d} \left(\frac{\partial f}{\partial p_k} \frac{\partial}{\partial q_k} - \frac{\partial f}{\partial q_k} \frac{\partial}{\partial p_k} \right)$. The volume element, or Liouville measure, on M is denoted by $dx := \omega^{\wedge d}$. A map $\chi : M \to M$ is called symplectic (or area preserving for $d = 1$) if it preserves ω and, consequently, the Liouville measure. The map χ acts on functions $f : M \to \mathbb{R}$ by $(\chi^* f)(x) = f(\chi(x))$. The dual of its inverse, χ^{-1*}, is called the Frobenius–Perron operator. A little computation,

$$(\chi^{-1*} f)(x) = f(\chi^{-1}(x))$$

$$= \int_M dy \, \delta(\chi^{-1}(x) - y) f(y) = \int_M dy \, \delta(x - \chi(y)) f(y) \, ,$$

shows that the Frobenius–Perron operator has the integral kernel

$$(\chi^{-1*})(x, y) = \delta(x - \chi(y)) \, ,$$

which identifies χ^{-1*} as the propagator (for one time step) of the classical dynamics obtained by iterating χ.

Quantization turns every smooth function $f : M \to \mathbb{R}$ into a self–adjoint operator $\mathcal{Q}_N(f) = F_N$ on a Hilbert space $\mathcal{H}_N$ of finite dimension $N = (2\pi\hbar)^{-d} \int_M dx$. The quantization of a smooth phase space function is called a pseudodifferential operator. The quantum analog of the classical map χ is a unitary operator $U_N : \mathcal{H}_N \to \mathcal{H}_N$, whose eigenvalues we denote by $e^{i\theta_n}$ ($n = 1, \ldots, N$). The θ_n's are sometimes called quasienergies. Let $\mathcal{D}_N = \mathcal{H}_N \otimes \mathcal{H}_N^* \simeq \mathrm{End}(\mathcal{H}_N)$ be the space of linear operators on $\mathcal{H}_N$. The unitary operator U_N acts on $\mathcal{D}_N$ by conjugation. More precisely, the quantum time evolution (by one time step) of an element $F_N \in \mathcal{D}_N$ is given by

$$\mathrm{Ad}(U_N)^{-1} F_N = U_N^{-1} F_N U_N \, .$$

(The notation $\mathrm{Ad}(U)$ for conjugation by U is taken from Lie algebra theory.) In the classical limit $\hbar \to 0$, $\mathrm{Ad}(U_N)^{-1}$ acting on a pseudodifferential operator $F_N = \mathcal{Q}_N(f)$ approaches the Frobenius–Perron operator:

$$\mathrm{Ad}(U_N)^{-1} F_N \xrightarrow{\hbar \to 0} \mathcal{Q}_N(\chi^{-1*} f) \, .$$

Examples illustrating these general definitions are furnished by the cat maps [6]. The phase space in this case is the two–torus T^2, coordinatized by a pair of canonically conjugate position and momentum functions q and p, which are defined modulo 1. For a concrete example, consider the cat map $\chi : \mathrm{T}^2 \to \mathrm{T}^2$ that acts on q and p as

$$\begin{aligned} q \circ \chi &= \ q + 2p \quad (\mathrm{mod}\ 1) \, , \\ p \circ \chi &= 2q + 5p \quad (\mathrm{mod}\ 1) \, , \end{aligned}$$

which entails a Frobenius–Perron operator expressed by

$$\chi^{-1*} = \exp\left(-2q\partial/\partial p\right) \exp\left(-2p\partial/\partial q\right)$$

on differentiable functions. [Note that $\exp\left(-\partial/\partial q\right) = \exp\left(-\partial/\partial p\right) = 1$, so that $\exp\left(-2p\partial/\partial q\right)$ and $\exp\left(-2q\partial/\partial p\right)$ are globally well–defined, although $p\partial/\partial q$ and $q\partial/\partial p$ are not.] The locally defined vector fields $\Xi_f = 2q\partial/\partial p$ and $\Xi_g = 2p\partial/\partial q$ derive from locally defined functions $f = -q^2$ and $g = p^2$, and quantization of χ yields a unitary operator

$$U_N = \exp\left(iP_N^2/\hbar\right) \exp\left(-iQ_N^2/\hbar\right) \, ,$$

where $\hbar^{-1} = 2\pi N \int_0^1 dp \int_0^1 dq = 2\pi N$. In a basis where $\exp\left(2\pi i P_N\right)$ is diagonal with the eigenvalues being the Nth roots of unity $\exp\left(2\pi i n/N\right)$ ($n = 1, 2, \ldots, N$), U_N has the following matrix

$$(U_N)_{nn'} = \frac{1}{\sqrt{N}} \exp \frac{2\pi i}{N} \left(n^2 + (n - n')^2\right) \, .$$

3. BGS CONJECTURE FORMULATED PRECISELY

There exist many statistical measures of level correlations. The one we will focus on here is the so-called pair correlation function, defined on a test function $f : \mathrm{U}(1) \to \mathbb{C}$ by

$$C(f) = \frac{2\pi}{N^2} \sum_{m,n=1}^{N} f\left(e^{i(\theta_m - \theta_n)}\right) - \int_0^{2\pi} f(e^{i\theta}) \, \mathrm{d}\theta .$$

If we take the liberty of choosing for f the Dirac δ–distribution centered at $e^{i\varphi}$, $\delta_\varphi(e^{i\theta}) = (2\pi)^{-1} \sum_{l \in \mathbb{Z}} e^{il(\theta - \varphi)}$, we get the two–level correlation function

$$R_2(\theta) = C(\delta_\theta) = \frac{2}{N^2} \sum_{l=1}^{\infty} \cos\left(l\theta\right) \left|\mathrm{Tr}\, U^l\right|^2 .$$

For our purposes, it is technically convenient to consider in addition to R_2 the following correlator of determinants:

$$\Omega(\gamma_{+0}, \gamma_{+1}; \gamma_{-0}, \gamma_{-1}) = \int_0^{2\pi} \frac{\mathrm{d}\theta}{2\pi} \frac{\mathrm{Det}(1 - \gamma_{+1} e^{i\theta} U_N)\mathrm{Det}(1 - \gamma_{-1} e^{-i\theta} U_N^{-1})}{\mathrm{Det}(1 - \gamma_{+0} e^{i\theta} U_N)\mathrm{Det}(1 - \gamma_{-0} e^{-i\theta} U_N^{-1})} ,$$

where $\gamma_{\pm 0}, \gamma_{\pm 1}$ are complex parameters. Note that the two–level correlation function can be extracted from Ω by taking two derivatives:

$$R_2(\theta) = -\frac{1}{2N^2} \frac{\partial^2}{\partial\varphi \partial\varphi'} \mathrm{Re}\, \Omega\left(e^{i\theta + i\varphi}, e^{i\theta - i\varphi}; e^{i\varphi'}, e^{-i\varphi'}\right)\Bigg|_{\varphi = \varphi' = 0} .$$

In Sects. 4 and 6, I will sketch the derivation of a formula for Ω that underlies the supersymmetric method for quantized maps, and I will go over some of the basic mathematics needed. Before doing so, it is proper to motivate the reader by giving an outline of the results that can be obtained.

The first and very important message is that our method *gives no answer at all* for the pair correlation function, or any other correlation function, of an *individual* quantized map. (This is to be contrasted with what has been claimed by Andreev et al.) For reasons that will be explained later on, we can make a statement only about the *expected value* of the correlations for an *ensemble* of quantized maps over a well–chosen probability space.

Let me therefore digress and describe this ensemble. We pick a finite number s of Hamiltonian functions $x_k : M \to \mathbb{R}$, with associated Hamiltonian vector fields Ξ_k, and Hamiltonian operators $X_{k,N} : \mathcal{H}_N \to \mathcal{H}_N$. The latter we multiply by uncorrelated Gaussian random variables ξ_k with zero mean and $\hbar$–dependent variances $\langle \xi_k^2 \rangle = 2\epsilon(\hbar)$, and we consider instead of a single U_N an s–parameter family of unitary operators

$$U_N(\xi) := \exp\left(i \sum_{k=1}^{s} \xi_k X_{k,N}/\hbar\right) U_N .$$

The choice of Hamiltonians x_k is constrained by the requirement that the weighted sum of squares $\sum_{k=1}^{s} \Xi_k^2$ be an *elliptic* operator.

The good news is that we can get an answer for the expected correlation functions $\langle \Omega \rangle$ and $\langle R_2 \rangle$ by taking $\epsilon(\hbar) = \mathrm{const} \times \hbar^\alpha$ where α is a number between 0 and 1 (and we may choose $\alpha = 1/2$, say). This dependence of the variances on $\hbar$ means that, roughly speaking, all members of the ensemble have the *same* classical limit. In other words, the ensemble we use is "quantum", and we could say that what we are averaging over are different quantizations of the same classical map.

Now, let the classical map be mixing, and arrange for the quantized map to have no unitary or anti–unitary symmetries. We will argue that the expected correlation functions in this case have the following semiclassical limits:

$$\lim_{N\to\infty} \left\langle \Omega\left(e^{ic_{+0}/N}, e^{ic_{+1}/N}; e^{-ic_{-0}/N}, e^{-ic_{-1}/N}\right)\right\rangle =$$

$$1 - \frac{(c_{+0} - c_{+1})(c_{-0} - c_{-1})}{(c_{+0} - c_{-0})(c_{+1} - c_{-1})}\left(1 - e^{i(c_{+1}-c_{-1})}\right), \tag{1}$$

and

$$\lim_{N\to\infty} \left\langle R_2(2\pi x/N)\right\rangle = \delta(x) - \sin^2(\pi x)/(\pi x)^2 .$$

To put this result into context, we observe that our expressions for $\langle\Omega\rangle$ and $\langle R_2\rangle$ coincide with those for a random unitary matrix in the large–N limit. What we propose, then, is a *precise* formulation of the BGS conjecture: random matrix theory applies to chaotic maps *on average* over a quantum ensemble of the kind specified. More specifically, although there exist counterexamples such as the cat maps, these are a set of measure zero in parameter space, and the random matrix result does obtain on averaging.

4. SUPERSYMMETRIC FORMULA FOR MAPS (I)

In what follows we shall construct a novel and useful integral representation of the correlator Ω. Our first step will be to express the determinants $\mathrm{Det}(1 - \gamma U_N)^{\pm 1}$ as traces over Fock space. To that end, let $\mathcal{F}_N$ be the 2^N–dimensional fermionic Fock space that is generated by acting with creation operators $f_n^\dagger$ ($n = 1, 2, \ldots, N$) on a particle vacuum $|0\rangle$. The operators $f_n^\dagger$ and their adjoints f_n satisfy the canonical anticommutation relation for fermions: $f_n^\dagger f_{n'} + f_{n'} f_n^\dagger = \delta_{nn'}$.

Now we claim that determinants such as those appearing in the numerator of the defining expression for Ω can be written as Fock space traces:

$$\mathrm{Det}(1 - U_N) = \mathrm{Tr}_{\mathcal{F}_N}(-1)^{N_F} \exp \sum_{n,n'=1}^{N} f_n^\dagger (\ln U_N)_{nn'} f_{n'}$$

where $N_F = \sum_n f_n^\dagger f_n$ is the fermion number. To prove that claim, we may assume the matrix U_N to be diagonal as both sides of the equation are invariant under a change of the single particle basis. Put $(U_N)_{nn'} = \lambda_n \delta_{nn'}$. Then the above equation is verified by the following simple computation:

$$\prod_{n=1}^{N}(1 - \lambda_n) = \prod_{n=1}^{N}\left(e^{0\times\ln\lambda_n} - e^{1\times\ln\lambda_n}\right)$$

$$= \prod_{n=1}^{N}\mathrm{Tr}_{\mathcal{F}_1}(-1)^{f_1^\dagger f_1}\exp\left(f_1^\dagger f_1 \ln\lambda_n\right)$$

$$= \mathrm{Tr}_{\mathcal{F}_N}(-1)^{N_F}\exp\sum_{n=1}^{N} f_n^\dagger f_n \ln\lambda_n .$$

The multivaluedness of the logarithm does not cause any ambiguity here as $\exp 2\pi i f_n^\dagger f_n = 1$.

A similar expression for the inverse of a determinant can be obtained by substituting bosons for fermions. Let $\mathcal{B}_N$ be the bosonic Fock space generated by canonical boson operators $b_n^\dagger$, b_n with commutation relations $b_{n'}b_n^\dagger - b_n^\dagger b_{n'} = \delta_{nn'}$. For $|\gamma| < 1$ we have the identity

$$\mathrm{Det}(1 - \gamma U_N)^{-1} = \mathrm{Tr}_{\mathcal{B}_N} \exp \sum_{n,n'=1}^{N} b_n^\dagger (\ln \gamma U_N)_{nn'} b_{n'} \,.$$

Again, this is verified by diagonalizing U_N and computing a single boson trace,

$$\mathrm{Tr}_{\mathcal{B}_1} e^{b_1^\dagger b_1 \ln \lambda} = e^{0 \times \ln \lambda} + e^{1 \times \ln \lambda} + \cdots + e^{n \ln \lambda} + \cdots = (1 - \lambda)^{-1} \,.$$

The inequality $|\gamma| < 1$ is needed in order for the sum on the right hand side to converge.

Let us now assemble the various pieces to build a formula for Ω. There are two determinants in both the numerator and the denominator, so we introduce two fermionic and two bosonic Fock spaces $\mathcal{F}_N^\pm$ and $\mathcal{B}_N^\pm$, and write

$$\frac{\mathrm{Det}(1 - \gamma_{+1} U_N)\mathrm{Det}(1 - \gamma_{-1} U_N^{-1})}{\mathrm{Det}(1 - \gamma_{+0} U_N)\mathrm{Det}(1 - \gamma_{-0} U_N^{-1})}$$

$$= \mathrm{Tr}_{\mathcal{B}_N^+ \otimes \mathcal{F}_N^+ \otimes \mathcal{B}_N^- \otimes \mathcal{F}_N^-} (-1)^{N_F} \exp \sum_{nn'} \Big(b_{+n}^\dagger (\ln \gamma_{+0} U_N)_{nn'} b_{+n'}$$

$$+ b_{-n}^\dagger (\ln \gamma_{-0} U_N^{-1})_{nn'} b_{-n'}$$

$$+ f_{+n}^\dagger (\ln \gamma_{+1} U_N)_{nn'} f_{+n'}$$

$$+ f_{-n}^\dagger (\ln \gamma_{-1} U_N^{-1})_{nn'} f_{-n'} \Big) \,. \tag{2}$$

Here $N_F = \sum_{n=1}^{N} \left(f_{+n}^\dagger f_{+n} + f_{-n}^\dagger f_{-n} \right)$, and convergence of the sums requires $|\gamma_{\pm 0}| <$ 1. To obtain Ω we need to replace U_N by $e^{i\theta} U_N$ and integrate over θ. If $N_\pm = \sum_n \left(b_{\pm n}^\dagger b_{\pm n} + f_{\pm n}^\dagger f_{\pm n} \right)$ counts the number of retarded $(+)$ or advanced $(-)$ particles, this produces an extra factor

$$\int_0^{2\pi} \frac{d\theta}{2\pi} (e^{i\theta})^{N_+ - N_-} =: P$$

under the trace. The operator P gives unity when applied to a state $|\psi\rangle$ with $(N_+ - N_-)|\psi\rangle = 0$, and zero otherwise, and thus projects the super Fock space $\mathcal{B}_N^+ \otimes \mathcal{F}_N^+ \otimes \mathcal{B}_N^- \otimes \mathcal{F}_N^-$ onto the subspace with equal retarded and advanced particle numbers. Hence, on writing $\sum_{nn'} b_{+n}^\dagger (\ln \gamma_{+0} U_N)_{nn'} b_{+n'} =: b_+^\dagger \ln (\gamma_{+0} U_N) b_+$ etc. to simplify the notation, we obtain

$$\Omega(\gamma_{+0}, \gamma_{+1}; \gamma_{-0}, \gamma_{-1}) =$$

$$\mathrm{Tr}_{N_+ = N_-} (-1)^{N_F} \exp \Big(b_+^\dagger \ln (\gamma_{+0} U_N) b_+ + b_-^\dagger \ln (\gamma_{-0} U_N^{-1}) b_-$$

$$+ f_+^\dagger \ln (\gamma_{+1} U_N) f_+ + f_-^\dagger \ln (\gamma_{-1} U_N^{-1}) f_- \Big) \,. \tag{3}$$

The final step in the construction of the supersymmetric formula for Ω is to pass from the trace over super Fock space to a Berezin integral, by the introduction of generalized coherent states. This step is perhaps unfamiliar and deserves a certain amount of explanation.

Let me first illustrate the basic idea at the example of $\gamma_{+0} = \gamma_{-0} = 0$, in which case the bosonic degrees of freedom are absent. Thus, we start from the relation

$$\Omega(0, \gamma_{+1}; 0, \gamma_{-1}) = \mathrm{Tr}_{\mathcal{F}_N^+ \otimes \mathcal{F}_N^-} P_F \exp \left(f_+^\dagger \ln (\gamma_{+1} U_N) f_+ + f_-^\dagger \ln (\gamma_{-1} U_N^{-1}) f_- \right)$$

where $P_F = (2\pi)^{-1} \int_0^{2\pi} d\theta \exp i\theta \sum_n (f_{+n}^\dagger f_{+n} - f_{-n}^\dagger f_{-n})$. Consider then the BCS (or pair) coherent states

$$|Z\rangle = \exp\left(\sum_{n,n'=1}^{N} f_{+n}^\dagger Z_{nn'} f_{-n'}^\dagger\right)|0\rangle$$

with complex amplitudes $Z_{nn'}$. Being made from pairs of one retarded and one advanced particle, these states satisfy the constraint $P_F|Z\rangle = |Z\rangle$. Moreover, they enjoy the key property of providing a resolution of the projector P_F:

$$P_F = \int_{\mathbb{C}^{N\times N}} d\mu_N(Z, \bar{Z})\, |Z\rangle\langle Z| \, ,$$

where $d\mu_N(Z, \bar{Z}) = \mathrm{const} \times \mathrm{Det}(1 + Z^\dagger Z)^{-2N-1} \prod_{n,n'=1}^{N} dZ_{nn'}\, d\bar{Z}_{nn'}$. The proof proceeds via Schur's lemma, by demonstrating that $\int d\mu_N(Z, \bar{Z})\, |Z\rangle\langle Z|$ commutes with all of the operators $f_{+n}^\dagger f_{-n'}$ and $f_{-n}^\dagger f_{+n'}$ $(n, n' = 1, \ldots, N)$. A standard computation on coherent states [7] yields

$$\langle Z| \exp \sum_{nn'} \left(f_+^\dagger \ln\left(\gamma_{+1} U_N\right) f_+ + f_-^\dagger \ln\left(\gamma_{-1} U_N^{-1}\right) f_- \right) |Z\rangle$$

$$= \mathrm{Det}(1 + \gamma_{+1}\gamma_{-1} Z^\dagger U_N Z U_N^{-1}) \, .$$

Hence, by inserting the coherent state resolution of P_F into the expression for the correlator $\Omega(0, \gamma_{+1}; 0, \gamma_{-1})$, we arrive at the formula

$$\Omega(0, \gamma_{+1}; 0, \gamma_{-1}) = \int_{\mathbb{C}^{N\times N}} d\mu_N(Z, \bar{Z})\, \mathrm{Det}\left(1 + \gamma_{+1}\gamma_{-1} Z^\dagger U_N Z U_N^{-1}\right) \, .$$

Our aim now is to extend this formula to allow for a nonvanishing first and third argument of Ω, in which case the Fock space expression involves both fermions and bosons. This requires a supergeneralization of the notion of coherent states and their integration measure. The purpose of the next section is to review some of the basic mathematics needed. We will return to the task of expressing Ω as a coherent state integral in Sect. 6.

5. BASIC NOTIONS OF ANALYSIS ON SUPERMANIFOLDS

Familiarity with the elementary theory of differentiable manifolds is assumed. In the present section the summation convention is in force.

Let $\Lambda\mathbb{R}^q$ be the Grassmann algebra over $\mathbb{R}$ with q generators $\psi^1, \psi^2, \ldots, \psi^q$. By definition, the Grassmann generators anticommute: $\psi^i\psi^j + \psi^j\psi^i = 0$. They are known to physicists from their use in the construction of path integrals for fermions. Note in particular $\psi^i\psi^i = -\psi^i\psi^i = 0$.

Now pick some domain $U \subset \mathbb{R}^p$. A map F from U into the Grassmann algebra,

$$F: \quad U \to \Lambda\mathbb{R}^q$$
$$a \mapsto F_0(a) + F_i(a)\psi^i + F_{i_1 i_2}(a)\psi^{i_1}\psi^{i_2} + \cdots + F_{i_1\cdots i_q}(a)\psi^{i_1}\cdots\psi^{i_q} \, ,$$

will be referred to as a *superfunction* (or simply a function) on U. F_0 is called the *number part* of F. We distinguish between *even* and *odd* superfunctions (the former are even and the latter are odd in the Grassmann generators). This distinction endows the algebra of superfunctions with a $\mathbb{Z}_2$-*grading*.

We can now describe what is meant by a (p,q)–dimensional real–analytic super-manifold in the sense of Berezin, Kostant and Leites [8]. Take a (real–analytic) manifold M of dimension p and cover it by domains $U_1, U_2, \ldots$. Let p even and q odd coordinate superfunctions be given on each domain. [For the purpose of illustration, we denote these by

$$x^1, \ldots, x^p;\ \xi^1, \ldots, \xi^q \quad \text{on } U_1,$$
$$\text{and by}\ \ y^1, \ldots, y^p;\ \eta^1, \ldots, \eta^q \quad \text{on } U_2.]$$

Assume that transition functions exist on overlapping domains and are consistent with the $\mathbb{Z}_2$–grading. [For example, on $U_1 \cap U_2$ this means that there are relations

$$y^i = f^i(x, \xi) = f^i_0(x) + f^i_{kl}(x)\xi^k\xi^l + \cdots,$$
$$\eta^j = \varphi^j(x, \xi) = \varphi^j_k(x)\xi^k + \varphi^j_{klm}(x)\xi^k\xi^l\xi^m + \cdots$$

(and the corresponding inverse relations) with f^i_0, f^i_{kl}, φ^j_k, φ^j_{klm} etc. being functions on a subset of $\mathbb{R}^p$.] If the transition functions are analytic functions, we call the algebra of superfunctions on M generated by the coordinate superfunctions a *real–analytic supermanifold* of *dimension* (p,q). Complex–analytic supermanifolds are defined in an analogous manner (just replace $\Lambda\mathbb{R}^q$ by $\Lambda\mathbb{C}^q$, the Grassmann algebra over $\mathbb{C}$ with q generators, and require the transition functions to be holomorphic). The manifold M is called the *base* of the supermanifold.

Of course, a proper definition avoids any reference to specific coordinates, and in a more mathematical exposition [9] one would define a supermanifold to be a "sheaf of algebras of superfunctions ...". However, the coordinate description given here is good enough for our purposes.

The standard example of a complex–analytic supermanifold is $\mathbb{CP}^{1|1}$. Let z^1, z^2 be canonical coordinates of $\mathbb{C}^2$, and consider the algebra of superfunctions: $\mathbb{C}^2 \setminus \{0\} \to \Lambda\mathbb{C}^1$, with the complex Grassmann generator of $\Lambda\mathbb{C}^1$ denoted by ζ^1. Then focus on the subalgebra of superfunctions which are homogeneous of degree zero in these generators, i.e. are invariant under rescalings $z^1 \to cz^1$, $z^2 \to cz^2$, $\zeta^1 \to c\zeta^1$ with $c \in \mathbb{C} \setminus \{0\}$. This algebra is what is called the complex projective superspace $\mathbb{CP}^{1|1}$. Its base is the ordinary complex projective space $\mathbb{CP}^1$. Because the null element has been removed from $\mathbb{C}^2$, at least one of the two coordinates z^1, z^2 is nonzero. Let U_1 be the domain where $z^1 \neq 0$. Then homogeneity means

$$f(z^1, z^2, \zeta^1) = f(1, z^2/z^1, \zeta^1/z^1),$$

so that f is a function of two variables $z := z^2/z^1$ and $\zeta := \zeta^1/z^1$, which are taken to be the coordinate (super)functions on U_1. Similarly on U_2, defined by $z^2 \neq 0$, we have

$$f(z^1, z^2, \zeta^1) = f(z^1/z^2, 1, \zeta^1/z^2),$$

and here we put $z' := z^1/z^2$ and $\zeta' := \zeta^1/z^2$. The transition functions on $U_1 \cap U_2$ are then given by

$$z' = \frac{1}{z}, \quad \zeta' = \frac{\zeta}{z}.$$

The domains U_1 and U_2 cover the base $\mathbb{CP}^1$, and the transition functions are holomorphic, so $\mathbb{CP}^{1|1}$ is a complex–analytic supermanifold, the complex dimension being (1,1).

Of the many structures that exist on supermanifolds, the most important one for our purposes is the Berezin integral. To define it, we must first introduce the notion of

Berezin form, also called superintegration measure or integral superform. If $\mathcal{A}$ denotes the algebra of superfunctions of our supermanifold and $\Omega^* M$ the space of volume forms on the base M, a Berezin form ω is a linear and local map

$$
\begin{aligned}
\omega : \quad & \mathcal{A} \to \Omega^* M \,, \\
& f \mapsto \omega[f] \,,
\end{aligned}
$$

i.e. a linear and local rule for converting superfunctions into volume forms on M. This process of conversion is sometimes called the "Fermi integral" or "integration over the Grassmann variables".

For example, an interesting class of Berezin forms on $\mathbb{CP}^{1|1}$ is

$$
\omega^{(n)} = \mathrm{d}z \wedge \mathrm{d}\bar{z} \circ \frac{\partial^2}{\partial\zeta\partial\bar{\zeta}} \circ (1 + \bar{z}z + \bar{\zeta}\zeta)^{-n}
$$

for $n \in N$. The meaning hereof is this: to compute $\omega^{(n)}[f]$, one first multiplies the superfunction f by $(1 + \bar{z}z + \bar{\zeta}\zeta)^{-n}$, then one takes two derivatives $\partial^2/\partial\zeta\partial\bar{\zeta}$ (thereby removing the Grassmann generators), and finally one multiplies by $\mathrm{d}z \wedge \mathrm{d}\bar{z}$ to obtain a two–form on $\mathbb{CP}^1$. To illustrate, we take

$$
f = f_0(z, \bar{z}) + f_{01}(z, \bar{z})\bar{\zeta} + f_{10}(z, \bar{z})\zeta + f_{11}(z, \bar{z})\bar{\zeta}\zeta \,,
$$

in which case we get

$$
\omega^{(n)}[f] = \left(\frac{f_{11}(z, \bar{z})}{(1 + \bar{z}z)^n} - n\frac{f_{00}(z, \bar{z})}{(1 + \bar{z}z)^{n+1}} \right) \mathrm{d}z \wedge \mathrm{d}\bar{z} \,.
$$

A special role is played by the Berezin form $\omega^{(1)}$. Converting $\omega^{(1)}$ from one chart to another by using the transition functions $z = 1/z'$ and $\zeta = \zeta'/z'$, one finds that $\omega^{(1)}$ operates by the same expression in both charts. (Moreover, $\omega^{(1)}$ is invariant under an action of the Lie supergroup $SU(2|1)$ on $\mathbb{CP}^{1|1}$.)

The definition of Berezin forms immediately leads to a notion of integration on supermanifolds called the Berezin integral, or superintegral. Since a Berezin form ω converts a superfunction f into a volume form $\omega[f]$ on M, the natural thing to do is to integrate $\alpha := \omega[f]$ in the ordinary fashion to produce the number $\int_M \alpha$. Thus a Berezin integral is defined to be a composite map

$$
f \mapsto \omega[f] \mapsto \int_M \omega[f] \,,
$$

taking superfunctions into the complex numbers. Note that this definition decrees the Berezin integral to be a two–step process: first the "integration" (or, rather, differentiation) of the Grassmann generators is done, and afterwards the differential form $\omega[f]$ is integrated in the usual sense. The advanced user occasionally finds it convenient to deviate from this rule and perform part of the ordinary integrals first. However, in case of doubt the definition recited [9] above is the one to go back to.

Ordinary integration is coordinate independent, i.e. one has the option of changing integration variables by the substitution rule. It turns out that this option also exists for superintegrals, albeit with a certain caveat. To describe the relevant statement, which is due to Berezin, we introduce the short hand notation

$$
D(x, \xi) := \mathrm{d}x^1 \wedge \ldots \wedge \mathrm{d}x^p \frac{\partial}{\partial\xi^1} \cdots \frac{\partial}{\partial\xi^q} \,,
$$

and consider a change of variables (consistent with the $\mathbb{Z}_2$-grading) from x, ξ to y, η by

$$
\begin{aligned}
y^i &= f^i(x^1, \ldots, x^p; \xi^1, \ldots, \xi^q) \qquad (i = 1, \ldots, p)\,, \\
\eta^j &= \varphi^j(x^1, \ldots, x^p; \xi^1, \ldots, \xi^q) \qquad (j = 1, \ldots, q)\,.
\end{aligned}
$$

The role of the Jacobian of ordinary analysis is taken by the *Berezinian*,

$$
\mathrm{Ber}\left(\frac{y, \eta}{x, \xi}\right) := \mathrm{SDet}\left(\begin{array}{cc} \partial f^i/\partial x^j & \partial f^i/\partial \xi^j \\ \partial \varphi^i/\partial x^j & \partial \varphi^i/\partial \xi^j \end{array}\right)\,,
$$

where the *superdeterminant* SDet of a supermatrix $\left(\begin{array}{cc} A & B \\ C & D \end{array}\right)$ (with the matrix elements of A, D being even and those of B, C odd) is defined by

$$
\mathrm{SDet}\left(\begin{array}{cc} A & B \\ C & D \end{array}\right) = \frac{\mathrm{Det}(A - BD^{-1}C)}{\mathrm{Det}\,D} = \frac{\mathrm{Det}\,A}{\mathrm{Det}(D - CA^{-1}B)}\,.
$$

Berezin's theorem [9] states that the substitution rule is valid, i.e.

$$
\int D(y, \eta) f = \int D(x, \xi)\, \mathrm{Ber}\left(\frac{y, \eta}{x, \xi}\right) f\,,
$$

provided that f is compactly supported.

The condition of compact support, unfamiliar from ordinary analysis, invites some explanation. For an instructive example [10], consider the noncompact supermanifold of functions: $]0, 1[\to \Lambda\mathbb{R}^2$, with canonical coordinates x, ξ^1, ξ^2. If we choose to integrate with the Berezin form $\mathrm{d}x\, \partial^2/\partial\xi^2\partial\xi^1$, a superfunction f that depends only on the combination $x + \xi^1\xi^2$ has the Berezin integral

$$
\int_0^1 \mathrm{d}x \frac{\partial^2}{\partial\xi^2\partial\xi^1} f(x + \xi^1\xi^2) = \int_0^1 \mathrm{d}x\, f'(x) = f(1) - f(0)\,.
$$

On the other hand, if we made a change of integration variables

$$
y = x + \xi^1\xi^2, \quad \eta^1 = \xi^1, \quad \eta^2 = \xi^2\,,
$$

which is easily seen to have unit Berezinian, we would get

$$
\int_0^1 \mathrm{d}x \frac{\partial^2}{\partial\xi^2\partial\xi^1} f(x + \xi^1\xi^2) \stackrel{??}{=} \int_0^1 \mathrm{d}y \frac{\partial^2}{\partial\eta^2\partial\eta^1} f(y) = 0\,.
$$

Because the Grassmann derivatives $\partial^2/\partial\eta^2\partial\eta^1$ have nothing to act on, the right hand side vanishes although for equality with the left hand side, it ought to be $f(1) - f(0)$. However, both sides vanish, and the discrepancy disappears, if we take f to be compactly supported (which implies $f(0) = f(1) = 0$), as is required in order for Berezin's theorem to apply. An alternative way of fixing the problem is to pass from the noncompact interval $]0, 1[$ to the compact circle S^1, by imposing periodic boundary conditions on x, so that $f(1) - f(0) = 0$.

The above example signals a general complication, which one has to confront when changing variables in a superintegral. While ordinary volume forms transform simply by the Jacobian, the transformation law for Berezin forms is *not* given only by the Berezinian, but involves an additional, "anomalous" term β:

$$
D(y, \eta) = D(x, \xi)\, \mathrm{Ber}\left(\frac{y, \eta}{x, \xi}\right) + \beta\,,
$$

whenever some of the even coordinates are shifted by nilpotents. In our simple example,

$$D(y, \eta)f = D(x, \xi)f + \mathrm{d}x \frac{\partial}{\partial x} f \Big|_{\xi^1 = \xi^2 = 0} .$$

The anomaly β always has the property of being exact (in the sense of differential forms), i.e. $p[f] = \mathrm{d}\alpha[f]$ where $\mathrm{d}\alpha[f]$ is the exterior derivative of some $(p-1)$–form $\alpha[f]$. Therefore, by Stokes' theorem the anomaly can be integrated and converted into an integral over the boundary of the base of the supermanifold (or the boundary of the chart used, if there exist coordinate singularities). Rothstein [10] has given an explicit expression for the anomaly in terms of the vector field generating the transformation from x, ξ to y, η.

6. SUPERSYMMETRIC FORMULA FOR MAPS (II)

After this brief tour of some basic supermathematics, we return to the task of setting up a supersymmetric formula for the pair correlations of quantized symplectic maps. We left off at Eq. (3) expressing Ω as a trace over S, the subspace of super Fock space $\mathcal{B}_N^+ \otimes \mathcal{F}_N^+ \otimes \mathcal{B}_N^- \otimes \mathcal{F}_N^-$ determined by the condition $N_+ = N_-$ (equal number of retarded and advanced particles).

As was motivated at the example of the purely fermionic case in Sect. 4, our goal is now to trade the trace over S for an integral over generalized coherent states. To that end, we observe that every Fock state $|\psi\rangle$ satisfying $(N_+ - N_-)|\psi\rangle = 0$ can be obtained by repeatedly acting on the vacuum with the pair creation operators $b_{+n}^\dagger b_{-n'}^\dagger$, $b_{+n}^\dagger f_{-n'}^\dagger$, $f_{+n}^\dagger b_{-n'}^\dagger$, and $f_{+n}^\dagger f_{-n'}^\dagger$ $(n, n' = 1, \ldots, N)$. This leads us to consider coherent states of the pairing form

$$|Z\rangle := \exp \sum_{nn'} \left(b_{+n}^\dagger Z_{nn'}^{\mathrm{BB}} b_{-n'}^\dagger + b_{+n}^\dagger Z_{nn'}^{\mathrm{BF}} f_{-n'}^\dagger + f_{+n}^\dagger Z_{nn'}^{\mathrm{FB}} b_{-n'}^\dagger + f_{+n}^\dagger Z_{nn'}^{\mathrm{FF}} f_{-n'}^\dagger \right) |0\rangle ,$$

where mathematical consistency requires taking complex numbers for the matrix elements $Z_{nn'}^{\mathrm{BB}}$, $Z_{nn'}^{\mathrm{FF}}$ and Grassmann generators for $Z_{nn'}^{\mathrm{BF}}$, $Z_{nn'}^{\mathrm{FB}}$. It is convenient to assemble these into a supermatrix

$$Z = \begin{pmatrix} Z^{\mathrm{BB}} & Z^{\mathrm{BF}} \\ Z^{\mathrm{FB}} & Z^{\mathrm{FF}} \end{pmatrix} .$$

Then a short hand notation for the pair coherent states is

$$|Z\rangle = \exp \left(c_+^\dagger Z c_-^\dagger \right) |0\rangle$$

where the creation operators $c_\pm^\dagger$ stand for both $b_\pm^\dagger$ and $f_\pm^\dagger$. We will also need the dual of a pair coherent state

$$\langle \tilde{Z}| := \langle 0| \exp \left(c_- \sigma \tilde{Z} c_+ \right), \quad \tilde{Z} = \begin{pmatrix} \tilde{Z}^{\mathrm{BB}} & \tilde{Z}^{\mathrm{BF}} \\ \tilde{Z}^{\mathrm{FB}} & \tilde{Z}^{\mathrm{FF}} \end{pmatrix}, \quad \sigma = \begin{pmatrix} 1_N & 0 \\ 0 & -1_N \end{pmatrix} .$$

In [11] it was shown that, with a suitable choice of Berezin form $D(Z, \tilde{Z})$, the pair coherent states provide a resolution of the projector P,

$$P = \int_0^{2\pi} \frac{d\theta}{2\pi} \exp i\theta(N_+ - N_-) = \int D(Z, \tilde{Z}) \, \langle \tilde{Z}|Z\rangle^{-1} \, |Z\rangle\langle \tilde{Z}| ,$$

where the domain of integration is fixed by the conditions

$$\tilde{Z}^{\mathrm{FF}} = -Z^{\mathrm{FF}\dagger} , \quad \tilde{Z}^{\mathrm{BB}} = +Z^{\mathrm{BB}\dagger} , \quad \text{and} \quad 1 - Z^{\mathrm{BB}\dagger}Z^{\mathrm{BB}} > 0 .$$

This resolution follows from Schur's lemma, once it has been
demonstrated that the right hand side commutes with all bilinears in Fock operators that commute with the projector P. The crucial relation needed for the latter statement is the invariance of the Berezin integral under a left translation through
$g = \begin{pmatrix} A & B \\ C & D \end{pmatrix} \in \mathrm{GL}(2N|2N)$,

$$\int D(Z,\tilde{Z})f(Z,\tilde{Z}) =$$
$$\int D(Z,\tilde{Z})f\Big((AZ+B)(CZ+D)^{-1},(C+D\tilde{Z})(A+B\tilde{Z})^{-1}\Big) ,$$

which determines the Berezin form $D(Z,\tilde{Z})$ to be the invariant superintegration measure on Efetov's model space II (or, using the terminology of [12], on a Riemannian symmetric superspace of type AIII/AIII). The invariant Berezin form $D(Z,\tilde{Z})$ turns out to be flat as a result of cancellations due to supersymmetry:

$$D(Z,\tilde{Z}) = \prod_{n,n'} \mathrm{d}Z_{nn'}^{\mathrm{BB}}\mathrm{d}\tilde{Z}_{nn'}^{\mathrm{BB}}\mathrm{d}Z_{nn'}^{\mathrm{FF}}\mathrm{d}\tilde{Z}_{nn'}^{\mathrm{FF}} \frac{\partial^4}{\partial Z_{nn'}^{\mathrm{BF}}\partial\tilde{Z}_{nn'}^{\mathrm{BF}}\partial Z_{nn'}^{\mathrm{FB}}\partial\tilde{Z}_{nn'}^{\mathrm{FB}}} + \dots$$

modulo boundary anomalies, which are indicated by the dots. The normalization of $D(Z,\tilde{Z})$ is fixed by $1 = \langle 0|P|0\rangle = \int D(Z,\tilde{Z})\langle\tilde{Z}|Z\rangle^{-1}$.

Given the coherent state resolution of the projector P, the trace of any operator O on Fock space can be converted into a Berezin integral:

$$\mathrm{Tr}_{N_+=N_-}(-1)^{N_F}O = \mathrm{STr}PO = \int D(Z,\tilde{Z})\langle\tilde{Z}|O|Z\rangle/\langle\tilde{Z}|Z\rangle .$$

Application to Eq. (3), followed by a computation of $\langle\tilde{Z}|Z\rangle$ and $\langle\tilde{Z}|\dots|Z\rangle$, results in the desired formula for Ω:

$$\Omega(\gamma_+;\gamma_-) = \int D(Z,\tilde{Z})\,\mathrm{SDet}(1-\tilde{Z}Z)\,\mathrm{SDet}^{-1}\Big(1-\tilde{Z}(\gamma_+\otimes U_N)Z(\gamma_-\otimes U_N^{-1})\Big) . \quad (4)$$

Here $\gamma_\pm = \mathrm{diag}(\gamma_{\pm 0},\gamma_{\pm 1})$ are diagonal 2×2 matrices, and $\gamma_+\otimes U_N$ is meant to be a supermatrix:

$$\gamma_+\otimes U_N = \begin{pmatrix} \gamma_{+0}U_N & 0 \\ 0 & \gamma_{+1}U_N \end{pmatrix} .$$

A closely related derivation of the formula for Ω is provided by the so–called color–flavor transformation ([11] and [13]). The two–level correlation function R_2 follows from Ω by taking two derivatives, as before.

7. SEMICLASSICAL LIMIT?

So far, we have achieved no more (and no less) than an exact reformulation of the original problem. As should be clear from its derivation, the supersymmetric formula for the correlator Ω applies to *any* unitary matrix U_N, independently of whether this matrix arises from quantization of a map or not. In order to go further and extract nonempty information from our formalism, we need to exploit the fact that U_N has a semiclassical limit and, in particular, we will have to make a distinction between chaotic and integrable systems.

164

To prepare these steps, we write $\Omega = \int D(Z, \tilde{Z})e^{-S}$ where, borrowing the terminology from quantum field theory, the function

$$S = -\mathrm{STr}\ln(1 - \tilde{Z}Z) + \mathrm{STr}\ln\left(1 - \tilde{Z}(\gamma_+ \otimes U_N)Z(\gamma_- \otimes U_N^{-1})\right) \tag{5}$$

will be called the "action functional". Now recall that the BB– and FF–blocks of Z are complex $N \times N$ matrices. Viewing them as linear operators on the Hilbert space $\mathcal{H}_N$, we might say that they are similar to the density matrix of a quantum mechanical state. And, in fact, an expression like $Z^{\mathrm{BB}'} := U_N Z^{\mathrm{BB}} U_N^{-1}$, which occurs under the second logarithm in S, can be interpreted as being the "density matrix" Z^{BB} evolved by one (inverse) time step. One may therefore be tempted [4] to transform to a Wigner representation, postulate a semiclassical limit for Z and subject the action functional to a semiclassical expansion. However, such an expansion is *entirely uncontrolled* in the present context and is, in fact, false. Unfortunately, a lot of confusion has been created among the community (including myself) concerning this point. Let me therefore change the style of presentation and give a detailed exposition of the issues as I see them. The reader is warned that, given the current level of understanding, the following has to be somewhat qualitative.

If A is a linear operator (on some quantum mechanical Hilbert space) with kernel $A(q, q')$ in, say, the position representation, one defines the Wigner transform (or Weyl symbol) of A by

$$\sigma_A(q, p) = \sum_{q'} A(q + q'/2, q - q'/2)\, e^{ipq'/\hbar}$$

whenever the transform exists. (In our case there is a technical complication due to the fact that the "position" on a compact phase space is not globally defined, but this is a peripheral issue and we are not going to worry about it here.) Now if A and B are pseudodifferential operators with Wigner transforms $\sigma_A(q, p)$ and $\sigma_B(q, p)$, their operator product has a semiclassical expansion

$$\sigma_{AB} = \sigma_A \sigma_B + (\hbar/2i)\{\sigma_A, \sigma_B\} + \mathcal{O}(\hbar^2), \tag{6}$$

where $\sigma_A \sigma_B$ is a product of functions, and the curly brackets denote the Poisson bracket.

The question now is whether this kind of expansion applies to products of the supermatrix Z. To be sure, there is nothing that prevents us from transforming Z to a Wigner-like representation, but are we really allowed to make a semiclassical expansion of the action functional?! The question is a valid one, for the semiclassical expansion (6) does require A and B to be *pseudodifferential operators*. Without the smoothness property provided by this condition, the Wigner transform of the operator product AB does *not* separate into the product of Wigner transforms $\sigma_A \sigma_B$ in the limit $\hbar \to 0$. For the standard operators appearing in quantum mechanics (position, momentum, energy etc.) the smoothness condition is of course always satisfied. However, there is *no* principle that guarantees smoothness of the Wigner transform σ_Z in the present case. Indeed, Z is not a fixed observable but is a *variable of integration*. While a small fraction of the Z that are integrated over do correspond to smooth σ_Z, the vast majority do not. (Recall that the integral is over all complex supermatrices $Z, \tilde{Z}$ with the integration measure being that of a Riemanian symmetric superspace.) This invalidates the error estimate in (6). One may still cherish the hope that the *dominant* contributions to the Z–integral do come from smooth configurations. Admittedly, this is what happens for Feynman's path integral in the limit $\hbar \to 0$, which is dominated by paths that are extrema of the action functional. However, the present situation is quite different. *There exists no saddle–point or stationary–phase or other principle here that would favor smoothness* in general.

To recognize the severity of the problem, recall that the unitary operator U_N has eigenvalues $e^{i\theta_n}$ and eigenfunctions ψ_n, and there exist N of them. As a result, the conjugation $\mathrm{Ad}(U_N) : \rho \mapsto U_N \rho U_N^{-1}$ has N eigendensities $\rho = \psi_n \otimes \bar{\psi}_n$ with eigenvalues $e^{i(\theta_n - \theta_n)} = 1$. To compute the pair correlation function in the microlocal scaling limit, we need to set $\gamma_\pm = e^{ic_\pm/N}$ and let N go to infinity. In this limit the action functional acquires N *zero modes* given by $Z = \sum_{n=1}^{N} C^{(n)} \psi_n \bar{\psi}_n$, where the $C^{(n)}$ are 2×2 supermatrices and are otherwise arbitrary. Thus there exist N directions through the point $Z = \tilde{Z} = 0$ along which the integrand is *exactly neutral* – a circumstance that surely causes very–large–amplitude fluctuations about this point. Moreover, the zero mode densities $\psi_n \bar{\psi}_n$ are never smooth. Indeed, in the integrable limit they are sharply localized on invariant sets, while in the chaotic limit they are highly irregular functions that vary on the shortest scales permitted by the uncertainty principle. In other words, the action functional (5) is unstable with respect to a large number N of zero modes and none of these has a smooth Wigner transform. In view of all this, we had better abandon the hope that the dominant contributions to the Z-integral come from configurations that possess the smoothness of a pseudodifferential operator. Quite on the contrary, the most important configurations are those where σ_Z varies with the shortest wave length possible! As $\hbar$ is lowered, the dominance of the N zero modes keeps introducing an increasing number of relevant field configurations that vary on finer and finer scales. In order for the semiclassical asymptotics (6) to set in, we would have to intervene and *limit* the variations of σ_Z to wave lengths that are greater than some fixed (i.e. $\hbar$-independent) minimal scale, and then let $\hbar$ go to zero. Without such an intervention, the Poisson bracket (and higher) terms in (6) fail to be $\hbar$-independent but *diverge* (with increasing powers) as $\hbar \to 0$, so that truncation of the semiclassical expansion of S is totally unjustified. Consequently, any conclusion resulting from the use of this expansion in (5) is at risk to be false.

Given the invalidity of the semiclassical step, how can we use the supersymmetric formula (4) to develop the theory further? Not surprisingly, the answer depends on what we are trying to achieve. Here we are pursuing no more than the modest goal (modest from a physicist's perspective) of formulating a precise version of the BGS conjecture and substantiating it. As will be shown in the sequel, this restricted goal offers us the option of postponing the semiclassical step until the very end of the calculation.

8. REGULARIZATION

Notwithstanding the fact that the formula (4) is well-defined and, in fact, mathematically rigorous, we are facing a key difficulty: the existence of N one–parameter groups along which the integrand lacks stability prevents us from investing semiclassical input and makes it difficult if not impossible to compute something from (4) as it stands. To make any progress at all, we must first solve this stability problem. Consider therefore, in the microlocal scaling limit $\gamma_\pm \to 1$, the quadratic part of the action functional at $Z = \tilde{Z} = 0$:

$$S^{(2)} = \mathrm{STr}\,\tilde{Z}Z - \mathrm{STr}\,\tilde{Z}U_N Z U_N^{-1} = \mathrm{STr}\,\tilde{Z}(1 - \mathrm{Ad}(U_N))Z \,,$$

which identifies the Hessian at this point as the operator $1 - \mathrm{Ad}(U_N)$. Now recall that the classical limit of $\mathrm{Ad}(U_N)$ is χ^*, the Frobenius–Perron operator of the inverse map χ^{-1}. To prepare the discussion of the quantum mechanical operator $\mathrm{Ad}(U_N)$ in the limit $N \to \infty$, we shall summarize what we know about χ^*.

First of all, because χ^* is unitary with respect to the Liouville L^2–measure on phase space, the spectrum of χ^* lies on the unit circle in $\mathbb{C}$. (If the map is mixing, the spectrum is known to be absolutely continuous [6].) Second, a salient feature of this spectrum is its *instability* with respect to regularization by smoothing. More precisely, if we project χ^* from the full L^2-space to a subspace of smooth functions, by coarse graining the phase space or imposing an ultraviolet (UV) cutoff, unitarity is lost and the spectrum moves to the interior of the unit circle. Third, if the map χ is mixing, the spectrum of the projected operator consists of one isolated *nondegenerate* eigenvalue at $+1$ (with the corresponding eigenfunction being the invariant density $\rho_0 = \mathrm{const}$) and so–called Ruelle resonances [14] inside the unit circle. It is important that this feature persists, i.e. the unitary spectrum is *not* recovered, when the UV cutoff is lowered to zero. (This is a statement of non–interchangability of limits.) Note also the difference between the quantum and classical cases: the quantum operator $1 - \mathrm{Ad}(U_N)$ always has N zero modes, whereas its classical limit $1 - \chi^*$ on smooth functions only has a *single* zero mode for a mixing system, namely the invariant density.

The usual approach taken in the study of Ruelle resonances is to regularize in the ultraviolet by imposing a smoothness or differentiability condition on the phase space functions acted upon by the Frobenius–Perron operator. A similar effect is achieved if we leave the functions unchanged and, instead, implement the UV regularization on the Frobenius–Perron operator *itself*. For our purposes the latter procedure is more convenient. What we do is to make the replacement

$$\chi^* \longrightarrow r^\epsilon \chi^* ,$$

where the operator r^ϵ is chosen to be the exponential of a sum of squares of Hamiltonian vector fields Ξ_k,

$$r^\epsilon = \exp \, \epsilon \sum_{k=1}^{s} \Xi_k^2 .$$

For example, for $M = \mathrm{S}^2$ we take for Ξ_k the generators of rotations $(k = 1, 2, 3)$, in which case $\sum_{k=1}^{3} \Xi_k^2$ is simply the Laplacian on S^2. More generally, the choice of Hamiltonian vector fields is constrained by the requirement that $\sum_k \Xi_k^2$ be an elliptic operator. The effect of r^ϵ, then, is to regularize in the ultraviolet by suppressing fluctuations on short scales. It is important that r^ϵ is the exponential of a differential operator of *second* order, whereas the Frobenius–Perron operator can be thought of as the exponential of a first–order differential operator. Therefore, r^ϵ is always relevant in the ultraviolet, no matter how small is the value of ϵ. As a result, r^ϵ remains effective as a regulator even in the limit $\epsilon \to 0$ and, again, the unitary spectrum is not retrieved when the UV cutoff is removed.

In the following we will make use of the fact that, if the map χ is mixing, the operator $1 - r^\epsilon \chi^*$ is strictly positive on the orthogonal complement of the uniform state ρ_0. The physical reason for such behavior is this. Consider the long time dynamics obtained by iterating $r^\epsilon \chi^*$ and, for simplicity of the argument, assume χ to be uniformly hyperbolic, which means that χ is everywhere stretching along unstable manifolds and contracting along stable ones. Thanks to the ellipticity of $\sum_{k=1}^{r} \Xi_k^2$, the operator r^ϵ counteracts the contraction along the stable manifolds via diffusion. After many iterations, the combination of stretching by χ and smoothing by r^ϵ will attract any initial state to the uniform state. Thus, as time l goes to infinity, the operator $(r^\epsilon \chi^*)^l$ approaches the projector onto the uniform state ρ_0. This implies strict positivity of the real part of the spectrum of $1 - r^\epsilon \chi^*$ operating on the orthogonal complement of ρ_0.

Let us now turn from the space of functions on M, acted upon by χ^*, to the quantum mechanical space $\mathcal{H}_N \otimes \mathcal{H}_N^*$, acted upon by the operator $\mathrm{Ad}(U_N)$. Crudely speaking, the quantum space differs from the classical one by a finite resolution, or cut-off, set by Planck's constant. When this cutoff is much smaller than the regularization scale set by ϵ, we have a good semiclassical expansion, (6), and the regularized $\mathrm{Ad}(U_N)$ spectrum looks qualitatively similar to the regularized Frobenius–Perron spectrum. On the other hand, if $\hbar$ is kept fixed at a finite value, the destruction of the unitary spectrum of $\mathrm{Ad}(U_N)$ with increasing ϵ is not abrupt but is rounded off. If we take $\epsilon \sim \hbar^\alpha$ with $\alpha > 2d$, the strength of the perturbation caused by the regulator becomes smaller than the level spacing of $\mathrm{Ad}(U_N)$ [which is $2\pi/N^2 \sim \hbar^{2d}$], and is therefore negligible, in the limit $\hbar \to 0$. In contrast, if we choose the scaling exponent α in the range $0 < \alpha < 1$, then by power counting the regulator remains relevant for $\hbar \to 0$. This is good news: it suggests that we can kill the unitary spectrum of $\mathrm{Ad}(U_N)$, and thereby get rid of the zero mode problem if the classical map is mixing, by a regulator with strength ϵ that *vanishes* in the classical limit.

The basic strategy to follow should now be clear: our aim must be to substitute for the unitary operator $\mathrm{Ad}(U_N)$ in the quadratic functional $S^{(2)}$ the regularized operator $R_N^\epsilon \mathrm{Ad}(U_N)$, where R_N^ϵ is a quantization of the diffusion operator r^ϵ. Such a substitution is expected to turn the unitary spectrum of $\mathrm{Ad}(U_N)$ into something akin to the Frobenius–Perron spectrum of the classical operator χ^*. If χ is mixing and ϵ depends on $\hbar$ as $\epsilon \sim \hbar^\alpha$ with $0 < \alpha < 1$, only the invariant zero mode of $1 - \mathrm{Ad}(U_N)$ remains neutral. The remaining $N - 1$ zero modes turn into positive directions of the regularized Hessian and therefore become amenable to perturbative treatment. From experience with the application of the supersymmetric method to disordered electron systems [5], we then anticipate that a careful integration over the invariant zero mode will give the random matrix answer for the correlation functions.

What is unsatisfactory about this strategy is the absence of any explanation of *where the UV regularization came from*. So far, regularization seems to be an ad hoc procedure, or "dirty trick", introduced so as to yield the result one wants to get. The situation would be different if we were working in a quantum field theoretic context. In that case, UV regularization would be well justified by the fundamental and inevitable existence of ultraviolet singularities (signaling incompleteness of the field theory, or "new physics" at short scales), which must be tamed by the introduction of a cutoff. In contrast, our formula (4) has no singularities whatsoever but is perfectly well–defined – we just don't know how to evaluate it – and there is no a priori need for regularization. In passing we mention that, if we ignored the arguments of Sect. 7 forbidding us to apply the semiclassical expansion (6) to the action functional S, and applied the expansion anyway, then we would, in fact, end up with an ultraviolet singular field theory [4]. However, the UV divergencies so introduced are entirely artificial – they simply confirm the fact that we have made improper use of the semiclassical expansion – and do not justify the postulate of any regulator.

To reiterate, it is illegal just to postulate UV regularization as a technical device or trick. For a convincing argument, one needs to explain exactly what modification of the original problem is implied by regularization, and one has to do that in a frame that precedes the supersymmetric quantum field theory formalism. As will be argued in the remainder of this section, regularization of the Hessian of (5) at $Z = \tilde{Z} = 0$ amounts to averaging over the "quantum" ensemble described in Sect. 3.

Recall that, if ξ_k ($k = 1, ..., s$) are uncorrelated Gaussian random variables with zero mean and variance $\langle \xi_k^2 \rangle = 2\epsilon$, the ensemble is specified by

$$U_N(\xi) = \exp\left(i \sum_{k=1}^{s} \xi_k X_{k,N}/\hbar \right) U_N \, .$$

By averaging the Hessian $1 - \mathrm{Ad}\,(U_N(\xi))$ over the probability space of the parameters

ξ_k, we get

$$\langle 1 - \mathrm{Ad}\,(U_N(\xi))\rangle = 1 - R_N^\epsilon \mathrm{Ad}(U_N)\,,$$

where the operator R_N^ϵ is identified as

$$R_N^\epsilon = \left\langle \mathrm{Ad}\left(\exp\, i\sum_{k=1}^{s} \xi_k X_{k,N}/\hbar\right)\right\rangle$$

$$= \left\langle \exp\, i\sum_k \xi_k \mathrm{ad}(X_{k,N})/\hbar\right\rangle = \exp\, -\epsilon\sum_k \mathrm{ad}^2(X_{k,N})/\hbar^2 + \dots\,.$$

[Note that, since the integrand in (4) depends on U_N in a nonlinear way, later we will have to confront averages of powers of $\mathrm{Ad}\,(U_N(\xi))$. For the time being, we concentrate on the Hessian.] The terms indicated by dots in the last expression become negligible in the semiclassical limit $\hbar \to 0$. Taking the commutator with $X_{k,N}$ in this limit corresponds to applying the Hamiltonian vector field Ξ_k:

$$\mathrm{ad}(X_{k,N}) = [X_{k,N}, \bullet] \xrightarrow{\hbar\to 0} i\hbar\Xi_k\,,$$

so R_N^ϵ is in fact a quantization of the classical diffusion operator $e^{\epsilon\sum_{k=1}^{s}\Xi_k^2} = r^\epsilon$, as desired.

To conclude, averaging over the specified ensemble regularizes the Hessian in the ultraviolet. As was argued earlier, if the classical map χ is mixing the real part of $1 - r^\epsilon\chi^*$ is positive on all states that are orthogonal to the uniform state ρ_0. In view of the semiclassical limits $R_N^\epsilon \to r^\epsilon$ and $\mathrm{Ad}(U_N) \to \chi^*$, we expect the same to be true for the regularized Hessian $1 - R_N^\epsilon\mathrm{Ad}\,(U_N)$ on $\mathcal{D}_N^\perp$, the subspace of $\mathcal{H}_N \otimes \mathcal{H}_N^*$ orthogonal to the invariant element. This key observation allows us to proceed as follows.

9. ASYMPTOTIC EXPANSION

We set $\gamma_\pm = e^{\pm ic_\pm/N}$ where $c_\pm = \mathrm{diag}(c_{\pm 0}, c_{\pm 1})$ are 2×2 matrices and, to abbreviate the notation, we put $Z_U := (1_2 \otimes U_N)Z(1_2 \otimes U_N^{-1})$. Then, since $c_\pm$ are held fixed while N is sent to infinity, we may expand the action functional:

$$S = -\mathrm{STr}\ln(1 - \tilde{Z}Z) + \mathrm{STr}\ln(1 - \tilde{Z}Z_U)$$

$$-\frac{i}{N}\mathrm{STr}\left(c_+ Z_U\tilde{Z}(1 - Z_U\tilde{Z})^{-1} - c_-\tilde{Z}Z_U(1 - \tilde{Z}Z_U)^{-1}\right) + \mathcal{O}(1/N^2)\,.$$

To isolate the invariant zero mode, we make the substitution

$$Z = (\zeta + Z_0)(1 + \tilde{Z}_0\zeta)^{-1}\,,$$
$$\tilde{Z} = (\tilde{\zeta} + \tilde{Z}_0)(1 + Z_0\tilde{\zeta})^{-1}\,.$$

The supermatrices ζ and $\tilde{\zeta}$ are taken to be traceless in each block (i.e. $\mathrm{Tr}\,\zeta^{\sigma\tau} = \mathrm{Tr}\,\tilde{\zeta}^{\sigma\tau} = 0$ for $\sigma,\tau = \mathrm{B}, \mathrm{F}$), and the zero mode is parameterized by $Z_0 = z_0 \otimes 1_N$, $\tilde{Z}_0 = \tilde{z}_0 \otimes 1_N$ with 2×2 matrices

$$z_0 = \begin{pmatrix} z_0^{\mathrm{BB}} & z_0^{\mathrm{BF}} \\ z_0^{\mathrm{FB}} & z_0^{\mathrm{FF}} \end{pmatrix}\,, \qquad \tilde{z}_0 = \begin{pmatrix} \tilde{z}_0^{\mathrm{BB}} & \tilde{z}_0^{\mathrm{BF}} \\ \tilde{z}_0^{\mathrm{FB}} & \tilde{z}_0^{\mathrm{FF}} \end{pmatrix}\,.$$

Because the transformation

$$Z \mapsto (AZ + B)(CZ + D)^{-1}\,,$$
$$\tilde{Z} \mapsto (C + D\tilde{Z})(A + B\tilde{Z})^{-1}\,,$$

is an isometry of the Riemannian symmetric superspace, the Berezinian of the change of variables from Z to (ζ, Z_0) is unity and $D(Z, \tilde{Z})$ factors into a product of Berezin forms $D(z_0, \tilde{z}_0) D(\zeta, \tilde{\zeta})$. (We pay no attention to possible boundary anomalies here.) Now, by inserting the expansion of the action S into (4), changing variables as indicated, and using identities such as

$$(1 - Z_U \tilde{Z})^{-1} = (1 + Z_0 \tilde{\zeta})(1 - \zeta_U \tilde{\zeta})^{-1}(1 + \zeta_U \tilde{Z}_0)(1 - Z_0 \tilde{Z}_0)^{-1} \, ,$$

we obtain

$$\Omega(e^{ic_+/N}; e^{-ic_-/N}) = \int D(z_0, \tilde{z}_0) \, \Omega_1(c_+/N, c_-/N; z_0, \tilde{z}_0)$$
$$\times \exp i\mathrm{STr}_{\mathbb{C}^{1|1}} \left(c_+ z_0 \tilde{z}_0 (1 - z_0 \tilde{z}_0)^{-1} - c_- \tilde{z}_0 z_0 (1 - \tilde{z}_0 z_0)^{-1} \right)$$

with Ω_1 defined by

$$\Omega_1(c_+/N, c_-/N; z_0, \tilde{z}_0)$$
$$= \int D(\zeta, \tilde{\zeta}) \, \mathrm{SDet}(1 - \tilde{\zeta}\zeta) \mathrm{SDet}^{-1}(1 - \tilde{\zeta}\zeta_U)$$
$$\times \exp \left(\frac{i}{N} \mathrm{STr} \, c_+ (1 + Z_0 \tilde{\zeta})(1 - \zeta_U \tilde{\zeta})^{-1} \zeta_U (\tilde{\zeta} + \tilde{Z}_0)(1 - Z_0 \tilde{Z}_0)^{-1} \right.$$
$$\left. - \frac{i}{N} \mathrm{STr} \, c_- (1 + \tilde{Z}_0 \zeta_U)(1 - \tilde{\zeta}\zeta_U)^{-1} \tilde{\zeta}(\zeta_U + Z_0)(1 - \tilde{Z}_0 Z_0)^{-1} \right) .$$

Note that convergence of the last integral is ensured by the condition $\mathrm{Im}\,c_{+0} > 0 > \mathrm{Im}\,c_{-0}$ resulting from $|\gamma_{\pm 0}| < 1$.

Our next goal is to establish an asymptotic $1/N$ expansion for the function Ω_1. To that end, we temporarily set $c_+/N = c_-/N = 0$ and consider the integral

$$\Omega_1(0, 0; z_0, \tilde{z}_0) = \int D(\zeta, \tilde{\zeta}) \, \mathrm{SDet}(1 - \tilde{\zeta}\zeta) \mathrm{SDet}^{-1} \left(1 - \tilde{\zeta}\zeta_U \right) .$$

To compute this, we observe that both the Berezin form $D(\zeta, \tilde{\zeta})$ and the two superdeterminants are invariant under transformations

$$\zeta \mapsto A\zeta D^{-1}, \quad \tilde{\zeta} \mapsto D\tilde{\zeta}A^{-1}$$

with $A = a \otimes 1_N$ and $D = d \otimes 1_N$. (Indeed, such transformations commute with $\mathrm{Ad}(U_N)$ and are isometries of the symmetric superspace.) This invariance results in the value of the integral being equal to the value of the integrand at the origin $\zeta = \tilde{\zeta} = 0$:

$$\Omega_1(0, 0; z_0, \tilde{z}_0) = 1 \, ,$$

by what is often called the Parisi–Sourlas–Efetov–Wegner theorem in disordered electron physics. The mechanism underlying the theorem is called "localization" in the mathematics literature. In a setting closely related to equivariant cohomology, localization of superintegrals has recently been discussed by Schwartz and Zaboronsky [15]. With reference to their results we can argue as follows. On setting

$$A = \exp \begin{pmatrix} 0 & \alpha \\ \bar{\alpha} & 0 \end{pmatrix} \otimes 1_N \, , \qquad D = \exp \begin{pmatrix} 0 & \delta \\ \bar{\delta} & 0 \end{pmatrix} \otimes 1_N \, ,$$

and differentiating with respect to the Grassmann parameters $\alpha, \bar{\alpha}, \delta, \bar{\delta}$, the transformation $\zeta \mapsto A\zeta D^{-1}$, $\tilde{\zeta} \mapsto D\tilde{\zeta}A^{-1}$ gives rise to four odd vector fields, each of which leaves the integrand invariant. The joint zero locus of these vector fields is $\zeta = \tilde{\zeta} = 0$, and the superdeterminant of the Hessian at this point is unity by cancellation due to supersymmetry. Moreover, the vector fields are *compact* in the sense of Schwartz and Zaboronsky. Therefore, by the version of the Parisi–Sourlas–Efetov–Wegner theorem proved in [15], the integral equals unity as claimed.

Given the result $\Omega_1(0,0;z_0,\tilde{z}_0) = 1$, one would like to proceed and make a saddle point approximation around $\zeta = \tilde{\zeta} = 0$, to produce an asymptotic expansion for $\Omega_1(c_+/N, c_-/N; z_0, \tilde{z}_0)$ of the form $1 + N^{-1}f(c_+, c_-; z_0, \tilde{z}_0) + \mathcal{O}(N^{-2})$. However, this is impossible as it stands. The reason is the notorious problem of $N-1$ zero modes: there exists an $(N-1)$-dimensional maximal torus of group actions that commute with $\mathrm{Ad}(U_N)$ and leave $\mathrm{SDet}(1-\tilde{\zeta}\zeta)\mathrm{SDet}^{-1}(1-\tilde{\zeta}\zeta_U)$ invariant, thereby causing a lack of stability of the candidate saddle point $\zeta = \tilde{\zeta} = 0$. Fortunately, given all the preparations that were made in Sect. 8, we immediately know how to fix this stability problem.

We substitute $U_N(\xi)$ for U_N and average over the probability space of the ensemble, replacing Ω by $\langle\Omega\rangle$, and Ω_1 by $\langle\Omega_1\rangle$. The integral for Ω_1 converges and, as a matter of fact, converges *uniformly* in the parameters ξ_k, if $\mathrm{Im}c_{+0} \geq 0 \geq \mathrm{Im}c_{-0}$. Therefore, we may interchange the order of integration and ensemble averaging. By the same localization argument as before,

$$\langle\Omega_1(0,0;z_0,\tilde{z}_0)\rangle = \int D(\zeta,\tilde{\zeta}) \left\langle \mathrm{SDet}(1-\tilde{\zeta}\zeta)\mathrm{SDet}^{-1}(1-\tilde{\zeta}\zeta_{U_N(\xi)}) \right\rangle = 1 \ .$$

What has improved in comparison with the situation without ensemble averaging is that now we do have a good asymptotic expansion. The reason is that integrals such as

$$\int D(\zeta,\tilde{\zeta}) \Big\langle \mathrm{SDet}(1-\tilde{\zeta}\zeta)\mathrm{SDet}^{-1}(1-\tilde{\zeta}\zeta_{U_N(\xi)})$$

$$\times \mathrm{STr}\ c_+(1-\zeta_{U_N(\xi)}\tilde{\zeta})^{-1}\zeta_{U_N(\xi)}\tilde{\zeta}(1-Z_0\tilde{Z}_0)^{-1} \Big\rangle$$

are expected to exist and remain finite in the limit $N \to \infty$. This can be checked by expanding around the saddle point $\zeta = \tilde{\zeta} = 0$ in the usual manner. By doing the resulting Gaussian integrals over $\zeta, \tilde{\zeta}$, one obtains traces involving powers of the inverse of the regularized Hessian $1 - R_N^\epsilon \mathrm{Ad}(U_N)$ acting on $\mathcal{D}_N^\perp$. We have argued earlier that the regularized Hessian remains strictly positive in the semiclassical limit if the classical map is mixing. Therefore, the inverse of $1 - R_N^\epsilon\mathrm{Ad}(U_N)$ exists on $\mathcal{D}_N^\perp$ and is uniformly bounded in N. Moreover, as follows from semiclassical trace formulas, the traces encountered are bounded uniformly in N if the fixed points of the classical map are isolated.

What all this amounts to is that, unlike Ω_1, the ensemble average $\langle\Omega_1\rangle$ does have an asymptotic expansion:

$$\langle\Omega_1(c_+/N, c_-/N; z_0, \tilde{z}_0)\rangle = 1 + N^{-1}f(c_+, c_-; z_0, \tilde{z}_0) + \mathcal{O}(N^{-2}) \ .$$

Using this in the formula for the correlator $\langle\Omega\rangle$ we obtain

$$\lim_{N\to\infty} \left\langle \Omega(e^{ic_+/N}; e^{-ic_-/N}) \right\rangle$$

$$= \int D(z_0,\tilde{z}_0) \exp\ i\mathrm{STr}\left(c_+ z_0\tilde{z}_0(1-z_0\tilde{z}_0)^{-1} - c_- \tilde{z}_0 z_0(1-\tilde{z}_0 z_0)^{-1}\right) \ .$$

Computation of this definite superintegral is a standard exercise and gives the random matrix answer (1). The result for the pair correlation function follows by taking two derivatives. The dependence on $z_0, \tilde{z}_0$ of the $1/N$ correction to $\langle\Omega_1\rangle$ can be shown to be such that the integral over $z_0, \tilde{z}_0$ exists and the coefficient of $1/N$ in $\langle\Omega\rangle$ is finite.

In conclusion, I believe that the arguments presented here are strong evidence in favor of the precise version of the BGS conjecture proposed. To turn them into a proof that is rigorous by mathematical standards, three improvements are required.

1. One needs to do a spectral analysis of the operator $R_N^\epsilon \mathrm{Ad}(U_N)$ to support our intuition of what happens in the limit $N \to \infty$.

2. The transformation of Berezin forms $D(Z, \tilde{Z}) \to D(z_0, \tilde{z}_0)D(\zeta, \tilde{\zeta})$ needs to be scrutinized carefully, to rule out the possible existence of boundary anomalies (Sect. 5) that might interfere with our argument.

3. A complete computation of the $1/N$ corrections is called for. In particular, one must resolve the puzzle why the calculation of Andreev et al, which is closely related to ours, fails to reproduce the diagonal approximation to the double sum over periodic orbits.

Acknowledgment. While participating in the workshop on "Disordered Systems and Quantum Chaos" at the Newton Institute, Cambridge, where these lectures were delivered, I enjoyed a number of illuminating discussions with Steve Zelditch, who got me straightened out on the subject of semiclassical limits and suggested to compose the deterministic map with a stochastic Hamiltonian flow that averages to a diffusion operator.

REFERENCES

1. O. Bohigas, M.J. Giannoni, and C. Schmit, Phys. Rev. Lett. **52**, 1 (1984).
2. M.V. Berry, Proc. R. Soc. London A **400**, 229 (1985).
3. E.B. Bogomolny and J.P. Keating, Phys. Rev. Lett. **77**, 1472 (1996).
4. A.V. Andreev, B.D. Simons, O. Agam, and B.L. Altshuler, Nucl. Phys. B **482**, 536 (1996).
5. K.B. Efetov, Adv. Phys. **32**, 53 (1983).
6. V.I. Arnold and A. Avez, *Ergodic Problems of Classical Mechanics* (Benjamin, New York, 1968).
7. A.M. Perelomov, *Generalized coherent states and their applications* (Springer–Verlag, Berlin, 1986).
8. F.A. Berezin and D.A. Leites, Sov. Math. Dokl. **16**, 1218 (1975); B. Kostant, Lect. Notes Math. **570**, 177 (1977).
9. F.A. Berezin, *Introduction to Superanalysis* (Reidel, Dordrecht, 1987).
10. M.J. Rothstein, Trans. Amer. Math. Soc. **299**, 387 (1987).
11. M.R. Zirnbauer, J. Phys. A **29**, 7113 (1996).
12. M.R. Zirnbauer, J. Math. Phys. **37**, 4986 (1996).
13. M.R. Zirnbauer, ICMP97 proceedings; chao-dyn/9810016.
14. D. Ruelle, Phys. Rev. Lett. **56**, 405 (1988).
15. A. Schwartz and O. Zaboronsky, Commun. Math. Phys. **183**, 463 (1997).

SEMICLASSICAL QUANTIZATION OF MAPS AND SPECTRAL CORRELATIONS

Uzy Smilansky

The Isaac Newton Institute of Mathematical Sciences
20 Clarkson Road, Cambridge, CB3 0EH, England, UK
and
Department of Physics of Complex Systems,
The Weizmann Institute of Science, Rehovot, 76100, ISRAEL

INTRODUCTION

Maps appear naturally in the analysis of classical chaotic systems: The Poincaré map, which maps the Poincaré section onto itself, is a powerful tool which was designed to simplify the description of complex multidimensional flows, without loosing their essential features. Its most important property is that the phase space measure of the section is preserved, thus maintaining the Hamiltonian character of the original system. Maps with similar properties arise also when the Hamiltonian is a periodic function of time. The stroboscopic description – where the dynamics is recorded and analyzed only at times which are integer multiples of the period – is carried out in terms of an area preserving classical map. We shall confine our attention to measure preserving maps which act on a compact phase space in two dimensions. These are the simplest, yet non trivial maps, which correspond to the simplest flows that display classical chaos.

The quantum mechanical analogues of the area preserving maps introduced above are unitary operators which act on Hilbert spaces of finite dimension, N. They are often referred to as evolution, scattering or Floquet operators, depending on the physical context where they are used. Their eigenvalues consist of N points on the unit circle (eigenphases). The eigenphases statistics was shown to depend on the nature of the underlying classical dynamics. The quantum analogues of classically integrable maps display Poissonian statistics. This was found numerically in many systems, and for a few particular cases, there exist some rigorous results 1, 2. Numerical studies of various classically chaotic systems suggest that the eigenphases statistics conform quite accurately with the results of Random Matrix Theory (RMT) for Dyson's "circular ensembles" 3, 4, 5, 6. This universal behavior which is directly linked with the underlying classical dynamics is one of the central issues in quantum chaology, and is the subject of the present paper.

There are three reasons why the study of quantum maps gained considerable interest during the past years:

Supersymmetry and Trace Formulae: Chaos and Disorder
Edited by Lerner *et al.*, Kluwer Academic / Plenum Publishers, New York, 1999

- The quantum energies (either exact or semiclassical) of Hamiltonian systems, can be derived from secular equations which involve the eigenvalues of certain unitary matrices, which, in the classical limit, correspond to Poincaré maps on a properly defined section. Notable examples are Bogomolny's semiclassical T matrix method, 7 and the scattering method developed by the present author and his coworkers 8, 9, 10. In the semiclassical limit, the statistics of the *energy levels* can be expressed in terms of the statistics of the *eigenphases* of the corresponding unitary operator. This correspondence enables one to take advantage of the structural simplicity of unitary matrices and their spectrum.

- Even though the original development and applications of Random Matrix Theory were focussed on the Gaussian ensembles of *Hermitian* matrices, it was soon realized by Dyson 11 that the circular ensembles of *unitary* matrices have properties which make them more convenient and natural to study, such as e.g., the fact that the spectra are confined to the unit circle and not to the entire real line as for the Hermitian operators.

- Recently, important progress was achieved by Zirnbauer and his coworkers 12, 13, who extended the field theoretical methods of Efetov to problems involving unitary operators and the corresponding circular ensembles.

At present, there exist two distinct theoretical approaches to the study of spectral statistics of unitary operators whose classical analogues are chaotic: The first is the field theoretical approach which is introduced and explained in M. Zirnbauer's contribution to this volume. The second is the semiclassical approximation, which will be the topic of this article. The material will be presented in the following way. The next section will introduce the statistical measures which we shall study, and summarize the predictions of RMT for these quantities. Preparing the background for the semiclassical approximation, a few topics from the theory of classical maps will be reviewed. Then, the semiclassical expression of the quantum evolution operator will be discussed, and the semiclassical trace formula appropriate to the study of unitary operators will be presented. In order to study spectral statistics, the "semiclassical ensemble" will be introduced and will be used to obtain expressions for the statistical measures which can be directly compared with the RMT predictions. In particular, two-points correlation function will be discussed using two complementary approaches. The first is an adaptation of a formalism introduced recently by Bogomolny and Keating 14, and the other, which emphasizes the rôle of periodic orbits correlations, follows the ideas developed by Argaman *et al.* 15 and Cohen *et. al.* 17. The summary section is dedicated to a critical discussion of the present status of the semiclassical approach, and it includes a list of open problems.

SPECTRAL STATISTICS - DEFINITIONS AND RESULTS FROM RMT

Consider an arbitrary $N \times N$ unitary matrix U with a spectrum $\left\{e^{i\theta_l}\right\}_{l=1,\ldots N}$. The spectral density can be written as

$$d_U(\theta) \equiv \sum_{l=1}^{N} \delta(\theta - \theta_l) = \frac{N}{2\pi} + \frac{1}{2\pi} \sum_{n=1}^{\infty} \left(e^{-in\theta}\mathrm{tr}U^n + e^{in\theta}\mathrm{tr}U^{-n}\right) \tag{1}$$

The corresponding counting (staircase) function is

$$N_U(\theta) \equiv \frac{N\theta - \Theta}{2\pi} + \frac{i}{2\pi} \sum_{n=1}^{\infty} \frac{1}{n} \left(e^{-in\theta}\mathrm{tr}U^n - e^{in\theta}\mathrm{tr}U^{-n}\right), \tag{2}$$

where $\det(-U) \equiv e^{i\Theta}$, and we set $N(0) = 0$. From now on, we shall use the notation $t_n = \mathrm{tr}U^n$. We write the characteristic polynomial as

$$p_U(z) \equiv \det(I - zU) = \sum_{n=0}^{N} a_n z^n \tag{3}$$

and its roots are $z_l = e^{-i\theta_l}$. An important consequence of the unitarity of U is the *inversive symmetry* of the coefficients a_n,

$$a_n = e^{i\Theta} a_{N-n}^* \ . \tag{4}$$

Starting from

$$\det(I - xU) = \sum_{n=0}^{N} a_n x^n = \exp\left(-\sum_{k=1}^{\infty} \frac{x^k}{k} \mathrm{tr}U^k\right) \ , \tag{5}$$

one can derive Newton's identities which relate the traces $t_n = \mathrm{tr}U^n$ and the coefficients of the secular polynomial

$$a_n = -\frac{1}{n}\left(t_n + \sum_{k=1}^{n-1} a_k t_{n-k}\right) \ . \tag{6}$$

Since $a_n = 0$ for $n > N$, the t_n for all $n > N$ depend linearly on the lower $n \leq N$ traces.

The two-point spectral statistics of interest here are defined as follows. Consider the oscillatory part of the number counting function:

$$\tilde{N}_U(\theta) \equiv N_U(\theta) - \frac{N\theta - \Theta}{2\pi}. \tag{7}$$

The number-number correlation function is

$$\begin{aligned}
\mathcal{N}(x, y) &\equiv \int_0^{2\pi} \frac{d\omega}{2\pi} \left\langle \tilde{N}_U(\omega + x)\tilde{N}_U(\omega + y)\right\rangle \\
&= \frac{1}{2\pi^2} \sum_{n=1}^{\infty} \frac{\langle|t_n|^2\rangle}{n^2} \cos[n(x - y)]
\end{aligned} \tag{8}$$

where we denote by $\langle\cdot\rangle$ the average over an ensemble of Unitary matrices which can be e.g., one of Dyson's circular ensembles or any other ensemble such as the semiclassical ensemble which will be introduced in the sequel. Due to the averaging over the variable ω, $\mathcal{N}(x, y)$ depends on its arguments only through the difference $\eta = x - y$. The more frequently used density-density correlation function can be derived from the number-number correlators by

$$R_2(\eta) \equiv \left(\frac{2\pi}{N}\right)^2 \int_0^{2\pi} \frac{d\omega}{2\pi} \left\langle \tilde{d}_U(\omega + \eta/2)\tilde{d}_U(\omega - \eta/2)\right\rangle = -\left(\frac{2\pi}{N}\right)^2 \frac{d^2}{d\eta^2}\mathcal{N}(\eta). \tag{9}$$

Another statistical measure which we shall study is the auto-correlation function of the characteristic polynomial 6, 18

$$\begin{aligned}
C(\eta) &\equiv e^{in\frac{N}{2}} \int_0^{2\pi} \frac{d\omega}{2\pi} \left\langle p_U(e^{-i(\omega+\eta/2)})p_U^*(e^{-i(\omega-\eta/2)})\right\rangle \\
&= \sum_{n=0}^{N} \langle|a_n|^2\rangle \cos[(n - N/2)\eta]
\end{aligned} \tag{10}$$

where the phase factor $e^{i\eta\frac{N}{2}}$ is introduced to keep the correlation function real. The inversive symmetry (4) implies that the Fourier components of $C(\eta)$ are symmetric about $N/2$. This statistical measure contains in it correlation between more than just pairs of eigenvalues, as can be easily shown by writing the a_n in terms of the eigenvalues. Hence, $C(\eta)$ tests aspects of the eigenvalues distribution which are not accessible by the study of the two-points functions defined previously.

There are a few important identities which will be used throughout this work, and which provide alternative expressions for the spectral functions which were introduced above. Using Cauchy's theorem one can calculate the number of eigenphases in the interval $[0, \theta]$ on the unit circle

$$N(\theta) = \frac{1}{2\pi i}\left[\log p_U(e^{i\theta+\epsilon}) - \log p_U(e^{i\theta-\epsilon})\right]_{\epsilon\downarrow 0} - \frac{1}{2\pi i}\left[\log p_U(e^{\epsilon}) - \log p_U(e^{-\epsilon})\right]_{\epsilon\downarrow 0} \quad (11)$$

The unitarity of U implies the "functional equation"

$$p_U(z) = z^N e^{-i\Theta}\left(p_U(\frac{1}{z^*})\right)^* . \quad (12)$$

Substituting in (11) one can get an expression for the ϵ smoothed stair-case function

$$N_\epsilon(\theta) = \frac{\theta N - \Theta}{2\pi} + \frac{1}{2\pi i}\log\left[\frac{\sum_{n=0}^{N} a_n^* e^{-in\theta}e^{-n\epsilon}}{\sum_{n=0}^{N} a_n e^{in\theta}e^{-n\epsilon}}\right] , \quad (13)$$

and $\lim_{\epsilon\downarrow 0} N_\epsilon(\theta) = N(\theta)$. The argument of the log in (13) does not have zeros or poles as long as $\epsilon > 0$. Hence, the phase

$$\phi_\epsilon(\theta) = \frac{1}{i}\log\left[\frac{\sum_{n=0}^{N} a_n^* e^{-in\theta}e^{-n\epsilon}}{\sum_{n=0}^{N} a_n e^{in\theta}e^{-n\epsilon}}\right] \quad (14)$$

is a well defined function for real θ and it assumes values in $[0, 2\pi]$. The limit $\epsilon \downarrow 0$ is not always possible to follow. Thus, we are sometimes forced to proceed with a small but finite value of ϵ, for which the number function $N_\epsilon(\theta)$ is smooth, and retains the stair-case character in an approximate way 14. This is realized by observing that $N_\epsilon(\theta)$ for ϵ which is smaller than the mean spacing $2\pi/N$ maintains some of the staircase characteristics, namely, it assumes approximately constant integer values which are interrupted by steeply increasing sections, located in the neighborhood of the eigenphases θ_l. One can obtain an approximate spectrum as the solutions $\theta_l(\epsilon)$ of the equation

$$N_\epsilon(\theta_l(\epsilon)) = l - \frac{1}{2} \quad \text{for} \quad l = 1\cdots N . \quad (15)$$

It is tacitly assumed that $N_\epsilon(\theta)$ is a monotonic increasing function, because otherwise the above equation might have more than N solutions. The corresponding sharpened spectral density is

$$d_\epsilon^\sharp(\theta) \equiv \sum_{l=1}^{N} \delta\left(N_\epsilon(\theta) - (l - \frac{1}{2})\right)\frac{\mathrm{d}N_\epsilon(\theta)}{\mathrm{d}\theta} . \quad (16)$$

Using the definitions (13,14), one gets

$$N_\epsilon^\sharp(\theta) = N_\epsilon(\theta) + \frac{1}{2\pi}\sum_{m\neq 0}\frac{(-1)^m}{m}e^{im(\phi_\epsilon(\theta)+\theta N-\Theta)} . \quad (17)$$

The staircase function, $N_\epsilon^\sharp(\theta)$, approximates the true counting function better and better as $\epsilon \downarrow 0$. It is hoped that even for finite but sufficiently small $\epsilon \approx \mathcal{O}(N^{-1})$, the spectral statistics calculated for $N_\epsilon^\sharp(\theta)$ imitates the exact spectral statistics. This equation will be the starting point for the semiclassical theory for the two-point correlations $\mathcal{N}(\eta)$.

To end this section, we shall quote the expressions provided by Random Matrix Theory for $\langle |t_n|^2 \rangle$ and $\langle |a_n|^2 \rangle$, for Dyson's circular ensembles COE and CUE, and for the circular Poisson ensemble 19.

$$
\textbf{COE}: \qquad \langle |t_n|^2 \rangle =
\begin{cases}
2n - n \sum_{m=1}^{n} \frac{1}{m + \frac{N-1}{2}} & \text{for } n\langle N \\[2em]
2N - n \sum_{m=1}^{N} \frac{1}{m+n - \frac{N+1}{2}} & \text{for } n \geq N
\end{cases}
$$

$$
\langle |a_n|^2 \rangle = 1 + \frac{n(N-n)}{N+1} \tag{18}
$$

$$
\textbf{CUE}: \qquad \langle |t_n|^2 \rangle =
\begin{cases}
n & \text{for } n\langle N \\[1em]
N & \text{for } n \geq N
\end{cases}
$$

$$
\langle |a_n|^2 \rangle = 1 \tag{19}
$$

$$
\textbf{POISSON}: \qquad \langle |t_n|^2 \rangle = N \qquad \text{for all } n
$$

$$
\langle |a_n|^2 \rangle = \binom{N}{n} \tag{20}
$$

Substituting these expressions in (8),(9) or (10), one can get the RMT expression for $\mathcal{N}(\eta)$, $R_2(\eta)$ or $C(\eta)$. As was mentioned previously, the numerical studies of the spectral statistics of quantum maps show that chaotic maps which obey time reversal symmetry adhere to the predictions of the COE ensemble, those which violate time reversal reproduce the predictions of the CUE ensemble, while for integrable maps, the statistical measures are given by the predictions of the Poisson ensemble. The main purpose of this paper is to show how these observations can be explained by the semiclassical theory of quantized maps, and to identify the circumstances where one would expect systematic deviations from the predictions of RMT.

SEMICLASSICAL QUANTIZATION OF MAPS

Classical maps

The quantum unitary operator U which we consider, is assumed to be the analogue of an area preserving map $\mathcal{F}$ acting on a finite phase space domain $\mathcal{M}$ with area $|\mathcal{M}|$. (For the present discussion we shall confine our attention to maps with the twist property. The semiclassical treatment can be extended to the general case.) The phase space coordinates are denoted by $\gamma = (q,p)$ and γ is mapped to $\gamma' \equiv \mathcal{F}(\gamma)$. Because of the area–preserving property, the map can be effected in terms of a generating function (action) $\Phi(q,q')$

$$
p = -\frac{\partial \Phi(q,q')}{\partial q} \quad ; \quad p' = \frac{\partial \Phi(q,q')}{\partial q'}. \tag{21}
$$

The explicit mapping function $\gamma' = \mathcal{F}(\gamma)$ is obtained by solving the implicit relations (21). The twist condition ensures that the implicit equations (21) have a unique solution. As an example, consider an integrable map, and denote by (I, ϕ) the action-angle variables, where I is the invariant momentum under the action of the map. The domain of the mapping is the annulus $I \in [I_{min}, I_{max}]$, $\phi \in [0, 2\pi)$. In this representation, the generating function must take the form $\Phi(\phi, \phi') = \Phi(\phi' - \phi)$. The explicit map is

$$I' = I \quad ; \quad \Delta\phi = \phi' - \phi = f(I) \tag{22}$$

Here, $f(I)$ (the angular velocity) is the inverse of the generating relation $I = \frac{d\Phi(\Delta\phi)}{d\Delta\phi}$, which gives $\Delta\phi$ in terms of I. The twist condition is fulfilled if $\frac{d^2\Phi(\Delta\phi)}{d\Delta\phi^2} \neq 0$.

A classical trajectory is obtained by applying the map to an arbitrary initial point in $\mathcal{M}$, and the corresponding action is accumulated along the trajectory.

Classical dynamics can be also viewed as the evolution of phase space densities in time. For maps, the evolution is mediated by the Perron-Frobenius operator,

$$W(\gamma, \gamma') \equiv \delta(\gamma' - \mathcal{F}(\gamma)), \tag{23}$$

and a phase space density $\rho(\gamma)$ evolves under the map as

$$\rho'(\gamma') = \int_{\mathcal{M}} d\gamma \, \rho(\gamma) W(\gamma, \gamma') \, . \tag{24}$$

The Ruelle ζ function of the map $\mathcal{F}$ is defined as

$$\zeta_{cl}(z) \equiv [\det(I - zW)]^{-1} = \exp\left[-\log\det(I - zW)\right] = \exp\left[\sum_{n=1}^{\infty} \frac{w_n}{n} z^n\right], \tag{25}$$

where, for hyperbolic maps,

$$w_n = \mathrm{tr}W^n = \sum_{p \in \mathcal{P}_n} \frac{n_p g_p}{|\det(I - T_p^r)|}. \tag{26}$$

This sum extends over the set $\mathcal{P}_n$ of all the primitive periodic orbits of $\mathcal{F}$, with periods n_p which are divisors of n, so that $n = n_p r$. Orbits which are related by a discrete symmetry are counted once, and their multiplicity is denoted by g_p. T_p is the linearization of the map $\mathcal{F}^{n_p}$ about the p periodic orbit, and its eigenvalues are λ_p and λ_p^{-1} with $|\lambda_p|\rangle 1$. The traces w_n can be interpreted as the probability to perform n-periodic motion. The ergodicity of the map dynamics implies that w_n approaches 1 as $n \to \infty$. Thus, the infinite sum in (25) converges absolutely for $|z| < 1$. One can substitute (26) in (25) and perform the summation over repetitions, to get the well known expression

$$\zeta_{cl}(z) = \left[\prod_{m=1}^{\infty} \prod_{p \in \mathcal{P}} \left(1 - \left(\frac{z^{n_p}}{\lambda_p^m}\right)\right)^{g_p m}\right]^{-1} \tag{27}$$

The summation extends over the set of primitive periodic orbits $p \in \mathcal{P}$. The function $\zeta_{cl}(z)$ is analytic in the open unit circle, with a pole at $z = 1$. An alternative expression for $\zeta_{cl}(z)$ in terms of the spectrum $\{w_n\}$ of W reads

$$\zeta_{cl}(z) = \left[\prod_{n=1}^{\infty}(1 - zw_n)\right]^{-1} \tag{28}$$

The rate of mixing induced by the map is determined by the gap between the main eiegnvalue ($w_1 = 1$) and the next one. In the limit of infinitely fast mixing, $w_n \to 0$ for $n > 1$. Hence,

$$\zeta_{cl}^{(m)}(z) \approx \frac{1}{1 - z} \, . \tag{29}$$

This is as much classical mechanics as we need, and the semiclassical theory follows in the next subsection.

The semiclassical approximation for $\mathrm{tr}U^n$

The semiclassical expression for the matrix elements of U in the q representation is 21, 23.

$$\langle q|U|q'\rangle = \left(\frac{1}{2\pi\hbar i}\right)^{\frac{1}{2}} \left[\frac{\partial^2\Phi(q,q')}{\partial q\partial q'}\right]^{\frac{1}{2}} e^{i\Phi(q,q')/\hbar} \tag{30}$$

In the semiclassical limit, N, the dimension of the Hilbert space where U acts, is the integer part of $\frac{|\mathcal{M}|}{2\pi\hbar}$.

We have seen above, that the main building blocks of the spectral statistics are the traces $t_n = \mathrm{tr}U^n$. The semiclassical approximation for t_n involves the periodic manifolds of the classical map. For hyperbolic maps 22, 23,

$$t_n \equiv \mathrm{tr}U^n \approx \sum_{p\in\mathcal{P}_n} \frac{g_p n_p e^{ir(\Phi_p/\hbar - \nu_p\frac{\pi}{2})}}{|\det(I - T_p^r)|^{\frac{1}{2}}} \tag{31}$$

The semiclassical approximation for t_n involves the same periodic orbits as the classical trace $\mathrm{tr}W^n$ (26), namely, $\mathcal{P}_n$ is the set of all primitive periodic orbits of $\mathcal{F}$, with periods n_p which are divisors of n, so that $n = n_p r$. T_p is the monodromy matrix, and orbits which are related by a discrete symmetry and multiplicity g_p are counted once. Each periodic orbit contribution is endowed with a phase which is the sum of the action along the periodic orbit,

$$\Phi_p = \sum_{j=1}^{n_p} \Phi(q_j, q_{j+1}) \quad (\text{with} \quad q_{n_p+1} = q_1), \tag{32}$$

and the Maslov contribution $-\frac{\nu_p\pi}{2}$.

For integrable maps, we use the phase space variables (I, ϕ) where I is the classical invariant. In the quantum picture, I is quantized to integer multiples of $\hbar$ so that $I_j = j\hbar$ and $1 \leq j \leq N$. The matrix U is diagonal in the j representation. The semiclassical approximation for the eigenphases can be carried out directly,

$$\theta_j(\hbar) \approx \frac{1}{\hbar}\left[\Phi(f(I_j)) - I_j f(I_j)\right] . \tag{33}$$

where $f(I)$ is the angular frequency (22). One can use the above equation to calculate $\mathrm{tr}U^n$,

$$t_n \equiv \mathrm{tr}U^n \approx \sum_{m=1}^{n}\left[\frac{2\pi}{n\hbar f'(I_{n,m})}\right]^{1/2} e^{i\left[n\Phi(\Delta\phi=2\pi\frac{m}{n})/\hbar - (n+\frac{1}{2})\frac{\pi}{2}\right]} \tag{34}$$

Where the summation is carried over the periodic manifolds of period n and winding number m. They occur at values of I for which the angular frequency is rational $\frac{f(I_{n,m})}{2\pi} = \frac{m}{n}$.

The expressions for t_n in the classically integrable and classically chaotic cases are the necessary input for the computation of the spectral measures which were introduced in the previous section.

A SEMICLASSICAL THEORY FOR THE SPECTRAL STATISTICS

Before we can proceed any further, we must clarify an essential point. In contrast with the RMT, the semiclassical theory deals with a *single* system, and not with an ensemble of systems. However, averaging is mandatory in order to get a meaningful theory since the quantities we calculate fluctuate and do not have a proper limit when $\hbar \to 0$. We generate the "semiclassical ensemble" by considering the inverse Planck constant $\beta = \hbar^{-1}$ as a parameter, and different realizations of the ensemble are distinguished by the value of the parameter β. We restrict β to the interval $|\beta - \beta_0| \leq \Delta/2$ and $\Delta = \frac{2\pi}{|\mathcal{M}|}$. This way, the matrices in the ensemble have the same dimension $N(\beta_0)$. The mean value β_0 is assumed to be sufficiently large to justify the use of the semiclassical approximation. That is, for typical orbits p and p' $\beta_0 |\Phi_p - \Phi_{p'}| >> 1$. The interval Δ is taken to be small on the scale of β_0, but sufficiently large so that $\Delta |\Phi_p - \Phi_{p'}| > 2\pi$. In this way, the phases (mod 2π) of the semiclassical expressions can be considered random. The averaging over the "semiclassical ensemble" is effected by

$$\langle A \rangle_\hbar \equiv \frac{1}{\Delta} \int_{\beta_0 - \Delta/2}^{\beta_0 + \Delta/2} \mathrm{d}\beta A(\beta). \tag{35}$$

With this definition of the ensemble average, we get for both classically integrable and chaotic maps,

$$\langle t_n \rangle_\hbar = 0. \tag{36}$$

The variance for the classically chaotic case reads,

$$\langle |t_n|^2 \rangle_\hbar \approx \sum_{p \in P_n} \frac{g_p^2 n_p^2}{|\det(I - T_p^r)|} = \langle g_p n_p \rangle_{cl(n)} w_n . \tag{37}$$

The RHS of (37) defines $\langle g_p n_p \rangle_{cl(n)}$, which, for large n approachs a well defined limit since averaging with respect to the classical phase space measure can be approximated by a periodic point average. If the repetitions of primitive orbits are neglected, we can write $\langle g_p n_p \rangle_{cl(n)} w_n = g n w_n$, which defines the mean multiplicity, and it is further assumed for simplicity that the mean multiplicity is independent of n.

For integrable maps we get

$$\langle |t_n|^2 \rangle_\hbar \approx \frac{2\pi(I_{max} - I_{min})}{2\pi\hbar} = N . \tag{38}$$

Thus, the β averaging provides the well know diagonal (random phase) approximation. The diagonal approximation is not valid uniformly in n, and it leads to completely wrong results for $n > N$. This issue will be explained and discussed further in the chapter on action correlations. We shall use the diagonal approximation in the restricted range $n < N(\beta_0) = \frac{\mathcal{M}\beta_0}{2\pi}$ where it can be justified. For integrable maps, the variances of t_n are independent of n. The result $\langle |t_n|^2 \rangle_\hbar \approx N$ for integrable systems implies that the spectral two-point correlation function in this case is Poisson 20. Let us consider higher moments of the t_n distribution, such as e.,g., $\langle (t_n)^k (t_m^*)^l \rangle_\hbar$. If n and m are relatively prime, the actions Φ_t which contribute to t_n and t_m are sufficiently different. Thus, all the terms in the product $(t_n)^k (t_m^*)^l$ are oscillatory and will yield a vanishing result upon averaging. If n and m have a common divisor, j, choose $k = m/j, l = n/j$, and all the amplitudes which involve repetitions of the primitive orbits of length j will contribute non oscillatory terms to the correlator. However, for hyperbolic maps, the number of periodic orbits which involve repetitions is exponentially smaller than the total number of periodic orbits, and the statistical independence of the variables t_n and t_m is ensured. If we check all other correlators using the approximation that repetitions can be neglected, we find that for $n < N/2$, t_n are Gaussian random variables. The corresponding approximation is more difficult to justify for integrable maps because the proliferation of periodic manifolds is only algebraic.

In summary, the approximation in which **repetitions of periodic orbits are neglected**, implies that we may replace the β averaging by averaging over the ensemble of the t_n variables, with $n \leq N/2$, whose distribution is defined as follows:

I . The semiclassical ensemble of $\{t_n\}$, $(n < N/2)$ is an ensemble of independent random Gaussian variables.

II . For classical chaotic systems, the variances $\langle |t_n|^2 \rangle_\hbar \approx g n w_n$, where w_n are the traces of the classical evolution operator (26), and g is the mean multiplicity.

III. For classical integrable systems, the variances are $\langle |t_n|^2 \rangle_\hbar \approx N$

This set of rules will be now used to represent the averaging with respect to the "semiclassical ensemble". In particular, they will be used to calculate the semiclassical expressions for the spectral statistics of interest here. As long as one stays strictly within the domain of validity of these rules, one obtains expressions which are very similar to the results derived by other groups using completely different methods.

Finally, it should be noted that in RMT, the traces t_n are indeed random Gaussian variables 19 in the limit where n is fixed and $N \to \infty$. This property, which is shared by the statistical and the semiclassical ensembles, has far reaching consequences, and it constitutes a strong link between RMT and quantum chaos.

A. The auto-correlation function of the characteristic polynomial. The Fourier coefficients of the auto-correlation function $\langle |a_n|^2 \rangle_\hbar$ will be derived from the generating function

$$
\begin{aligned}
G_\hbar(x, y) &= \left\langle \det(I - xU) \det(I - yU^\dagger) \right\rangle_\hbar \\
&= \left\langle \exp\left(-\sum_{k=1}^{\infty} \frac{1}{k}(x^k t_k + y^k t_k^*) \right) \right\rangle_\hbar \\
&= \exp\left(\sum_{k=1}^{\infty} (xy)^k \frac{\langle |t_k|^2 \rangle_\hbar}{k^2} \right)
\end{aligned}
\tag{39}
$$

so that

$$
\langle a_n a_m^* \rangle_\hbar = \delta_{n,m} \frac{1}{n!} \frac{\partial^n}{\partial^n v} Z_\hbar(v) \Big|_{v=0} ,
\tag{40}
$$

with

$$
Z_\hbar(v) = \exp\left(\sum_{k=1}^{\infty} \langle |t_k|^2 \rangle_\hbar \frac{v^k}{k^2} \right) .
\tag{41}
$$

The $\langle |a_n|^2 \rangle_\hbar$ can be obtained from the recursion relations

$$
\langle |a_l|^2 \rangle_\hbar = \frac{1}{l} \sum_{k=1}^{l} \langle |a_{l-k}|^2 \rangle_\hbar \frac{\langle |t_k|^2 \rangle_\hbar}{k} .
\tag{42}
$$

It is important to remember that for the purpose of calculating the autocorrelation function (10), it suffices to obtain the $\langle |a_n|^2 \rangle_\hbar$ for $n \leq N/2$. The rest are provided by the inversive relation $\langle |a_n|^2 \rangle_\hbar = \langle |a_{N-n}|^2 \rangle_\hbar$. Note that the iteration procedure (42) is constructed in such a manner, that the only input necessary for the calculation of the $N/2$ lower $\langle |a_n|^2 \rangle_\hbar$ consists only of the lowest $N/2$ values of $\langle |t_n|^2 \rangle_\hbar$, for which the working hypothesis which underlies the present derivation can be justified.

The semiclassical expressions for $\langle |t_k|^2 \rangle_\hbar$, depend on the nature of the corresponding classical dynamics (37),(38). For chaotic classical dynamics we have

$$
\langle |t_n|^2 \rangle_\hbar \approx g n w_n
\tag{43}
$$

¿From which it follows that (41) can be written as

$$Z_\hbar(v) = \exp\left(\sum_{k=1}^{\infty} \langle |t_k|^2 \rangle_\hbar \frac{v^k}{k^2}\right)$$

$$\approx \exp\left(g \sum_{k=1}^{\infty} w_k \frac{v^k}{k}\right) \qquad (44)$$

$$= (\zeta_{cl}(v))^g$$

Where $\zeta_{cl}(z)$ is the Ruelle ζ function for the classical mapping $\mathcal{F}$ defined previously in (25). This can be derived by substituting (43) in (42) and the recursion relations take the form

$$\langle |a_m|^2 \rangle_\hbar = \frac{g}{m} \sum_{k=1}^{m} \langle |a_{m-k}|^2 \rangle_\hbar w_k \qquad (45)$$

which should be solved with the initial condition $\langle |a_0|^2 \rangle_\hbar = 1$.

Let us consider systems which are strongly mixing and for which all transients die out on a short time scale. In such cases, $\zeta_{cl}(v) \approx = \zeta_{cl}^{(m)}(v) = \frac{1}{1-v}$ and $Z_\hbar(v) \approx \frac{1}{(1-v)^g}$. Direct substitution gives the following results:

Chaotic systems which *violate* time reversal symmetry (TRS) have $g = 1$, and

$$\langle |a_m|^2 \rangle_\hbar = 1 \ . \qquad (46)$$

Thus, the semiclassical result coincides with the prediction of RMT for the CUE, (see (19)), $\langle |a_m|^2 \rangle_{CUE} = 1$.

Chaotic systems which *respect* TRS, have $g = 2$ and

$$\langle |a_l|^2 \rangle_\hbar \approx \begin{cases} 1+l & \text{for } 1 < l \le N/2 \\ 1+N-l & \text{for } N > l \ge N/2 \end{cases} . \qquad (47)$$

This expression does not reproduce the RMT result for the COE case (18). However, for large N, where the semiclassical approximation is justified, the semiclassical result agrees with the exact expression in a domain of l values of size $\sqrt{N}$ in the vicinity of the end points of the l interval, $l = 0$ and $l = N$. The deterioration of the quality of the agreement between the semiclassical and the RMT expressions when TRS is imposed is typical, and its explanation remains one of the open problems in quantum chaos 20.

So far, we discussed systems for which all transients die out on a fast time scale, which was imposed by setting $w_k = \text{tr} W^k = 1$ for all k. This is possible only when one eigenvalue of W is 1 and all the rest vanish. In generic systems, the spectrum is not distributed in this extreme way. Rather, beside the eigenvalue 1 which, for ergodic systems is not degenerate, and which corresponds to the invariant phase–space measure, the spectrum is in the interval $[0, 1)$, and it accumulates at 0. The rate of decay of transients is determined by the magnitude of the gap in the spectrum of W. To get the leading correction due to the non vanishing eigenvalues of W, one can expand the recursion relation (45) to first order in $\mu = w_1 - 1$. One obtains in this way recursion relations for the correction to $\langle |a_m|^2 \rangle_\hbar$. They are particularly simple for the cases with $g = 1, 2$ and the corrected coefficients are

$$\langle |a_l|^2 \rangle_\hbar = 1 + \mu \qquad \text{for } g = 1 \qquad (48)$$

$$= 1 + l + 2\mu l \qquad \text{for } g = 2$$

The symmetry $\langle |a_l|^2 \rangle_\hbar = \langle |a_{N-l}|^2 \rangle_\hbar$ should be implemented for $l > N/2$. Recently, the Essen group studied numerically the variances of the coefficients of the characteristic polynomial for the quantum kicked top 19. They checked systems with and without TRS, and their numerical results show systematic deviations from the RMT predictions which are consistent with the expressions (48). The numerical results for the case without TRS are particularly convincing. Other numerical tests which involve $S(k)$ matrices which appear in the theory of quantized graphs 24, were recently carried out. For these systems, the strong mixing limit is not justified, and indeed, the resulting $\langle|a_l|^2\rangle_\hbar$ deviate appreciably from the predictions of RMT. However, the semiclassical theory which uses the classical Ruelle ζ as input, reproduces the main features of the numerical data.

For integrable maps we have (38)

$$\langle|t_n|^2\rangle_\hbar \approx N \tag{49}$$

The resulting recursion relations for the coefficients $\langle|a_m|^2\rangle_\hbar$ are

$$\langle|a_m|^2\rangle_\hbar = \frac{N}{m} \sum_{k=1}^{m} \frac{\langle|a_{m-k}|^2\rangle_\hbar}{k} \tag{50}$$

We were not able to find a closed form for the solution of this equation. However, to leading order $\langle|a_m|^2\rangle_\hbar \approx \frac{N^m}{m!}$ which coincides with the leading term of the result for the Poisson ensemble (20).

B. The spectral two point correlation functions.   Encouraged by the results of the previous section, we shall try to use the same strategy to compute the two-point correlation functions. The essential point in this approach is to express all quantities of interest in terms of quantities which involve the short periodic orbits. This will be achieved here by following the ideas presented in the first section, namely, one generates a *synthetic* point spectrum which might not coincide with the true spectrum in detail, but is expected to reproduce its statistical measures when $\epsilon \approx 1/N$. In this spirit, we calculate the number-number correlation function of the "sharpened" number function (17)

$$N_\epsilon^\sharp(\theta) = N_\epsilon(\theta) + \frac{1}{2\pi} \sum_{m\neq 0} \frac{(-1)^m}{m} e^{im(\phi_\epsilon(\theta)+\theta N - \Theta)} \tag{51}$$

where (14),

$$\phi_\epsilon(\theta) = \frac{1}{i} \log \left[\frac{\sum_{n=0}^{N} a_n^* e^{-in\theta} e^{-n\epsilon}}{\sum_{n=0}^{N} a_n e^{in\theta} e^{-n\epsilon}} \right] . \tag{52}$$

The spectral statistics will be performed for the semiclassical ensemble defined previously. It will be shown bellow, that setting $\epsilon = c/N$ amounts to the introduction of a smooth cut-off which suppresses the dependence of the present theory on t_n with $n > N/2c$.

The oscillatory part of $N_\epsilon^\sharp(\theta)$ comes from the oscillatory part of the first term,

$$\tilde{N}_\epsilon(\theta) = N_\epsilon(\theta) - \frac{\theta N - \Theta}{2\pi} , \tag{53}$$

and from the infinite sum

$$\tilde{N}_{\epsilon,m\neq 0}(\theta) = \frac{1}{2\pi} \sum_{m\neq 0} \frac{(-1)^m}{m} e^{im(\phi_\epsilon(\theta)+\theta N - \Theta)} . \tag{54}$$

$\tilde{N}_{\epsilon,m\neq0}(\theta)$ involves much higher frequencies then those involved in $\tilde{N}_\epsilon(\theta)$, and therefore there are no cross correlations. Accordingly, we shall write

$$\mathcal{N}^\hbar(\eta) = \mathcal{N}^{diag}(\eta) + \mathcal{N}^{off}(\eta) \tag{55}$$

where $\mathcal{N}^{diag}$ and $\mathcal{N}^{off}$ stand for the contributions from correlations in $\tilde{N}_\epsilon(\theta)$ and from $\tilde{N}_{\epsilon,m\neq0}(\theta)$, respectively. Starting with $\mathcal{N}^{diag}(\eta)$, we write

$$\begin{aligned}
\tilde{N}_\epsilon(\theta) &= \frac{1}{2\pi i}\left[\log \sum_{n=1}^{N} a_n^* e^{(-i\theta-\epsilon)n} - \log \sum_{n=1}^{N} a_n e^{(i\theta-\epsilon)n}\right] \\
&= \frac{1}{2\pi i}\sum_{l=1}^{\infty}\frac{e^{-\epsilon l}}{l}\left(t_l^* e^{-i\theta l} - t_l e^{i\theta l}\right),
\end{aligned} \tag{56}$$

and get,

$$\begin{aligned}
\mathcal{N}^{diag}(\eta) &= \int_0^{2\pi}\frac{d\omega}{2\pi}\left\langle \tilde{N}_\epsilon(\omega+\eta/2)\tilde{N}_\epsilon(\omega-\eta/2)\right\rangle_\hbar \\
&= \frac{g}{(2\pi)^2}\left(\sum_{l=1}^{\infty}\frac{w_l}{l}(e^{(i\eta-2\epsilon)l} + e^{(-i\eta-2\epsilon)l})\right) = \frac{1}{2\pi^2}\Re e\log\left(\zeta_{cl}(e^{-2\epsilon+i\eta})\right)^g.
\end{aligned} \tag{57}$$

Above we used the approximate relation $\langle|t_l|^2\rangle_\hbar = glw_l$. Had we chosen $c = 1$, we would have obtained a damping factor $e^{-\frac{2}{N}l}$ which suppresses the $l > N/2$ terms in (58), in a way which justifies *a posteriori* the use of the semiclassical ensemble as defined above. Unfortunately, one has to choose $c = 1/2$ in order that the off–diagonal contribution agrees with the CUE result, rendering the damping less effective. This will be shown in the sequel.

To compute $\mathcal{N}^{off}(\eta)$, we have to evaluate integrals of the type

$$\int_0^{2\pi}\frac{d\omega}{2\pi}\left\langle e^{im(\phi_\epsilon(\omega+x)+(\omega+x)N-\Theta)}e^{im'(\phi_\epsilon(\omega+y)+(\omega+y)N-\Theta)}\right\rangle_\hbar. \tag{58}$$

For this purpose we write

$$\begin{aligned}
\left\langle e^{im\phi_\epsilon(\omega+x)}e^{im'\phi_\epsilon(\omega+y)}\right\rangle_\hbar &= \left\langle \left(\frac{\exp[-\sum_{l=1}^{\infty}\frac{t_l^*}{l}(\psi^*)^l]}{\exp[-\sum_{l=1}^{\infty}\frac{t_l}{l}(\psi)^l]}\right)^m\left(\frac{\exp[-\sum_{l=1}^{\infty}\frac{t_l^*}{l}(\chi^*)^l]}{\exp[-\sum_{l=1}^{\infty}\frac{t_l}{l}(\chi)^l]}\right)^{m'}\right\rangle_\hbar \\
&= \exp\left[-\sum_{l=1}^{\infty}\frac{\langle|t_l|^2\rangle_\hbar}{l^2}\left|m\psi^l + m'\chi^l\right|^2\right]
\end{aligned} \tag{59}$$

where, $\psi = e^{i(\omega+x)-\epsilon}$ and $\chi = e^{i(\omega+y)-\epsilon}$. The last line in (59) is independent of ω, and therefore, when the ω integration is performed in (58), only terms with $m = -m'$ contribute. Recalling the approximate relation $\langle|t_l|^2\rangle_\hbar = lgw_l$ and the definition of the Ruelle ζ function,

$$\begin{aligned}
\mathcal{N}^\hbar(\eta) &= \mathcal{N}^{diag}(\eta) + \mathcal{N}^{off}(\eta) \\
&= \frac{1}{2\pi^2}\Re e\log\left(\zeta_{cl}(e^{-2\epsilon+i\eta})\right)^g \\
&\quad + \frac{1}{2\pi^2}\sum_{m=1}^{\infty}\frac{\cos(Nm\eta)}{m^2}\left(\frac{\zeta_{cl}(e^{-2\epsilon+i\eta})\zeta_{cl}(e^{-2\epsilon-i\eta})}{\zeta_{cl}{}^2(e^{-2\epsilon})}\right)^{gm^2}
\end{aligned} \tag{60}$$

The two-point correlation function is obtained by taking the second derivative of (60). Expressing it in terms of the "unfolded" phase difference $s = \eta\frac{N}{2\pi}$, and neglecting corrections which are of order $1/N$ we get

$$R_2^{(\hbar)}(s) = R^{diag}(s) + R^{off}(s)$$

$$= -\frac{1}{2\pi^2}\frac{d^2}{ds^2}\left[\Re\log\left(\zeta_{cl}(e^{-2\epsilon+i\frac{2\pi}{N}s})\right)^g\right]$$

$$+ 2\left[\sum_{m=1}^{\infty}\cos(2\pi ms)\left(\frac{\zeta_{cl}(e^{-2\epsilon+i\frac{2\pi}{N}s})\zeta_{cl}(e^{-2\epsilon-i\frac{2\pi}{N}s})}{\zeta_{cl}^{\,2}(e^{-2\epsilon})}\right)^{gm^2}\right] \tag{61}$$

This is the central result of the present chapter, expressing the two point statistics in terms of the Ruelle ζ function of the classical map. Analogueous expressions were previously derived for the Gaussian ensembles using field theoretical methods 25, 26, and using the semiclassical trace formula 14. The partitioning of R_2 to its "diagonal" and "off-diagonal" parts conforms with the notations used in 14. M. Zirnbauer presents a field theoretical derivation for maps in his contribution to this volume. To bring our result into a form which is closer to the expressions derived in previous studies, we note that for $\pi|s| > 1$, the m sum in $R^{off}(s)$ converges very rapidly, so that the $m = 1$ term is sufficient. This is not true in the vicinity of $s = 0$, since there, the entire m sum is necessary to reproduce the $\delta(s)$ singularity.

We shall now test to what extent (61) reproduces the RMT limit when we approach the strong mixing limit, for which $\zeta_{cl}(z) = \zeta_{cl}^{(m)}(z) = (1-z)^{-1}$ and assume that TRS is violated ($g = 1$) (see (29)). This should be compared with the CUE result in the semiclassical limit $N \to \infty$. We shall set $\epsilon = \frac{c}{N}$ and the yet unknown constant c will be chosen such that the best agreement with the CUE expression is achieved. For the diagonal term we get

$$R^{diag}(s) \approx \frac{1}{2}\frac{c^2-(\pi s)^2}{\left(c^2+(\pi s)^2\right)^2}\xrightarrow[\pi s\gg 1]{} - \frac{1}{2(\pi s)^2}. \tag{62}$$

For $\pi s \gg c$ this coincides with the CUE expression. $R^{diag}(s)$ can also be written as

$$R^{diag}(s) = \frac{1}{N}\sum_{n=1}^{\infty}\frac{n}{N}e^{-2c\frac{n}{N}}\left(e^{i2\pi s\frac{n}{N}} + e^{-i2\pi s\frac{n}{N}}\right). \tag{63}$$

Thus, the Fourier coefficients of $R^{diag}(s)$ almost coincide with the CUE coefficient for $n < \frac{N}{2c}$, and for $n > \frac{N}{2c}$ the Fourier coefficients approach 0 exponentially.

As was indicated above, the calculation of the "off diagonal" term requires different approximations depending on whether $|\pi s|$ is larger or smaller than 1. For $|\pi s| > 1$ one can truncate the series at the $m = 1$ term. The $m = 1$ contribution is

$$R^{off}_{m=1}(s) = 2\cos 2\pi s\,\frac{c^2}{c^2+(\pi s)^2} \tag{64}$$

To get agreement with the CUE expression for $\pi s \gg 1$, one must choose the regularization constant to be $c = 1/2$. This is disappointing, because it implies that the high n terms are less effectively damped, and this is not consistent with the assumptions which underlie the semiclassical ensemble.

In the domain $\pi s \leq 1$ one can approximate $(1+(\frac{\pi s}{c})^2)^{-1} \approx e^{-(\frac{\pi s}{c})^2}$, taking $c = 1/2$ and using the Poisson summation formula to re-sum the m series we obtain

$$R^{off}(s) = -1 + \frac{1}{\sqrt{4\pi s^2}}\sum_{m=-\infty}^{\infty}e^{-(\frac{s-m}{2s})^2} \tag{65}$$

Since for $t \to 0$

$$\frac{1}{\sqrt{4\pi t}} \sum_{m=-\infty}^{\infty} e^{-(\frac{x-m}{2\sqrt{t}})^2} \quad \to \quad \delta(x) \tag{66}$$

We can identify $x = s$ and $t = s^2$ and get

$$R^{off}(s) \quad \approx -1 + \delta(s) \quad \text{for} \quad |\pi s| << 1. \tag{67}$$

which shows that the expected $\delta(s)$ singularity is reproduced.

The reconstruction of the main features of the CUE expression manifests a few of the difficulties encountered in this approach. They stem from the regularization procedure, which requires a specific choice of the regularization parameter to get the right answer. The smoothing parameter $\epsilon = \frac{1}{2N}$ which should be used, is not large enough to damp the contribution of the $n > N/2$. This pushes the "semiclassical ensemble" beyond its strict domain of validity. The same difficulty arises also in the work of Keating and Bogomolny 14, where the "cut-off" time must be taken as the Heisenberg time, and not its half.

The purpose of the approach presented above was to circumvent the need to use the semiclassical approximation for the higher traces t_n for $n > N$. It is very successful for the calculation of the autocorrelations of the characteristic polynomial. However, it meets with the difficulties mentioned above, when unitarity (pure point spectrum on the unit circle) is to be restored within the semiclassical approximation. In the next chapter we shall present another approach which does not suffer from these problems, but which introduces a new concept which still requires much more study - periodic orbits correlations.

PERIODIC ORBITS CORRELATIONS

In the previous sections we derived the semiclassical theory of spectral fluctuation using semiclassical information on the periodic orbits with period $n \leq N$. In other words, we based the theory on the evolution of the classical system during times which are shorter than the relevant Heisenberg time. However, this by itself is not sufficient to reproduce the most distinctive property of the quantum spectrum - the fact that it consists of N points on the unit circle. This feature was incorporated by generating the "synthetic" spectrum (15) out of the ϵ smoothed spectrum, which served the desired end at the cost of obtaining a theory which depends critically on the smoothing parameter.

In the present section we shall take a completely different route. We shall use the semiclassical expression (31) or (34) for $t_n = \mathrm{tr} U^n$ for classically chaotic or integrable systems, respectively. Averaging $|t_n|^2$ over the domain of β values (35), and *retaining the off diagonal terms* we shall be able to express the two point quantum correlations in terms of correlations in the spectrum of periodic orbits of the classical system. In a way, we do not solve the problem which we have set to solve, but defer the problem to another unknown function which is the classical correlation function. However, recent detailed numerical studies actually confirmed the existence of such classical correlations. 15,16,17. Once the classical correlations are studied and theoretically confirmed, the present formalism will provide the desired semiclassical theory of spectral statistics.

Periodic orbit correlations appear also in the discussion of integrable systems (where one should refer to periodic tori). To be able to treat the two types of systems in a uniform way, we shall introduce a short-hand notation and rewrite (31) and (34) as

$$t_n(\beta) \equiv \sum_{l=1}^{N(\beta)} e^{in\theta_l(\beta)} \approx \sum_{p \in \mathcal{P}_n} A_p e^{i\beta\Phi_p} . \tag{68}$$

The summation over p goes over the set of unstable n–periodic orbits (primitive and repetitions) for chaotic dynamics, and over the set of n–periodic tori for integrable dynamics. The complex coefficients A_p are given in (31), (34) and their phase is determined by the Maslov index. The dependence of the quantum spectrum on the value of $\hbar = \beta^{-1}$ is explicitly indicated. Also, $N(\beta)$ is the integer part of $\left[\frac{|\mathcal{M}|\beta}{2\pi}\right]$ and $|\mathcal{M}|$ is the classical phase space area.

We define a classical density

$$d_{cl}(x; n) \equiv \sum_{p \in \mathcal{P}_n} A_p \delta(x - \Phi_p) , \tag{69}$$

where x has the dimension of action. Recalling the quantum density of eigenphases on the unit circle

$$d_{qm}(\omega; \beta) \equiv \sum_{l=1}^{N} \delta(\omega - \theta_l(\beta)) \tag{70}$$

we can write (68) as

$$\int_0^{2\pi} d\omega e^{i\omega n} d_{qm}(\omega; \beta) \approx \int_{\Phi_{min}(n)}^{\Phi_{max}(n)} dx e^{i\beta x} d_{cl}(x; n) \tag{71}$$

In this way, the trace formulae (31) or (34) are interpreted as a relationship between a strictly quantum density and a strictly classical density. The quantum density involves N points on the unit circle, which have equal (positive) weights. $d_{qm}(\omega; \beta)$ is a function (distribution) in ω and it depends *parametrically* on β. The classical density corresponds to unstable n - periodic orbits (n - periodic tori) whose number increases exponentially (algebraically) with n, and which are weighted by *complex* coefficients which are different from each other in generic systems. $d_{cl}(x; n)$ is a function (distribution) in x and it depends *parametrically* on n. The actions of n-periodic orbits (tori) are bounded from bellow and from above so that $\Phi_{min}(n) \leq x \leq \Phi_{max}(n)$.

Defining the oscillatory part of the classical density,

$$\tilde{d}_{cl}(x; n) \equiv d_{cl}(x; n) - \frac{1}{\Phi_{max}(n) - \Phi_{min}(n)} \sum_{p \in \mathcal{P}_n} A_p , \tag{72}$$

we can construct the *classical* two point correlation function

$$\rho_{cl}(\xi; n) \equiv \frac{\sin \frac{\xi \pi}{|\mathcal{M}|}}{\frac{\xi \pi}{|\mathcal{M}|}} \int_{\Phi_{min}(n)}^{\Phi_{max}(n)} dx \ \tilde{d}_{cl}(x + \xi/2; n)\tilde{d}_{cl}(x - \xi/2; n) \tag{73}$$

The factor in front of the integral comes from the β averaging, and it limits the range of action correlations to $|\xi| < |\mathcal{M}|$. As we shall see above, the correlations of interest here occur on much smaller action differences.

In the present context, it is convenient to normalize the *quantum* two point correlation function (9) in a different way

$$\rho_{qm}(\eta; N) \equiv \int_0^{2\pi} \frac{d\omega}{2\pi} \langle \tilde{d}_{qm}(\omega + \eta/2; \beta)\tilde{d}_{qm}(\omega - \eta/2; \beta)\rangle_\hbar = \left(\frac{N}{2\pi}\right)^2 R_2(\eta). \tag{74}$$

Comparing (73) and (74) we get

$$\int_0^{2\pi} d\eta e^{-i\eta n} \rho_{qm}(\eta; N) = \frac{1}{2\pi} \int_{\Phi_{min}(n)}^{\Phi_{max}(n)} d\xi e^{i(N+1/2)\frac{\pi\xi}{\mathcal{M}}} \rho_{cl}(\xi; n) \equiv \rho(n; N) \tag{75}$$

The function $\rho(n; N)$ expresses the duality between the quantum and classical spectra in the clearest way, because it generates the two-point correlations of the two spectra by taking Fourier transforms with respect to either variable:

$$\rho_{qm}(\eta; N) = \frac{1}{2\pi} \sum_n e^{i\eta n} \rho(n; N) \tag{76}$$

and

$$\rho_{cl}(\xi; n) = \frac{2\pi}{|\mathcal{M}|} \sum_N e^{-i\frac{\xi\pi}{|\mathcal{M}|}(2N+1)} \rho(n; N). \tag{77}$$

It is easy to show that

$$\rho(-n; N) = \rho(n; -N) = \rho(-n; -N) = \rho(n; N) , \tag{78}$$

which completes the definition of $\rho(n; N)$. $\rho(n; N)$ can be considered as a function of n for a fixed N, and then it stands for the *quantum* spectral form factor for the system which is quantized with a Planck constant $\hbar = \frac{|\mathcal{M}|}{2\pi N}$. If n is kept fixed then $\rho(n; N)$ is the *classical* form factor for the *classical* spectrum of n-periodic orbits (tori).

Since both the classical and the quantum spectral densities consist of isolated δ functions, one can extract their diagonal parts,

$$\rho_{qm}(\eta; N) = \frac{N}{2\pi}\left[\delta(\eta) - p_{qm}(\eta; N)\right] \quad \text{with} \quad \int p_{qm}(\eta; N)\mathrm{d}\eta = 1 . \tag{79}$$

Similarily,

$$\rho_{cl}(\xi; n) = \left[\sum_{p \in \mathcal{P}_n} A_p^2\right][\delta(\xi) - p_{cl}(\xi; n)] \quad \text{with} \quad \int p_{cl}(\xi; n)\mathrm{d}\xi = 1 . \tag{80}$$

The normalization of the functions p_{qm} and p_{cl} follows from the fact that the correlators ρ_{qm} and ρ_{cl} were constructed from the oscillatory parts of the corresponding densities.

The implementation of these relations to systems which are integrable classically will be discussed first, as a transparent example of the duality idea. The quantum spectrum in this case is known to be Poisson, hence $\rho_{qm}^{Integ}(\eta; N) = \frac{N}{2\pi}\delta(\eta)$, which implies $\rho^{Integ}(n; N) = \frac{N}{2\pi}$. For classically integrable systems, we know that $\sum_{p\in\mathcal{P}_n} A_p^2 = N$ (38). When this is substituted in (80) and in (75), one obtains $\rho^{Integ}(n; N) = \frac{N}{2\pi}$ if the classical correlation $p_{cl}^{Integ}(\xi; n) = 0$. That is, the classical–quantum spectral duality implies that both spectra have to be Poisson! This result can be also substantiated on different grounds. The quantum eigenphases (in the semiclassical approximation) are obtained by quantizing the action variables to integer multiples of $\hbar$, and with $I_j = j\hbar$,

$$\theta_j(\hbar) \approx \frac{1}{\hbar}\left(\Phi(f(I_j)) - I_j f(I_j)\right) . \tag{81}$$

$f(I)$ is the angular frequency (22). The reason why this series of phases is Poissonian is because the correlations are lost when the phases are considered mod 2π. This happens in the semiclassical limit, and when e.g., $\Phi(f(I)) \approx I^\mu$, with $\mu \geq 2$. The same argument applies also for the spectrum of actions of n- periodic tori

$$\Phi_m(n) = n\Phi(2\pi\frac{m}{n}) \quad ; \quad 1 \leq m \leq n . \tag{82}$$

These actions are related to the eigenphases by a Legendre transformation, and will have a Poisson distribution on the unit circle provided n is large enough.

For system which are chaotic in the classical limit, we can write $gnw_n = \sum_{p\in\mathcal{P}_n} A_p^2$ so that

$$\rho_{cl}(n; N) = gnw_n \left[\delta(\xi) - p_{cl}(\xi; n)\right]. \tag{83}$$

To obtain the *universal* behavior of $p_{cl}(\xi; n)$ we use the strong mixing limit $w_n \approx 1$. We also assume that the quantum two points correlation function converges as $N \to \infty$ to a well defined limit when it is expressed in terms of the unfolded variable $s = \eta\frac{N}{2\pi}$. This is a strong requirement which leads automatically to a scaling form for $\rho(n; N)$

$$\rho_{universal}(n; N) = \frac{N}{2\pi}q(n/N) \tag{84}$$

and $q(x) \to 1$ for $x \gg 1$. With this information and using (77), $p_{cl}(\xi; n)$ can be extracted. Replacing the N summation by an integral,

$$p_{cl}(\xi; n) = \frac{1}{g}\frac{2}{(|\mathcal{M}|/n)}\int_0^\infty \cos\left[2\pi s\frac{\xi}{(|\mathcal{M}|/n)}\right]\left[g - sq(\frac{1}{s})\right]ds = \frac{1}{g}\frac{1}{\lambda(n)}Q\left(\frac{\xi}{\lambda(n)}\right) \tag{85}$$

This is the main result of the present section. It expresses the *classical* correlation function in terms of the function Q which is an appropriate Fourier transform of the quantum function q. The action correlation length $\lambda(n) = \frac{|\mathcal{M}|}{n}$ can be interpreted as the mean phase space area per point on the periodic orbit. It has the dimension of action, and it exceeds by far the mean spacing between actions in the classical spectrum, which are exponentially small in n. This is consistent with the results of numerical studies which show that the action spectrum is Poisson when studied on the scale of a few mean spacings. For systems which violate time reversal symmetry, $g = 1$ and RMT provides $q(x) = \min(x, 1)$. Then,

$$Q_{CUE}(x) = \left(\frac{\sin \pi x}{\pi x}\right)^2 \tag{86}$$

which coincides with the RMT expression for the quantum two point correlation function. The expression for Q_{COE} was derived in 15.

Numerical tests of the Baker map 15, 16 and the deformed cat map 15 confirm the existence of classical correlations in these systems, and reproduce the scaling behavior. However, the numerical correlation functions deviate from the predicted universal functions by up to a factor 2. In 17 the classical quantum duality was investigated for the Sinai billiard in two and three dimensions. For these systems, the classical origin of the action correlations was investigated, and a possible mechanism was identified and was shown to be consistent with the numerical data. This is, however, a mechanism which is particular to these billiards. The origin of the correlations in generic hyperbolic systems remains an enigma.

DISCUSSION AND CONCLUSIONS

In the previous sections we described a few methods to substantiate the empirical finding that the spectral correlations of unitary matrices which represent quantum evolution of classical maps reflect the underlying classical dynamics. In particular, we have shown under what conditions, the spectral correlation functions approach the predictions of random matrix theory.

It should have been made clear from the outset, that typical evolution operators which pertain to actual physical systems can rarely reproduce the full joint probability distribution of the N eigenphases, as given by RMT. (a non trivial exception for $N = 2$ is discussed in 27.) The reason for this observation is that short time evolution usually does not reflect details on the forces and constraints which characterize the system, and which dominate the long time and non trivial evolution. This feature is mediated in our formalism by the appearance of the Ruelle ζ function in the semiclassical expressions of the two spectral measures which we derived. Only if the actual ζ_{cl} is replaced by the function $\zeta_{cl}^{(m)}$ which reflects the long time - and hence the strong mixing - properties of the system, do we recover the RMT results. In other words, the only limit in which the RMT can be reproduced to any desired accuracy is the semiclassical limit $(N \to \infty)$ where short time effects can be neglected.

Once we agree that the universal behavior can be reproduced only in the semiclassical limit, we can now ask what are the main attributes which characterize the approach to this limit. The short time evolution in bound systems which approach complete mixing limit via phase space diffusion were the first systems which were shown to reveal systematic deviations from the RMT results 28. The classical evolution operator on the relevant time scale can be approximated by the diffusion evolution operator, which gives, for the 1-d diffusion discussed in 28, $w_n \approx 1/\sqrt{nD}$ for $n < L^2/D$ where D is the diffusion constant and L is the length of the system. Only for $n > L^2/D$ one recovers the strong mixing limit $w_n = 1$, and hence the RMT attributes. In most systems, however, one cannot identify a systematic approach to the strong mixing limit, and the approach to the RMT limit is affected by the idiosyncratic behavior of short periodic orbits. A few examples will clarify this point. Consider the billiard bounce maps. This map has no fixed points in the interior of the phase space. (The phase space boundaries form a continuous set of fixed points which do not appear in the semiclassical treatment). Hence, for any billiard map, $t_1 = 0$, which is at variance with the RMT result $\langle |t_1|^2 \rangle \approx g$. Billiards with parallel boundary sections support families of bouncing ball orbits. They and their repetitions affect all the $\langle |t_n|^2 \rangle$, and should be subtracted away if any degree of agreement with RMT is to be expected.

The appearance of "freak" short periodic orbits affect not only the spectral measures which are related to short time dynamics, (such as e.g., the lowest $\langle |t_n|^2 \rangle$), but also the other statistical measures. Even if we do not consider their repetitions, they appear in other indirect ways. The quantities which are most affected by the short periodic orbits are the coefficients of the characteristic polynomial. This can be easily understood from the Newton identities through which they are constructed (6). The traces t_k appear in the relations which define all a_n with $n \geq k$, and any possible anomaly which is due to a k - periodic orbit will be noticed. This is why the auto-correlation of the characteristic polynomial is a statistical measure which is most sensitive to the deviation from universality. Another venue through which short periodic orbits become noticeable, is also connected with the Newton identities, and the self inversive symmetry (4)which is imposed by unitarity. Consider

$$a_k = -\frac{1}{k}\left(t_k + \sum_{l=1}^{k-1} a_l t_{k-l}\right) \tag{87}$$

and

$$a_{N-k} = -\frac{1}{N-k}\left(t_{N-k} + \sum_{l=1}^{N-k-1} a_l t_{N-k-l}\right) \tag{88}$$

Since $a_k = e^{i\Theta} a_{N-k}^*$, we must have

$$\frac{1}{N-k}t_{N-k} = \frac{1}{k}e^{i\Theta} t_k^* + \cdots \tag{89}$$

which shows how the inversive symmetry may also affect the 2-point spectral statistics.

The universal classical correlation function is also expected to be modified due to the deviations from the strong mixing limit at short times. This can be clearly seen in derivation of the scaling properties of $\rho_{cl}(\xi; n)$ which could be derived only when the w_n in (83) were set to their strong mixing limit $w_n = 1$.

Finally let me point out a few problems which remain open, and deserve in my opinion further work. The classical correlations which were discussed above are far from being understood. These correlations are at the heart of any theory which attempts to explain a phenomenon which depends crucially on quantum interference. Eckhardt 30 have shown that periodic orbit correlations are necessary to explain the distribution of the dwell times in chaotic scattering. Doron Cohen 29 used the concept of classical correlations to provide a semiclassical explanation of the scaling theory for Anderson localization. However, to achieve this goal, the classical correlation function was not derived from first principles, but was supported by physical arguments and intuition.

Spectral correlations were discussed in this paper from the semiclassical point of view. The results obtained by using the notion of the semiclassical ensemble are formally similar to the results obtained using the field theoretical approach. The semiclassical theory is based on one important assumption, namely, that repetitions of short periodic orbits can be neglected. Can this assumption be reformulated in a way which will agree with the approximations which underlie the field theoretical approach? The comparison of the two methods calls for further study to elucidate this point.

ACKNOWLEDGEMENTS

I am indebted to the Newton Institute for providing comfortable and stimulating environment where much of the research reported here was carried out. I would like to thank Doron Cohen, Harel Primack and Holger Schanz for helping me to better understand the concept of classical correlations, and Shmuel Fishman, Bernhard Mehlig, Fritz Haake, Zeev Rudnick, John Keating and Richard Prange for many discussions suggestions and comments. This work was supported in part by the Minerva Center for Nonlinear Physics of Complex Systems, and by a grant from the Israel Science Foundation.

REFERENCES

1. Z. Rudnick and P. Sarnak. The pair correlation function for fractional parts of polynomials *Comm. Math. Phys* (in press) (1997)

2. S. Zelditch, Level spacings for quantum maps in genus zero preprint, Isaac Newton Inst. Cambridge 1997)

3. F. M. Izrailev. Simple models of Quantum Chaos: Spectrum and eigenfunctions Phys. Rep. **196**, 299 (1990)

4. R. Blümel and U. Smilansky. Random matrix description of chaotic scattering: Semiclassical Approach. *Phys. Rev. Lett.* **64** (1990) 241.

5. F. Haake. *Quantum signatures of Chaos* Springer, Berlin (1991).

6. U. Smilansky. Quantum Chaos and Random Matrix Theory - Some New Results *Physica D* (To appear Nov. 1997)

7. E.B. Bogomolny Semiclassical quantization of multidimensional systems. *Nonlinearity* **5** (1992) 805.

8. E. Doron and U. Smilansky. Semiclassical Quantization of Chaotic Billiards - a Scattering Theory Approach. *Nonlinearity* **5** (1992) 1055.

9. B. Dietz and U. Smilansky. A Scattering Approach to the Quantization of Billiards - The Inside-Outside Duality. *Chaos* **3**(1993) 581-590.

10. J. P. Eckmann and C. A. Pillet. Spectral duality for planar biliards, *Comm. Math. Phys.* **170** (1995) 283.

11. F.J. Dyson. Statistical Theory of the Energy Levels of Complex Systems *J. Math. Phys.* **3** (1962) 140.

12. M. R. Zirnbauer. Riemannian symmetric superspaces and their origin in Random Matrix Theory *Jour. Math. Phys.* **37** (1996) 4986.

13. M. R. Zirnbauer. Supersymmetry for systems with unitary disorder: Circular ensembles. *J. Phys. A***29** (1996) 7113.

14. E.B. Bogomolny and J. Keating Gutzwiller's trace formula and spectral statistics: Beyond the diagonal approximation. *Phys. Rev. Lett.* **77** (1996) 1472.

15. N. Argaman, F. Dittes , E. Doron, J. Keating, A. Kitaev, M. Sieber and U. Smilansky. Correlations in the Acions of Periodic Orbits Derived from Quantum Chaos. *Phys. Rev. Letters* **71**, (1993) 4326-4329

16. F.M. Dittes, E. Doron and U. Smilansky. Long time Behavior of the Semiclassical Baker's Map. *Phys Rev E* **49**,(1994) R963-R966

17. D. Cohen, H. Primack and U. Smilansky. Quantal- classical duality and the semiclassical trace formula. *Ann. of Phys.***264** (1998), 108-170.

18. S. Ketemann, D. Klakow and U. Smilansky. Characterization of Quantum Chaos by the Autocorrelation Function of Spectral Determinants. *J. Phys. A***30** (1997) 3643.

19. F. Haake, M. Kuś, H.-J. Sommers, H. Schomerus, and K. Zyczkowski. Secular determinants of random unitary matrices. *J. Phys. A* **29** (1996) 3641.

20. M.V. Berry. Semiclassical Theory of Spectral Rigidity. *Proc. Royal Soc. Lond* **A 400** (1985) 229.

21. W. H. Miller. Classical-limit quantum mechanics and the theory of molecular collisions. *Adv. Chem. Phys.* **25**(1974) 69.

22. M. Tabor A Semiclassical Quantization of Area–Preserving Maps. *Physica* **D6**, 195 (1983)

23. U. Smilansky. Semiclassical Quantization of Chaotic Billiards - A Scattering Approach. in *Proc. of the Les Houches Summer School on Mesoscopic Quantum Physics.* Elsevier Science Publ. (1995) Ed. E. Akkermans, G. Montambaux and J. L. Pichard.

24. T. Kottos and U. Smilansky. Quantum Chaos on Graph. *Phys. Rev. Lett.* **79** (1997), 4794.

25. O. Agam, B.L. Altshuler, and A.V. Andreev. Spectral statistics: from disordered to chaotic systems. *Phys. Rev. Lett.* **75** (1995) 4389.

26. B. A. Muzykantskii and D. E. Khmelnitskii. Effective action in Theory of quasi–ballistic disordered conductors. *JETP Lett.* **62**, 76 (1995).

27. E. Doron and U. Smilansky. Some Recent Developments in the Quantum Theory of Chaotic Scattering. *Nuclear Physics A* **545**, (1992) 455c.

28. T. Dittrich and U. Smilansky. Spectral properties of systems with dynamical localization I and II *Nonlinearity* **4**, 59-84 (1991) and *ibid*, 85-101 (1991).

29. D. Cohen. Periodic orbits, breaktime and localization. *J. Phys. A* **31** (1998) 277.

30. B. Eckhardt. Correlations in quantum time delay *Chaos* **3**, (1993) 613–617

WAVE FUNCTIONS, WIGNER FUNCTIONS AND GREEN FUNCTIONS OF CHAOTIC SYSTEMS

Shmuel Fishman

Department of Physics, Technion, Haifa 32000, Israel
and
The Isaac Newton Institute For Mathematical Sciences,
University of Cambridge
20 Clarkson Road, Cambridge, CB3 OEH, U.K.

1 INTRODUCTION

The nature of the spectrum of systems that are chaotic in the classical limit has been extensively studied[1, 2, 3, 4, 5] and was the subject of several talks during the conference. The main result is that over a wide energy range the energy level correlations are described by random matrix theory (RMT)[6, 7], but there are deviations that are important in some regimes[8, 9, 10, 11, 12, 13, 14]. Some of these studies rely on semiclassical approximations of various degrees of sophistication that started from the Gutzwiller trace formula[15] and were followed by more sophisticated methods for obtaining meaningful results from the trace formula[16, 17, 18] and for analyzing level-level correlations[10, 11]. Field theoretical methods were very instructive for the exploration of global spectral correlations and for the understanding of the regimes where random matrix theory holds and where deviations are important[12, 13, 14]. These deviations were found to be related to classical quantities such as periodic orbits and decay rates of disturbances to the invariant density.

For most physical problems the knowledge of the energy levels is not sufficient for the understanding of the relevant physical process. Very often transitions between energy states are measured and matrix elements are required. For atoms and molecules these are usually the matrix elements of the dipole operator. The eigenfunctions are useful for the understanding of these matrix elements[19]. For example for mesoscopic systems[20], such as quantum dots, transport is dominated by the wave functions with energy around the Fermi level, that are large in the vicinity of the leads. For typical chaotic systems eigenfunctions look random[21, 22]. Some of these eigenfunctions are much larger in the vicinity of classical unstable periodic orbits, than statistically expected for a function that is randomly distributed in space. These are known as "scars"[23, 24, 25]. In this contribution, semiclassical properties of wave functions of classically chaotic systems will be described. It turns out that these are also of importance for the understanding of the relation between spectral correlations and the properties of the underlying classical system.

The most elementary quantum object is the wave function $\psi(\mathbf{r})$. Its relation to classical objects is not transparent. Therefore the Wigner function

$$W_\psi(\mathbf{x}) = \frac{1}{h^{d_f}} \int d\mathbf{r}' e^{-\frac{i}{\hbar}\mathbf{P}\cdot\mathbf{r}'} \psi^*(\mathbf{r} - \frac{1}{2}\mathbf{r}')\psi(\mathbf{r} + \frac{1}{2}\mathbf{r}') , \tag{1}$$

corresponding to the classical density, was introduced. Here $\mathbf{x} = (\mathbf{r}, \mathbf{p})$ where $\mathbf{r}$ is the position, $\mathbf{p}$ is the momentum and d_f is the number of degrees of freedom. For a system with the Hamiltonian $\mathcal{H}$ the wave functions are propagated by the evolution operator

$$U_t = e^{-\frac{i}{\hbar}\mathcal{H}t}. \tag{2}$$

An evolution operator can be defined for the Wigner function (see Sec. 2). It corresponds to the evolution of the classical phase space densities. The time integral of the evolution operator is the resolvent

$$\hat{\mathcal{R}} = \frac{-i}{\hbar} \int_0^\infty dt e^{\frac{i}{\hbar}(E-\hat{\mathcal{H}})t - \frac{\varepsilon t}{\hbar}} = \frac{1}{E + i\varepsilon - \hat{\mathcal{H}}}. \tag{3}$$

Its representation in coordinate space is the Green function

$$G(\mathbf{r}, \mathbf{r}'; E) = \langle \mathbf{r}|\hat{\mathcal{R}}|\mathbf{r}'\rangle = \sum_\alpha \frac{\psi_\alpha^*(\mathbf{r}')\psi_\alpha(\mathbf{r})}{E + i\epsilon - E_\alpha} , \tag{4}$$

where E_α and ψ_α are the eigenenergies and the eigenfunctions respectively. The resolvent Wigner function is the Weyl transform of the Green function,

$$\begin{aligned}
W(\mathbf{x}; E) &= \int d\mathbf{r}' e^{-\frac{i}{\hbar}\mathbf{P}\cdot\mathbf{r}'} G(\mathbf{r} + \frac{1}{2}\mathbf{r}', \mathbf{r} - \frac{1}{2}\mathbf{r}'; E) \\
&= h^{d_f} \sum_\alpha \frac{W_\alpha(\mathbf{x})}{E + i\varepsilon - E_\alpha}
\end{aligned} \tag{5}$$

where $W_\alpha(\mathbf{x})$ is the Wigner function of the eigenstate ψ_α. In Sec. 3 expressions for the Green function and the resolvent Wigner function are developed for chaotic systems. The eigenfunctions can be calculated from these expressions with the help of (4) and (5). As in the case of the trace formula for the spectrum[15], one encounters sums that are not absolutely convergent. Resummation methods were used for the calculation of the semiclassical spectral determinant[16, 17, 18]

$$\Delta(E) = e^{-i\pi\bar{N}(E)} \prod_{p=p.p.o} \prod_j \left(1 - t_p e^{-ju_p}\right) \tag{6}$$

with

$$t_p = e^{\frac{i}{\hbar}S_p(E) - i\gamma_p - \frac{1}{2}u_p}. \tag{7}$$

Here the subscript p denotes a primitive periodic orbit, $S_p(E)$ is its corresponding action, γ_p is a phase determined by the focusing paths close to p, and u_p is the instability exponent. The mean energy staircase (mean number of levels with energy less than E) is $\bar{N}(E)$.

In Sec. 4, "scars", namely eigenfunctions that are large in the vicinity of unstable classical periodic orbits[23, 24, 25] are discussed. Various definitions of this concept are

194

studied and a criterion[26] for their existence based on the expressions of the Green function and the Wigner function, developed in Sec. 3, is presented.

2 THE CLASSICAL LIMIT OF QUANTUM EVOLUTION IN CHAOTIC SYSTEMS

The singular behavior of quantum evolution in the classical limit is of particular interest for chaotic systems [1,2,3]. In this section, the limit of long time behavior will be studied for such systems. It is of special interest in the view of recent calculations of energy correlations[12,10,13,14,11], where deviations from Random Matrix Theory (RMT)[6,7] related to the long time behavior of the classical systems were found. The specific question that will be studied here is what is the semiclassical limit of the evolution operator of the Wigner function for bound systems. For such systems the energy shell is bounded and consequently for a non-vanishing value of the Planck constant $\hbar$ the evolution operator of the wave functions is effectively a finite dimensional matrix, and so is the evolution operator of the Wigner functions. The Liouville operator, that is the evolution operator of the classical densities, on the other hand, is infinite dimensional, since classical dynamics is defined on arbitrarily small scales in phase space. In the $\mathcal{L}^2$ Hilbert space it is a unitary operator [27,28]. A similar situation holds for maps on the torus, where the quantum evolution operator is given by a finite dimensional matrix, while the corresponding classical evolution operator is an infinite dimensional matrix. If one assumes that the basis states of this matrix are ordered by increasing resolution, then truncation of the matrix is equivalent to limited resolution or coarse graining. This truncated matrix is no more unitary and since this is a projection of a unitary matrix, its eigenvalues must lie inside the unit circle in the complex plane. Projecting the quantum evolution operator on the basis of the classical evolution operator, that is ordered by increasing resolution and truncated by the dimension of the quantum evolution operator, will lead to the projection of the quantum evolution operator on this finite dimensional space. It can be considered also as a coarse graining of the quantum evolution operator by finite resolution in the classical phase space. It is expected to approximate the classical truncated evolution operator for small values of $\hbar$. The question is what happens in the limit where the dimension of all these matrices tends to infinity. In this section it is argued in general, and tested numerically for the baker map[27,28], that in this limit the spectrum of both evolution operators consists of the Ruelle resonances, that are related to relaxation rates of disturbances of the classical invariant density[29,30,31]. These lie inside the unit circle. The operations of taking the spectrum of a sequence of finite dimensional matrices and the diagonalization of the infinite limiting matrix do not commute in this case. In recent field theoretical calculations of the level-level correlation functions, the results are expressed in terms of the Ruelle resonances[12,14]. Results that are similar within some approximation were found using direct semiclassical methods[10,11]. In all these calculations some degree of coarse graining is involved.

The classical evolution operator for the map

$$\mathbf{x}' = \mathbf{F}(\mathbf{x}), \tag{8}$$

where $\mathbf{x}$ and $\mathbf{x}'$ are on a torus in phase space, is the Liouville or Frobenius-Perron operator:

$$M\rho(\mathbf{x}') = \int d\mathbf{x}\, \delta(\mathbf{x}' - \mathbf{F}(\mathbf{x}))\rho(\mathbf{x}). \tag{9}$$

For an area preserving and invertible map, corresponding to a Hamiltonian system, M is unitary, in $\mathcal{L}^2$, the space of square integrable functions. It is instructive to introduce the resolvent,

$$R(z) = \frac{1}{z - M} = \frac{1}{z} \sum_{n=0}^{\infty} (Mz^{-1})^n. \tag{10}$$

Since M is unitary its resolvent is singular on the unit circle. The function $\rho = const.$ is an eigenfunction with the eigenvalue $z = 1$. In analogy with continuous flow of Hamiltonian systems[32] there are many "eigenfunctions" that are concentrated on periodic orbits that are not normalizable and therefore lie outside of the $\mathcal{L}^2$ Hilbert space. The terms "eigenvalue" and "eigenfunction" are used here for Λ and ψ_Λ if for the operator $\mathcal{O}$ they satisfy $\mathcal{O}\psi_\Lambda = \Lambda\psi_\Lambda$, but ψ_Λ is not necessarily in the Hilbert space where $\mathcal{O}$ is defined. Let $\{\mathbf{x}_1...\mathbf{x}_n\}$, where $\mathbf{x}_{j+1} = \mathbf{F}(\mathbf{x}_j)$ and $\mathbf{x}_1 = \mathbf{F}(\mathbf{x}_n)$ be periodic orbits of period n. The function

$$\rho_\alpha(\mathbf{x}) = \sum_j \delta(\mathbf{x} - \mathbf{x}_j)\, e^{i2\pi\alpha j}, \tag{11}$$

with $\alpha = \frac{l}{n} \bmod 1$, is an "eigenfunction" of M with the eigenvalue $\Lambda_\alpha = e^{-i2\pi\alpha}$ for all integer values of l. In the familiar case of quantum evolution, the physical resolvent for the Hamiltonian $\mathcal{H}$ is defined as (3), so that the Fourier transform of the evolution operator converges in the upper half plane. In the present case M is unitary, and the corresponding definition is the extrapolation of the resolvent from outside of the unit circle where the sum in (10) converges. On the unit circle the resolvent is singular but an analytic continuation from the region $|z| > 1$ to $|z| < 1$ can still be meaningful. It is useful to introduce the orthonormal basis $\varphi_k(\mathbf{x})$ in phase space, where the basis functions are ordered by increasing phase space resolution, and define the infinite dimensional matrix $M_{kk'} = \langle k|M|k'\rangle$ where $\langle \mathbf{x}|k\rangle = \varphi_k(\mathbf{x})$. Finite truncations of this matrix correspond to finite resolution in phase space and elimination of structures generated on arbitrary fine scales. The truncated matrix is not unitary. One can diagonalize it or at least transform it to the Jordan form. The "eigenvalues" (or the entries on the diagonal of the Jordan form) are inside the unit circle. What happens when the dimension of the matrix is increased? The eigenvalues do *not* approach the unit circle, they stay inside and the corresponding eigenfunctions approach some limiting expression. Since for a unitary matrix all normalizable eigenfunctions correspond to eigenvalues on the unit circle, this limiting expression must be some type of distribution[28]. In summary two types of operations are possible:

1. Diagonalization of the infinite matrix M (corresponding to long time evolution) with infinite precision.

2. Diagonalization of the finite dimensional matrix corresponding to finite resolution and long time evolution. Then the limit of infinitely fine resolution, corresponding to an infinite matrix M, is taken.

These two operations do not lead to the same result. The reason is that truncation (or coarse graining) and diagonalization do not commute. The second operation leads to the Ruelle resonances. Quantization, *combined with coarse graining*, provides a truncation of the evolution operator resulting in operation (2). Quantization by itself does not lead to identification of the Ruelle resonances since the quantum evolution is unitary. The coarse graining is required to break this unitarity. The quantization, however, sets the scale of coarse graining, since it is not reasonable to consider resolution that is finer than the quantization. The distributions found in operation (2) are uniform in the unstable direction and are of infinite complexity in the stable direction. This reflects the fact that stretching enhances uniformity while folding enhances modulations[28].

We turn now to the behavior of the quantum systems. For notational simplicity we confine ourselves to two dimensional maps, corresponding to a system with one degree of freedom. The function corresponding to the phase space density is the Wigner function (for one degree of freedom, $d_f = 1$, studied in this section)

$$W_\psi(q,p) = \frac{1}{h} \int dq' e^{-iq'p/\hbar} \langle q + \frac{1}{2} q' | \psi \rangle \langle \psi | q - \frac{1}{2} q' \rangle. \tag{12}$$

The inverse of this transform is

$$\tilde{W}_\psi(q_1, q_2) = \psi(q_1)\psi^*(q_2) = \int dp\, e^{ip(q_1-q_2)/\hbar}\, W_\psi(\frac{1}{2}(q_1+q_2), p). \tag{13}$$

The function $\tilde{W}_\psi(q_1, q_2)$ that is the product of the wave function with its complex conjugate at different points can be considered as a coordinate representation of the Wigner function. For maps on the torus quantization implies that the wave functions and Wigner functions are defined on an $N \times N$ lattice, where $N = 1/h$, with positions $q_n = \frac{n}{N}$ and momenta $p_m = \frac{m}{N}$, therefore a modification of (12) and (13) is required[33, 34]. The evolution operator U is an $N \times N$ unitary matrix, therefore the evolution operator of the Wigner function is the $N^2 \times N^2$ matrix,

$$U^{(W,N)} = U^\dagger \otimes U \tag{14}$$

with elements $\left(U^{(W,N)}\right)_{ij,i'j'} = U^\dagger_{jj'} U_{ii'}$. In order to explore the relation between this propagator and the classical evolution operator, a discretized version of the basis functions $\varphi_k(\mathbf{x} = (p, q))$ is defined,

$$\varphi_k^{(N)}(n, m) = \frac{1}{\sqrt{N}} \varphi_k\left(\frac{n}{N}, \frac{m}{N}\right) \tag{15}$$

These functions form an orthonormal basis in the limit $N \to \infty$. Since the Wigner functions are related to phase space densities, in the large N limit, a discrete transform corresponding to (13) is introduced for these functions,

$$\tilde{W}_{\varphi_k}(n_1, n_2) = \frac{1}{N} \sum_m e^{i2\pi(n_1-n_2)m/N}\, \varphi_k^{(N)}\left(\frac{1}{2}(n_1+n_2), m\right) \equiv V\, \varphi_k^{(N)} \tag{16}$$

where V is the linear transformation between the phase space wave functions that depend on $(\frac{n}{N}, \frac{m}{N})$ corresponding to position and momentum, and the functions $\tilde{W}_\psi$, that depend only on the position space coordinates (n_1, n_2). It is a matrix with elements $V_{n'm} = \frac{1}{N} e^{i2\pi n'm/N}$ and the RHS of (16) is $\sum_m V_{n'm}\varphi_k^{(N)}(n, m)$ with the identification $n' = n_1 - n_2$ and $n = \frac{1}{2}(n_1 + n_2)$. In the basis (15) the matrix elements of the evolution operator of the Wigner function are:

$$U_{k\,k'}^{(W,N)} = \langle \varphi_k^{(N)} | V^\dagger \left(U^\dagger \otimes U\right) V | \varphi_{k'}^{(N)} \rangle. \tag{17}$$

In the limit $N \to \infty$ (corresponding to $\hbar \to 0$) these are expected to approach $M_{kk'}$. For large but finite N, these approximate $M_{kk'}$ but only for $k', k \ll N$. Functions φ_k with resolution higher than the mesh size (of order $1/N$) cannot be involved in the representation of $U^{(W,N)}$. This results in a truncation of M dictated by the value of N.

The matrix elements of the corresponding resolvent are

$$R_{kk'}^{(W,N)} = \langle \varphi_k^{(N)} | V^\dagger \frac{1}{z - U^\dagger \otimes U} V | \varphi_{k'}^{(N)} \rangle. \tag{18}$$

When analytically continued from $|z| > 1$ these are expected to be singular for $|z| \leq 1$ and in the limit $N \to \infty$ the singularities should approach the singularities of the matrix elements of the resolvent of the classical evolution operator R of (10) when continued in the same way. These singularities are the Ruelle resonances.

The arguments presented will be tested numerically for the baker map where analytic expressions for both M and U are known[28, 35, 36]. The baker map is defined in the $\mathbf{x} = (x, y)$ phase plane as

$$(x', y') = \mathbf{F}(x, y) = \begin{cases} (2x, y/2) & for \;\; 0 \leq x < 1/2 \\ (2x - 1, (y + 1)/2) & for \;\; 1/2 \leq x < 1 \end{cases} \tag{19}$$

Convenient basis functions are

$$\varphi_{k=(i,j)}(\mathbf{x} = (x, y)) = \tilde{P}_i(x)\tilde{P}_j(y) \tag{20}$$

where

$$\tilde{P}_i(x) = \sqrt{2i + 1}P_i(1 - 2x) \tag{21}$$

are the normalized (modified) Legendre polynomials while P_i are the Legendre polynomials. The matrix elements of the classical evolution operator are[28]

$$M_{ij,i'j'} = \langle ij|M|i'j'\rangle = \frac{1}{2}\left[1 + (-1)^{i+i'+j+j'}\right]I_{ii'}I_{j'j} \tag{22}$$

where $I_{ii'} = \int_0^1 \tilde{P}_i(x)\tilde{P}_{i'}(x/2)\,dx$ vanish for $i > i'$. This results from the fact that the map is stretching in the x direction. Consequently the weight of the basis states with smaller values of i increases, because of increasing uniformity. The opposite holds for the stable direction y where application of the map results in increasing complexity. The non-recurrence property of M, namely the fact that all the matrix elements that satisfy $i > i'$ and $j < j'$ vanish, results in a simple form of the diagonal matrix elements of the classical resolvent[28]

$$R_{ij,ij} = \frac{1}{z - M_{ij,ij}} = \frac{1}{z - 2^{-(i+j)}} \tag{23}$$

where the explicit values of $I_{ii'}$ and $I_{j'j}$ were used in the last equality. The fact that for the baker map the stable and unstable manifolds are parallel to the x and y axes enables to find easily the basis where the non-recurrence property manifests itself. For more generic maps such a property is expected to be found in some basis when the resolution is refined.

For the quantum baker map the evolution operator U, that was introduced by Balazs and Voros[35], will be used here (a different version was introduced by Saraceno[36]). Using the evolution operator U, the matrix elements (17), $U_{ij,i'j'}^{(W,N)}$ (here $k = ij$ and $k' = i'j'$) of the evolution operator of the Wigner function were computed numerically for various values of N. In App. A the corner of the matrices M and $U^{(W,N)}$ where $i + j, i' + j' \leq 2$ is presented. The deviation from the classical values in the corner of the matrix where $i + j \leq \bar{N}$

$$Dm(N, \bar{N}) = \sum_{i+j \leq \bar{N}, i'+j' \leq \bar{N}} \left(U_{ij,i'j'}^{(W,N)} - M_{ij,i'j'}\right)^2 \tag{24}$$

was calculated for various values of N. The results are plotted in Fig. 1 and one sees that the difference decreases with N, but it is hard to determine exactly at what rate. To test whether the evolution operator $U^{(W,N)}$ of the Wigner function exhibits the

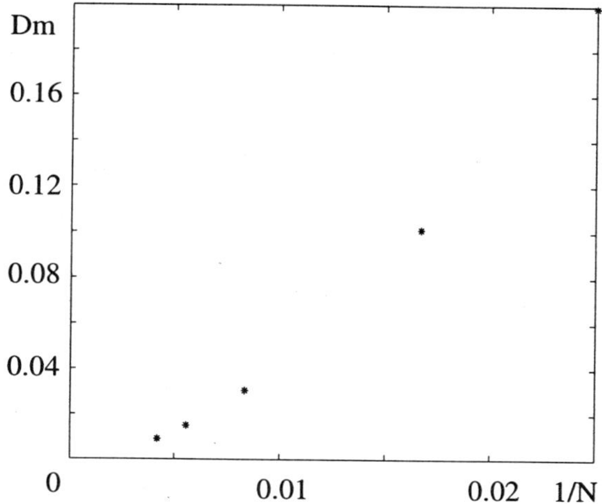

Figure 1. The deviation $Dm(N, \bar{N} = 2)$ for $40 \leq N \leq 240$.

Ruelle resonances, the diagonal matrix elements (18) of the resolvent were calculated for $|z| > 1$ and compared to the classical result (23). In Fig. 2, the inverses of the matrix elements,

$$f_{ij}^{(Q)}(z) = \frac{1}{R_{ij,ij}^{(W,N)}(z)} \tag{25}$$

are plotted and compared to the corresponding classical ones

$$f_{ij}^{(C)}(z) = \frac{1}{R_{ij,ij}(z)} = z - 2^{-(i+j)} \tag{26}$$

for $(i,j) = (0,0), (1,0), (2,0), (1,2)$. Note the relatively large deviations between these in the vicinity of $|z| = 1$. The values of the poles obtained from best fit of $f^{(Q)}$ in

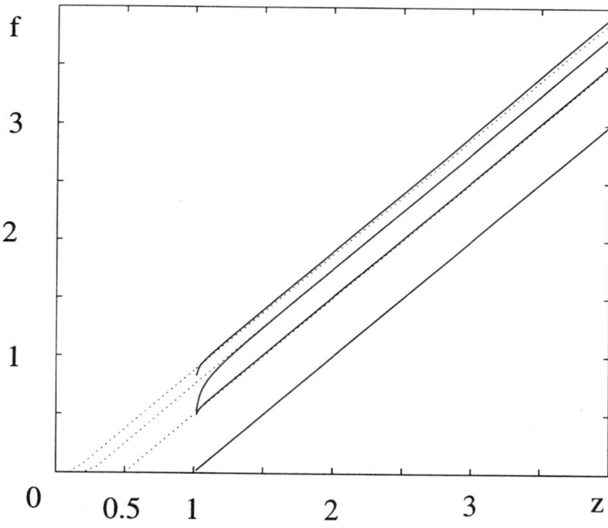

Figure 2. The functions $f_{ij}^{(Q)}$ (solid line) and $f_{ij}^{(C)}$ (dotted line) for $(i,j) = (0,0), (1,0), (2,0), (1,2)$ (from bottom to top) for $N = 400$.

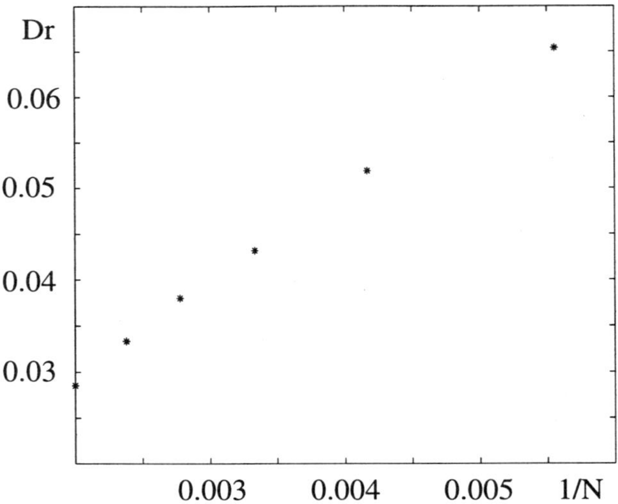

Figure 3. The integrated difference Dr between the quantum and classical matrix elements for $180 \le N \le 500$.

the region $2.5 < z < 7$ are $1.00, 0.488, 0.249, 0.116$ that are very close to $2^{-(i+j)}$. The integrated difference

$$Dr = \int_{z_1}^{z_2} dz |f_{ij}^{(Q)}(z) - f_{ij}^{(C)}(z)| \tag{27}$$

is presented in Fig. 3 for $z_1 = 1.02$ and $z_2 = 3.9$ as a function of $1/N$.

In this section it was argued in general and demonstrated for the baker map that for large but finite N the quantum evolution operator of the Wigner function of a chaotic system approaches the classical evolution operator when both operators are restricted to finite resolution. When matrix elements of the resolvent, that is the operator relevant for the long time evolution are calculated and then the classical limit $(N \to \infty)$ is taken, these are found to be singular for $|z| \le 1$. The singularities are at the "eigenvalues" of the Perron-Frobenius operator, $z = 1$ is related to the invariant density while the $|z| < 1$ singularities give the relaxation rates to it. The fact that these are found for the quantum evolution operator is a manifestation of the classical chaotic behavior. The essential operations in obtaining this relation are the coarse graining and the extrapolation from $|z| > 1$. Preliminary results for the kicked top[2] support the generality of the arguments presented in this section[37]. In recent calculations the level-level correlations were expressed in terms of the determinant[12, 14]

$$\zeta_q^{-1}(z) = Det_\epsilon(z - U^\dagger \otimes U), \tag{28}$$

where some coarse graining of magnitude ϵ or ensemble averaging was applied. It was *assumed* that in the limit $\epsilon \to 0$ it can be replaced by the classical function

$$\zeta_c^{-1}(z) = \prod_j (z - \mu_j), \tag{29}$$

where μ_j are the Ruelle resonances and the product is truncated at j of order N [14]. This work provides some justification for the replacement of ζ_q by ζ_c.

3 THE RATIO FORM OF THE GREEN FUNCTION AND THE WIGNER FUNCTION

The purpose of this section is to transform the Green function to the form which reveals its poles,

$$G(\mathbf{r}, \mathbf{r}'; E) = \frac{\mathcal{N}_G(\mathbf{r}, \mathbf{r}'; E))}{\Delta(E)} \tag{30}$$

where Δ is the spectral determinant and all the coordinate dependence is in the numerator $\mathcal{N}_G$. The corresponding form of the resolvent Wigner function is,

$$W(\mathbf{x}; E) = \frac{\mathcal{N}_W(\mathbf{x}; E)}{\Delta(E)} \tag{31}$$

where $\mathbf{x} = (\mathbf{p}, \mathbf{r})$ are the phase space coordinates. The expression has poles where the spectral determinant $\Delta(E)$ vanishes, namely for the eigenvalues of $\mathcal{H}$. The form (30) or (31) enables one to find the eigenfunctions with the help of (4) and (5). For most chaotic systems it is impossible to obtain exact expressions for Δ, $\mathcal{N}_G$ and $\mathcal{N}_W$ and approximations are required. In this section various approximations are presented.

In subsection 3.1 Fredholm theory[38] is used to obtain the form (30) with meaningful convergent expressions for $\mathcal{N}_G$ and Δ[39]. It holds whenever Fredholm theory is applicable as will be explained in what follows. Various levels of the semiclassical approximation for $\mathcal{N}_G$ and Δ are introduced and discussed. The main advantage of this method is that the semiclassical approximation is applied at a late stage of the calculation and if not applicable one can resort to numerical methods at this late stage. In subsection 3.2 the resolvent Wigner function in the form (31) will be obtained from (30) with the help of (5)[39]. An alternative derivation[19] will be outlined as well. Exact expressions for $\mathcal{N}_G$ and Δ for a billiard on a surface of constant negative curvature are presented in subsection 3.3[40]. These are used as a test for the approximate formulas obtained in subsections 3.1 and 3.2. In subsections 3.1 and 3.2 the theory will be developed for a system of d_f degrees of freedom. Specific formulas will be written in many cases for $d_f = 2$ for the sake of simplicity.

3.1 The Fredholm Theory for the Green Function

The Green function $G(\mathbf{r}, \mathbf{r}'; E)$ can be considered as a sum over paths from $\mathbf{r}'$ to $\mathbf{r}$. These contributions are the time Fourier transforms of the contributions of the Feynman paths to the evolution operator. It is convenient to define a Poincaré surface of section (PSS) and let $\mathbf{T}$ be the transfer operator that transforms the surface of section on itself. Let $\mathbf{V}_+$ ($\mathbf{V}_-$) be the operator that transforms points $\mathbf{r}'$ to the PSS (from the PSS to points $\mathbf{r}$). The contributions of paths from $\mathbf{r}'$ to $\mathbf{r}$ that cross the PSS exactly $n + 1$ times can be expressed as

$$G_{n+1}(\mathbf{r}, \mathbf{r}'; E) = \langle \mathbf{r} | \mathbf{V}_- \mathbf{T}^n \mathbf{V}_+ | \mathbf{r}' \rangle. \tag{32}$$

For a billiard a convenient PSS is the boundary and $\mathbf{T}$ transforms the boundary to itself at constant E[41]. In general it is not clear that an exact unitary transfer operator can be defined. Such an operator can always be defined in the semiclassical approximation[17]. It will be assumed in this section that $\mathbf{T}$ is a kernel of a *Fredholm integral equation* of the second kind, defined by the operator equation[38],

$$x = x_0 + \lambda \mathbf{T} x \tag{33}$$

which is equivalent to

$$x(q) = x_0(q) + \lambda \int dq' T(q, q') x(q') \tag{34}$$

The unknown function $x(q)$ and the known functions x_0 and T are defined on some finite domain, [e.g. an interval of reals]. If the functions, $x_0(q)$, $T(q, q')$ are sufficiently nice, [e.g. continuous, or square integrable], the *Fredholm alternative* holds. That is, *either* there is a unique solution of Eqs. (33) and (34), with the same nice properties, *or* the homogeneous version of Eq. (33) and (34) [$x_0 \equiv 0$] has a solution. There is a discrete set of complex eigenparameters $\lambda = \lambda_n$ for which the solution $x(q)$ is not unique.

In operator language, x and x_0 are elements of a Hilbert or Banach space, and $\mathbf{T}$ is an $\mathcal{L}^2$ operator on that space. In this terminology, it is said that the inverse operator $[\mathbf{1} - \lambda \mathbf{T}]^{-1}$ exists except for a discrete set of λ's.

The kernel $T(q, q')$ can also be regarded as a continuous or infinite dimensional matrix (for $d_f = 2$ the PSS is one dimensional and the PSS coordinates are scalars. The explicit equations will be presented for $d_f = 2$ for the sake of simplicity). Most of the schemes for numerical solution of Eq. 34 exploit this by making a discretization or truncation, which reduces the problem to inversion of a finite matrix. In this case, $[\mathbf{1} - \lambda \mathbf{T}]^{-1}$ can be expressed as a ratio

$$\frac{1}{1 - \lambda \mathbf{T}} = \frac{\mathbf{N}(\lambda)}{D(\lambda)} \tag{35}$$

If $\mathbf{T}$ is approximated as an $N \times N$ matrix, the determinant $D(\lambda) = \det(\mathbf{1} - \lambda \mathbf{T})$ is an N'th order polynomial in λ whose N zeroes give an approximation to the λ_n. The numerator is a polynomial of order $N - 1$.

The main result of Fredholm theory[38] is that, if $\mathbf{T}$ is an $\mathcal{L}^2$ operator, that may be infinite dimensional, the expression (35) continues to hold, but with $D(\lambda)$ an *entire* function of λ i.e. it is a series *absolutely convergent* for all $|\lambda| < \infty$ rather than a polynomial. Similarly, the numerator is an *operator valued* entire function of λ. We give the explicit expressions for the Fredholm determinant D and the numerator operator $\mathbf{N}$ in terms of the kernel $\mathbf{T}$ below.

The Green function takes the form

$$\langle \mathbf{r} | G | \mathbf{r}' \rangle = \langle \mathbf{r} | G_0 | \mathbf{r}' \rangle + \sum_n \langle \mathbf{r} | \mathbf{V}_- \mathbf{T}^n \mathbf{V}_+ | \mathbf{r}' \rangle \tag{36}$$

or explicitly

$$G(\mathbf{r}, \mathbf{r}'; E) = G_0(\mathbf{r}, \mathbf{r}'; E) + \sum_{n=0}^{\infty} \int V_-(\mathbf{r}, q) T^n(q, q') V_+(q', \mathbf{r}') dq dq' \tag{37}$$

where G_0 corresponds to the direct trajectories between $\mathbf{r}$ and $\mathbf{r}'$, which do not cross the PSS. We say that an orbit coming from the term T^n has *PSS length* $n + 1$. Direct orbits have PSS length zero, and orbits from the term $\mathbf{V}_- \mathbf{V}_+$ have PSS length one. We denote the PSS length of an orbit j by the integer L_j. The expression (37) contains the sum of a geometric series $\sum \mathbf{T}^n$ which is formally equal to $[\mathbf{1} - \mathbf{T}]^{-1}$.

The *Fredholm determinant* $D(\lambda)$ for $[\mathbf{1} - \lambda \mathbf{T}]$ is given by the absolutely convergent series[38]

$$D(\lambda) = \sum_{n=0}^{\infty} \lambda^n D_n. \tag{38}$$

where

$$D_n = \frac{(-1)^n}{n!} \int\int_{PSS} dq_1 \ldots dq_n \det \begin{vmatrix} T(q_1, q_1) & \cdots & T(q_1, q_n) \\ \vdots & \ddots & \vdots \\ T(q_n, q_1) & \cdots & T(q_n, q_n) \end{vmatrix} \tag{39}$$

Similarly, one expresses

$$\mathbf{N}(\lambda) = \sum_{n=0}^{\infty} \lambda^n \mathbf{N}_n, \tag{40}$$

where $\mathbf{N}_n$ is an operator corresponding to the kernel

$$N_{n+1}(q, q') = D_{n+1}\delta(q - q') + \tag{41}$$

$$+ \frac{(-1)^n}{n!} \int\int_{PSS} dq_1 \ldots dq_n \det \begin{vmatrix} T(q, q') & T(q, q_1) & \cdots & T(q, q_n) \\ T(q_1, q') & T(q_1, q_1) & \cdots & T(q_1, q_n) \\ \vdots & \vdots & \ddots & \vdots \\ T(q_n, q') & T(q_n, q_1) & \cdots & T(q_n, q_n) \end{vmatrix}.$$

As $D = \det(\mathbf{1} - \lambda\mathbf{T})$, the theory shows that the well-known equality between the logarithm of the determinant of a matrix and the trace of its logarithm generalizes to this case, i.e.

$$D_n = -\frac{1}{n}\sum_{r=1}^{n} \sigma_r D_{n-r}. \tag{42}$$

where

$$\sigma_r = \mathrm{Tr}\mathbf{T}^r = \int\int_{PSS} dq_1 \ldots dq_r T(q_1, q_2)\ldots T(q_r, q_1), \tag{43}$$

Also using the fact that $\mathbf{N}(\lambda) = D(\lambda)/(\mathbf{1} - \lambda\mathbf{T})$ one has:

$$\mathbf{N}_n = \sum_{r=0}^{n} D_r \mathbf{T}^{n-r}. \tag{44}$$

If $\mathbf{T}$ is of finite rank, that is, it can be expressed as a bilinear sum of N functions, it is equivalent to a finite matrix of size N. Then the series will be truncated after the term of index N for D, and after the term of index $N - 1$ for $\mathbf{N}$.

Using the Fredholm result, (36) and (37) take the form,

$$G(\mathbf{r}, \mathbf{r}'; E) = G_0(\mathbf{r}, \mathbf{r}'; E) + \frac{1}{D(E)} \int V_-(\mathbf{r}, q)N(q, q'; E)V_+(q', \mathbf{r}')dqdq'. \tag{45}$$

where $D(E)$ is the Fredholm determinant and $N(q, q'; E)$ is the numerator of (35) at $\lambda = 1$, regarded as a function of E and the PSS coordinates. The resulting expression for the Green function is

$$G(\mathbf{r}, \mathbf{r}'; E) = G_0(\mathbf{r}, \mathbf{r}'; E) + \frac{1}{D(E)} \sum_{n=0}^{\infty} \sum_{s=0}^{n} D_{n-s}(E)G_{s+1}(\mathbf{r}, \mathbf{r}'; E). \tag{46}$$

where G_s is defined by (32). Taking a common denominator one finds the form (30) with

$$\mathcal{N}_G = \overline{\mathcal{N}}_G + \tilde{\mathcal{N}}_G \tag{47}$$

where

$$\overline{\mathcal{N}}_G = G_0(\mathbf{r}, \mathbf{r}'; E)\Delta(E) \tag{48}$$

and

$$\tilde{\mathcal{N}}_G = \frac{1}{\Delta(E)} \sum_{n=0}^{\infty} \sum_{s=0}^{n} \Delta_{n-s}(E) G_{s+1}(\mathbf{r}, \mathbf{r}'; E). \tag{49}$$

with

$$\Delta_n = D_n \, e^{-i\pi \bar{N}}. \tag{50}$$

The spectral determinant is:

$$\Delta = D \, e^{-i\pi \bar{N}}. \tag{51}$$

At the most elementary level of the semiclassical approximation the various operators are approximated by their semiclassical expressions. The semiclassical transfer operator is

$$T(q, q') = \frac{1}{(2\pi i\hbar)^{\frac{1}{2}}} \left| \frac{\partial^2 S}{\partial q \partial q'} \right|^{\frac{1}{2}} e^{\frac{i}{\hbar} S(q, q', E) - i\gamma \frac{\pi}{2}}, \tag{52}$$

that is unitary in the framework of this approximation. The action is S while γ is the Maslov index. The operators connecting an arbitrary point to the PSS are

$$V_+(q', \mathbf{r}') = \frac{1}{\sqrt{i\hbar}} \frac{1}{\sqrt{2\pi i\hbar}} \frac{1}{\sqrt{v'}} \left| \frac{\partial^2 S}{\partial q' \partial y'} \right|^{\frac{1}{2}} e^{\frac{i}{\hbar} S(q', \mathbf{r}', E) - i\gamma \frac{\pi}{2}}, \tag{53}$$

and

$$V_-(\mathbf{r}, q) = \frac{1}{\sqrt{i\hbar}} \frac{1}{\sqrt{2\pi i\hbar}} \frac{1}{\sqrt{v}} \left| \frac{\partial^2 S}{\partial y \partial q} \right|^{\frac{1}{2}} e^{\frac{i}{\hbar} S(\mathbf{r}, q, E) - i\gamma \frac{\pi}{2}}, \tag{54}$$

where v is the velocity. Within this approximation the convergence of the expressions for the numerator and denominator is expected to hold since these rely on the Hadamard inequality[38] that guarantees that for an $n \times n$ matrix with elements bounded by B, the determinant is bounded by $n^{n/2}B$. Consequently the terms (39) and (41) fall off as $n^{-n/2}$ and the series (38) and (40) absolutely converge. The Hadamard inequality relies on exact cancellation between the various $n!$ permutations constituting the determinant. A higher level in the semiclassical approximation involves the calculation of the integrals in (43) and (44) semiclassically, leading to an explicit formula for the G_s that appear in (46),

$$G_s(\mathbf{r}, \mathbf{r}'; E) = \frac{1}{(2\pi i\hbar)^{\frac{1}{2}}} \frac{1}{i\hbar} \sum_{j, L_j = s} \frac{1}{\sqrt{vv'}} \sqrt{\frac{\partial^2 S_j}{\partial y \partial y'}} e^{\frac{i}{\hbar} S_j(\mathbf{r}, \mathbf{r}', E) - i\gamma \frac{\pi}{2}} \tag{55}$$

where the sum is now over classical trajectories that cross the PSS exactly $L_j = s$ times. With this approximation the Hadamard inequality may be violated and Fredholm theory does not guarantee anymore the convergence of the expressions for $\mathcal{N}_G$ and D. In order to obtain meaningful expressions for $\mathcal{N}_G$ and Δ, when the last approximation is made, resummation, that makes use of the unitarity of $\mathbf{T}$ or the hermiticity of $\mathcal{H}$, is required [16−19,39].

3.2 The Resolvent Wigner Function

The Wigner transform of (46) is[39]

$$W(\mathbf{x}; E) = W_0(\mathbf{x}; E) + \frac{1}{D(E)} \sum_{n=0}^{\infty} \sum_{s=0}^{n} D_{n-s}(E) W_{s+1}(\mathbf{x}; E). \tag{56}$$

Here W_j is the Weyl transform (5) of G_j. If the G_j are calculated in the semiclassical approximation also the Wigner transform should be calculated in this approximation, that will be assumed throughout this subsection. From (56) the resolvent Wigner function in the form (31) is obtained with

$$\mathcal{N}_W = \overline{\mathcal{N}}_W + \mathcal{N}_W^{(po)}. \tag{57}$$

The contribution of short trajectories is

$$\overline{\mathcal{N}}_W = \Delta(E)\overline{W}(\mathbf{x}; E) \tag{58}$$

with

$$\overline{W}(\mathbf{x}; E) = -i\pi \left\{ A(\mathbf{x}; E) - iG^{(i)}(\mathbf{x}; E) \right\} \tag{59}$$

where

$$A(\mathbf{x}; E) = \frac{2}{\hbar \left| \frac{\ddot{\mathbf{x}} \wedge \dot{\mathbf{x}}}{\hbar} \right|^{\frac{1}{3}}} \operatorname{Ai}\left(\frac{2[\mathcal{H}(\mathbf{x}) - E]}{(\hbar^2 \ddot{\mathbf{x}} \wedge \dot{\mathbf{x}})^{\frac{1}{3}}} \right) \tag{60}$$

and

$$G^{(i)}(\mathbf{x}; E) = \frac{2}{\hbar \left| \frac{\ddot{\mathbf{x}} \wedge \dot{\mathbf{x}}}{\hbar} \right|^{\frac{1}{3}}} \operatorname{Gi}\left(\frac{2[\mathcal{H}(\mathbf{x}) - E]}{(\hbar^2 \ddot{\mathbf{x}} \wedge \dot{\mathbf{x}})^{\frac{1}{3}}} \right) \tag{61}$$

The Airy function $\operatorname{Ai}(z)$ and the function $\operatorname{Gi}(z)$ are defined as the real and the imaginary parts of the integral

$$\frac{1}{\pi} \int_0^\infty dt\, e^{i\left(\frac{1}{3}t^3 + zt\right)} \tag{62}$$

respectively. This term describes the pattern of Airy fringes as $\mathbf{x}$ moves off the energy surface. The contribution from the periodic orbits is of the approximate form

$$\mathcal{N}_W^{(po)} \approx 4\pi \sum_{p=p.p.o} \Delta^{(p)}(E) A(\mathbf{x}; E) e^{\frac{i}{\hbar}\tilde{\mathbf{X}}\mathbf{R_p}\mathbf{X}} \tag{63}$$

where the sum is over primitive periodic orbits. The coordinate on the Poincaré surface is $\mathbf{X}$, associated with the primitive periodic orbit p, for the canonical variables

$$\mathcal{H}, \quad t, \quad \text{and} \quad \mathbf{X}(Q, P) = (Q_1, \cdots Q_{d_f-1}, P_1, \cdots P_{d_f-1}) \tag{64}$$

where $\mathcal{H}$ is the Hamiltonian, $\mathbf{X}$ are the $2d_f - 2$ coordinates on the Poincarè surface of section, and t is the time along the periodic orbit. The exponent term in equation (63) describes a structure of quadratic fringes as $\mathbf{X}$ moves off the closed orbit. The Matrix $\mathbf{R}_p$ is related to the monodromy matrix and is defined in App. B. The functions $\Delta^{(p)}(E)$ are similar to the spectral determinant except that the factor $(1 - t_p)$ in (6) is replaced by $-it_p$ for $j = 0$ and all factors with $j \neq 0$ are ignored (the t_p are defined by (7)). Thus,

$$\Delta^{(p)}(E) = -ie^{-i\pi\bar{N}(E)} \prod_{p' \neq p} (1 - t_{p'})\, t_p \tag{65}$$

A more accurate formula that takes into account the $j \neq 0$ terms is presented in App. B. The form (31) with (57) was obtained originally[19] starting from the expression[42]

$$W(\mathbf{x}; E) = \overline{W}(\mathbf{x}; E) + \sum_P W_P^{(po)}(\mathbf{x}; E) \tag{66}$$

where $\overline{W}$ is given by (59) while the periodic orbits' contribution consists of terms

$$W_P^{(po)}(\mathbf{x}; E) = -i\pi \frac{2^{d_f}}{\sqrt{\det(M_P + I)}} \times \tag{67}$$

$$A(\mathbf{x}; E) \exp\left\{ i\left(\frac{1}{\hbar}[S_P + \tilde{\mathbf{X}} J \frac{M_P - I}{M_P + I}\mathbf{X}] - \gamma_P \right) \right\}$$

where is T_P is the period of the periodic trajectory P, while M_P is the corresponding monodromy matrix, S_P is its action, γ_P is the Maslov phase and J is the unit symplectic matrix,

$$J = \begin{pmatrix} 0 & I \\ -I & 0 \end{pmatrix}. \tag{68}$$

Taking the leading order in the expansion of $\sqrt{\det(M_P + I)}$ in e^{-u_P}, one finds

$$W_P^{(po)}(\mathbf{x}; E) = -4\pi i A(\mathbf{x}; E) e^{-\frac{1}{2}u_P} e^{\frac{i}{\hbar}\left(S_P + \tilde{\mathbf{X}}\mathbf{R_p}\mathbf{X}\right) - i\gamma_P} \tag{69}$$

Here P is an arbitrary periodic orbit. If it consists of r repetitions of the primitive orbit p then $S_P = S_p r$ and $u_P = u_p r$. The sum over repetitions is a geometric sum leading to

$$W_p^{(po)}(\mathbf{x}; E) = -4\pi i A(\mathbf{x}; E) \frac{e^{-\frac{1}{2}u_p} e^{\frac{i}{\hbar}\left(S_p + \tilde{\mathbf{X}}\mathbf{R_p}\mathbf{X}\right) - i\gamma_p}}{1 - e^{-\frac{1}{2}u_p} e^{\frac{i}{\hbar}S_p - i\gamma_p}} \tag{70}$$

and now (66) takes the form

$$W(\mathbf{x}; E) = \overline{W}(\mathbf{x}; E) + \sum_p W_p^{(po)}(\mathbf{x}; E) \tag{71}$$

where the sum is only over primitive periodic orbits. Taking a common denominator in the sum (71) one obtains the form (57) with (63) or more accurately with (109). The advantage of the direct derivation is the fact that in this derivation the role of the summation over repetitions is transparent. The spectral determinant is the product of the denominators of the various primitive periodic orbits (70). Since the product (6) is not absolutely convergent the zeros are *not* the zeros of the various factors, and therefore not directly related to single primitive periodic orbits. One should be able to obtain the ratio form (30) for the Green function starting from a semiclassical formula for the Green function given in[43], by the same route used to obtain the ratio form (31) for the Wigner function from the semiclassical formula (66) derived in Ref. 42. Resummation is required to obtain meaningful expressions for $\mathcal{N}_W$ and Δ [16-19,39].

3.3 The Green Function for Billiards on Surfaces of Constant Negative Curvature

The systems considered in this subsection are tiling billiards on the pseudo-sphere. There is extensive mathematical literature concerning these systems. Various aspects of geometry can be found in Ref. 44, while many of the results about quantum mechanical properties of these systems, can be found in Refs. 45, 46 and 47. Useful reviews are [48] and [49]. An example of such a billiard is presented in Fig. 4.

The pseudo-sphere is a surface of constant negative Gaussian curvature $-\frac{1}{R_{ps}^2}$. Setting $R_{ps} = 1$, all distances are measured in units of R_{ps}. Because the surface is two dimensional, it can be projected onto the complex plane, so that the two dimensional coordinates are given by complex numbers. Two very useful projections are the Poincaré disk, in which the surface is projected onto the interior of the unit disk, and the Poincaré half plane, in which it is projected onto the upper half of the complex plane. In both of these models the geodesics on the surface turn out to be arcs of circles perpendicular to the boundary (the circle or the real axis).

The invariant measure in the half plane coordinates, that will be used here is:

$$d\mu(\zeta) = \frac{dxdy}{y^2}, \tag{72}$$

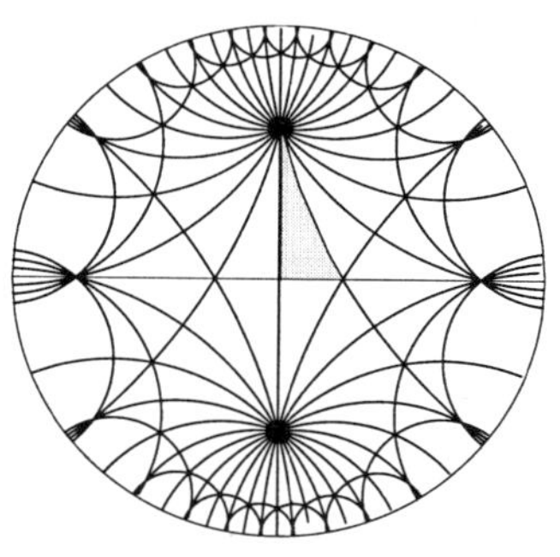 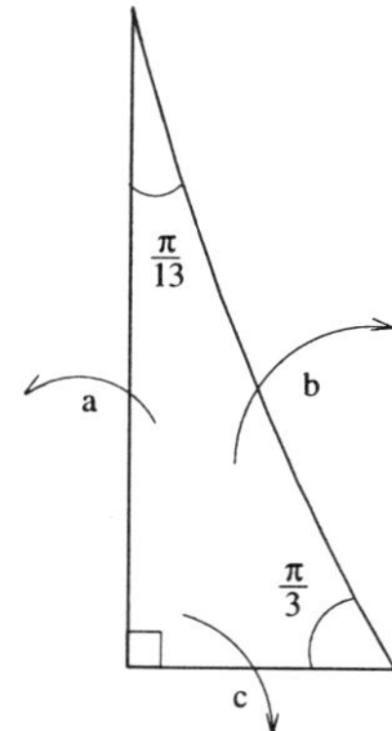

Figure 4. Left - the action of some isometries of the group generated by reflections at the sides of the shaded triangle on it. One can see that the copies tile the pseudo-sphere. **Right** - the billiard chosen for the numerical calculations (a magnification of the shadowed area on the left).

where $\zeta = x + iy$. The distance between two points is:

$$\cosh d_{\zeta,\zeta'} = 1 + \frac{1}{2} \frac{|\zeta' - \zeta|^2}{y' \, y} \ . \tag{73}$$

There are several kinds of isometries that are going to be encountered here:

1. <u>Even boosts</u>: these isometries are the analogues of Euclidean translations. They have two fixed points at infinity, which lie on the boundary of both Poincaré models. Any two points on the surface are connected by a unique geodesic. The unique geodesic that connects these fixed points is called the boost's *invariant geodesic*.

2. <u>Rotations</u>: these isometries are the analogues of Euclidean rotations. They have one fixed point which does not lie at infinity.

3. <u>Inversions</u>: reflections across the invariant geodesics of boosts.

4. <u>Odd boosts</u>: an even boost followed by an inversion through it's invariant geodesic.

The distance a point is transferred by a boost is closely related to the it's distance from the boost's invariant geodesic. If g is an even boost[44]:

$$\sinh \frac{1}{2} d_g(\zeta) = \cosh d_g^{\perp}(\zeta) \ \sinh \frac{1}{2} L_g \ , \tag{74}$$

where $d_g(\zeta) \equiv d_{\zeta,g(\zeta)}$ is the distance of a point from it's image under g, L_g is the distance along the geodesic, while $d_g^{\perp}(\zeta)$ is the distance of the point ζ from the invariant geodesic of g measured along a perpendicular geodesic. A relation of similar nature holds for odd boosts.

The Green function is calculated exactly by the method of images, from the free Green function, by virtue of the underlying discrete group of isometries. In half-plane coordinates, the Schrödinger equation reads[48]:

$$-\frac{\hbar^2}{2m} \, y^2 \left(\frac{\partial^2}{\partial x^2} + \frac{\partial^2}{\partial y^2} \right) \Psi = E\Psi \ . \tag{75}$$

Distance is measured in units of R_{ps}, and from this point on in this subsection (and in App. C), energy will be measured in units of $\frac{\hbar^2}{2mR_{ps}^2}$ while $m = \frac{1}{2}$ and $\hbar = 1$. In these units, it turns out that the free retarded Green function is given by[48]:

$$G^0(\zeta, \zeta'; E) = \frac{-1}{2\pi} Q_{-\frac{1}{2}-ik} [\cosh d_{\zeta,\zeta'}] \, , \tag{76}$$

where $E = k^2 + \frac{1}{4}$ and $Q_l[x]$ is the Legendre function of the second kind[50].

The billiards under consideration are the fundamental domains of discrete groups of isometries. Therefore, all the copies of the billiard tile the surface perfectly. In addition, as all the copies are identical, a solution of the Schrödinger equation in one, solves it in all the other tiles, but for a coordinate transformation. This means that the Green function has to be invariant under the action of the group isometries ($g_i \in \Gamma$). A simple way to obtain such a function is to use the method of images and free Green's function:

$$G(\zeta, \zeta'; E) = \sum_{g \in \Gamma} \chi_g G^0(\zeta, g(\zeta'); E) \, . \tag{77}$$

where $\chi_g = (\pm 1)^{parity\ of\ g}$ (+1 is the choice for Neumann and periodic boundary conditions and -1 for the Dirichlet case).

Now the *exact* retarded Green function for fundamental domain billiards can be written using (76):

$$G(\zeta, \zeta'; E) = \frac{-1}{2\pi} \sum_{g \in \Gamma} \chi_g Q_{-\frac{1}{2}-ik} \left[\cosh d_{\zeta, g(\zeta')} \right] \, . \tag{78}$$

This equation gives the Green function between ζ and ζ' as a sum over all the *classical* orbits between them. This is fundamentally different from other quantum mechanical problems, where all Feynman paths are required for the calculation of the Green function. This property, that holds when tiling is possible, makes the semiclassical approximation exact for the problems studied here. An exact solution is possible, although the classical motion for the system is chaotic!!

The sum over the group members of (78) can be arranged in a suggestive manner. Any group can be divided into *conjugacy classes*[51]. A conjugacy class is defined as the collection of all different group members that are similar to one another. All the members of a conjugacy class can be generated from one representative via:

$$g' = h^{-1}gh \quad h \in \Gamma/\Gamma_g \, , \tag{79}$$

where Γ/Γ_g is the sub-group of Γ, whose members are unity and the members that do not commute with g. The reason for excluding Γ_g (the sub-group of the group members that commute the g) is that each member of a conjugacy class should be accounted for only once.

It may be shown that there is a one-to-one correspondence between the conjugacy classes of boosts and the periodic orbits[45, 48]. This means that the list of boost conjugacy classes is the list of periodic orbits. Therefore, it has the form:

$$\{\ldots, p_i^r, \ldots\} \, ,$$

where $r = 1, 2, 3, \ldots$ are repetitions and $\{p_i\}$ is the list of primitive periodic orbits of the billiard. Note that p and p^{-1} can only be conjugate if there are reflections in the group, *i.e* if the periodic orbit is self retracing.

The exact retarded Green function is given by (78). Rearranging the sum over boosts in complete boost conjugacy classes, one obtains:

$$G(\zeta, \zeta'; E) = \frac{-1}{2\pi} \sum_{\substack{g \\ \text{not a boost}}} \chi_g Q_{-\frac{1}{2}-ik} \left[\cosh d_{\zeta, g(\zeta')} \right] +$$

$$+ \frac{-1}{2\pi} \sum_{p} \sum_{r=1}^{\infty} \sum_{h \in \Gamma/\Gamma_p} \chi_{p^r} Q_{-\frac{1}{2}-ik} \left[\cosh d_{\zeta, h^{-1}p^r h(\zeta')} \right] , \qquad (80)$$

where $\sum_p \sum_{r=1}^{\infty}$ is the sum over all the different boost conjugacy classes, and $\sum_{h \in \Gamma/\Gamma_p}$ is the sum over the different members of each class.

For a billiard with periodic boundary conditions, the group of isometries, whose fundamental domain is the billiard, contains only the identity and even boosts. The retarded Green function is given, in this case, by:

$$G(\zeta, \zeta'; E) = \frac{-1}{2\pi} Q_{-\frac{1}{2}-ik} \left[\cosh d_{\zeta, \zeta'} \right] +$$

$$+ \frac{-1}{2\pi} \sum_{\{p\}} \sum_{r=1}^{\infty} \sum_{h \in \Gamma/\Gamma_p} Q_{-\frac{1}{2}-ik} \left[\cosh d_{\zeta, h^{-1}p^r h(\zeta')} \right] , \qquad (81)$$

where $\sum_{\{p\}} \sum_{r=1}^{\infty}$ is the sum over all the different even boost conjugacy classes.

In order to obtain a transparent formula for the Green function, the Legendre functions, which constitute it, are expanded in powers of e^{-L_g} (see App. C):

$$Q_{-\frac{1}{2}-ik} \left[\cosh d_g(\zeta) \right] = \sum_{m=0}^{\infty} \mathcal{F}_m \left[k, d_g^{\perp}(\zeta) \right] e^{-\left(\frac{1}{2}+m-ik \right) L_g} , \qquad (82)$$

where

$$\mathcal{F}_m \left[k, d_g^{\perp}(\zeta) \right] \equiv \sum_{n=0}^{\left[\frac{m}{2} \right]} \frac{(2n-1)!!}{(2n)!!} B \left[n + \frac{1}{2} - ik, \frac{1}{2} \right] \times$$

$$\times a_{m-2n} \left[\frac{1}{\cosh^2 d_g^{\perp}(\zeta)}, \partial\eta \right] \left(\frac{e^{-\eta}}{\cosh^2 d_g^{\perp}(\zeta)} \right)^{\frac{1}{2}+2n-ik} \Bigg|_{\eta=0} , \qquad (83)$$

in which $B(x, y) = \frac{\Gamma(x)\Gamma(y)}{\Gamma(x+y)}$ is the Beta function, and $a_m[x, y]$ is the product two real polynomials of order m, one of x and one of y (see App. C).

All the members of a conjugacy class are related to one periodic orbit:

$$L_{h^{-1}p^r h} = L_{p^r} = r L_p . \qquad (84)$$

Since p^r and p have the same invariant geodesic,

$$d_{h^{-1}p^r h}^{\perp}(\zeta) = d_{h^{-1}ph}^{\perp}(\zeta) . \qquad (85)$$

The set of distances $\left\{ d_{h^{-1}ph}^{\perp}(\zeta) \right\}_{h \in \Gamma/\Gamma_p}$ is the set of distances of ζ from all different images of the invariant geodesic of p. Therefore, the summation over the distances from the invariant geodesics of members of a boost conjugacy class takes into account the distance of a point from all of the images of the invariant geodesic of the class representative.

The next aim is to derive Eq. 30 with (47) for this system in terms of (89) and (90), that express the exact Green function in terms of sums over periodic orbit lengths. This will be achieved by substituting (84) and (85) in Eq. 82:

$$G(\zeta, \zeta; E) = \frac{-1}{2\pi} Q_{-\frac{1}{2}-ik} [1] +$$

$$+ \frac{-1}{2\pi} \sum_{\{p\}} \sum_{r=1}^{\infty} \sum_{h \in \Gamma/\Gamma_p} \sum_{m=0}^{\infty} \mathcal{F}_m \left[k, d_{h^{-1}ph}^{\perp}(\zeta) \right] e^{-\left(\frac{1}{2}+m-ik\right)rL_p} . \tag{86}$$

Note that the contribution of a single periodic orbit (p) has been decoupled into two separate parts - the perpendicular distance, $d_{h^{-1}ph}^{\perp}(\zeta)$, and the length of the periodic orbit, L_p. To achieve this in a typical chaotic system one usually has to resort to the semiclassical approximation, while here it is exact.

Exchanging the order of summation, a geometric series is encountered:

$$\mathcal{S}^{(m)}(L_p, k) \equiv \sum_{r=1}^{\infty} \left[e^{-\left(\frac{1}{2}+m-ik\right)L_p} \right]^r = \frac{e^{-\left(\frac{1}{2}+m-ik\right)L_p}}{1 - e^{-\left(\frac{1}{2}+m-ik\right)L_p}} . \tag{87}$$

This enables one to write the retarded Green function (86) in the form:

$$G(\zeta, \zeta; E) = \frac{-1}{2\pi} Q_{-\frac{1}{2}-ik} [1] + \frac{-1}{2\pi} \sum_{\{p\}} \sum_{h \in \Gamma/\Gamma_p} \sum_{m=0}^{\infty} \mathcal{F}_m \left[k, d_{h^{-1}ph}^{\perp}(\zeta) \right] \mathcal{S}^{(m)}(L_p, k) , \tag{88}$$

that is given in terms of the lengths of the primitive periodic orbits of the system. Taking a common denominator and multiplying the numerator and denominator by $e^{-i\pi \bar{N}(E)}$, one obtains a convenient *exact* expression for the Green function, in the form (30) with (47) where,

$$\mathcal{N}_G(E) \equiv \frac{-1}{2\pi} Q_{-\frac{1}{2}-ik} (1) \Delta(E) , \tag{89}$$

in which $\Delta(E)$ is the spectral determinant, as it is given *exactly* by the Selberg trace formula. In addition:

$$\tilde{\mathcal{N}}_G = \mathcal{N}_G^{(po)}(\zeta; E) \equiv \frac{-1}{2\pi} \sum_{\{p\}} \sum_{h \in \Gamma/\Gamma_p} \sum_{m=0}^{\infty} \mathcal{F}_m \left[k, d_{h^{-1}ph}^{\perp}(\zeta) \right] \Delta^{(p,m)}(E) , \tag{90}$$

where $\Delta^{(p,m)}(E)$ is defined in (110). In the present case there are exactly three sorts of contributions to (30). The first is from short non-periodic orbits, and it is given by $\overline{\mathcal{N}}_G(E)$, as it results from the identity. The second and third contributions are related to periodic orbits and are given by $\mathcal{N}_G^{(po)}(\zeta; E)$. The first contribution to this function is related to the actual periodic orbits in the billiard, while the second is related to their images. The reason is that a "real" periodic orbit is made up from only a few segments of invariant geodesics of a conjugacy class's elements. The invariant geodesics of the rest of the conjugacy class make up "images" of the periodic orbit, as they never intersect with the original billiard. "Real" periodic orbit segments generally have relatively small $d_{h^{-1}ph}^{\perp}(\zeta)$, while the images of them tend to have larger $d_{h^{-1}ph}^{\perp}(\zeta)$. Note that the formula takes into account the contributions of *all* trajectories.

Eq. 30 is a manifestation of the results of Fredholm's theory, as presented in Subsection 3.1. The precise nature of the connection with Fredholm's theory has not been worked out yet for this case. Once this is done, Fredholm's theory will supply a prescription as to the correct reordering of terms in the series, so that they converge absolutely to the exact result. In absence of such ordering, resummation has to be used

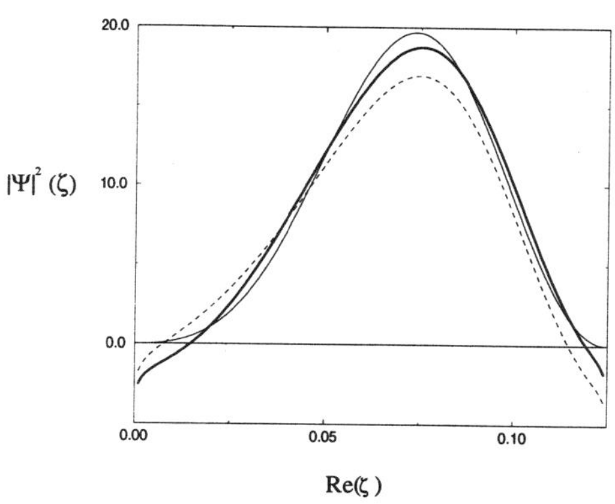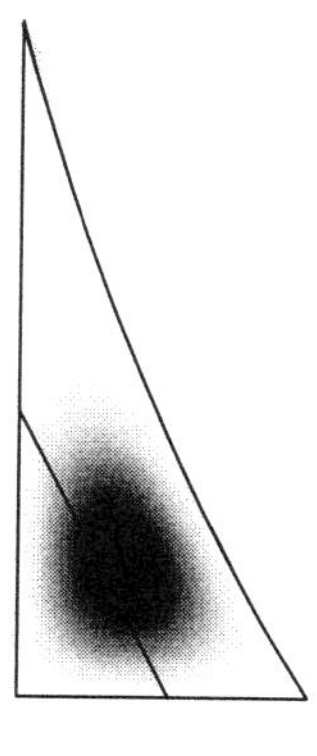

Figure 5. Comparison of an exact cut through the ground-state density (thin line) with the semiclassical result, before including periodic orbits (dashed line) and after including them (thick line). On the right is the exact ground-state density in the whole billiard, with the cut through it denoted by a line. The average square deviation is $\delta_1^{(app)} \approx 0.023$ and $\delta_2^{(app)} \approx 0.005$.

to get meaningful results from this formula. This is also the case for the (exact) Selberg trace formula.

For hard wall billiards (Dirichlet boundary conditions) a formula similar to (90) for the periodic orbit contribution holds. The contribution from short non-periodic orbits is for hard wall billiards of the form:

$$\mathcal{N}_G\left(\zeta; E\right) = \frac{-1}{2\pi} \left[Q_{-\frac{1}{2}-ik}\left[1\right] \; - \sum_{\substack{Inversions \\ \mathcal{I}}} Q_{-\frac{1}{2}-ik}\left[\cosh d_{\mathcal{I}}(\zeta)\right] \; + \right.$$
$$\left. + \sum_{\substack{Rotations \\ \mathcal{R}}} Q_{-\frac{1}{2}-ik}\left[\cosh d_{\mathcal{R}}(\zeta)\right] \right] \times \Delta(E) \ . \tag{91}$$

The main difference between (89) and this result, is that the former is a constant function of position, whereas the latter is not.

Eq. 30 (with (47)) for billiards with periodic boundary conditions with (89) and (90), and the corresponding ones for hard wall billiards, are exact. These cannot be used, however, for the calculation of eigenfunctions, since both the numerator and the denominator are expressed as series that are not absolutely convergent. In what follows, resummed expressions that approximate the above equations will be calculated and compared with exact eigenfunctions found numerically. In particular, an arbitrary slice through the two lowest eigenfunctions of the hard wall billiard defined in Fig. 4 is calculated. This billiard tiles the pseudo-sphere perfectly by reflections, because all of it's angles are of integer fractions of π [52]. The results are shown in Figs. 5 and 6. Presented are the exact cut through the eigenfunction density, the contribution of the non-boost isometries of (91), and the corrections to this due to periodic orbits (the counterpart of (90) for hard wall billiards). The contribution of the non-boost isometries accounts for most of the features of the eigenfunction density, while the addition of the periodic orbits provides only a small correction.

One can see that even after the inclusion of all of the isometries in the sums that are truncated as resummation requires, the eigenfunctions are still not reproduced perfectly. A measure of the quality of an approximation of the function $f(x)$ by $f_{approx}(x)$ is given

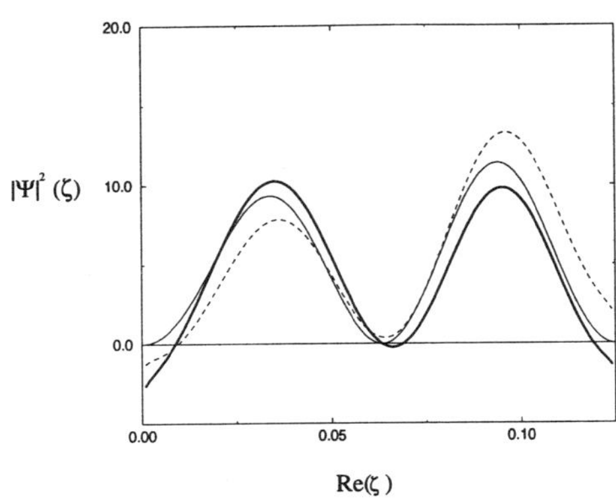
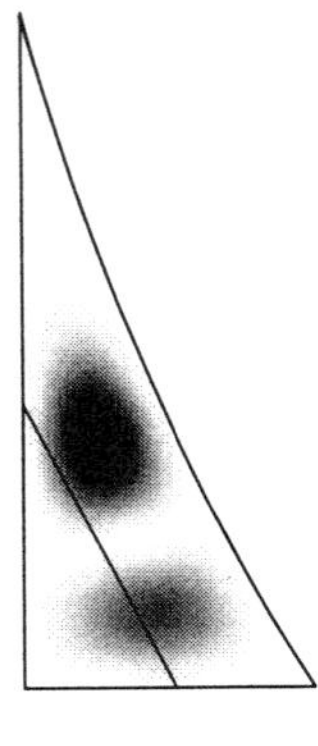

Figure 6. Same as Fig. 5 but for the first excited state. The average square deviation is $\delta_1^{(app)} \approx 0.07$ and $\delta_2^{(app)} \approx 0.03$.

by the average squares deviation:

$$\delta^{(app)} \equiv \frac{\int \, dx \, |f(x) - f_{approx}(x)|^2}{\int \, dx \, |f(x)|^2} \, , \tag{92}$$

where in this case $f(x) = |\Psi(\zeta)|^2$, and the integral is taken along the cut (see Figs. 5 and 6). It is interesting to see the difference in the quality of the approximation without including the periodic orbit contribution, $\delta_1^{(app)}$, and including it, $\delta_2^{(app)}$.

These results show that the quality is fairly good even without including the periodic orbits' contribution. Including this last contribution improves the approximation appreciably. The reason the fit is not perfect is that at such low energies of the wave functions presented here the number of periodic orbits that contribute to the truncated sums is very small, and the contribution of the orbits in the vicinity of the truncation point is not negligible.

4 SCARS

In this section "scars" will be discussed. The discussion will be confined to systems with two degrees of freedom ($d_f = 2$), mainly to billiards. Scars are defined by (Ref. 24, p. 636): "A quantum eigenstate of a classically chaotic system has a *scar* of a periodic orbit if its density on the classical invariant manifolds near the periodic orbit differs significantly from the statistically expected density." Such states are indeed found, for example for the truncated hyperbola billiard depicted in Fig. 7. Some eigenfunctions with the scarring periodic orbits are presented in Fig. 8 (boundaries of the regions where these are expected to be large are drawn to guide the eye).

It is obvious that for this example there is a clear relation between the wave functions and the classical unstable periodic orbits and the issue is, not whether scars exist, but rather how to define them quantitatively and how to derive conditions for their existence. It was already commented in Ref. 24 that the term "significant" in the definition is loose. Therefore a more quantitative definition is called for.

The following argument for the existence of scars was put forward by Heller[24, 25]. Let ϕ be a coherent state in the vicinity of an orbit. Projection of the state propagated a time t on the initial one, $\langle\phi|U_t|\phi\rangle$ is large if ϕ is centered on a periodic orbit. It

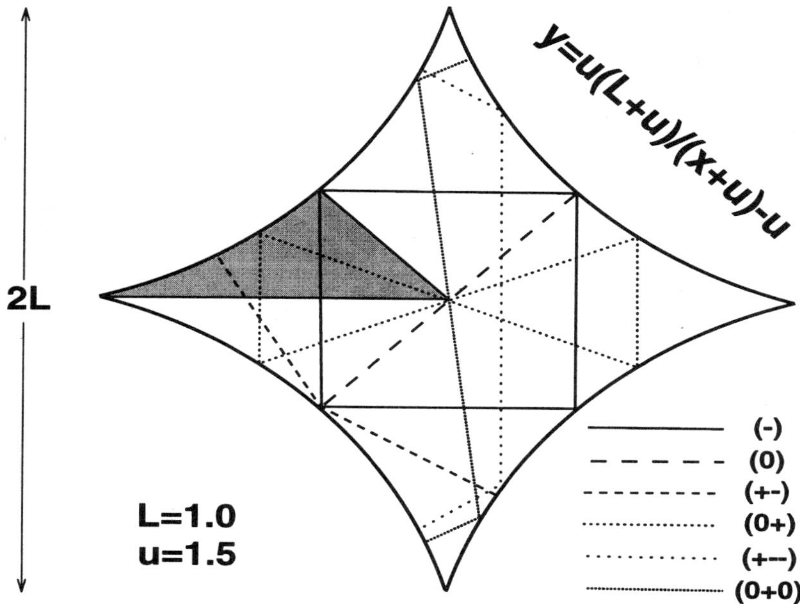

Figure 7. The truncated hyperbola billiard together with 6 of its periodic orbits. The shadowed area marks the fundamental domain.

will decay as a function of time with the classical Lyapunov exponent of the orbit, $\bar{u}_p = u_p/T_p$, where u_p is the instability exponent and T_p is the period. Consequently the local density of states on ϕ

$$R_\phi(E) = \sum_\alpha |\langle \phi|\alpha\rangle|^2 \delta(E - E_\alpha), \qquad (93)$$

that is the time Fourier transform of $\langle \phi|U_t|\phi\rangle$, exhibits peaks of width proportional to $\bar{u}_p$ (and height inversely proportional to it). Since $R_\phi(E)$ is the local density of states related to a wave packet started on a periodic orbit p, it is a measure of scarring of eigenfunctions with eigenenergies near E, by this orbit. One can probably turn it to a quantitative measure of average scarring of groups of states. However, it does not have predictive power for individual states.

To develop a scar theory for individual states one should define a quantity that measures the strength of the scar of a *particular state* by a *particular periodic orbit*. Such a measure, $Y_p(E_\alpha)$ is the integral of the Wigner function $W_\alpha(\mathbf{x})$ corresponding to the wave function ψ_α, with energy E_α, in a phase space tube around the periodic orbit p or an integral over $|\psi_\alpha(\mathbf{r})|^2$ in a coordinate space tube around the periodic orbit. The size and the shape of this tube is somewhat arbitrary, and may depend on the orientation of the stable and unstable manifolds, but its order of magnitude is determined by the nature of the semiclassical theories. For systems with two degrees of freedom it is of the order of $\hbar$ in phase space as will be argued in what follows. It is of the order of $\delta y = \sqrt{2\lambda_{db}L}$ in real space, where λ_{db} is the de-Broglie wavelength while L is characteristic classical length of the system, such as the size of a billiard. This estimate can be easily derived for a billiard, noting that for a trajectory of length L between the walls, trajectories of length smaller than $L + \lambda_{db}$ contribute terms with similar phases

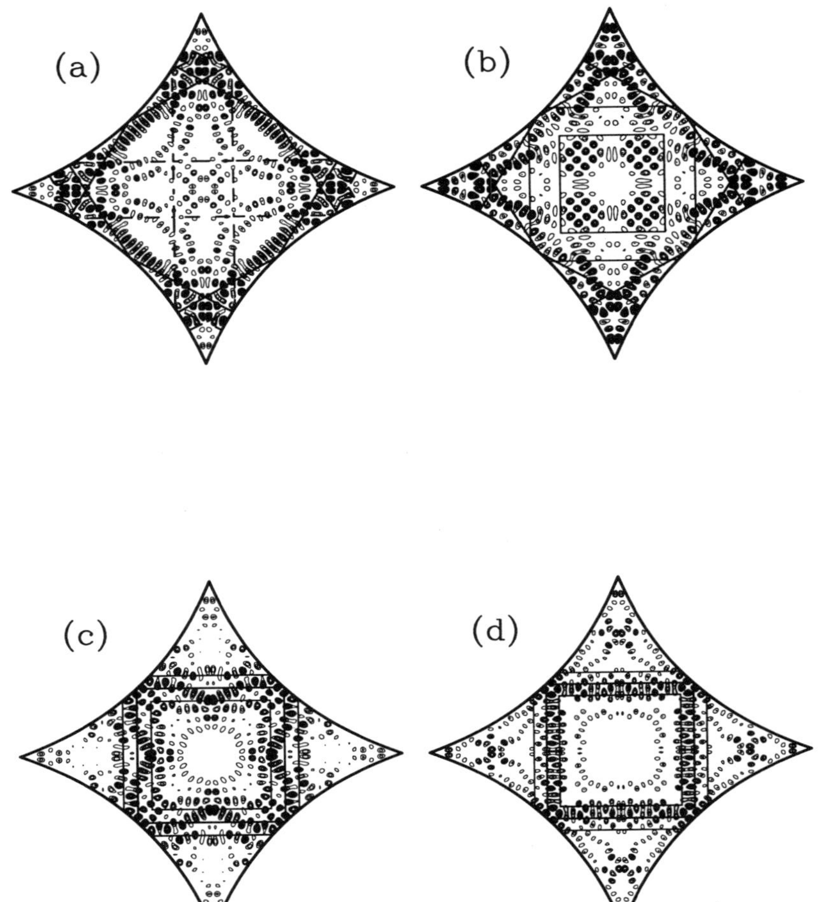

Figure 8. The densities of the wave functions: (a) No. 94 scarred by the orbits $(+-)$ and $(+--)$; (b) No. 73 scarred by $(+-)$ and anti-scarred by $(-)$; (c) 87, and (d) 107 both scarred by the orbit $(-)$. The corresponding weights are indicated by arrows in Fig. 9.

to the semiclassical formulas. Therefore only trajectories with separation δy from the original one, that satisfy $\sqrt{L^2 + \delta y^2} \leq L + \lambda_{db}$ have similar phases and contribute significantly.

With the help of a resummed version of (31), using (57), (58) and (63) one finds[19]

$$W_\alpha(\mathbf{x}) \simeq \frac{4\pi A(\mathbf{x}; E_\alpha)}{h^2 \Delta'(E_\alpha)} \left\{ \frac{1}{4} \Delta_i(E_\alpha) + \Re \sum_p \Delta^{(p)}(E_\alpha) e^{\frac{i}{\hbar} \tilde{\mathbf{X}} \mathbf{R_P} \mathbf{X}} \right\}, \tag{94}$$

or more accurately (111). The explicit form of the function $A(\mathbf{x}; E)$ is given by (60), Δ is a resummed version of the spectral determinant (6) that is manifestly real, $\Delta'(E) = \frac{d\Delta(E)}{dE}$ and the function $\Delta_i(E)$ is the imaginary part of the expansion of (6) as a sum over periodic orbits. One sees from (94) that the contribution associated with a given periodic orbit is concentrated in a small region surrounding the orbit, since the contributions from more distant parts average out to zero. The size of this region on the Poincarè surface is of order h. Therefore, if the tube is wide enough, the integral over the contribution to $W_\alpha(\mathbf{x})$ associated with the p-th orbit can be calculated semiclassically. Denoting this contribution by $Y_p(E_\alpha)$ one obtains:

$$Y_p(E_\alpha) \simeq \frac{T_p}{\hbar \Delta'(E_\alpha)} \Re \Delta^{(p)}(E_\alpha) \tag{95}$$

or more accurately (112). Here T_p is the period of the primitive periodic orbit p. This expression does not measure the actual weight of the scar, since in the same tube there may be contributions from other orbits which are not taken into account. However, due to the normalization of the Wigner function, if $Y_p(E_\alpha)$ is of order unity, ψ_α is expected to be scarred along the p-th orbit. In this case the scar weight may be approximated by $Y_p(E_\alpha)$. In the more general case, ψ_α is predicted to be scarred if a considerable part of the weight is concentrated on *few* $Y_p(E_\alpha)$ that are much higher than the rest. Then the scarring pattern is expected to be more complicated.

The contributions from the periodic orbits as well as those associated with the background are interrelated. This is a consequence of the normalization of the Wigner function which leads to the sum rule:

$$\sum_p Y_p(E_\alpha) + \pi \rho_0(E_\alpha) \frac{\Delta_i(E_\alpha)}{\Delta'(E_\alpha)} = 1 \tag{96}$$

where $\rho_0(E)$ is the mean density of states. The scar weight is measured relative to the background (that is proportional to $\Delta_i(E_\alpha)$). A large negative value of $Y_p(E_\alpha)$, therefore, corresponds to a situation in which the probability density in the vicinity of the periodic orbit is small, i.e. to an anti-scar.

For billiards, (94) is inapplicable for the calculation of Wigner functions[53]. Nevertheless, the expression for the scar weight (95) holds in this case as well. It may be derived from an extension of Ref. 42 to billiards[53] following the steps of the derivation presented here.

The weights $Y_p(E_\alpha)$, calculated for the orbits $(-)$ and $(+-)$, of the truncated hyperbola billiard of Fig. 7 are plotted in Fig. 9. A clear feature of this figure is the nearly periodic structure of the peaks along the energy axis. These peaks may be intuitively associated with the set of energies $\{\bar{E}_n\}$ which allow "standing waves" along the periodic orbit, i.e. those which satisfy the quantization condition $S_p(\bar{E}_n) = 2\pi\hbar n + \gamma_p\hbar$, where n is an integer and γ_p is the Maslov phase. The densities corresponding to several wave functions, numbered and indicated by arrows in Fig. 9, are shown in Fig. 8. States are indeed scarred if the corresponding $Y_p(E_\alpha)$ are large. Agreement was found for all states of this billiard that were calculated.

The amplitudes of the peaks of the least unstable orbits may be estimated to be

$$Y_p(E_\alpha) = \frac{1}{1+C}, \quad \text{with} \quad C \approx \frac{u_p \tau_H}{2T_p}, \tag{97}$$

where $\tau_H = 2\pi\hbar\rho_o(E)$ is the Heisenberg time, while ρ_0 is the mean density of states. This estimate was obtained from (95), substituting the sum rule (96), using the diagonal approximation and the Hannay and Ozorio de Almeida sum rule[54]. For an orbit p that is not very unstable the approximations $1 - t_p \approx u_p/2$ and $1 - t_{p'} \approx 1$ for $p' \neq p$ were used. The test of this approximation is presented in Fig. 9. For short orbits with Lyapunov exponent, $\bar{u}_p$, satisfying $T_p \ll \tau_H$,

$$Y_p(E_\alpha) \approx 2\frac{1}{u_p}\frac{T_p}{\tau_H} = 2\frac{1}{\bar{u}_p \tau_H}, \tag{98}$$

that agrees with the statistical estimate presented following (93).

Calculation of $Y_p(E_\alpha)$ given the tube, is a well defined problem. In many cases scarred wave functions are visually observed when $Y_p(E_\alpha)$ is sufficiently large. Other situations may be encountered as well. This is the case for the billiard on a surface of constant negative curvature presented in subsection 3.3. Integration over $|\psi_\alpha|^2$ in tubes around periodic orbits that can be done analytically, in the semiclassical approximation, starting from the resummed version of (30) with (47), and the hard wall counterpart of (89) and (90), leads to results presented in Fig. 10 for the periodic orbits plotted in Fig. 11. Comparing Fig. 10 with Fig. 9 one realizes that in Fig. 10 the plots are very wiggled compared those of Fig. 9. Consequently each periodic orbit scars many states and in turn each state is scarred by many periodic orbits. Therefore visual observation of scars is expected to be difficult in this case as can indeed be seen from Fig. 12 and Fig. 13. The eigenfunctions plotted in Figs. 12a, 12b, 12c and 13 are expected, from the results presented in Fig. 10 to be strongly scarred by the orbits drawn in the corresponding figures, while the eigenfunction of Fig. 12d is expected to be moderately scarred. This expectation relied on the assumption that each eigenfunction is scarred only by few orbits. However only the eigenfunctions plotted in Figs. 12b and 13b, look strongly scarred by these orbits. The reason probably is that here each eigenfunction is scarred or anti-scarred by many periodic orbits, all having the same Lyapunov exponent. Such a mechanism is conceivable for the eigenfunction of Fig. 13a, that although predicted to be scarred by the shortest orbit, it is also anti-scarred by several other orbits in its vicinity (see Fig. 10 and 11), leading to destruction of the scar. Eigenfunctions for various billiards on surfaces of constant negative curvature were studied extensively and no clear visually observable scars, that are significantly related to periodic orbits were found, although correlations that differ from those of the random wave function hypothesis[21, 22] were obtained[55]. What is special for billiards on surfaces of constant negative curvature? All orbits have the same Lyapunov exponent $\bar{u}_p = u_p/T_p$. If various orbits have different values of $\bar{u}_p$, the ones with smaller $\bar{u}_p$, namely the less unstable ones, are more likely to scar. For surfaces of constant negative curvature, all orbits "scar" with similar strength resulting in the wiggly shape of the Y_p plots in Fig. 10 compared the plots of Fig. 9. Therefore the scarring patterns by specific orbits may not be obvious.

The theory presented so far is a linear theory, but it takes into account other orbits than the scarring orbit p via Δ and $\Delta^{(p)}$ (or more accurately $\Delta^{(p,n)}$ of App. B). In this way nonlinear effects, that determine the periodic orbits in vicinity of the orbit p are taken into account. Nonlinear theories that use the stable and unstable manifolds, rather then periodic orbits, as the underlying classical objects were introduced in

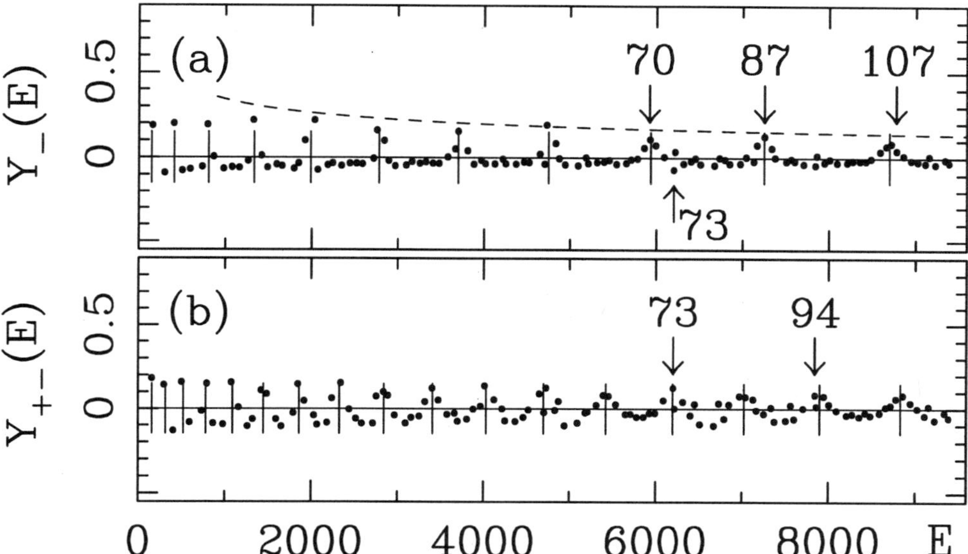

Figure 9. The scar weights of the orbits $(-)$ and $(+-)$ calculated from (112) that is a more accurate version of (95). The vertical bars are located at the energies $\bar{E}_n$ which allow standing waves along the orbits. The dashed line is the approximation (97) for the peak amplitude.

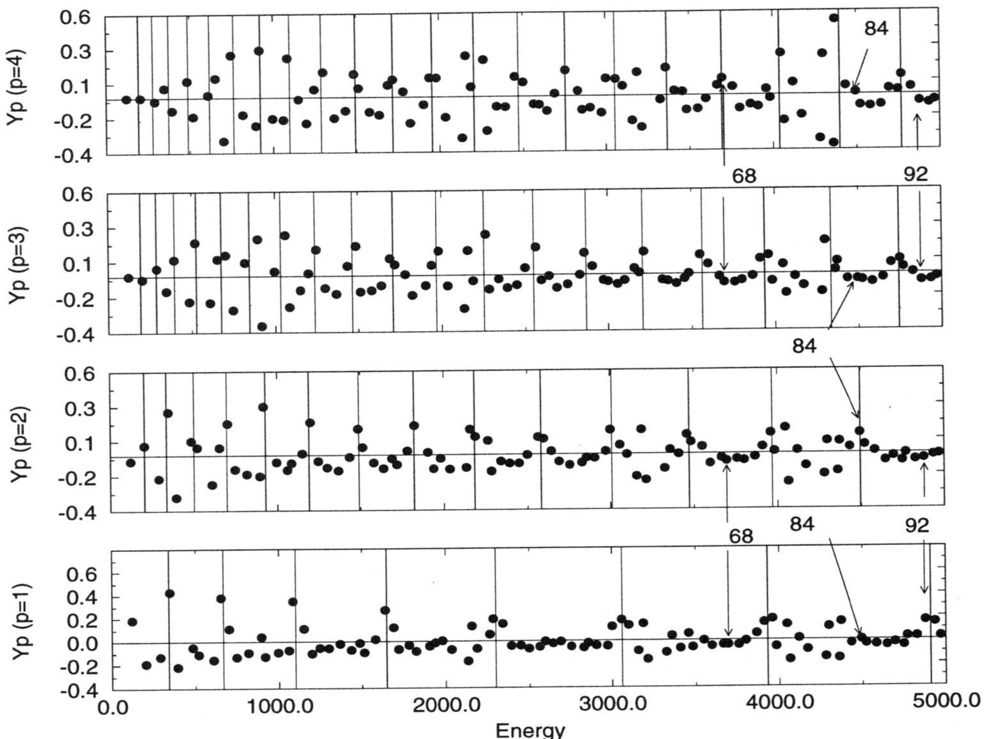

Figure 10. The scar weight on the shortest four primitive periodic orbits ($p = 1$ denotes the shortest, $p = 2$ the next shortest, etc.) calculated on the zeros of the approximate spectral determinant. In analogy with Fig. 9 the vertical lines are the locations of the eigenenergies which allow a standing wave along the orbit.

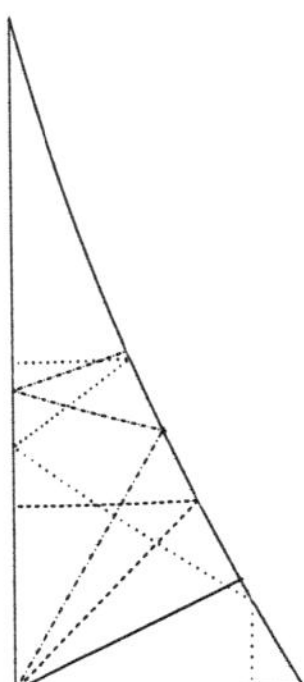

Figure 11. The four shortest periodic orbits: solid line ($p = 1$) - the shortest orbit, dashed ($p = 2$) - next in length, dot-dashed ($p = 3$)- longer still and the fourth shortest ($p = 4$) is dotted.

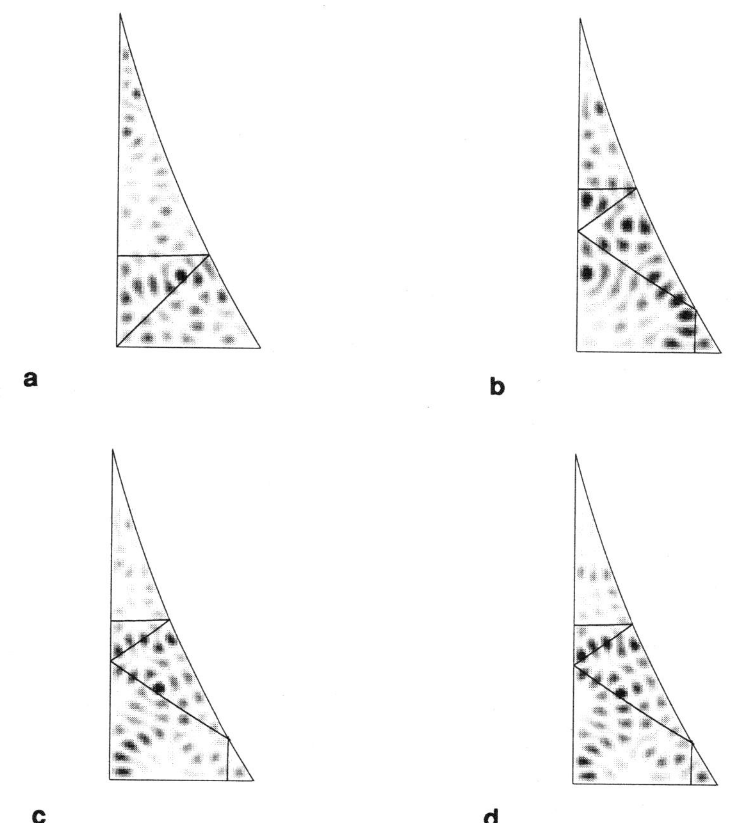

Figure 12. (a) - Density of eigenfunction No. 84 and the second shortest periodic orbit ($p = 2$). The fourth shortest periodic orbit ($p = 4$) on the density of eigenfunction No. 68 (b), No. 90 (c) and No 91 (d).

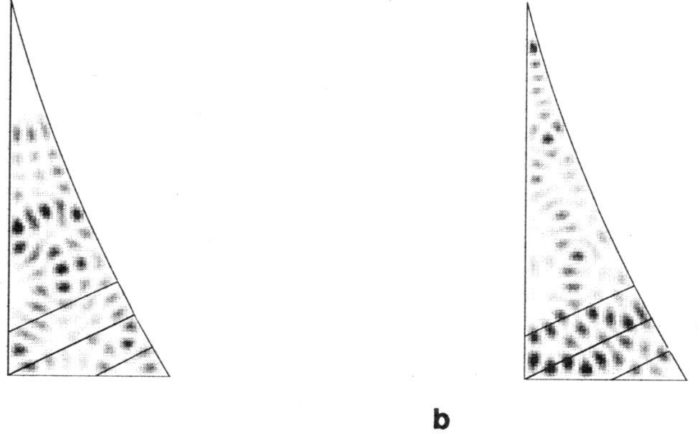

Figure 13. Density of eigenfunction No. 92 (a), and density of eigenfunction No. 93 (b). On both figures the shortest periodic orbit ($p = 1$) is presented, as well as a tube of width $\sqrt{\lambda_{db}} = \frac{1}{\sqrt{k}}$ around it.

recent years[56, 57]. These take into account the effect of periodic orbits other than the scarring one. Scar weights and profiles were obtained also with the help of a scattering approach[58, 59].

In the asymptotic limit of high energies the scar weight Y_p is predicted to vanish as $1/\sqrt{E}$ for two dimensional billiards. It follows from (98) since $\rho_0(E)$ and therefore τ_H do not depend on energy while T_p is inversely proportional to the velocity. It was proved rigorously[60] that for arithmetic billiards there are no scars in the limit $E \to \infty$, where $\lambda_{db} = 0$. If there are scars in such a limit in the sense that $Y_p > 0$, the eigenfunctions must be singular and such a singularity was ruled out rigorously[60]. This is also the situation for generic chaotic systems, where the scarred eigenfunction or the corresponding Wigner function, is significantly larger near the scarring orbit than the microcanonical average, but in the framework of the heuristic theories it is not expected to diverge in the semiclassical limit. Contributions from various orbits can conspire to result in a contribution with a weight that is approximately constant over a wide range of energy[61].

Scars were found experimentally in microwave cavities where the electric field plays the role of the wave function[62]. Many of the eigenfunctions of the hydrogen atom in a magnetic field were found to be scarred[63]. It is possible that the effects of scars were observed in the conductance through quantum dots. Sequences of correlated peaks in plots of the conductance versus energy (Fig. 2 of Ref. 64) remind one the peaks in Fig. 9 and it is tempting to speculate that these result from scarred states, with wave functions that are large on the leads. In absence of any calculations in this direction it is merely a speculation at this stage. It was shown, in the framework of density functional theory, that scars can survive the effects of electron-electron interactions[65].

Appendix A: The Matrix Elements of M and $U^{(W,N)}$ for the Baker Map.

One would like to compare the matrices M and $U^{(W,N)}$ of the baker map, so that one can see how $U^{(W,N)}$ approaches M as N is increased. It is convenient to order the states by increasing $J_M = i + j$ and within states with the same J_M by increasing i. The corner with $J_M = 0, 1, 2$ of M is (ignoring vanishing matrix elements),

$$M = \begin{pmatrix} 1 & & & & & \\ & 0.5 & 0.75 & & & \\ & & 0.5 & 0 & & \\ & & & 0.25 & 0.839 & \\ & & & & 0.25 & 0.839 \\ & & & & & 0.25 \end{pmatrix} \tag{99}$$

The corresponding corner of $U^{(W,N)}$ for $N = 120$ is,

$$U^{(W,120)} = \begin{pmatrix} 1.000 & -0.014 & -0.014 & & & \\ -0.014 & 0.474 & 0.725 & & -0.011 & 0.024 \\ -0.014 & 0.005 & 0.474 & -0.021 & & -0.008 \\ & -0.023 & -0.023 & 0.245 & 0.780 & \\ & 0.003 & 0.004 & & 0.209 & 0.753 \\ & -0.020 & -0.031 & 0.010 & & 0.246 \end{pmatrix} \tag{100}$$

for $N = 180$ it is,

$$U^{(W,180)} = \begin{pmatrix} 1.000 & -0.010 & -0.010 & & & \\ -0.010 & 0.481 & 0.733 & & -0.007 & 0.016 \\ -0.010 & 0.003 & 0.481 & -0.014 & & -0.005 \\ -0.015 & -0.016 & 0.247 & 0.797 & & \\ 0.002 & 0.003 & & 0.220 & 0.779 \\ -0.014 & -0.021 & 0.005 & & 0.247 \end{pmatrix} \tag{101}$$

while for $N = 240$ it is,

$$U^{(W,240)} = \begin{pmatrix} 1.000 & -0.007 & -0.007 & & & \\ -0.007 & 0.485 & 0.738 & & -0.005 & 0.012 \\ -0.007 & 0.002 & 0.485 & -0.010 & & -0.004 \\ -0.012 & -0.012 & 0.248 & 0.807 & & \\ 0.002 & 0.002 & & 0.226 & 0.793 \\ -0.013 & -0.016 & 0.003 & & 0.248 \end{pmatrix} \tag{102}$$

Entries smaller than 10^{-3} were ignored. We see that indeed the elements become closer to those of M as N is increased.

Appendix B: A Formula for $\mathcal{N}_W^{(po)}$

In this Appendix a formula that is more accurate than (63) will be presented. First define the matrix $\mathbf{R_p}(\mathbf{x})$,

$$\mathbf{R_p}(\mathbf{x}) = \frac{1}{R_p^+ - R_p^-} \begin{pmatrix} -2R_p^+ R_p^- & R_p^+ + R_p^- \\ R_p^+ + R_p^- & -2 \end{pmatrix} \tag{103}$$

with,

$$R_p^{\pm} = \frac{v_{p2}^{\pm}}{v_{p1}^{\pm}} \tag{104}$$

where $\mathbf{v}_p^{\pm}$ are the eigenvectors of the monodromy matrix for the p-th orbit. Then define

$$g^{(n)}(b_p) = \sum_{l=0}^{n} (-1)^{n-l} f_p^{(l)} \tag{105}$$

where the functions $f_p^{(l)}$ are,

$$f_p^{(l)} = e^{-\frac{i}{\hbar}\tilde{\mathbf{X}}\mathbf{R_p}\mathbf{X}} \frac{1}{l!} \left(\frac{\partial}{\partial z} \right)^l e^{\frac{i}{\hbar}\frac{1-z}{1+z}\tilde{\mathbf{X}}\mathbf{R_p}\mathbf{X}} \bigg|_{z=0} \tag{106}$$

with

$$b_p = b_p(\mathbf{x}) = \frac{i}{\hbar}\tilde{\mathbf{X}}\mathbf{R_p}\mathbf{X} \tag{107}$$

The first few $g^{(n)}$ are,

$$\begin{aligned} g^{(0)}(b_p) &= 1 \\ g^{(1)}(b_p) &= -1 - 2b_p \\ g^{(2)}(b_p) &= 1 + 4b_p + 2b_p^2 \end{aligned} \tag{108}$$

With the help of these, the periodic orbit contribution to (57) is written in the form, that is obtained after summation over repetitions is performed, namely[19],

$$\mathcal{N}_W^{(po)} = 4\pi \sum_{p.p.o} \sum_n \Delta^{(p,n)}(E) A(\mathbf{x}; E) g^{(n)}(b_p) e^{\frac{i}{\hbar} \tilde{\mathbf{X}} \mathbf{R_p} \mathbf{X}} \tag{109}$$

where the functions $\Delta^{(p,n)}(E)$ are similar to the spectral determinant except that the factor $(1 - t_p e^{-u_p n})$ is replaced by $-it_p e^{-u_p n}$. Thus,

$$\Delta^{(p,n)}(E) = -ie^{-i\pi \bar{N}(E)} \prod_{p' \neq p} \prod_j \left(1 - t_{p'} e^{-u_{p'} j}\right) \times$$
$$\prod_{j \neq n} \left(1 - t_p e^{-u_p j}\right) t_p e^{-u_p n} \tag{110}$$

and $\Delta^{(p)} = \Delta^{(p,0)}$.

The resulting more accurate versions of (94) and (95) are[19]:

$$W_\alpha(\mathbf{x}) \simeq \frac{4\pi A(\mathbf{x}, E_\alpha)}{h^2 \Delta'(E_\alpha)} \left\{ \frac{1}{4} \Delta_i(E_\alpha) + \Re \sum_{p,n} \Delta^{(p,n)}(E_\alpha) g_p^{(n)} e^{\frac{i}{\hbar} \tilde{\mathbf{X}} \mathbf{R_p} \mathbf{X}} \right\} \tag{111}$$

and

$$Y_p(E_\alpha) \simeq \frac{T_p}{\hbar \Delta'(E_\alpha)} \Re \sum_{n=0}^{\infty} \Delta^{(p,n)}(E). \tag{112}$$

Appendix C: Expanding Legendre Functions in e^{-L_g}

To prove (82) one starts with the integral representation of the relevant Legendre function[50]:

$$Q_{-\frac{1}{2}-ik}[\cosh d] = \frac{e^{ikd}}{\sqrt{2}} \int_0^\infty \frac{e^{iky} dy}{\sqrt{\cosh(d+y) - \cosh(d)}} \quad \text{for } d > 0 \; .$$

Expanding the denominator and doing the integrals term by term, one finds:

$$Q_{-\frac{1}{2}-ik}[\cosh d] = \sum_{n=0}^{\infty} \frac{(2n-1)!!}{(2n)!!} B\left[n + \frac{1}{2} - ik, \frac{1}{2}\right] e^{(ik - \frac{1}{2} - 2n)d} \; . \tag{113}$$

Since the d's to be used in the above expression are the $d_g(\zeta)$'s satisfying Eq. 74, one can write, after some straight-forward algebra:

$$e^{d_g(\zeta)} = \cosh^2 d_g^{\perp}(\zeta) e^{L_g} \cdot f\left[x, d_g^{\perp}(\zeta)\right]\Big|_{x=e^{-L_g}} \; ,$$

where

$$f\left[x, d_g^{\perp}(\zeta)\right] \equiv (1 - x)^2 \frac{1}{2} \left[1 + \sqrt{1 + \frac{4}{\cosh^2 d_g^{\perp}(\zeta)} \frac{x}{(1-x)^2} + \frac{2}{\cosh^2 d_g^{\perp}(\zeta)} \frac{x}{(1-x)^2}}\right] \; .$$

Now $\left[e^{d_g(\zeta)}\right]^{ik - \frac{1}{2} - 2n}$ can be expanded in powers of $x = e^{-L_g}$:

$$\left[e^{d_g(\zeta)}\right]^{ik - \frac{1}{2} - 2n} = \left[\frac{e^{-L_g}}{\cosh^2 d_g^{\perp}(\zeta)}\right]^{\frac{1}{2} + 2n - ik} \cdot \left\{f\left[x, d_g^{\perp}(\zeta)\right]\right\}^{ik - \frac{1}{2} - 2n} =$$
$$= \sum_{m=0}^{\infty} a_m \left[\frac{1}{\cosh^2 d_g^{\perp}(\zeta)}, ik - \frac{1}{2} - 2n\right] e^{-mL_g} \; , \tag{114}$$

where the first few $a_m[x,y]$'s are:

$$a_0[x,y] = 1 \; ,$$
$$a_1[x,y] = -2y + 2yx \; .$$

Since the $a_m[x,y]$'s are polynomials of y, one may write:

$$a_m[x,y] = a_m[x, \partial_\eta] e^{\eta y} \Big|_{\eta=0} \; ,$$

and bring (114) to a more convenient form:

$$\left[e^{d_g(\zeta)} \right]^{ik-\frac{1}{2}-2n} =$$
$$= \sum_{m=0}^{\infty} a_m \left[\frac{1}{\cosh^2 d_g^\perp(\zeta)}, \partial_\eta \right] \left(\frac{e^{-\eta}}{\cosh^2 d_g^\perp(\zeta)} \right)^{\frac{1}{2}+2n-ik} \times e^{ikL_g} e^{-(\frac{1}{2}+m+2n)L_g} \Bigg|_{\eta=0} \; . \qquad (115)$$

Inserting (115) into (113) one obtains:

$$Q_{-\frac{1}{2}-ik}[\cosh d_g(\zeta)] = \sum_{n=0}^{\infty} \sum_{m=0}^{\infty} \frac{(2n-1)!!}{(2n)!!} B\left[n + \frac{1}{2} - ik, \frac{1}{2} \right] \times$$
$$\times a_m \left[\frac{1}{\cosh^2 d_g^\perp(\zeta)}, \partial_\eta \right] \left(\frac{e^{-\eta}}{\cosh^2 d_g^\perp(\zeta)} \right)^{\frac{1}{2}+2n-ik} \times e^{ikL_g} e^{-(\frac{1}{2}+m+2n)L_g} \Bigg|_{\eta=0} \; . \qquad (116)$$

Changing summation variables in (116):

$$\sum_{n=0}^{\infty} \sum_{m=0}^{\infty} A_n B_m \xrightarrow{m'=m+2n} \sum_{n=0}^{\infty} \sum_{m'=2n}^{\infty} A_n B_{m'-2n} \longrightarrow \sum_{m'=0}^{\infty} \sum_{n=o}^{\left[\frac{m'}{2}\right]} A_n B_{m'-2n} \; , \qquad (117)$$

leads to Eq. 82.

Acknowledgments

The work presented in this contribution was done in collaboration with Oded Agam, Ophir Auslaender, Bertrand Georgeot and Richard E. Prange. It is my great pleasure to thank also M.V. Berry, D. Cohen, P. Cvitanović, B. Eckhardt, F. Haake, E. Heller, L. Kaplan, J. Keating, B. Mehlig, U. Smilansky and M. Zirnbauer for stimulating and informative discussions. I would like to thank in particular Ophir Auslaender, that helped to bring this manuscript to its final form. This work was supported in part by the U.S.-Israel Binational Science Foundation (BSF), the fund for the Promotion of Research at the Technion and the Minerva Center for Nonlinear Physics of Complex Systems.

REFERENCES

1.   M.C. Gutzwiller, *Chaos in Classical and Quantum Mechanics* (Springer, NY 1990).
2.   F. Haake, *Quantum Signatures of Chaos* (Springer, NY 1991).
3.   *Chaos and Quantum Physics*, Proc. of the Les-Houches Summer School, Session LII, 1989, Edited by M. J. Giannoni, A. Voros and J. Zinn-Justin (North Holland, Amsterdam, 1991).
4.   *Quantum Chaos*, Proc. of the International School of Physics "Enrico Fermi", Varenna, July 1991, Edited by G. Casati, I. Guarneri and U. Smilansky (North-Holland, N.Y., 1993).

5.   *Quantum Dynamics of Simple Systems*, Proc. of the 44th Scottish Universities Summer School in Physics, Stirling, August 1994, edited by G.L. Oppo, S.M. Barnett, E. Riis and M. Wilkinson.

6.   M. L. Mehta, *Random Matrices*, (Academic Press, New York, 1991).

7.   O. Bohigas, in Ref. 3.

8.   M.V. Berry, in Ref. 3.

9.   M.V. Berry, Proc. R. Soc. Lond. **A 400**, 229 (1985).

10. E. Bogomolny and J. Keating, Phys. Rev. Lett. **77**, 1472 (1996).

11. S. Ketemann, D. Klakow and U. Smilansky, J. Phys. **A 30**, 3643 (1997); U. Smilansky, Physica **D** (in press) (1997); and this volume.

12. A. V. Andreev and B. L. Altshuler, Phys. Rev. Lett. **75**, 902 (1995); O. Agam, B. L. Altshuler, and A. V. Andreev, Phys. Rev. Lett. **75**, 4389 (1995); A. V. Andreev, O. Agam, B. D. Simons and B. L. Altshuler, Phys. Rev. Lett. **76**, 3947 (1996); Nuclear Physics **B482**, 536 (1996); Phys. Rev. Lett. **79**, 1778 (1997); O. Agam, A. V. Andreev, and B. D. Simons, Chaos, Solitons & Fractals **8**, 1099 (1997); J. Math. Phys. **38**, 1982 (1997).

13. B.A. Muzykantskii and D.E. Khmelnitskii, JETP Lett. **62**, 76 (1995).

14. M.R. Zirnbauer, to be published.

15. M. C. Gutzwiller, J. Math. Phys. **8**, 1979 (1967); **10**, 1004 (1969); **11**, 1791 (1970); **12**, 343 (1971).

16. M. V. Berry and J. P. Keating, Proc. Roy. Soc. (London) **A 437**, 151 (1992).

17. E. B. Bogomolny, Nonlinearity **5**, 805 (1992).

18. E. Doron and U. Smilansky, Phys. Rev. Letters **68**, 1255 (1992); Nonlinearity **5**, 1055 (1992).

19. O. Agam and S. Fishman, J. Phys. **A 26**, 2113 (1993); corrigendum J. Phys. **A 26**, 6595 (1993).

20. C. M. Marcus, A. J. Rimberg, R. M. Westervelt, P. F. Hopkins, and A. C. Gossard, Phys. Rev. Lett. **69**, 509 (1992); A.M. Chang, H.U. Baranger, L.N. Pfeiffer and K.W. West, Phys. Rev. Lett. **73**, 2111 (1994); T. M. Fromhold, L. Eaves, F.W. Sheard, T. J. Foster, M. L. Leadbeater, and P. C. Main, Physica B **201**, 367 (1994) ; T.S. Monteiro, D. Delande, A.J. Fisher and G.S. Boebinger, Phys. Rev. **B 56**, 3913 (1997).

21. M.V. Berry, J. Phys. **A 10**, 2083 (1977).

22. S.W. McDonald and A.N. Kaufman, Phys. Rev. **A 37**, 3067 (1988).

23. E.J. Heller, Phys. Rev. Lett. **53,** 1515 (1984).

24. E.J. Heller in Ref. 3.

25. E.J. Heller, Lecture Notes in Physics **263**, 162 (1986).

26. O. Agam and S. Fishman, Phys. Rev. Lett. **73**, 806 (1994).

27. V.I. Arnold and A. Avez, *Ergodic Problems of Classical Mechanics*, (Addison-Wesley NY, 1989).

28. H.H. Hasegawa and W.C. Saphir, Phys. Rev. **A 46**, 7401 (1992).

29. D. Ruelle, *Statistical Mechanics, Thermodynamic Formalism*, (Addison-Wesley, Reading MA, 1978).

30. P. Cvitanović and B. Eckhardt, J. Phys. **A 24**, L237 (1991).

31. P. Gaspard, Phys. Rev. **E 53**, 4379 (1996).

32. M.V. Berry, in *New Trends in Nuclear Collective Dynamics*, eds: Y. Abe, H. Horiuchi, K. Matsuyanagi Springer proceedings in Physics, vol **58** pp183-186 (1992).

33. O. Agam and N. Brenner, J. Phys. **A 28**, 1345 (1995).

34. M.V. Berry, Phil. Trans. R. Soc. **A 287**, 237 (1977).

35. N.L. Balazs and A. Voros, Ann. Phys. (NY) **190**, 1 (1989).

36. M. Saraceno, Ann. Phys. **199**, 37 (1990).

37. F. Haake and J. Weber, private communication.

38. F. Smithies, *Integral Equations*, Cambridge Tracts in Mathematics and Mathematical Physics **49**, (Cambridge University Press, Cambridge, 1962).

39. S. Fishman, B. Georgeot and R.E. Prange, J. Phys. **A29**, 919 (1996).

40. O. M. Auslaender and S. Fishman, Technion preprint, 1997, *chao-dyn/9707021*, to be published in Physica **D**.

41. B. Georgeot and R. E. Prange, Phys. Rev. Lett. **74**, 2851 (1995).

42. M. V. Berry, Proc. R. Soc. Lond. **A423,** 219 (1989); **A424,** 279 (1989).

43. E. B. Bogomolny, Physica **D31**, 169 (1988).

44. A.F. Beardon, *The Geometry of Discrete Groups*, Springer-Verlag, New York, 1983.

45. A. Selberg, J. Indian Math. Soc. **20**, 47 (1956).

46. D.A. Hejhal, Duke Math. J.**43**, 441 (1976).

47. D.A. Hejhal, *The Selberg Trace Formula for PSL(2,R)*, volume *548 & 1001*, Springer, Berlin, 1976 & 1983.

48. N.L. Balazs and A. Voros, Phys. Rep.**143**, 109 (1986).

49. E.B. Bogomolny, in Ref. 5.

50. I.S. Gradshteyn and I.M. Ryzhik, *Table of Integrals, Series, and Products.* Academic Press, New-York, 1980.

51. M. Hamermesh, *Group theory and it's application to physical problems*, Addison-Wesley, 1962.

52. H.S.M. Coxeter and W.O.J. Moser, *Generators and relations for discrete groups*, Springer-Verlag, 1965.

53. M. Feingold, R. G. Littlejohn, S. B. Solina and J. S. Pehing, Phys. Lett. A **146**, 199 (1990); M. Feingold, Z. Phys. B **95**, 121 (1994), O. Agam, unpublished.

54. J. H. Hannay and A. M. Ozorio de Almeida, J. Phys. A **17**, 3429 (1984).

55. R. Aurich and F. Steiner, Physica D *48*, 445 (1991); *64*, 185 (1993); Chaos, Solitons & Fractals, **5**, 229(1995).

56. S. Tomsovic and E.J. Heller, Phys. Rev. Lett. **67**, 664 (1991); Phys. Rev. E **47**, 282 (1993); *Physics Today*, **46**(7), 38 (1993); P.W. O'Connor, S. Tomsovic and E.J. Heller, Physica D **55**, 340 (1992).

57. L. Kaplan and E.J. Heller, Ann. Phy. **264**, 171 and this volume.

58. D. Klakow and U. Smilansky, J. Phys. A **29**, 3213 (1996).

59. T.M. Antonsen, Jr., E. Ott, Q. Chen, and R.N. Oerter, Phys. Rev. E **51**, 111 (1995).

60. Z. Rudnick and P. Sarnak, Comm. Math. Phys **161**, 195 (1994).

61. B. Li, Phys. Rev. E **55**, 5376 (1997).

62. J. Stein and H. -J. Stöckman, Phys. Rev. Lett. **68**, 2867 (1992); S. Sridhar and E. J. Heller, Phys. Rev. A **46**, 1728 (1992).

63. K. Müller and D. Wintgen, J. Phys. B **27**, 2693 (1994).

64. J.A. Folk, S.R. Patel, S.F. Godijn, A.G. Huibers, S.M. Cronenwelt, C.M. Marcus, K. Campman and A.C. Gossard, Phys. Rev. Lett. **76**, 1699 (1996).

65. O. Agam, Phys. Rev. B **54**, 2607 (1996).

WAVE FUNCTIONS IN CHAOTIC BILLIARDS: SUPERSYMMETRY APPROACH

K.B. Efetov

Fakultät für Physik und Astronomie, Ruhr-Universität Bochum,
Germany and
Landau Institute for Theoretical Physics, Moscow, Russia

INTRODUCTION

Due to recent progress in semiconductor technology, experimental study of nanoscale electronic systems has become a very popular field of research. In small devices studied experimentally, electron motion at low temperatures is coherent and the quantum nature of carriers can become very important [1, 2]. During the previous decade, the quantum effects in transport have been mainly explored in systems where they give small corrections to values of physical quantities expected on the basis of classical mechanics. Weak localization effects [3] and mesoscopic universal conductance fluctuations [4] are the best known examples.

In more recent studies [5] of smaller structures put into a weak (tunneling) contact to the bulk electrodes and cooled down to the temperatures of the order of tens of mK, the resonant tunneling regime of the transport through a single discrete level in a quantum dot has been achieved. Since the value of the resonance tunneling conductance is determined by the amplitudes of the resonance state wave functions in the vicinity of the contact, fluctuations and spatial structure of the single particle eigenstates in a dot become important observables, especially regarding a rich experimental information that is already available in the literature [6, 7].

So, studying experimentally statistics of peak heights in the resonance regime one can get an information about fluctuations of the wave functions in the quantum dot. Usually this regime is achieved when Coulomb effects are important and transport is basically possible only at certain values of the gate voltage. A Breit-Wigner type formula used to describe the resonance conductance [8, 9] contains directly the amplitudes of the wave functions near the contacts, and the distribution function of the conductances is directly related to a joint distribution function of the amplitudes at different space points.

At present, it is quite clear that the character of the fluctuations in the conductance peaks series depends on whether the corresponding classical motion of a carrier in quantum dot is chaotic or regular. In the latter case calculation of the wave functions should follow a straightforward procedure, whereas a description of the wave

Supersymmetry and Trace Formulae: Chaos and Disorder
Edited by Lerner *et al.*, Kluwer Academic / Plenum Publishers, New York, 1999

functions in the regime of chaotic dynamics demands considerable efforts. By studying experimentally the conductance fluctuations for "chaotic" dots, one can check theoretical results for statistics of the amplitudes of the chaotic electron waves in mesoscopic dots—at least, would a single-particle description work being applied to the system of interacting electrons.

Besides the measurements on the quantum dots, it is relevant to mention a series of experiments on microwave scattering in metallic cavities in which the local amplitudes of electromagnetic waves were measured and their statistics were analyzed [10, 11]. Being confined to a thin slab, the electromagnetic waves are strongly polarized and obey the same equation as the Schrödinger equation for the electron. One can make either ballistic cavities where the electromagnetic waves are scattered by the walls or add into the cavity small pieces of a good metal, which models impurities. Changing an impurity configuration or sweeping the frequency and passing from one resonance to another, one can collect statistics of the eigenmodes intensities at some fixed point of a sample.

Both types of the experiments (which can be classified as mesoscopic) provide a good reason for studying in great detail fluctuations of wave functions in chaotic or disordered confined systems. This question has been addressed quite long ago in nuclear physics where it was suggested to describe fluctuations of wave functions by fluctuations of eigenvectors of random matrices [12, 13]. Using this hypothesis one comes to the result that depending on the presence or absence of the time reversal invariance in the system (i.e., of a magnetic field), the eigenstates of the random matrices can be described as either real or complex vectors and have a purely Gaussian distribution of the amplitudes of their projections onto some arbitrarily chosen direction in the basis (Porter-Thomas distribution in the random matrix theory (RMT)), which is equivalent to the universal distribution of local densities of single particle wave functions in the chaotic regime. Intensive numerical studies of a large number of high-lying eigenstates of confined systems, such as "quantum billiards" [14, 15], have confirmed the Gaussian distribution of local wave function amplitudes, and this has been observed directly in the microwave experiments [10].

Another phenomenological approach related to the RMT that also confirms the Gaussian distribution of the wave functions amplitudes has been proposed by Berry. It is based on the assumption that the local structure of the eigenstates can be represented as a superposition of an infinite number of plane waves with random phases and equal wave number [16]. Originally, the applicability of such a conjecture was justifies by classical ergodicity of chaotic systems. For ergodic systems one can assume that relevant classical trajectories are typical orbits that pass close to all points on the energy surface corresponding to the energy ε of the state with the wave function φ and, in the semiclassical approximation, this leads to random phases. Using this approach one can describe spatial correlations of the amplitudes of the wave functions of a chaotic billiard [17] and show that their behavior is similar to that of the Friedel oscillation [18].

Both the advantage and disadvantage of the phenomenology is related to the statistical equivalence of eigenstates that is built into the construction of the random matrix substituting the real dynamics or stands behind Berry's conjecture. This reveals a set of universal features of chaos in the limits where they do exist but hides peculiarities of physically different systems in the cases when universality is broken. In particular, the phenomenology can be helpful in the limiting cases of the orthogonal and unitary ensembles corresponding to time reversal invariant systems or to systems where this invariance is completely broken but not in the crossover regime (quantum billiard in a weak magnetic field). Moreover, they cannot be used in situations where localization effects due to real disorder become important and the assumption of the statistical equivalence of the eigenstates is no longer valid.

As a result of the disadvantages of the phenomenological approaches, a derivation of statistics of amplitudes of the wave functions directly from a well defined model without using additional hypotheses (i.e., from the first principles) is desirable, and this requires development of other analytical methods. Fortunately, this goal can be achieved by a modification of the supersymmetry technique [19]. In the present lecture, it is demonstrated how the supersymmetry technique can be applied to the problem of the statistics of wave functions in disordered and chaotic confined systems. The presentation is based on previous works [9, 20-26]. More information can be found in the review [27] and the book [28].

WAVE FUNCTIONS AND NON-LINEAR SUPERMATRIX σ-MODEL

Below, wave functions in models with disorder will be studied. In all cases consideration is restricted to systems of non-interacting particles in a finite volume. The assumption is very well justified if we consider electromagnetic or sound waves in a box. At the same time, the validity of it is less clear for electrons in a quantum dot where both electron-electron and electron-phonon interactions can be quite important. Nevertheless, at low temperatures the inelastic mean free path can be large. Even if the electron-electron interaction is strong one can argue that instead of considering electrons one can do calculations for quasi-particles using the Landau theory of Fermi-liquids. So, very often the one-particle approximation serves as a good description of interesting physical phenomena.

To simplify the discussion one can speak about electrons in a box and study solutions of the Schrödinger equation although most of the results obtained below are also applicable to electromagnetic and sound waves. The basic equation can be written in the form

$$\hat{H}\varphi_\alpha\left(\mathbf{r}\right) = \varepsilon_\alpha\varphi_\alpha\left(\mathbf{r}\right), \qquad \hat{H} = \hat{H}_0 + U\left(\mathbf{r}\right) \tag{1}$$

where $U\left(\mathbf{r}\right)$ is an interaction with impurities. In principle, an external magnetic field can be applied and the potential can include magnetic and spin-orbit interactions. In Eq. (1), ε_α are the eigenenergies measured from the energy ε fixed by an external observer (e.g., the Fermi energy in the bulk electrodes); $\varphi_\alpha\left(\mathbf{r}\right)$ are the corresponding eigenfunctions, and

$$\hat{H}_0 = \varepsilon\left(-i\nabla - \frac{e}{c}\mathbf{A}\right) - \varepsilon \tag{2}$$

with $\varepsilon\left(\mathbf{p}\right)$ being the energy of free motion. Below the vector potential $\mathbf{A}$ is used in the London gauge

$$div\mathbf{A} = 0, \qquad \mathbf{An}|_S = 0 \tag{3}$$

where $\mathbf{n}$ is the unit vector normal to the surface S of the sample.

The random potential $U\left(\mathbf{r}\right)$ is assumed to be Gaussian and satisfy the following relations

$$\langle U\left(\mathbf{r}\right)\rangle = 0, \qquad \langle U\left(\mathbf{r}\right)U\left(\mathbf{r}'\right)\rangle = \frac{1}{2\pi\nu}\delta\left(\mathbf{r} - \mathbf{r}'\right) \tag{4}$$

where τ is the mean free time and ν is the average density of states at the energy ε. The angular brackets stand for the averaging procedure.

The eigenstates problem, Eq. (1), should be complemented by the boundary conditions for the wave functions. It is assumed that the electron is confined in a finite region of space and therefore the energy spectrum is discrete.

The most complete information about statistical and correlation properties of the wave functions is contained in the N-point correlation functions $f_N(p_1, ..., p_N)$ defined as

$$f_N(p_1, ..., p_N; \mathbf{r}_1, ..., \mathbf{r}_N) = \Delta \left\langle \sum_a \delta(\varepsilon - \varepsilon_a) \prod_{n=1}^{N} \delta\left(p_n - |\varphi_\alpha(\mathbf{r}_n)|^2\right) \right\rangle \tag{5}$$

In Eq. (5), $\Delta = (\nu V)^{-1}$ is the mean level spacing which is finite in a confined system. In this definition, spin degrees of freedom are not included. Those, who are interested in a generalization to the case when the spin degrees of freedom are important can find some details in Ref. [25].

The function $f_N(p_1, ..., p_N)$ measures the probability of given amplitudes of the wave function corresponding to the energy ε at N different coordinates $\mathbf{r}_N$. When speaking about the problem of the distribution of irradiation in a chaotic cavity, this function describes a probability to detect a given configuration of local electromagnetic field intensities. In principle, using the supersymmetry technique one can try to compute the function $f_N(p_1, ..., p_N)$ for arbitrary N, but the functions f_1 and f_2 are of special importance for physical applications. For example, the distribution of conductance peaks in the resonance regime can be directly related to f_2 [8], whereas $f_1(p)$ has been directly measured in the microwave experiments [10]. So, let us concentrate on calculation of these two functions. They are related to each other by a simple formula

$$f_1(p, \mathbf{r}_1) = \int_{-\infty}^{\infty} f_2(p, p'; \mathbf{r}_1, \mathbf{r}') \, dp' \tag{6}$$

In some cases it is interesting to calculate the so-called inverse participation numbers (IPN) P_m defined as

$$P_m = \left\langle \sum_\alpha |\varphi_\alpha(\mathbf{r})|^{2m} \delta(\varepsilon - \varepsilon_\alpha) \right\rangle \equiv \int_0^{\infty} p^m f_1(p, \mathbf{r}) \, dp \tag{7}$$

To apply the supersymmetry technique to the calculation of physical quantities, one has to express these quantities in terms of retarded G_ε^R and advanced G_ε^A Green functions. These functions are solutions of equations

$$\left(\hat{H} \pm i\frac{\gamma}{2}\right) G_\varepsilon^{R,A}(\mathbf{r}, \mathbf{r}') = \delta(\mathbf{r} - \mathbf{r}') \tag{8}$$

Each of the functions $G_\varepsilon^{R,A}(\mathbf{r}, \mathbf{r}')$ is analytic in the upper (lower) half plane of complex ε. The broadening γ is introduced here artificially and is assumed to be the same for all quantum levels ε_α. The spectral expansion of the Green functions $G_\varepsilon^{R,A}$ can be written as

$$G_\varepsilon^{R,A}(\mathbf{r}, \mathbf{r}') = \sum_\alpha \frac{\varphi_\alpha(\mathbf{r}) \varphi_\alpha^*(\mathbf{r}')}{\varepsilon - \varepsilon_\alpha \pm i\gamma/2} \tag{9}$$

Using Eq. (9) one can express without difficulties the distribution function f_2 in terms of the Green functions.

First, one should use the following equality

$$\lim_{\gamma \to 0, \beta \to 1} \int_0^1 dt \frac{d}{d\beta} \left[\beta \frac{(i\gamma\beta t)^m}{(x + i\gamma/2)^m} \frac{(-i\gamma\beta(1-t)^n)}{(x - i\gamma/2)^n} \frac{\gamma}{x^2 + (\gamma/2)^2} \right] = 2\pi\delta(x) \tag{10}$$

To prove Eq. (10) one can integrate, first, over t which gives

$$\int_0^1 t^m(1-t)^n \, dt = \frac{m! \, n!}{(m+n+1)!} \tag{11}$$

and then over x using the formula

$$\gamma^{m+n+1} i^{m-n} \int_{-\infty}^{\infty} \frac{dx}{(x + i\gamma/2)^{m+1} (x - i\gamma/2)^{n+1}} = 2\pi \frac{(m+n)!}{m!n!} \qquad (12)$$

Therefore, using Eq. (9) one can rewrite the distribution function f_2 in the form

$$f_2 (p_1, p_2; \mathbf{r}_1, \mathbf{r}_2) = \frac{\Delta}{\pi} \lim_{\gamma \to 0, \beta \to 1} \frac{d}{d\beta} < \beta \int_0^1 dt \int d\mathbf{r} Im \left[G_\varepsilon^A (\mathbf{r}, \mathbf{r}) \right]$$

$$\times \delta \left(p_1 + iY_A G_\varepsilon^A (\mathbf{r}_1, \mathbf{r}_1) \right) \delta \left(p_2 - iY_R G_\varepsilon^R (\mathbf{r}_2, \mathbf{r}_2) \right) > \qquad (13)$$

where $Y_A = \gamma \beta V t$ and $Y_R = \gamma \beta V (1 - t)$.

As soon as the physical quantity of interest has been expressed in terms of the Green functions one can try to express this quantity as an integral over supervectors containing both commuting s and anticommuting χ elements [28, 29]. To rewrite Eq. (13) in such a form, one can expand the δ-functions in a series in the Green functions $G_\varepsilon^{R,A}$ and use the representation of the latter in terms of functional integrals over the supervectors. To do this, one can use a 4-component supervector for each of the retarded and advanced Green functions, so that the minimal size of the superspace is 8 and the structure of the supervectors can be written as

$$\psi = \begin{pmatrix} \psi_1 \\ \psi_2 \end{pmatrix}, \ \psi_m = \begin{pmatrix} \Delta^m \\ v^m \end{pmatrix}, \ \Delta^m = \frac{1}{\sqrt{2}} \begin{pmatrix} \chi^{m*} \\ \chi^m \end{pmatrix}, \ v^m = \frac{1}{\sqrt{2}} \begin{pmatrix} s^{m*} \\ s^m \end{pmatrix} \qquad (14)$$

$$\bar{\psi} = \left(\bar{\psi}_1, -\bar{\psi}_2 \right), \ \bar{\Delta}^m = \frac{1}{\sqrt{2}} \left(\chi_m, \chi_m^* \right), \ \bar{s}_m = \frac{1}{\sqrt{2}} \left(s_m, s_m^* \right), \ m = 1, 2$$

In Eq. (14), the superscripts 1 and 2 relate to the advanced and retarded Green Functions, respectively.. The components χ and χ^* are anticommuting Grassmann variables whereas s and s^* are conventional complex numbers conjugate with respect to each other.

In principle, to obtain a proper expression for a product of a large number of Green functions related to different energies, one would have to increase the size of the superspace, which would make any further computation extremely difficult. Fortunately, the distribution functions f_N describing properties of a single quantum level are all expressed in terms of $G_\varepsilon^{R,A}$ taken at exactly the same energy. This enables us to operate with the supervectors ψ, Eq. (14), without increasing their dimensions. Within the supersymmetry technique, one should use the Lagrangian L in the form

$$L [\psi] = - \int d\mathbf{r} \bar{\psi} (\mathbf{r}) \left[\hat{H}_0 + U (\mathbf{r}) + \frac{i\gamma}{2} \Lambda \right] \psi (\mathbf{r}) \qquad (15)$$

where 8×8 supermatrix Λ is

$$\Lambda = \begin{pmatrix} 1 & 0 \\ 0 & -1 \end{pmatrix}, \qquad (16)$$

and represent any product of Green functions that would appear in the formal series expansion of Eq. (13)

$$B_{mn} = \left\langle G_\varepsilon^A (\mathbf{r}, \mathbf{r}) \left[G_\varepsilon^A (\mathbf{r}_1, \mathbf{r}_1) \right]^m \left[G_\varepsilon^R (\mathbf{r}_2, \mathbf{r}_2) \right]^n \right\rangle \qquad (17)$$

as

$$m!n!i^{n-m-1} B_{mn} = \left\langle \int \chi_1 (\mathbf{r}) \chi_1^* (\mathbf{r}) |s_1 (\mathbf{r}_1)|^{2n} |s_2 (\mathbf{r}_2)|^{2m} e^{-iL[\psi]} D\psi \right\rangle \qquad (18)$$

Performing the averaging in Eq. (18) one obtains an effective "interaction" $(\bar{\psi}\psi)^2$ in the Lagrangian which should be decoupled by Gaussian integration over 8×8 supermatrices Q slowly varying in space. After that one comes to an effective Lagrangian $L_{eff}[\psi, Q]$

$$L_{eff}[\psi, Q] = -\int \left[\bar{\psi}(\mathbf{r}) \left(\hat{H}_0 + \frac{i\gamma}{2}\Lambda + \frac{i}{2\tau}Q \right) \psi(\mathbf{r}) \right] dr \tag{19}$$

As usual, one can perform the Gaussian integration using the Wick theorem. In principle, not only pairing of the type $\langle \psi_\alpha(\mathbf{r}_n) \bar{\psi}_\alpha(\mathbf{r}_n) \rangle$, such that both ψ are taken at the same point is possible, but also pairings like $\langle \psi_\alpha(\mathbf{r}_n) \bar{\psi}_\alpha(\mathbf{r}_m) \rangle$, $n \neq m$, should be taken into account. The latter pairing determines spatial dependence of the correlations of the wave functions. They decay already at distances of the order of wave length and are even exponentially suppressed at distances longer than the mean free path l. As soon as one studies correlations at distances $|\mathbf{r} - \mathbf{r}'| > l$, only the pairings at coinciding points need to be considered and for the moment let us restrict ourselves to this case. Then, the Gaussian integration over ψ in Eq. (18) with the Lagrangian L replaced by the Lagrangian L_{eff}, Eq. (19), can be easily performed. After collecting all terms B_{mn} one can reduce the function f_2 to the form

$$f_2(p_1, p_2; |\mathbf{r}_1 - \mathbf{r}_2| \gg l) = \lim_{\gamma \to 0, \beta \to 1} \frac{d}{d\beta} < \frac{\beta}{2V} \int \frac{d\zeta_1 d\zeta_2}{(2\pi)^2} \int_0^1 dt \int d\mathbf{r} \left(Q_{11}^{11}(\mathbf{r}) - Q_{11}^{22}(\mathbf{r}) \right)$$

$$\times \delta\left(p_1 - \frac{t\gamma\beta}{2V\Delta}\bar{z}_1 Q(\mathbf{r}_1) z_1 \right) \delta\left(p_2 + \frac{(1-t)\gamma\beta}{2V\Delta}\bar{z}_2 Q(\mathbf{r}_2) z_2 \right) >_Q \tag{20}$$

where $\langle ... \rangle_Q$ stands for the integration over the supermatrices Q

$$\langle ... \rangle_Q = \int (\cdots) \exp(-F[Q]) \, dQ \tag{21}$$

with the free energy functional $F[Q]$,

$$F[Q] = \frac{\pi\nu}{8}STr \int \left[D\left(\nabla Q - \frac{ie}{c}\mathbf{A}[Q, \tau_3] \right)^2 - \gamma\Lambda Q \right] dr, \tag{22}$$

In Eqs. (20-22) STr is supertrace and D is the classical diffusion coefficient. The supermatrix Q obeys the constraint $Q^2 = 1$ and can be represented in the form

$$Q = U\Lambda\bar{U} \tag{23}$$

where U is a unitary matrix satisfying the condition $U\bar{U} = 1$. The overbar stands for an operation of conjugation. The definition of this operation as well as an explicit form of the supermatrix Q can be found in Refs.[29, 28]. The vectors $z_{1,2}$ in Eq. (20) have the form

$$\bar{z}_1 = \left(0, 0, e^{i\zeta_1}, e^{-i\zeta_1}, 0, 0, 0, 0 \right), \qquad \bar{z}_2 = \left(0, 0, 0, 0, 0, 0, e^{i\zeta_2}, e^{-i\zeta_2} \right) \tag{24}$$

and τ_α, $\alpha = 1, 2, 3$ are standard Pauli matrices. Eqs. (20-24) completely reduce the problem of computation of the distribution function f_2 to functional integration over supermatrices Q within the non-linear σ-model. One should remember, however, that this has become possible due to the discreteness of the spectrum.

The same can be repeated for the more simple function f_1. As a result, one obtains

$$f_1(p_1, \mathbf{r}_1) = \frac{1}{4V} \int \frac{d\zeta_1 d\zeta_2}{(2\pi)^2} \lim_{\gamma \to 0} < \int d\mathbf{r} \left(\bar{z}_1 (Q(\mathbf{r}) - \Lambda) z_1 \right)$$

$$\times \delta \left(p - \frac{\gamma}{2V\Delta} \bar{z}_2 Q\left(\mathbf{r}_1\right) z_2 \right) >_Q \tag{25}$$

In the next sections, some new results obtained with the supermatrix formulation of the problem, Eqs. (20-25), are presented.

RESULTS OBTAINED WITH THE ZERO-DIMENSIONAL σ-MODEL

As usual [29, 28], the most simple are calculations with the zero-dimensional ($0D$) version of the σ-model that can be obtained taking into account only the zero space harmonics of Q in Eqs. (21, 22). The $0D$ σ-model gives the same results for the level statistics as the RMT [29, 30, 28]. As concerns the wave functions, RMT predicts a Gaussian distribution of the amplitudes of the wave functions. The function $f_1\left(p,\mathbf{r}\right)$ corresponding to this result can be written as [12]

$$f_1\left(p,\mathbf{r}\right) = \begin{cases} \sqrt{\frac{V}{2\pi p}} \exp\left(-\tfrac{1}{2}Vp\right), & \textit{orthogonal ensemble} \\ V\exp\left(-Vp\right), & \textit{unitary ensemble} \end{cases} \tag{26}$$

The same approach leads to the result that the function f_2 taken at different space points is merely the product of the functions f_1, Eq. (26). The question about behavior of the distribution functions in the crossover regime has not been addressed until recently.

Using the $0D$ σ-model one can confirm the results of the RMT but also describe the crossover regime as well as calculate the function $f_2\left(p_1;p_2;\mathbf{r}_1,\mathbf{r}_2\right)$ at arbitrary distances. Moreover, one can go beyond the $0D$ σ-model and consider the wave functions statistics in a higher dimension provided these functions are localized and the spectrum is discrete. The possibility to go beyond the $0D$ σ-model is also important in finite samples if one wants to consider the probability of having very high amplitudes.

Now, let us present the basic steps of the calculations. Although explicit results will be obtained in this section for the $0D$ σ-model only, it is reasonable to derive here several formulae that will be also used for more general considerations.

Calculating quantities characterizing wave functions it is convenient to integrate first over the zero "diffuson" mode and obtain a new *reduced* σ-model. This procedure is important because one can take the limit $\gamma \to 0$ before doing further computations. The reduced σ-model obtained in this way contains the "cooperon" degrees of freedom and non-zero harmonics of space variations of "diffuson" elements. To distinguish between the diffuson and cooperon degrees of freedom it is convenient to use the parametrization of Refs. [31, 32, 28]. Within this parametrization the supermatrix Q, Eq. (23), is represented in the form

$$Q = U_d Q_c \bar{U}_d,$$

$$Q_c = \begin{pmatrix} u_c & 0 \\ 0 & v_c \end{pmatrix} \begin{pmatrix} \cos\hat{\theta}_c & i\sin\hat{\theta}_c C_0 \\ -i\sin\hat{\theta}_c C_0^T & -\cos\hat{\theta}_c \end{pmatrix} \begin{pmatrix} \bar{u}_c & 0 \\ 0 & \bar{v}_c \end{pmatrix} \tag{27}$$

$$U_d = \begin{pmatrix} u_d & 0 \\ 0 & v_d \end{pmatrix} \begin{pmatrix} \cos\left(\hat{\theta}_d/2\right) & -i\sin\left(\hat{\theta}_d/2\right) \\ -i\sin\left(\theta_d/2\right) & \cos\left(\hat{\theta}_d/2\right) \end{pmatrix} \begin{pmatrix} \bar{u}_{2d} & 0 \\ 0 & 1 \end{pmatrix}$$

where the matrix C_0 is

$$C_0 = \begin{pmatrix} -i\tau_2 & 0 \\ 0 & \tau_1 \end{pmatrix}$$

T stands for transposition, and the τ_α, $\alpha = 1, 2, 3$ are the standard Pauli matrices.

The supermatrix U_d in Eq. (27) commutes with the matrix τ_3 in Eq. (22) and corresponds to the diffuson degrees of freedom. The "eigenvalue" matrix $\hat{\theta}_{c,d}$ has the form

$$\hat{\theta}_{c,d} = \begin{pmatrix} \theta_{c,d} & 0 \\ 0 & i\theta_{1c,d} \end{pmatrix}, \qquad \begin{array}{l} 0 < \theta_c < \pi/2,\ 0 < \theta_d < \pi, \\ 0 < \theta_{1c,d} < \infty \end{array} \tag{28}$$

The explicit form of the matrices $u_{c,d}$, $v_{c,d}$ as well as the Jacobian corresponding to the parametrization, Eqs. (27, 28), can be found in Refs. [31, 32, 28].

To derive the reduced σ-model, one should represent the supermatrix field $Q(\mathbf{r})$ at any point $\mathbf{r}$ inside the sample in the form

$$Q(\mathbf{r}) = U_d(\mathbf{r}_1)\,\tilde{Q}(\mathbf{r})\,\bar{U}_d(\mathbf{r}_1) \tag{29}$$

The integration over the zero diffuson mode is equivalent to integration of Eqs. (20, 25) over all possible rotations of the sample as a whole in the space of the supermatrices U_d. Due to the δ-functional form of the pre-exponential part of the expressions involved, only large $\theta_d(\mathbf{r}_1)$, such that $\exp\theta_d \sim \Delta/\gamma \to \infty$, contribute to the integrals in Eqs. (20, 25). As a result, the general σ-model transforms into the reduced one with a modified free energy functional. It is easier to follow the derivation calculating the distribution function f_1. In particular, the second term $F_2[Q]$ in $F[Q]$ in Eq. (22) containing infinitely small γ converts after the integration of $\delta\left(p - (\gamma/2V\Delta)\,\bar{z}_1 Q(\mathbf{r}_1) z_1\right)$ over all elements of U_d into

$$F_2[Q] \to F_2\left[\tilde{Q}\right] = \frac{p}{2\left(\bar{z}_1\Lambda\Pi\tilde{Q}(\mathbf{r}_1)z_1\right)}\int STr\left(\Lambda\Pi\tilde{Q}(\mathbf{r})\right)d\mathbf{r} \tag{30}$$

where the matrix Π has the form

$$\Pi = \begin{pmatrix} \pi_b & \pi_b \\ \pi_b & \pi_b \end{pmatrix}, \qquad \pi_b = \begin{pmatrix} 0 & 0 \\ 0 & 1 \end{pmatrix}$$

The term described by Eq. (30) plays the role of the interaction with an "external field" breaking the symmetry between bosonic and fermionic degrees of freedom of the field Q. Note that now, for the field $\tilde{Q}(\mathbf{r})$ parametrized as $\tilde{Q} = \tilde{U}_d\tilde{U}_c\Lambda\overline{\tilde{U}}_c\overline{\tilde{U}}_d$ one has the new boundary condition

$$U_d(\mathbf{r}_1) = 1 \tag{31}$$

On the contrary, the gradient term $F_1[Q]$ of the free energy $F[Q]$ in Eq. (22) does not change its form

$$F_1[Q] \to F_1\left[\tilde{Q}\right] = \frac{\pi\nu D}{8}STr\int\left(\nabla\tilde{Q} - \frac{ie}{c}\mathbf{A}\left[\tilde{Q},\tau_3\right]\right)^2 d\mathbf{r} \tag{32}$$

As concerns the form of the pre-exponential factor in the generating functional of the reduced σ-model, it depends on the index N of the joint distribution and can be sufficiently simplified in several special cases, in particular, when studying the regime of crossover between the orthogonal and unitary symmetry classes in the approximation of the $0D$ σ-model. Within this scheme one can neglect spatial variations of the field $\tilde{Q}$, so that $\tilde{Q}(\mathbf{r}) \equiv \tilde{Q}(\mathbf{r}_1)$. Then, the free energy of the lowest "cooperon" mode is described as

$$F_1\left[\tilde{Q}\right] = -\left(\frac{X}{4}\right)^2 STr\left(\left[\tilde{Q},\tau_3\right]^2\right) = X^2\left(\lambda_{1c}^2 - \lambda_c^2\right), \tag{33}$$

$$X^2 = (2\pi)^3 \frac{D}{\Delta} \int \frac{d\mathbf{r}}{V} \left(\frac{\mathbf{A}(\mathbf{r})}{\phi_0} \right)^2 = \alpha_g \frac{\phi^2}{\phi_0^2} \frac{E_c}{\Delta}$$

where $\phi_0 = h/ce$ is the flux quantum, α_g is a sample geometry dependent factor, D is the classical diffusion coefficient, E_c is the Thouless energy proportional to the inverse time of flight through the system L^2/D.

Let us calculate now the function $f_2(p_1, p_2; \mathbf{r}_1, \mathbf{r}_2)$ at two points separated by the distance $|\mathbf{r}_1 - \mathbf{r}_2|$ much larger than the wave length. In principle, it can be even larger than the mean free path l. It is interesting to compare it with the product of two functions f_1 and therefore one should find the proper expression also for this function. Using Eqs. (30, 33) and carrying out all possible integrations one can represent the joint distribution function f_N, $N = 1, 2$ in the form [26, 27]

$$f_N = V^N \int_1^\infty B(X, x) \prod_{n=1}^{N} M(V p_n, x) \, dx, \tag{34}$$

where

$$B(X, x) = X^2 \left[\left(xX^2 - 1 \right) \Phi_2(X) + \Phi_1(X) \right] e^{-X^2(x-1)},$$

$$M(q, x) = \sqrt{x} \exp(-qx) I_0 \left(q\sqrt{x^2 - x} \right)$$

Here, $I_0(x)$ is the Bessel function, and

$$\Phi_1(X) = \frac{e^{-X^2}}{X} \int_0^X e^{y^2} \, dy, \qquad \Phi_2(X) = \frac{1 - \Phi_1(X)}{X^2}$$

Eq. (34) has a rather complicated form and therefore, it is interesting to discuss its limits. First of all, let us compare the result with the predictions of the RMT. The limit $X \to \infty$ should correspond to the unitary case. In this limit essential $(x - 1)$ in the integral in Eq. (34) are of the order of X^{-2} and the variable x in the function $M(q, x)$ can be replaced by 1. Calculation of the remaining integral of the function $B(X, x)$ is trivial and one obtains

$$f_{2unit}(p_1, p_2; \mathbf{r}_1, \mathbf{r}_2)|_{|\mathbf{r}_1 - \mathbf{r}_2| \to \infty} = f_{1unit}(p_1) \, f_{1unit}(p_2) \tag{35}$$

where the function $f_{1unit}(p)$ is the one-point distribution function for the unitary ensemble determined by the second line of Eq. (26).

In the opposite limit $X \to 0$ one can write approximately

$$B(x) \simeq \frac{X^2}{3} \left(1 + 2xX^2 \right) \exp\left(-X^2 x\right) \tag{36}$$

Using the asymptotic form of the Bessel function I_0 at large x

$$I_0 \left(q\sqrt{x^2 - x} \right) \simeq \frac{\exp\left(q(x - 1/2) \right)}{\sqrt{2\pi}} \tag{37}$$

one can see that essential x in the integral in Eq. (34) are of the order of X^{-2}. Substituting Eqs. (36, 37) into Eq. (34) one obtains after integration over x

$$f_{2orth}(p_1, p_2; \mathbf{r}_1, \mathbf{r}_2)|_{|\mathbf{r}_1 - \mathbf{r}_2| \to \infty} = f_{1orth}(p_1) \, f_{2orth}(p_2) \tag{38}$$

where $f_{1orth}(p)$ is the one-point distribution function for the orthogonal ensemble given by the first line of Eq. (26). (Calculating directly the one-point distribution functions f_1 from Eqs. (34) one comes also to the result of the RMT, Eq. (26).

In principle, one can also find using the supersymmetry technique the function $f_2(p_1, p_2; \mathbf{r}_1, \mathbf{r}_2)$ at distances comparable with the wave length $\lambda = 2\pi k^{-1}$, k being the wave vector, or the mean free path l. For the unitary ensemble, this calculation has been performed in Ref. [22, 23] and the final result reads

$$f_2(p_1, p_2; \mathbf{r}_1, \mathbf{r}_2) = \frac{V^2}{1 - g^2(r)} \exp\left(-\frac{V(p_1 + p_2)}{1 - g^2(r)}\right) I_0\left(\frac{2V\sqrt{p_1 p_2 g^2(r)}}{1 - g^2(r)}\right) \tag{39}$$

where $r = |\mathbf{r}_1 - \mathbf{r}_2|$ and the $g(r)$ depends on the dimensionality and has the form

$$g = \begin{cases} J_0(kr)\, e^{-r/l}, & 2D \\ (\sin(kr)/kr)\, e^{-r/l}, & 3D \end{cases}$$

In the limit $r \to \infty$, the function $g(r)$ vanishes and one comes again to Eq. (35). This behavior is in agreement with the semiclassical consideration of Ref. [16].

Eqs. (35, 38, 39) show that correlations between wave functions vanish at large distances in the limiting cases of the orthogonal and unitary ensembles. However, in the regime of the crossover between the ensembles the correlations exist even in the limit $r \to \infty$. This can be seen investigating Eq. (34) written for the functions f_1 and f_2.

For example, at small fluxes corresponding to $X \ll 1$ one has

$$[f_1(0)]^2 \simeq (4\pi/9)\, X^{-2} \tag{40}$$

whereas

$$f_2(0, 0) \simeq \frac{5}{3} X^{-2}$$

Although the quantity $[f_1(0)]^2 - f_2(0, 0)$ is numerically very small, it is finite. In fact, the correlations have numerically the order of 1%. More details can be found in Ref. [27].

Since the work by Berry [16] it has been believed that in classically ergodic systems the local wave function density can be represented as a superposition of an infinite number of plane waves with random phases and equal wave vectors. In the unitary case, this is a complete randomness. In the orthogonal ensemble, the pairs of time-reversed plane waves with wave vectors $\pm\mathbf{k}$ enter this representation with conjugate phases. The randomness of the phases results in the Gaussian randomness of the amplitude φ_α and vanishing of the correlations at large distances. The result given by Eq. (34) means that a weak magnetic flux through the sample area introduces some non-local correlations in φ_α related to the fact that the Aharonov-Bohm phases taken by an electron moving in a quantum billiard cannot be arbitrarily large and that the phases of the "plane waves" in the phenomenological picture are slightly correlated at each point, which has not been anticipated in previous semiclassical theories. These correlations are present in the entire crossover regime and can be suppressed only by localization effects.

Recently it was demonstrated using a random matrix approach that Eq. (34) holds for an arbitrary N [35].

To have true localization effects one needs an infinite sample. However, some pre-localized states can be seen in any finite sample. This states are studied in the next section.

PRE-LOCALIZED STATES

To study the localization effects on the wave function statistics one has to go beyond the 0D approximation of the σ-model. Depending on the system to be studied and problem to be considered, the inhomogeneous field $Q(\mathbf{r})$ can be treated in different ways. In the perturbation theory calculations in the weak localization regime, they appear as non-zero spatial diffusion modes [3]. In quasi-1D wires, one can come to localization using the transfer matrix technique [33]. Recently, it was suggested that the long-living current relaxation in mesoscopic conductors can be described by non-trivial saddle point solutions of the supermatrix σ-model with some boundary conditions [34]. The results were interpreted in terms of the existence of pre-localized states.

It turns out [24, 25] that the wave functions of the pre-localized states can be described by the ground state of the reduced σ-model, Eq. (30-32). Proper calculations can be performed for all symmetry classes but here the most simple unitary ensemble will be considered. The results for the orthogonal ensemble are practically the same.

Integrating Eq. (20) over the zero space harmonics, as it has been described in the preceding section, and using Eqs. (29-32) one can express the distribution function f_1 for the unitary ensemble in the form

$$f_1(p) = \frac{1}{V}\frac{d^2\Phi(p)}{dp^2}, \qquad \Phi(p) = \int_{\tilde{Q}(\mathbf{r}_0)=\Lambda} \exp\left(-F_{red}\left[\tilde{Q}\right]\right) D\tilde{Q} \qquad (41)$$

where the free energy $F_{red}\left[\tilde{Q}\right]$ is

$$F_{red}\left[\tilde{Q}\right] = \int STr\left[\frac{\pi\nu D}{8}\left(\nabla\tilde{Q}\right)^2 - \frac{p}{4}\Lambda\Pi\tilde{Q}(\mathbf{r})\right]d\mathbf{r} \qquad (42)$$

with the matrix Π defined in Eq. (30). The functional integral in Eq. (41) should be calculated with the boundary condition

$$\tilde{Q}(\mathbf{r}_0) = \Lambda \qquad (43)$$

where $\mathbf{r}_0$ is the coordinate of the observation point. The moments P_n of the distribution function f_1 can be written for $n \geqslant 2$ as

$$P_n = \frac{n(n-1)}{V}\int_0^\infty p^{n-2}\Phi(p)\,dp \qquad (44)$$

The function $\Phi(p)$ has several remarkable features. At $p = 0$, the functional $F_{red}[p]$ is invariant with respect to all rotations of the supermatrix $\tilde{Q}$ (is supersymmetric) and therefore, $\Phi(0) = 1$. This corresponds to the normalization of all wave functions

$$VP_1 = \Phi(0) = 1 \qquad (45)$$

On the other hand, for any finite p the symmetry of the reduced σ-model is broken, so that the minimum of the free energy in Eq. (42) corresponds to an inhomogeneous $\tilde{Q}_p(\mathbf{r})$. Indeed, the term $(p/4)\Lambda\Pi$ in the free energy functional, Eq. (42), resembles a field tending to align the matrix $\tilde{Q}$ along the non-compact "direction" in the Q-space (related to the angle θ_1), whereas the boundary condition at $\mathbf{r}_0$ together with the gradient term is analogous to a rigidity preventing that. The competition between these two tendencies results in a non-trivial minimum configuration of $\tilde{Q}$. To find this non-trivial vacuum one can represent Q as

$$\tilde{Q}(\mathbf{r}) = U_{vac}(\mathbf{r})\,\Lambda\frac{1+iP}{1-iP}\bar{U}_{vac}(\mathbf{r}), \qquad P = \begin{pmatrix} 0 & B \\ \bar{B} & 0 \end{pmatrix} \qquad (46)$$

where the supermatrix P stands for fluctuations around the minimum and the supermatrices B have the form [29]

$$B = \begin{pmatrix} a & \sigma \\ i\rho & ib \end{pmatrix}$$ (47)

a, b being complex numbers, σ, ρ-Grassmann anticommuting variables. The parametrization, Eq. (47), for the fluctuations is especially convenient since the Jacobian is equal to unity. The form of the saddle-point $\tilde{Q}_{vac} = U_{vac}\Lambda\bar{U}_{vac}$ should be found from the requirement of the absence of linear terms in the expansion of F into the series

$$F_{red}\left[\tilde{Q}\right] = F_{red}\left[\tilde{Q}_{vac}\right] + F^{(2)} + F^{(3)} + F^{(4)}$$ (48)

in the perturbation $B, \bar{B}$. This gives

$$U_{vac} = \exp\begin{pmatrix} 0 & \frac{1}{2}\theta_p e^{i\chi\tau_3}\pi_b \\ \frac{1}{2}\theta_p e^{i\chi\tau_3}\pi_b & 0 \end{pmatrix}$$ (49)

where the parameter θ_p satisfies the equation

$$\nabla^2\theta_p\left(\mathbf{r}\right) = -\frac{p}{\pi\nu D}e^{-\theta_p}, \qquad \chi = \pi$$ (50)

with the boundary conditions $\theta_p\left(\mathbf{r}_0\right) = 0$ and $\left(\mathbf{n}\nabla\right)\theta_p = 0$ at the surface of the sample. The symbol ∇^2 stands for the Laplacian in the real d-dimensional space, and the projection matrix π_b is defined in Eq. (30).

A similar analysis of the reduced σ-model can also be performed for the other symmetry classes and leads to the same Eq. (50) for the minimum. The difference between the unitary, orthogonal and symplectic ensembles results in different values of the coefficient β

$$\beta_o = 1, \qquad \beta_u = 2, \qquad \beta_s = 2$$ (51)

in the expression for the free energy functional F_p at the minimum

$$F_p = \frac{\beta}{2}\int\left[\frac{\pi\nu D}{2}\left(\nabla\theta_p\right)^2 + pe^{-\theta_p}\right]d\mathbf{r}$$ (52)

and in the higher order terms of the expansion of $F_{red}\left[\tilde{Q}\right]$ in the vicinity of the vacuum state Q_{vac}.

Eq. (50) for the minimum looks very similar to a saddle-point equation derived in Ref. [34] when considering the problem of long-living current relaxation. Nevertheless, the fact that now, a reduced σ-model derived for studying properties of a single quantum state in a confined system is considered, results in a different non-linearity and a different boundary condition. What is remarkable, Eq. (50) for the non-trivial vacuum of the reduced σ-model is exactly the Liouville equation known in the conformal theory of quantum gravity [36]. Within this model one has to calculate the functional integral over all θ with the free energy functional determined by Eq. (52). Although this can lead to helpful analogies [37], results that might be anticipated in this way can be used only as intermediate asymptotics. In conventional models of disorder, there are several relevant length scales, such as the mean free path l which is kept much smaller than the size of the system L and the localization length L_c. The latter is assumed to be much larger than L, which gives a possibility to neglect fluctuations near the minimum.

Due to the additional degrees of freedom present in the supermatrix Q, the conformal Liouville theory and the σ-model description of the disordered systems are different.

In 2D systems, the fluctuations near the minimum for the values of parameters $\pi\nu D \sim 1$ are not small and Eqs. (50-52) alone do not describe the disordered systems. At the same time, the Liouville theory is well defined for these values of parameters.

In the models of disorder, the fluctuations are small provided the inequality

$$(\pi\nu D)^{-1} \ll 1 \tag{53}$$

is fulfilled. Only under this condition the function f_1 can be calculated without difficulties.

The result is most interesting in 2D and, therefore, this case will be considered below. To simplify calculations let us assume that the sample has the form of a disk with the radius L and the observation point $\mathbf{r}_0$ is located at the center. Then, the solution of Eq. (50) can be sought in a rotationally symmetric form $\theta_p(r)$, and one obtains

$$\left(r^{-1}\partial_r r \partial_r\right)\theta_p = -\frac{p}{\pi\nu D}e^{-\theta_p}, \qquad \theta_p\left(r_0 \sim l\right) = 0, \qquad \partial_r\theta_p\left(L\right) = 0 \tag{54}$$

The σ-model has been derived under the assumption that the characteristic lengths of variations of the supermatrix Q are larger than the mean free path l. So, the shortest length in the system is l and the second relation in Eq. (54) corresponds to the first boundary condition written after Eq. (50). The third relation in Eq. (54) is the condition at the boundary of the sample.

Eq. (54) can be solved exactly. Introducing the notation

$$\rho = \sqrt{\frac{2\pi\nu D}{pl^2}}$$

one finds the solution

$$\exp\left(-\frac{\theta_p}{2}\right) = \frac{2\left(l/r\right)^{1-A}\left[\sqrt{\left(1/A\rho\right)^2 + 1} + 1\right]}{\left[\sqrt{\left(1/A\rho\right)^2 + 1} + 1\right]^2 - \left(1/A\rho\right)^2\left(r/l\right)^{2A}} \tag{55}$$

where A should be found from the boundary condition at the sample edge $r = L$,

$$\sqrt{A^2 + \rho^{-2}} + A = \frac{\left(L/l\right)^A}{\rho}\sqrt{\frac{1+A}{1-A}} \tag{56}$$

Substituting the minimum solution $\theta_p(r)$ from Eq. (55) into Eq. (52) one finds the free energy at the minimum

$$F_p = 2\beta\pi^2\nu D\left\{\ln\left(\frac{\left(L/l\right)^{\left(1+A^2\right)}}{\rho^2\left[1-A^2\right]}\right) + 2\left(1 - \sqrt{A^2 + \rho^{-2}}\right)\right\} \tag{57}$$

Eqs. (56, 57) are rather complicated for a further explicit analytical treatment but can be used for numerical analysis of the distribution function. Eq. (56) has positive roots only if $1 < \ln\left(L/l\right) < \rho$, which provides a reasonable limitation for the amplitudes of the wave functions that can be studied using the present method. This limits the density p of a splash by the value $\sim \left(\lambda l\right)^{-1}$ related to the density of an electron state bound to the forward-and-backward scattering trajectory between two impurities (in fact, a more strict condition $p < p_{\max} \sim \left(\lambda l\right)^{-1}/\ln\left(L/l\right)$, where λ is the wavelength, should be satisfied). At the same time, in the limit $\rho \gg 1$, the roots of Eq. (56) can be written as $A = 1 - \mu$ with $\mu \ll 1$. The same conditions enable us to write the solution in Eq. (55) as

$$e^{-\theta_p} \approx (l/r)^{2\mu} \tag{58}$$

One can show [25] that the characteristic shape of the envelope of $|\varphi_\alpha(\mathbf{r})|^2$ is proportional to $e^{-\theta_p(r)}$. Therefore, Eq. (58) demonstrates existence of states with wave functions decaying essentially in a power law. The parameter μ should be found from the equation

$$\mu = \frac{(L/l)^{2-2\mu}}{2(1-\mu)\rho^2},$$

which gives

$$\mu \approx \frac{z(T)}{2\ln(L/l)}, \qquad \text{where } ze^z = T \equiv \frac{pV\ln(L/l)}{2\pi^2\nu D} \tag{59}$$

For small values of p, the parameter μ is small. When p grows approaching $p_{\max} \sim (l\lambda)^{-1}/\ln(L/l)$, μ approaches the limiting value $\mu \approx 1$ which corresponds to $|\varphi_p(\mathbf{r})|^2 \propto r^{-2}$. With the same accuracy, the free energy of the vacuum state can be approximated by

$$F_p \approx 2\beta\pi^2\nu D\left\{\mu + \mu^2\ln(L/l)\right\} \tag{60}$$

Now, using Eq. (41) one can find the distribution function f_1. The entire curve can be drawn only numerically. Analytically, one can find asymptotic behavior for $T \ll 1$ and $T \gg 1$. In the limit of small $T \ll 1$, (which corresponds to the limit of small amplitudes $p \ll 2\pi\nu D\left(L^2\ln(L/l)\right)^{-1}$) one obtains

$$f_1 \approx V\exp\left(-\frac{\beta Vp}{2}\left[1 - \frac{T}{2} + ...\right]\right) \times \left\{\begin{array}{ll} \frac{1}{\sqrt{2\pi Vp}}, & \textit{orthogonal ensemble} \\ 1, & \textit{unitary ensemble} \end{array}\right. \tag{61}$$

Comparing Eq. (61) with Eq. (26) we see that the first term in Eq. (61) agrees with the result of RMT and Eq. (61) describes the first correction to the RMT.

In the opposite limit, $p \gg 2\pi\nu D\left(L^2\ln(L/l)\right)^{-1}$, the distribution function f_1 takes the form

$$f_1 \sim V\exp\left(-\frac{\beta\pi^2\nu D}{2\ln(L/l)}\ln^2 T\right) \tag{62}$$

Eq. (62) demonstrates that for different symmetry classes the appearance of high-amplitude splashes of wave functions is much more probable than one would expect from the Porter-Thomas formula, Eq. (26). The logarithmically-normal law in Eq. (62) coincides with the form of the asymptotics of the distribution of the local density of states in open systems discovered in Ref. [38], although now the theory presented is developed for closed systems and is based on a completely different scheme of calculations. As we have seen, Eq. (62) has been obtained non-perturbatively from a non-trivial minimum of the reduced Lagrangian whereas the authors of Ref. [38] came to this result after a rather complicated summation of terms of a perturbation theory.

Using Eq. (44) one can also calculate the inverse participation numbers P_n. To find the moments P_n accurately enough, one has to notice that, although T in Eq. (61) is the parameter of the expansion in the exponent, corrections to the Porter-Thomas distribution are small only when $pVT \ll 1$ ($pV \ll \sqrt{2\pi\nu D}$). Therefore, only first few numbers P_n, $2 \leq n \ll \sqrt{2\pi\nu D}$, can be estimated using an expansion of $f_1(p)$ in a series in T. Proceeding in this way one can reproduce corrections to the Porter-Thomas distribution found perturbatively in Ref. [39]. To derive higher order IPN one should calculate the integral in Eq. (44) using the saddle-point method. The final result can be written as

$$P_n \approx \frac{\min\{c(n), [2\pi\nu D/\ln(L/l)]^n\}}{l^{2\delta}V}\left(\frac{1}{V}\right)^{\tau(n)/2} \tag{63}$$

where

$$\tau(n) = (n-1)\,d^*(n), \qquad d^*(n) \approx 2 - \frac{n}{2\beta\pi^2\nu D} \tag{64}$$

In Eq. (63), $c(n) = n!$ for the unitary ensemble and $c(n) = (2n-1)!!$ for the orthogonal one.

The intermediate asymptotic scaling laws, Eqs. (63, 64), manifest the multifractal behavior of quantum states. The set of indices $\tau(n)$ determines the spectrum of fractal dimensions $d^*(n)$. The multifractality indicates that the density of a wave function can be imagined as distributed over a tree whose Hausdorf dimension decreases with increasing the length scale. In truly conformal theories [37], this property can be traced both by enlarging the system size L for a fixed amplitude of the wave function, or by looking at increasingly growing density peaks for a fixed L. Both ways of moving through the increasing density p or length scale L would lead to a selection of a fractal dimension $d^*(n_*)$ determined by a typical n_* corresponding to the typical wave functions at this scale. In the disordered systems under consideration the process of moving along the n-axis of the function $\tau(n)$ cannot continue indefinitely since one meets one of two obstacles: the limiting value of the density, $p_{\max} \sim (l\lambda)^{-1}/\ln^2(L/l)$, or the localization length. The effective dimensions at large n should start saturating in order to remain positive, such that $d^* > 0$.

Proceeding in the same way one can study the pre-localized states in quantum wires ($1D$ systems) [25, 27]. The results are in agreement with those one can obtain using the transfer matrix technique.

DISCUSSION

Recent results on statistical properties and spatial structure of chaotic wave functions in disordered systems are reviewed. The progress is achieved using the supermatrix non-linear σ-model. The use of the σ-model allows us not only recover the universal statistics of wave functions previously obtained by means of the random matrix theory and semiclassical considerations but also go beyond the universality limit. In particular, a full description of properties of chaotic wave functions in the regime of crossover between the fundamental classes (orthogonal and unitary) is given and localization effects are considered. The explicit computation for the wave functions given in the present lecture shows in a disordered cavity some sort of pre-localized states that should exist at any weak disorder. The existence of the pre-localized states follows from strong deviations of the large amplitude tails of the distribution function of the local wave function density from the Porter-Thomas distributions. In two dimensions, the pre-localized states have nearly critical properties and show multifractal behavior.

The present results are obtained with the help the supermatrix σ-model derived from models of disorder. Electron motion inside the billiard is assumed to be diffusive and, strictly speaking, one needs the condition $l \ll L$ to be fulfilled. At the same time, some results obtained are more general than the diffusive σ-model. This definitely concerns the Porter-Thomas distribution, Eq. (26), obtained with the $0D$ σ-model. If the concentration of impurities is very small such that the mean free path l exceeds the sample size L one can try to use the "ballistic σ-model" obtained averaging over impurities for a very low concentration in Ref. [40] or averaging over energy levels in

Ref. [41]. The $0D$ version of the ballistic σ-model is the same as that of the diffusive one and is applicable under very general conditions. Apparently, one can obtain in the ballistic case also Eq. (39), in which one should take the limit $l \to \infty$.

Eq. (34) for the crossover between the orthogonal and unitary ensembles is also quite general. Although Eq. (33) is written for the diffusive regime, the first line seems to be more general. Of course, the parameter X entering the first equation should be related to the vector-potential $\mathbf{A}$ by a more complicated formula than that in the second line of Eq. (33). This conjecture is supported by the derivation of Eq. (39) from a RMT [35], which is definitely not restricted to the diffusive regime.

The results of the preceding section about the behavior of the distribution function for large amplitudes of the wave functions are less general and applicable for the diffusive case only. The final formulae, Eqs. (61, 62), are written for a sample in a form of a disk with the observation point located at the center of the disk. These formulae can change if the observation point is located at another point or the sample has another shape. The ballistic case may be even more interesting because now classical periodic trajectories could play an important role. Possibly, studying the probability of high amplitudes within the ballistic σ-model one could see some kind of scars.

REFERENCES

1. *Mesoscopic Phenomena in Solids*, edited by B.L. Altshuler, P.A. Lee, and R.A. Webb (Elsevier, Amsterdam, 1991)

2. *Nanostructures and Mesoscopic Systems,* edited by W.P. Kirk and M.A. Reed (Academic, San Diego, 1992)

3. P.A. Lee and T.V. Ramakrishnan, Rev. Mod. Phys. **57**, 287 (1985), and references therein.

4. B.L. Altshuler, Pis'ma v ZhETF **51**, 530 (1985) (JETP Lett. **41**, 648 (1985); P.A. Lee and A.D. Stone, Phys. Rev. Lett. **55**, 1662 (1985)

5. M.A. Kastner, Rev. Mod. Phys. **64**, 849 (1992)

6. A.M. Chang, H.U. Baranger, L.N. Pfeifer, K.W. West, and T.Y. Chang, Phys. Rev. Lett. **76**, 1695 (1996)

7. J.A. Folk, S.R. Patel, S.F. Godijn, A.G. Huibers, S.M. Cronenwett, C.M. Marcus, K. Campman, and A.C. Gossard, Phys. Rev. Lett. **76**, 1699 (1996)

8. R.A. Jalabert, A.D. Stone, and Y. Alhassid, Phys. Rev. Lett. **68**, 3468 (1992)

9. V.N. Prigodin, K.B. Efetov, and S.Iida, Phys. Rev. Lett. **71**, 1230 (1993)

10. S. Sridar, Phys. Rev. Lett. **67**, 785 (1991)

11. A. Kudrolli, V. Kidambi, and S. Sridhar, Phys. Rev. Lett.**75**, 822 (1995); V.N. Prigodin, N. Taniguchi, A.Kudrolli, V. Kidambi, and S. Sridhar, ibid **75**, 2392 (1995)

12. T.A. Brody, J.Flores, J.B. French, P.A. Mello, A.Pendey, and S.S.M. Wong, Rev. Mod. Phys. **53**, 395 (1981)

13. M.L. Mehta, *Random Matrices* (Academic, San Diego, CA 1991)

14. E.J. Heller, in *Chaos abd Quantum Systems*, edited by M.-J. Giannoni, A. Voros, and J. Zinn-Justin (Elsevier, Amsterdam, 1991); M.C. Gutzwiller, *Chaos in Classical and Quantum Mechanics* (Springer, New York, 1990)

15. B. Li and M. Robnik, J. Phys. A**27**, 5509 (1994)

16. M.V. Berry, J. Phys. A **10**, 2083 (1977); M.V. Berry, Proc. R. Soc. London Ser. A**400**, 229 (1985)

17. M. Srednicki, Phys. Rev. E **54**, 954 (1996)

18. D. Pines and P. Noziéres, *The Theory of Quantum Liquids*, (Plenum, New York, 1966).

19. K. B. Efetov, Adv. in Phys, **32**, 53 (1983)

20. K.B. Efetov and V.N. Prigodin, Phys. Rev. Lett. **70**, 1315 (1993); Mod. Phys. Lett. B **7**, 981 (1993)

21. V.I. Falko and K.B. Efetov, Phys. Rev. B **50**, 11267 (1994)

22. V.N. Prigodin, Phys. Rev. Lett. **74**, 1566 (1995)

23. E.R. Mucciolo, V.N. Prigodin, and B.L. Altshuler, Phys. Rev. Lett. **75**, 1360 (1995)

24. V.I. Falko and K.B. Efetov, Europhys. Lett. **32**, 627 (1995)

25. V.I. Falko and K.B. Efetov, Phys. Rev. B **52**, 17413 (1995)

26. V.I. Falko and K.B. Efetov, Phys. Rev. Lett. **77**, 912 (1996)

27. V.I. Falko and K.B. Efetov, J. Math. Phys. **37**, 4935 (1996)

28. K.B. Efetov, *Supersymmetry in Disorder and Chaos* (Cambridge University Press, New York, 1997)

29. K.B. Efetov, Adv. in Phys. **32**, 53 (1983)

30. J.J.M. Verbaarschot, H.A. Weidenmüller, and M.R. Zirnbauer, Phys. Rep. **129**, 367 (1985)

31. A. Altland, S. Iida, and K.B. Efetov, J. Phys. A **26**, 3545 (1993)

32. K.B. Efetov, S. Iida, Phys. Rev. B **47**, 15794 (1993)

33. K.B. Efetov and A.I. Larkin, Zh. Eksp. Teor. Fiz. **85**, 763 (1983) [Sov. Phys. JETP **58**, 444 (1983)]

34. B.A. Muzykantskii and D.E. Khmelnitskii, Phys. Rev. B **51**, 5480 (1995)

35. S.A. van Langen, P.W. Brouwer, C.W.J. Beenakker, Phys. Rev. E **55**, R1 (1997)

36. N. Seiberg, Prog. Theor. Phys. **102** Suppl., 319 (1990)

37. A. Tsevelik, C. Mudry, and Y. Kogan, Phys. Rev. Lett. **77**, 707 (1996)

38. B.L. Altshuler, V.E. Kravtsov, and I.V. Lerner, in Ref. 1, p.449

39. Y.V. Fyodorov and A.D. Mirlin, JETP Letters **60**, 790 (1994)

40. B.A. Muzykantskii and D.E. Khmelnitskii, JETP Lett. **62**, 76 (1995)

41. A.V. Andreev, O. Agam, B.D. Simons, and B.L. Altshuler, Phys. Rev. Lett. **76**, 3947 (1996)

CORRELATIONS OF WAVE FUNCTIONS IN DISORDERED SYSTEMS

Alexander D. Mirlin *

Institut für Theorie der kondensierten Materie, Universität Karlsruhe, 76128 Karlsruhe, Germany

INTRODUCTION

Statistical properties of eigenfunction amplitudes in disordered and chaotic systems have attracted considerable research interest recently. Fluctuations of the wave function amplitudes are believed to determine statistical properties of conductance peaks in quantum dots in the Coulomb blockade regime [1, 2, 3, 4, 5], and can be directly measured in the microwave cavity experiments [6, 7]. On the theoretical side, the recent progress is based on application of the supersymmetry method to the problem of eigenfunction statistics [8, 2]. It was found that in so-called zero-mode approximation the distribution of eigenfunction amplitudes is correctly described by formulas of the random matrix theory (RMT). Deviations from the RMT predictions were studied in Refs.8, 9, 10, 11.

The present article addresses a problem of correlations of eigenfunction amplitudes. A description of correlations in amplitudes of a wave function in relatively close points of a chaotic billiard was proposed by Berry [12] within a RMT-like assumption that the wave function is a superposition of plane waves with random coefficients. More recently, these correlations were considered in a disordered system within the zero-mode approximation [13]; the result was later shown [14] to be equivalent to that of Ref.12. This result is valid for small separations of the two points (less than the mean free path l). Here we consider such correlations for arbitrary distances. These correlations determine, in particular, fluctuations of matrix elements of the (Coulomb) interaction, which are in turn important for statistical properties of spectra of quantum dots. Another topic addressed in the present paper is that of correlation of amplitudes of different eigenfunctions. Such correlations, while absent in RMT, appear in a disordered system. They become especially strong near the Anderson metal-insulator transition, where they play a crucial role in supporting the RMT-like level repulsion.

This article is based on recent works done in collaboration with Ya. M. Blanter, Y. V. Fyodorov, and B. A. Muzykantskii [9, 15, 16, 17, 18, 19].

---

*Also at Petersburg Nuclear Physics Institute, 188350 St. Petersburg, Russia.

Supersymmetry and Trace Formulae: Chaos and Disorder
Edited by Lerner *et al.*, Kluwer Academic / Plenum Publishers, New York, 1999

EIGENFUNCTION CORRELATIONS IN THE WEAK LOCALIZATION REGIME

In this section, we study the correlations of eigenfunctions in the regime of a good conductor [9, 15, 16]. The correlation function of amplitudes of one and the same eigenfunction with energy E can be formally defined as follows:

$$\alpha(\mathbf{r}_1, \mathbf{r}_2, E) = \left\langle |\psi_k(\mathbf{r}_1)\psi_k(\mathbf{r}_2)|^2 \right\rangle_E \equiv \Delta \left\langle \sum_k |\psi_k(\mathbf{r}_1)\psi_k(\mathbf{r}_2)|^2 \delta(E - \epsilon_k) \right\rangle, \tag{1}$$

where $\psi_k(\mathbf{r})$ and ϵ_k are eigenfunctions and eigenvalues of the Hamiltonian in a particular disorder configuration $U(\mathbf{r})$, the angular brackets $\langle \ldots \rangle$ denote averaging over the disorder potential $U(\mathbf{r})$, and $\Delta = \langle \sum_k \delta(E - \epsilon_k) \rangle^{-1}$ is the mean level spacing. In the case of a real or numerical experiment, calculation of the correlation function (1) would include averaging over all states in a relatively narrow energy window around E. An analogous correlation function for two different eigenfunctions is defined as

$$\sigma(\mathbf{r}_1, \mathbf{r}_2, E, \omega) = \left\langle |\psi_k(\mathbf{r}_1)\psi_l(\mathbf{r}_2)|^2 \right\rangle_{E,\omega}$$
$$\equiv \Delta^2 R_2^{-1}(\omega) \left\langle \sum_{k \neq l} |\psi_k(\mathbf{r}_1)\psi_l(\mathbf{r}_2)|^2 \delta(E - \epsilon_k)\delta(E + \omega - \epsilon_l) \right\rangle, \tag{2}$$

where $R_2(\omega)$ denotes the two-level correlation function,

$$R_2(\omega) = \Delta^2 \left\langle \sum_{k,l} \delta(E - \epsilon_k)\delta(E + \omega - \epsilon_l) \right\rangle. \tag{3}$$

Eq.(2) defines an overlap of the eigenfunctions ψ_k and ψ_l provided they have energies close to E with the energy difference equal to ω.

To evaluate $\alpha(\mathbf{r}_1, \mathbf{r}_2, E)$ and $\sigma(\mathbf{r}_1, \mathbf{r}_2, E, \omega)$ (Ref.16), we employ an identity

$$2\pi^2 \left[\Delta^{-1}\alpha(\mathbf{r}_1, \mathbf{r}_2, E)\delta(\omega) + \Delta^{-2}\tilde{R}_2(\omega)\sigma(\mathbf{r}_1, \mathbf{r}_2, E, \omega) \right] \tag{4}$$
$$= \mathrm{Re} \left[\left\langle G^R(\mathbf{r}_1, \mathbf{r}_1, E)G^A(\mathbf{r}_2, \mathbf{r}_2, E + \omega) - G^R(\mathbf{r}_1, \mathbf{r}_1, E)G^R(\mathbf{r}_2, \mathbf{r}_2, E + \omega) \right\rangle \right],$$

where $G^{R,A}(\mathbf{r}, \mathbf{r}', E)$ are retarded and advanced Green's functions and $\tilde{R}_2(\omega)$ is non-singular part of the level-level correlation function: $R_2(\omega) = \tilde{R}_2(\omega) + \delta(\omega/\Delta)$. A natural question, which arises at this point, is whether the r.h.s. of Eq.(4) cannot be simply found within the diffuson-Cooperon perturbation theory [20]. Such a calculation would, however, be justified only for $\omega \gg \Delta$ (more precisely, one has to introduce an imaginary part of frequency: $\omega \rightarrow \omega + i\Gamma$, and require that $\Gamma \gg \Delta$). Therefore, it would only allow to find a smooth in ω part of $\sigma(\mathbf{r}_1, \mathbf{r}_2, E, \omega)$ for $\omega \gg \Delta$. Evaluation of $\alpha(\mathbf{r}_1, \mathbf{r}_2, E)$, as well as of $\sigma(\mathbf{r}_1, \mathbf{r}_2, E, \omega)$ at $\omega \sim \Delta$ cannot be done within such a calculation. For this reason, we employ a non-perturbative supersymmetry approach below.

The r.h.s. of Eq.(4) can be expressed in terms of the supermatrix σ-model [21], yielding:

$$2\pi^2 \left[\Delta^{-1}\alpha(\mathbf{r}_1, \mathbf{r}_2, E)\delta(\omega) + \Delta^{-2}\sigma(\mathbf{r}_1, \mathbf{r}_2, E, \omega)\tilde{R}_2(\omega) \right]$$
$$= (\pi\nu)^2 \left[1 - \mathrm{Re}\langle Q_{bb}^{11}(\mathbf{r}_1)Q_{bb}^{22}(\mathbf{r}_2) \rangle_F - k_d(\mathbf{r}_1 - \mathbf{r}_2)\mathrm{Re}\langle Q_{bb}^{12}(\mathbf{r}_1)Q_{bb}^{21}(\mathbf{r}_1) \rangle_F \right], \tag{5}$$

where $k_d(\mathbf{r}) = (\pi\nu)^{-2}\langle \mathrm{Im}G^R(\mathbf{r}) \rangle^2$ is a short-range function explicitly given by

$$k_d(\mathbf{r}) = \exp(-r/l) \begin{cases} J_0^2(p_F r), & 2D \\ (p_F r)^{-2}\sin^2 p_F r, & 3D \end{cases}. \tag{6}$$

Here $\langle\ldots\rangle_F$ denotes the averaging with the action of the supermatrix sigma-model $F[Q]$:

$$\langle\ldots\rangle_F = \int DQ(\ldots)\exp(-F[Q]),$$

$$F[Q] = -\frac{\pi\nu}{4}\int d\mathbf{r}\, \mathrm{Str}[D(\nabla Q)^2 + 2i(\omega + i0)\Lambda Q], \tag{7}$$

where D is the diffusion coefficient, ν is the density of states, $Q = T^{-1}\Lambda T$ is a 4×4 supermatrix, $\Lambda = \mathrm{diag}(1,1,-1,-1)$, and T belongs to the supercoset space $U(1,1|2)/U(1|1)\times U(1|1)$. The symbol Str denotes the supertrace defined as $\mathrm{Str}B = B_{bb}^{11} - B_{ff}^{11} + B_{bb}^{22} - B_{ff}^{22}$. The upper matrix indices correspond to the retarded-advanced decomposition, while the lower indices denote the boson-fermion one. The action (7) is written for the case of so-called unitary ensemble (broken time reversal symmetry), which we consider below. Generalization to a system with time reversal symmetry (orthogonal ensemble) is straightforward, and the results are presented in the end of the section. Evaluating the σ-model correlation functions in the r.h.s. of Eq.(5) and separating the result into the singular the singular (proportional to $\delta(\omega)$) and regular at $\omega = 0$ parts, one can obtain the correlation functions $\alpha(\mathbf{r}_1, \mathbf{r}_2, E)$ and $\sigma(\mathbf{r}_1, \mathbf{r}_2, E, \omega)$. The two-level correlation function $R_2(\omega)$ entering Eq.(5) is given in this formalism by

$$R_2(\omega) = \frac{1}{2V^2}\mathrm{Re}\int d\mathbf{r}_1 d\mathbf{r}_2 \left[1 - \langle Q_{bb}^{11}(\mathbf{r}_1)Q_{bb}^{22}(\mathbf{r}_2)\rangle_F\right]. \tag{8}$$

In the metallic (weak localization) regime, the sigma-model correlation functions $\langle Q_{bb}^{11}(\mathbf{r}_1)Q_{bb}^{22}(\mathbf{r}_2)\rangle_F$ and $\langle Q_{bb}^{12}(\mathbf{r}_1)Q_{bb}^{21}(\mathbf{r}_2)\rangle_F$ can be calculated for relatively low frequencies $\omega \ll E_c$ with the use of a general method developed in Refs.22, 9, which allows one to take into account spatial variations of the field Q. The results are obtained in form of expansions in g^{-1}, where g is the dimensionless conductance. First, we restrict ourselves to the terms of order g^{-1}. Then, the result for the first correlation function reads as

$$\langle Q_{bb}^{11}(\mathbf{r}_1)Q_{bb}^{22}(\mathbf{r}_2)\rangle_F = -1 - 2i\frac{\exp(i\pi s)\sin\pi s}{(\pi s)^2} - \frac{2i}{\pi s}\Pi(\mathbf{r}_1,\mathbf{r}_2), \tag{9}$$

where $s = \omega/\Delta + i0$. Here the diffusion propagator Π is the solution to the diffusion equation

$$-D\nabla^2\Pi(\mathbf{r}_1,\mathbf{r}_2) = (\pi\nu)^{-1}[\delta(\mathbf{r}_1 - \mathbf{r}_2) - V^{-1}] \tag{10}$$

with the Neumann boundary condition (normal derivative equal to zero at the sample boundary), which can be presented in the form

$$\Pi(\mathbf{r}_1,\mathbf{r}_2) = (\pi\nu)^{-1}\sum_{\mathbf{q}}(Dq^2)^{-1}\phi_{\mathbf{q}}(\mathbf{r}_1)\phi_{\mathbf{q}}(\mathbf{r}_2), \tag{11}$$

with $\phi_{\mathbf{q}}$ being the eigenfunction of the diffusion operator corresponding to the eigenvalue Dq^2, $\mathbf{q} \neq 0$. The first two terms in Eq. (9) represent the result of the zero-mode approximation [21], which takes into account only the spatially constant configurations of the field $Q(\mathbf{r})$, so that the functional integral over $DQ(\mathbf{r})$ is reduced to an integral over a single matrix Q. The last term is the correction of order g^{-1}. An analogous calculation for the second correlator yields:

$$\langle Q_{bb}^{12}(\mathbf{r}_1)Q_{bb}^{21}(\mathbf{r}_2)\rangle_F = -2\left\{\frac{i}{\pi s} + \left[1 + i\frac{\exp(i\pi s)\sin\pi s}{(\pi s)^2}\right]\Pi(\mathbf{r}_1,\mathbf{r}_2)\right\}. \tag{12}$$

Now, separating regular and singular parts in r.h.s. of Eq. (5), we obtain the following result for the autocorrelations of the same eigenfunction:

$$V^2\langle|\psi_k(\mathbf{r}_1)\psi_k(\mathbf{r}_2)|^2\rangle_E - 1 = k_d(r)[1 + \Pi(\mathbf{r}_1,\mathbf{r}_1)] + \Pi(\mathbf{r}_1,\mathbf{r}_2), \tag{13}$$

and for the correlation of amplitudes of two different eigenfunctions

$$V^2 \langle |\psi_k(\mathbf{r}_1)\psi_l(\mathbf{r}_2)|^2 \rangle_{E,\omega} - 1 = k_d(r)\Pi(\mathbf{r}_1,\mathbf{r}_1), \quad k \neq l \tag{14}$$

In particular, for $\mathbf{r}_1 = \mathbf{r}_2$ we have

$$V^2 \langle |\psi_k(\mathbf{r})\psi_l\mathbf{r})|^2 \rangle_{E,\omega} - 1 = \delta_{kl} + (1 + \delta_{kl})\Pi(\mathbf{r},\mathbf{r}). \tag{15}$$

Note that the result (13) for $\mathbf{r}_1 = \mathbf{r}_2$ is the inverse participation ratio calculated in Ref. 9; on the other hand, neglecting the terms with the diffusion propagator (i.e. making the zero-mode approximation), we reproduce the result of Refs.12, 13, 14.

Eqs. (14), (15) show that the correlations between different eigenfunctions are relatively small in the weak disorder regime. Indeed, they are proportional to the small parameter $\Pi(\mathbf{r},\mathbf{r})$, which is equal in the case of 2D geometry to (L is the size of the system)

$$\Pi(\mathbf{r},\mathbf{r}) = (\pi g)^{-1}\ln L/l, \qquad 2D, \tag{16}$$

with $g = 2\pi\nu D$. For a quasi-1D wire or strip of the length L,

$$\Pi(r,r) = \frac{2}{g}\left[\frac{1}{6} + B_2\left(\frac{r}{L}\right)\right], \qquad 0 \leq r \leq L, \tag{17}$$

where $g = 2\pi\nu D/L$, and $B_2(x) = x^2 - x + 1/6$ is the Bernoulli polynomial.[†] The correlations are enhanced by disorder; when the system approaches the strong localization regime, the relative magnitude of correlations, $\Pi(\mathbf{r},\mathbf{r})$ ceases to be small. The correlations near the Anderson localization transition will be discussed in the next section of the paper.

Another correlation function, generally used for the calculation of the linear response of the system,

$$\gamma(\mathbf{r}_1,\mathbf{r}_2,E,\omega) = \langle \psi_k^*(\mathbf{r}_1)\psi_l(\mathbf{r}_1)\psi_k(\mathbf{r}_2)\psi_l^*(\mathbf{r}_2) \rangle_{E,\omega}$$

$$\equiv \Delta^2 R_2^{-1}(\omega)\left\langle \sum_{k \neq l} \psi_k^*(\mathbf{r}_1)\psi_l(\mathbf{r}_1)\psi_k(\mathbf{r}_2)\psi_l^*(\mathbf{r}_2)\delta(E - \epsilon_k)\delta(E + \omega - \epsilon_l) \right\rangle \tag{18}$$

can be calculated in a similar way; the result reads

$$V^2 \langle \psi_k^*(\mathbf{r}_1)\psi_l(\mathbf{r}_1)\psi_k(\mathbf{r}_2)\psi_l^*(\mathbf{r}_2) \rangle_{E,\omega} = k_d(r) + \Pi(\mathbf{r}_1,\mathbf{r}_2), \quad k \neq l. \tag{19}$$

As is seen from Eqs. (13), (14), (19), in the $1/g$ order the correlation functions $\alpha(\mathbf{r}_1,\mathbf{r}_2,E)$ and $\gamma(\mathbf{r}_1,\mathbf{r}_2,E,\omega)$ survive for the large separation between the points, $r \gg l$, while $\sigma(\mathbf{r}_1,\mathbf{r}_2,E,\omega)$ decays exponentially for the distances larger than the mean free path l. This is, however, an artifact of the g^{-1} approximation, and the investigation of the corresponding tails requires the extension of the above calculation to the terms proportional to g^{-2}. We find that the correlator $\langle Q_{bb}^{11}(\mathbf{r}_1)Q_{bb}^{22}(\mathbf{r}_2) \rangle_F$ gets the following correction:

$$\delta\langle Q_{bb}^{11}(\mathbf{r}_1)Q_{bb}^{22}(\mathbf{r}_2) \rangle_F = -f_1 + 2f_4 + \exp(2i\pi s)f_3 - 2i\frac{\exp(2i\pi s)}{\pi s}(f_2 - f_3)$$

$$- \frac{\exp(2i\pi s) - 1}{2(\pi s)^2}(f_1 - 4f_2 + 3f_3 - 2f_4). \tag{20}$$

[†]In the 3D geometry, the sum over the momenta q in Eq.(11) determining $\Pi(\mathbf{r},\mathbf{r})$ diverges at large q and is determined by the upper cut-off, $q \sim 1/l$, yielding $\Pi(\mathbf{r},\mathbf{r}) \sim g^{-1}L/l$. This reflects the fact that in 3D geometry the truly local ($\mathbf{r}_1 = \mathbf{r}_2$) correlations may not be given correctly by the diffusion approximation and can depend on microscopic structure of the random potential. For this reason, we do not consider local correlations in 3D geometry here. Note, however, that this concerns the global geometry of the sample; locally the system can be either of 2D or 3D nature, which determines the form of the function $k_d(r)$ (e.g, a wire is locally 3D, but has a quasi-1D geometry).

Here we defined the functions

$$f_1(\mathbf{r}_1, \mathbf{r}_2) = \Pi^2(\mathbf{r}_1, \mathbf{r}_2),$$

$$f_2(\mathbf{r}_1, \mathbf{r}_2) = (2V)^{-1} \int d\mathbf{r} \left[\Pi^2(\mathbf{r}, \mathbf{r}_1) + \Pi^2(\mathbf{r}, \mathbf{r}_2) \right],$$

$$f_3 = V^{-2} \int d\mathbf{r} d\mathbf{r}' \Pi^2(\mathbf{r}, \mathbf{r}'),$$

$$f_4(\mathbf{r}_1, \mathbf{r}_2) = V^{-1} \int d\mathbf{r} \Pi(\mathbf{r}, \mathbf{r}_1) \Pi(\mathbf{r}, \mathbf{r}_2). \tag{21}$$

Consequently, we obtain the following results for the correlations of different ($k \neq l$) eigenfunctions at $r > l$:

$$V^2 \langle |\psi_k(\mathbf{r}_1) \psi_l(\mathbf{r}_2)|^2 \rangle_{E,\omega} - 1 = \frac{1}{2}(f_1 - f_3 - 2f_4)$$

$$+ 2(f_2 - f_3) \left(\frac{\sin^2 \pi s}{(\pi s)^2} - \frac{\sin 2\pi s}{2\pi s} \right) \left(1 - \frac{\sin^2 \pi s}{(\pi s)^2} \right)^{-1}. \tag{22}$$

As it should be expected, the double integral over the both coordinates of this correlation function is equal to zero. This property is just the normalization condition and should hold in arbitrary order of expansion in g^{-1}.

The quantities f_2, f_3, and f_4 are proportional to g^{-2}, with some (geometry-dependent) prefactors of order unity. On the other hand, f_1 in $2D$ and $3D$ geometry depends essentially on the distance $r = |\mathbf{r}_1 - \mathbf{r}_2|$. In particular, for $l \ll r \ll L$ we find

$$f_1(\mathbf{r}_1, \mathbf{r}_2) = \Pi^2(\mathbf{r}_1, \mathbf{r}_2) \approx \begin{cases} \dfrac{1}{(\pi g)^2} \ln^2 \dfrac{L}{r} \,, & 2D \\[2ex] \dfrac{1}{(4\pi^2 \nu D r)^2} \,, & 3D \end{cases}$$

Thus, for $l < r \ll L$, the contributions proportional to f_1 dominate in Eq.(22), and we get

$$V^2 \langle |\psi_k(\mathbf{r}_1) \psi_l(\mathbf{r}_2)|^2 \rangle_{E,\omega} - 1 = \frac{1}{2} \Pi^2(\mathbf{r}_1, \mathbf{r}_2) \,, \quad k \neq l. \tag{23}$$

On the other hand, for the case of quasi-1D geometry (as well as in 2D and 3D for $r \sim L$), all quantities f_1, f_2, f_3, and f_4 are of order of $1/g^2$. Thus, the correlator $\sigma(\mathbf{r}_1, \mathbf{r}_2, E, \omega)$ acquires a non-trivial (oscillatory) frequency dependence on a scale $\omega \sim \Delta$ described by the second term in the r.h.s. of Eq.(22). In particular, in the quasi-1D case the function $f_2 - f_3$ determining the spatial dependence of this term has the form

$$f_2 - f_3 = -\frac{2}{3g^2} \left[B_4 \left(\frac{r_1}{L} \right) + B_4 \left(\frac{r_2}{L} \right) \right] \,, \tag{24}$$

where $B_4(x) = x^4 - 2x^3 + x^2 - 1/30$.

Let us remind the reader that the above derivation is valid for $\omega \ll E_c$. In order to obtain the results in the range $\omega \geq E_c$ one can calculate the sigma-model correlation functions entering Eqs. (5) by means of the perturbation theory [20]. We find then

$$V^2 \langle |\psi_k(\mathbf{r}_1) \psi_l(\mathbf{r}_2)|^2 \rangle_{E,\omega} = 1 + \mathrm{Re} \left\{ k_d(r) \Pi_\omega(\mathbf{r}_1, \mathbf{r}_2) \right.$$

$$\left. + \frac{1}{2} \left[\Pi_\omega^2(\mathbf{r}_1, \mathbf{r}_2) - \frac{1}{V^2} \int d\mathbf{r} d\mathbf{r}' \Pi_\omega^2(\mathbf{r}, \mathbf{r}') \right] \right\}, \tag{25}$$

$$V^2 \langle \psi_k^*(\mathbf{r}_1) \psi_l(\mathbf{r}_1) \psi_k(\mathbf{r}_2) \psi_l^*(\mathbf{r}_2) \rangle_{E,\omega} = k_d(r) + \mathrm{Re} \Pi_\omega(\mathbf{r}_1, \mathbf{r}_2),$$

where $\Pi_\omega(\mathbf{r_1}, \mathbf{r_2})$ is the finite-frequency diffusion propagator

$$\Pi_\omega(\mathbf{r_1}, \mathbf{r_2}) = (\pi\nu)^{-1} \sum_\mathbf{q} \frac{\phi_q(\mathbf{r_1})\phi_q(\mathbf{r_2})}{Dq^2 - i\omega}, \tag{26}$$

and the summation in Eq. (26) now includes $\mathbf{q} = 0$. As was mentioned, the perturbation theory should give correctly the non-oscillatory (in ω) part of the correlation functions at $\omega \gg \Delta$. Indeed, it can be checked that Eqs.(25) match the supersymmetric σ-model results in this regime. Furthermore, in the $1/g$ order [which means keeping only linear in Π_ω terms in (25) and neglecting $-i\omega$ in denominator of Eq.(26)] Eqs.(25), (26) reproduce the exact results (13), (19) even at small frequencies $\omega \sim \Delta$. We stress however that the perturbative calculation is not justified in this region and only the supersymmetry method provides a rigorous derivation of these results.

As was mentioned, the case of broken time reversal symmetry (unitary ensemble) was considered above. Generalization to a system with unbroken time reversal symmetry is straightforward [16]; in $1/g$-order Eqs.(13), (14), and (19) are modified as follows:

$$V^2 \langle |\psi_k(\mathbf{r_1})\psi_k(\mathbf{r_2})|^2 \rangle_E = [1 + 2k_d(r)]\,[1 + 2\Pi(\mathbf{r_1}, \mathbf{r_2})], \tag{27}$$

$$V^2 \langle |\psi_k(\mathbf{r_1})\psi_l(\mathbf{r_2})|^2 \rangle_{E,\omega} - 1 = 2k_d(r)\Pi(\mathbf{r_1}, \mathbf{r_2}). \tag{28}$$

$$V^2 \langle \psi_k^*(\mathbf{r_1})\psi_l(\mathbf{r_1})\psi_k(\mathbf{r_2})\psi_l^*(\mathbf{r_2}) \rangle_{E,\omega} = k_d(r) + [1 + k_d(r)]\,\Pi(\mathbf{r_1}, \mathbf{r_2}), \quad k \neq l. \tag{29}$$

Using the supersymmetry method, one can calculate also higher order correlation functions of eigenfunction amplitudes. In particular, the correlation function $\langle |\psi_k^4(\mathbf{r_1})||\psi_k^4(\mathbf{r_2})| \rangle_E$ determines fluctuations of the inverse participation ratio (IPR) $I_2 = \int d\mathbf{r} |\psi^4(\mathbf{r})|$. Details of the corresponding calculation can be found in Ref.9; the result for the relative variance of IPR, $\delta(I_2) = \mathrm{var}(I_2)/\langle I_2 \rangle^2$ being

$$\delta(I_2) = \frac{8}{\beta^2} f_3 = \frac{32 a_d}{\beta^2 g^2}, \tag{30}$$

with a numerical coefficient a_d depending on the sample dimensionality d and equal to $a_1 = 1/90$, $a_2 \approx 0.0266$ and $a_3 \approx 0.0527$ for quasi-1D, 2D, and 3D geometry respectively. The fluctuations (30) have the same relative magnitude as the famous universal conductance fluctuations. Note also that extrapolating Eq.(30) to the Anderson transition point, where $g \sim 1$, we find $\delta(I_2) \sim 1$, so that the magnitude of IPR fluctuations is of the order of its mean value (which is, in turn, much larger than in the metallic regime; see the next section).

Similarly, correlations of eigenfunction amplitudes determine fluctuations of matrix elements of an operator of some (say, Coulomb) interaction computed on eigenfunctions ψ_k of the one-particle Hamiltonian in a random potential. Such a problem naturally arises, when one wishes to study the effect of interaction onto statistical properties of excitations in a mesoscopic sample (see below).

Finally, the above consideration can be generalized to a ballistic chaotic system, by applying a recently developed ballistic generalization of the σ-model [23, 24]. The results are then expressed in terms of the (averaged over the direction of velocity) kernel $g(\mathbf{r_1}, \mathbf{n_1}; \mathbf{r_2}, \mathbf{n_2})$ of the Liouville operator $\hat{K} = v_F \mathbf{n}\nabla$ governing the classical dynamics in the system,

$$\Pi(\mathbf{r_1}, \mathbf{r_2}) = \int d\mathbf{n_1} d\mathbf{n_2}\, g(\mathbf{r_1}, \mathbf{n_1}; \mathbf{r_2}, \mathbf{n_2});$$

$$\hat{K} g(\mathbf{r_1}, \mathbf{n_1}; \mathbf{r_2}, \mathbf{n_2}) = (\pi\nu)^{-1} \left[\delta(\mathbf{r_1} - \mathbf{r_2})\delta(\mathbf{n_1} - \mathbf{n_2}) - V^{-1} \right]. \tag{31}$$

Here $\mathbf{n}$ is a unit vector determining the direction of momentum, and normalization $\int d\mathbf{n} = 1$ is used. Equivalently, the function $\Pi(\mathbf{r_1}, \mathbf{r_2})$ can be defined as

$$\Pi(\mathbf{r_1}, \mathbf{r_2}) = \int_0^\infty dt \int d\mathbf{n_1} \, \tilde{g}(\mathbf{r_1}, \mathbf{n_1}, t; \mathbf{r_2}) \, , \tag{32}$$

where $\tilde{g}$ is determined by the evolution equation

$$\left(\frac{\partial}{\partial t} + v_F \mathbf{n_1} \nabla_1 \right) \tilde{g}(\mathbf{r_1}, \mathbf{n_1}, t; \mathbf{r_2}) = 0 \, , \qquad t > 0 \tag{33}$$

with the boundary condition

$$\tilde{g}|_{t=0} = (\pi \nu)^{-1} [\delta(\mathbf{r_1} - \mathbf{r_2}) - V^{-1}]. \tag{34}$$

Eq.(31) is a natural "ballistic" counterpart of Eq.(10). In particular, generalization of Eqs.(13), (27) for the correlations of an eigenfunction amplitudes in two different points onto the ballistic case reads [19]

$$\alpha(\mathbf{r_1}, \mathbf{r_2}, E) = 1 + \frac{2}{\beta} \Pi(\mathbf{r_1}, \mathbf{r_2}). \tag{35}$$

Note that the ballistic σ-model approach is of semiclassical nature and thus valid on distances much larger than the wave length λ_F. For this reason, Eqs.(35) represent the smoothed correlation function, which is valid for $|\mathbf{r_1} - \mathbf{r_2}| \gg \lambda_F$. Indeed, the contribution to $\Pi(\mathbf{r_1}, \mathbf{r_2})$ from the straight line motion from $\mathbf{r_2}$ to $\mathbf{r_1}$ can be easily evaluated, yielding e.g. in the 2D case $\Pi^{(0)}(\mathbf{r_1}, \mathbf{r_2}) = 1/(\pi p_F |\mathbf{r_1} - \mathbf{r_2}|)$. This is nothing else but the smoothed version of the function $k_d(|\mathbf{r_1} - \mathbf{r_2}|) = J_0^2(p_F |\mathbf{r_1} - \mathbf{r_2}|)$ giving the leading contribution to the short-scale correlations. A formula for the variance of matrix elements closely related to Eq.(35) was obtained in the semiclassical approach in Ref.25. In a very recent paper [26] a similar generalization of the Berry formula for $\langle \psi_k^*(\mathbf{r_1}) \psi_k(\mathbf{r_2}) \rangle$ was proposed.

Eq.(35) shows that correlations in eigenfunction amplitudes in remote points are determined by the classical dynamics in the system. It is closely related to the phenomenon of scarring of eigenfunctions by the classical orbits [27, 28]. Indeed, if $\mathbf{r_1}$ and $\mathbf{r_2}$ belong to a short periodic orbit, the function $\Pi(\mathbf{r_1}, \mathbf{r_2})$ is positive, so that the amplitudes $|\psi_k(\mathbf{r_1})|^2$ and $|\psi_k(\mathbf{r_2})|^2$ are positively correlated. This is a reflection of the "scars" associated with this periodic orbits and a quantitative characterization of their strength in the coordinate space. Note that this effect gets smaller with increasing energy E of eigenfunctions. Indeed, for a strongly chaotic system and for $|\mathbf{r_1} - \mathbf{r_2}| \sim L$ (L being the system size), we have in the 2D case $\Pi(\mathbf{r_1}, \mathbf{r_2}) \sim \lambda_F/L$, so that the magnitude of correlations decreases as $E^{-1/2}$. The function $\Pi(\mathbf{r_1}, \mathbf{r_2})$ was explicitly calculated in Ref.19 for a circular billiard with diffusive surface scattering.

STRONG CORRELATIONS OF EIGENFUNCTIONS AND LEVEL REPULSION AT THE ANDERSON LOCALIZATION TRANSITION

In the preceding section, we considered the metallic regime, where a typical wavefunction $\psi_i(\mathbf{r})$ is extended and covers uniformly all the sample volume. When system approaches the point of Anderson transition E_c, these extended eigenfunctions become less and less homogeneous in space showing regions with larger and smaller amplitudes and eventually forming a multifractal structure in the vicinity of E_c. To characterize the degree of non-homogeneity quantitatively, it is convenient to use the inverse participation ratio (IPR) $\langle I_2(E) \rangle = \int dr \alpha(\mathbf{r}, \mathbf{r}, E)$ For extended states this quantity is inversely proportional to the system volume: $\langle I_2(E) \rangle = A(E) L^{-d}$, with L and d standing for the system size and spatial dimension, respectively. The coefficient A in this relation measures a fraction of the system volume where eigenfunction is appreciably non-zero. In the regime of a good conductor $A \approx 1 + 2/\beta \sim 1$, whereas close to the mobility edge $E = E_c$ it becomes large and diverges like $A(E) \propto |E - E_c|^{-\mu_2}$, with a critical index $\mu_2 > 0$ [29]. This means that eigenfunctions become more and more sparse, when the system approaches the critical point of the Anderson transition. Just at the mobility edge eigenfunctions occupy a vanishing fraction of the system volume and IPR scales like $\langle I_2(E) \rangle \propto L^{-d+\eta}$, with $\eta > 0$. Such a behavior reflects multifractal [30, 31, 32] structure of critical eigenstates. At last, in the insulating phase any eigenstate is localized in a domain of finite extension ξ and IPR remains finite in the limit of infinite system size $L \to \infty$.

This transparent picture serves as a basis for qualitative understanding of spectral properties of disordered conductors. Indeed, as long as eigenstates are well extended and cover the whole sample, they overlap substantially and corresponding energy levels repel each other in the same way as in RMT. As a result, the Wigner-Dyson (WD) statistics describes well energy levels in a good metal [21, 20, 22]. In contrast, in the insulating phase different eigenfunctions corresponding to levels close in energy are localized far apart from one another and their overlap is negligible. This is the reason for absence of correlations of energy levels in this regime – the so-called Poisson statistics.

However, a naive extrapolation of this argument to the vicinity of the transition point would lead to a wrong conclusion. Indeed, one might expect that sparse (multifractal in the critical point) eigenstates fail to overlap, that would result in essential weakening of level correlations close to the mobility edge and vanishing level repulsion at $E = E_c$. However, numerical simulations show [33, 34, 35, 36, 37] that even at the mobility edge levels repel each other strongly, though the whole statistics is different from the WD one. The purpose of this section is to explain how this apparent contradiction is resolved. We will show that critical eigenstates for nearby levels are so strongly correlated that they overlap well in spite of their sparse structure.

To calculate the overlap function $\sigma(\mathbf{r}, \mathbf{r}, E, \omega)$ in the critical regime [17], we will exploit an exactly solvable model of the Anderson transition – so-called sparse random matrix (SRM) model [38]. This model is essentially equivalent to a tight-binding model, which is locally of tree-like stricture (i.e has no small-scale loops), with matrix size N playing a role of the system volume (number of sites of the tight-binding model). The SRM model can be used to construct an effective mean-field theory of Anderson localization [39] corresponding to the limit $d = \infty$. The inverse participation ratio $\langle I_2(E) \rangle = N\alpha(r, r, E)$ is proportional to $1/N$ in the delocalized phase, as expected: $\langle I_2(E) \rangle = A(E)/N$. The coefficient $A(E)$ diverges close to the transition point, $|E - E_c| \ll E_c$, like $A(E) \propto \exp\left(\mathrm{const} |E - E_c|^{-1/2}\right)$ [39] which differs from the power-law behavior expected for conventional $d-$dimensional systems. The origin of such a critical dependence was explained in [39, 40] and stems from the fact that $A(E)$ is determined essentially by the "correlation volume" $V(\xi)$ (i.e. number of sites at a distance smaller than correlation length ξ), which is exponentially large, $V(\xi) \propto \exp(\mathrm{const}\,\xi)$, for tree-like structures, whereas $V(\xi) \propto \xi^d$ for a $d-$dimensional lattice. Having this difference in mind, one can translate the results obtained in the framework of $d = \infty$ models to their finite-dimensional counterparts [40].

The calculation of the inverse participation ration $N\alpha(r, r, E)$ can be extended onto the overlap correlation function $\sigma(r, r, E, \omega)$. Evaluating the r.h.s. of Eq.(4), we find the following relation [17]:

$$\sigma(r, r, E, \omega) = \frac{\beta}{\beta + 2}\alpha(r, r, E) \ . \tag{36}$$

This relation is valid *everywhere* in the phase of extended eigenstates, up to the mobility edge $E = E_c$, provided the number of sites (the system volume) exceeds the correlation volume. In particular, it is valid in the critical region $|E - E_c| \ll E_c$, where a typical eigenfunction is very sparse and $\alpha(r, r, E)$ grows like $\exp\left(\mathrm{const}|E - E_c|^{-1/2}\right)$.

Eq.(36) implies the following structure of eigenfunctions within an energy interval $\delta E = \omega < A^{-1}(E)$. Each eigenstate can be represented as a product $\Psi_i(r) = \psi_i(r)\Phi_E(r)$. Here the function $\Phi_E(r)$ is an eigenfunction envelope of "bumps and dips" It is the same for all eigenstates around energy E, reflects underlying gross (multifractal) spatial structure and governs the divergence of the inverse participation ratio (i.e. of the factor $A(E)$) at the critical point. In contrast, $\psi_i(r)$ is Gaussian white-noise component fluctuating in space on the scale of lattice constant. It fills in the envelope function $\Phi_E(r)$ in an individual way for each eigenfunction, but is not critical, i.e. is not sensitive to the vicinity of the Anderson transition. These Gaussian fluctuations are responsible for the factor $\beta/(\beta + 2)$ (which is the same as in the corresponding Gaussian Ensemble) in Eq.(36).

As was already mentioned, this picture is valid in the energy window $\delta E \sim A^{-1}(E)$ around the energy E; the number of levels in this window being large as $\delta E/\Delta \sim NA^{-1}(E) \gg 1$ in the thermodynamic limit $N \to \infty$. These states form a kind of Gaussian Ensemble on a spatially non-uniform (multifractal for $E \to E_c$) background $\Phi_E(r)$. Since the eigenfunction correlations are described by the formula (36), which has exactly the same form as in the Gaussian Ensemble, it is not surprising that the level statistics has the WD form everywhere in the extended phase [38].

We believe on physical grounds that the same picture should hold for a conventional d-dimensional conductor. First of all, the general mechanism of the transition is the same in $d < \infty$ and $d = \infty$ models. Furthermore, the sparsity (multifractality) of eigenstates near the transition point takes its extreme form for $d = \infty$ models [40], so that since the strong correlations (36) take place at $d = \infty$ it would be very surprising if they do not hold at finite d as well. Finally, Eq.(36) is supported by the results of the calculations in the weak localization regime. Keeping only the leading (in $d \geq 2$) terms proportional to the powers of $\Pi(\mathbf{r}, \mathbf{r})$ and considering the unitary ensemble for definiteness, we have up to the two-loop order (see Eq.(22) and Ref.9)

$$\sigma(\mathbf{r}, \mathbf{r}, E, \omega) = V^{-2}\left[1 + \Pi(\mathbf{r}, \mathbf{r}) + \frac{1}{2}\Pi^2(\mathbf{r}, \mathbf{r}) + \ldots\right] = \frac{1}{2}\alpha(\mathbf{r}, \mathbf{r}, E) \ , \tag{37}$$

in full agreement with Eq.(36).

Replacing $A(E)$ by the d-dimensional correlation volume $\sim \xi^d$, we conclude that for E close to E_c Eq.(36) should be valid for $\omega < \Delta_\xi$, where $\Delta_\xi \propto 1/\xi^d$ is the level spacing in the correlation volume. For larger ω, $\sigma(\mathbf{r}, E, \omega)$ is expected to decrease as $\omega^{-\eta/d}$ according to the scaling arguments [31, 32, 41], so that we find $\sigma(\mathbf{r}, E, \omega)/\alpha(\mathbf{r}, E) \sim (\omega/\Delta_\xi)^{-\eta/d}$, up to a numerical coefficient of order of unity. Again, for any value of the energy E in the delocalized phase, taking the system size L large enough, $L \gg \xi$, we have a large number of levels $\delta E/\Delta \sim \Delta_\xi/\Delta \propto (L/\xi)^d$ in the energy window δE where Eq.(36) holds, so that the level correlation will be of the WD form.

Finally, let us consider what happens when we go from the critical regime (ξ large, but $L \gg \xi$) to the critical point ($\xi \gg L$). For this purpose, let us keep the system size L fixed and change the energy toward E_c, so that ξ increases. When ξ gets comparable to the system size, $\xi \sim L$, we have $\Delta_\xi \sim \Delta$. This is the border of applicability of the above consideration. Correspondingly, we find

$$\sigma(\mathbf{r}, \mathbf{r}, E, \omega)/\alpha(\mathbf{r}, \mathbf{r}, E) \sim 1 , \qquad \omega < \Delta \tag{38}$$

and $\sigma(\mathbf{r}, \mathbf{r}, E, \omega)/\alpha(\mathbf{r}, \mathbf{r}, E) \sim (\omega/\Delta)^{-\eta/d}$ for $\omega > \Delta$. When E approaches further E_c, the correlation length $\xi \gg L$ gets irrelevant, so that these results will hold in the critical point ($\xi = \infty$). Of course, Eq.(38) is not sufficient to ensure the WD statistics in the critical point, since there is only of order of one level within its validity range $\delta E \sim \Delta$. Indeed, the numerical simulations show that the level statistics on the mobility edge is different from the WD one [33, 34, 35, 36, 37].

However, Eq.(38) allows us to make an important conclusion concerning the behavior of $R_2(\omega)$ at small $\omega < \Delta$, or, which is essentially the same, the behavior of the nearest neighbor spacing distribution $P(s)$, $s = \omega/\Delta$, at $s < 1$. For this purpose, it is enough to consider only two neighboring levels. Let their energy difference be $\omega_0 \sim \Delta$. Let us now perturb the system by a random potential $V(\mathbf{r})$ with $\langle V(\mathbf{r}) \rangle = 0$, $\langle V(\mathbf{r})V(\mathbf{r}') \rangle = \Gamma \delta(\mathbf{r} - \mathbf{r}')$. For the two-level system it reduces to a 2×2 matrix $\{V_{ij}\}$, $i, j = 1, 2$, with elements $V_{ij} = \int d^d\mathbf{r}\, V(\mathbf{r})\Psi_i^*(\mathbf{r})\Psi_j(\mathbf{r})$. The crucial point is that the variances of the diagonal and off-diagonal matrix elements are according to Eq.(38) equal to each other up to a factor of order of unity:

$$\langle V_{11}^2 \rangle / \langle |V_{12}^2| \rangle = \sigma(\mathbf{r}, \mathbf{r}, E, \omega)/\alpha(\mathbf{r}, \mathbf{r}, E) \sim 1 \tag{39}$$

The distance between the perturbed levels is given by $\omega = [(V_{11} - V_{22} + \omega_0)^2 + |V_{12}|^2]^{1/2}$. Choosing the amplitude of the potential in such a way that the typical energy shift $V_{11} \sim \Delta$ and using Eq.(39), we find $\langle |V_{12}|^2 \rangle \sim \Delta$. As a result, the probability density for the level separation ω is for $\omega \ll \Delta$ of the form $dP \sim (\omega/\Delta)^\beta d\omega/\Delta$, with some prefactor of order of unity. We thus conclude that in the critical point $P(s) \simeq c_\beta s^\beta$ for $s \ll 1$ with a coefficient c_β of order of unity, in agreement with the numerical findings [34, 35, 37].

FLUCTUATIONS IN THE ADDITION SPECTRA OF QUANTUM DOTS

In this section we discuss effects of the eigenfunction statistics on the spectral properties of quantum dots. The electron levels of a quantum dot can be resolved if the temperature T is less than the mean single-particle level spacing in a dot and can be studied in transport experiments (see the recent review [42] and references therein). In small dots containing only few electrons these levels show a regular structure familiar from the atomic physics [43]. What we have in mind here are however larger dots (containing in typical experiments from few hundred to few thousand conducting electrons). These dots are either disordered (diffusive) or, although being ballistic, are expected to have chaotic dynamics due to their irregular shape. As a result, it is natural to use a statistical description of the properties of energy levels and eigenfunctions in such dots.

There are two types of the quantum dot spectra studied experimentally via measuring their I–V characteristics: (i) excitation spectrum, when excited levels are probed in a dot with given number of electrons by increasing the source-drain voltage, and (ii) addition spectrum, when electrons are added one by one by changing the gate voltage. In the latter case, which will be the subject of our consideration here, one finds narrow conductance peaks separated the regions of (almost) zero current (Coulomb blockade). Statistics of the spacings between these peaks was studied in a number of recent experiments [44, 45, 46]; we will return to the experimental results below.

The simplest theoretical model which may be used to study distribution of the spacings is as follows. One considers a dot as a fixed size diffusive mesoscopic sample and assumes that changing a gate voltage by an amount δV_g simply reduces to a uniform change of the potential inside the dot by a constant $\gamma \delta V_g$, with certain numerical coefficient γ ("lever arm"). Such a model was used for numerical simulations of the addition spectra in Refs.[44, 47]. Below we consider the statistics of peak spacings within this model [18], and later return to the approximations involved. We will neglect the spin degree of freedom of electrons of first; inclusion of the spin will be also discussed in the end of the section.

The distance between the two consecutive conductance peaks is given by

$$
\begin{aligned}
S_N &= (E_{N+2} - E_{N+1}) - (E_{N+1} - E_N) \\
&= \mu_{N+1}^{N+2} - \mu_N^{N+1},
\end{aligned}
\tag{40}
$$

where E_N is the ground state of a sample with N electrons. In the second line of Eq.(40) we rewrote S_N in terms of the Hartree-Fock single electron energy levels, with μ_i^j denoting the energy of the state $\#j$ in the dot containing i electrons. It is convenient to decompose S_N in the following way

$$
\begin{aligned}
S_N &= (\mu_{N+1}^{N+2} - \mu_N^{N+2}) + (\mu_N^{N+2} - \mu_N^{N+1}) \\
&\equiv E_1 + E_2
\end{aligned}
\tag{41}
$$

The quantity E_2 is the distance between the two levels of the same one-particle (Hartree-Fock) Hamiltonian $\hat{H}_N$ (describing a dot with N electrons) and is expected to obey RMT; in particular $\langle E_2 \rangle = \Delta$ and r.m.s.$(E_2) = a\Delta$ with a numerical coefficient a of order of unity [$a = 0.52$ (0.42) for the orthogonal (resp. unitary) ensemble]. On the other hand, E_1 is a shift of the level $\#(N+2)$ due to the change of the Hamiltonian $\hat{H}_N \to \hat{H}_{N+1}$ accompanying addition of the electron $\#(N+1)$ to the system. It can be in turn decomposed into the following three contributions [18]

$$
\begin{aligned}
E_1 &= e^2/C + \int d\mathbf{r} \left(|\psi_{N+1}^2(\mathbf{r})| + |\psi_{N+2}^2(\mathbf{r})| \right) \delta U(\mathbf{r}) \\
&\quad + \int d\mathbf{r}d\mathbf{r}' |\psi_{N+1}^2(\mathbf{r})||\psi_{N+2}^2(\mathbf{r}')| U_\kappa(|\mathbf{r} - \mathbf{r}'|) \\
&= E_1^{(0)} + E_1^{(1)} + E_1^{(2)}
\end{aligned}
\tag{42}
$$

Here C is the dot capacitance, $\delta U(\mathbf{r})$ is the change of the self-consistent potential due to addition of one electron (i.e. difference in the self-consistent potential in the dots with N and $N+1$ electrons), and $U_\kappa(\mathbf{r})$ is the screened Coulomb interaction (with the subscript κ denoting the inverse screening length). In particular, in the experimentally most relevant 2D case (which we will consider below) and assuming a circular form of the dot with radius R, we have

$$
\delta U(\mathbf{r}) = -\frac{e^2}{2\kappa R}(R^2 - r^2)^{-1/2},
\tag{43}
$$

while U_κ is given in the Fourier space by $\tilde{U}_\kappa(\mathbf{q}) = 2\pi e^2/\epsilon(q + \kappa)$ with $\kappa = 2\pi e^2 \nu/\epsilon$ and ϵ being the dielectric constant. The first term in Eq.(42) (the charging energy) determines the average value $\langle E_1 \rangle$ and thus the average peak spacing $\langle S_N \rangle$ (since $e^2/C \gg \Delta$ for a large dot with $N \gg 1$). This is the only contribution to E_1 which is kept by so-called constant interaction model, which in addition neglects fluctuations of the capacitance C. Consequently, fluctuations of S_N in the constant interaction model are determined solely by fluctuations of the single-particle level spacing E_2 and thus should be described by RMT: $\text{r.m.s.}(S_N) = a\Delta$.

The term E_1 in Eq.(41) is however an additional source of fluctuations and is thus responsible for the enhancement of fluctuations in comparison with RMT. In principle, all three terms $E_1^{(0)}$, $E_1^{(1)}$, and $E_1^{(2)}$ in Eq.(42) contribute to this enhancement. Fluctuations of the first one, $E_1^{(0)} = e^2/C$ are due to the fact that the capacitance is slightly different from its purely geometric value because of a finite value of the screening length. The corresponding correction to C can be expressed in terms of the polarization operator $P(\mathbf{r}, \mathbf{r}')$ [16]. The latter is a fluctuating quantity (because of fluctuations of the eigenfunctions in the Fermi sea) and contains a random part $P_r(\mathbf{r}, \mathbf{r}')$ leading to the following expression for the random part of the charging energy:

$$(e^2/C)_r = 2 \int d\mathbf{r} d\mathbf{r}' \delta U(\mathbf{r}) P_r(\mathbf{r}, \mathbf{r}') \delta U(\mathbf{r}'). \tag{44}$$

Evaluating the fluctuations of the polarization operator [16], we find [‡]

$$\text{var}(E_1^{(0)}) = \frac{48}{\beta} \nu^2 \ln g \left[\frac{1}{V} \int d\mathbf{r_1} d\mathbf{r_2} \delta U(\mathbf{r_1}) \Pi(\mathbf{r_1}, \mathbf{r_2}) \delta U(\mathbf{r_2}) \right]^2$$

$$\propto \frac{1}{\beta} \ln g \left(\frac{\Delta}{g} \right)^2 \tag{45}$$

Now we consider fluctuations of the last term, $E_1^{(2)}$, in Eq.(42). Using Eqs.(13), (27) for the correlations of eigenfunction amplitudes in two remote points, the variance of $E_1^{(2)}$ is found to be

$$\text{var}(E_1^{(2)}) = \frac{4}{\beta^2 V^4} \int d\mathbf{r_1} d\mathbf{r_1'} d\mathbf{r_2} d\mathbf{r_2'} U_\kappa(|\mathbf{r_1} - \mathbf{r_1'}|) U_\kappa(|\mathbf{r_2} - \mathbf{r_2'}|) \Pi(\mathbf{r_1}, \mathbf{r_2}) \Pi(\mathbf{r_1'}, \mathbf{r_2'})$$

$$\approx \frac{4\Delta^2}{\beta^2 V^2} \int d\mathbf{r_1} d\mathbf{r_2} \Pi^2(\mathbf{r_1}, \mathbf{r_2})$$

$$\propto \frac{1}{\beta^2} \left(\frac{\Delta}{g} \right)^2. \tag{46}$$

Finally, fluctuations of the term $E_1^{(1)}$ can be also evaluated with help of Eqs.(13), (27), yielding

$$\text{var}(E_1^{(1)}) = \frac{4}{\beta V^2} \int d\mathbf{r_1} d\mathbf{r_2} \delta U(\mathbf{r_1}) \Pi(\mathbf{r_1}, \mathbf{r_2}) \delta U(\mathbf{r_2})$$

$$\propto \frac{1}{\beta} \frac{\Delta^2}{g}. \tag{47}$$

It is seen that for $g \gg 1$ all the contributions Eqs.(45)–(47) are parametrically small compared to the RMT fluctuations (which are $\sim \Delta$). Fluctuations of the term $E_1^{(1)}$ related to the change $\delta U(\mathbf{r})$ of the self-consistent potential represent parametrically leading contribution to the enhancement of the peak spacing fluctuations with respect to RMT.

Let us now discuss approximations made in the course of the above derivation:

[‡]We use the obvious notations for the variance and the root mean square deviations of a quantity X: $\text{var}(X) = \langle X^2 \rangle - \langle X \rangle^2$; $\text{r.m.s.}(X) = [\text{var}(X)]^{1/2}$.

i) It was assumed that changing of the gate voltage results in a spatially uniform change of the potential in the sample, that led us to the expression Eq.(43) for the change of the self-consistent potential $\delta U(\mathbf{r})$ accompanying the addition of one electron to the dot. This result would correspond to a gate located far enough from the sample. In a more realistic situation, when the gate is relatively narrow and located close to the sample, the potential change $\delta U(\mathbf{r})$ (as well as the additional electron density) will be mainly located on the side of the dot facing the gate. Furthermore, the total area of the dot is not fixed, so that the dot gets larger (and slightly deformed) with adding each electron to it. These effects lead to some increase of the fluctuations of $E_1^{(1)}$ in comparison with the model considered above. If the size of the gate and its distance to the dot are of the same order of magnitude as the dot size, then

$$\text{r.m.s.}(E_1^{(1)}) \propto \Delta/\sqrt{\beta g},$$

as in Eq.(47), with a geometry dependent-numerical prefactor. The upper bound for the magnitude of fluctuations of $E_1^{(1)}$ is given by the opposite limiting case, when additional electron density occupies an area $\sim \lambda_F \times \lambda_F$ near the gate, in which case one finds

$$\text{r.m.s.}(E_1^{(1)}) \sim \Delta/\sqrt{\beta}.$$

ii) The dot was supposed to be diffusive in the calculation. For a ballistic dot one should replace $\Pi(\mathbf{r},\mathbf{r}')$ by $\Pi_B(\mathbf{r},\mathbf{r}')$, as was explained above. This would mean that the parameter g is replaced by $\sim N^{1/2} \sim L/\lambda_F$, where N is the number of electrons in the dot and L the characteristic linear dimension. The numerical coefficient would depend, however, on "how strongly chaotic" is the dot. Role of the eigenfunctions fluctuations and correlations ("scars") in enhancement of the peak spacing fluctuations was studied in [48] via numerical simulations of a dot with $N \approx 100$ electrons.

iii) It was assumed that the dot energy and the measured gate voltage are related through a constant (or smoothly varying) coefficient γ. This "lever arm" γ depends, however on the dot-gate capacitance, which is also a fluctuating quantity. If the gate size and the distance to the gate is of the same order as the size of the dot, these fluctuations should be of the same order as fluctuations of the dot self-capacitance given by Eq.(45), and thus lead to additional fluctuations which are parametrically small compared to Δ. In a more general situation (thin gate located close to the sample) an additional analysis along the lines of Ref.16 is necessary.

iv) The calculation was done within the random phase approximation, which assumes that $r_s \equiv e^2/\epsilon v_F \ll 1$, with v_F being the Fermi velocity. In realistic dots however $r_s \sim 1$. Since this value is still considerably lower than the Wigner crystallization threshold, the calculations should be still valid, up to a numerical factor $\alpha(r_s)$ [depending on r_s only and such that $\alpha(r_s \ll 1) = 1$].

v) We considered the model of spinless electrons up to now. Let us briefly discuss the role of the spin degree of freedom. Within the constant interaction model, it would lead to a bimodal distribution [47] of peak spacings

$$\mathcal{P}(S_N) = \frac{1}{2}\left[\delta(S_N - e^2/C) + \frac{1}{2\Delta}P_{WD}\left(\frac{S_N - e^2/C}{2\Delta}\right)\right], \tag{48}$$

257

where $P_{WD}(s)$ is the Wigner-Dyson distribution and Δ denotes the level spacing in the absence of spin degeneracy. The value of the coefficient a in the relation r.m.s.$(S_N) = a\Delta$ is then increased (compared to the spinless case) and is equal to 1.24 (1.16) for the orthogonal (resp. unitary) ensemble. Taking into account fluctuations of eigenfunctions (and thus of E_1) however modifies the form of the distribution [18]. The value of the term $E_1^{(2)}$ representing the interaction between two electrons is larger in the case when ψ_{N+2} and ψ_{N+1} correspond to two spin-degenerate states (i.e. have the same spatial dependence of the wave function), since

$$\left\langle \int d\mathbf{r} d\mathbf{r}' |\psi_i^2(\mathbf{r})||\psi_i^2(\mathbf{r}')|U_\kappa(|\mathbf{r} - \mathbf{r}'|)\right\rangle - \left\langle \int d\mathbf{r} d\mathbf{r}' |\psi_i^2(\mathbf{r})||\psi_j^2(\mathbf{r}')|U_\kappa(|\mathbf{r} - \mathbf{r}'|)\right\rangle$$

$$= \frac{2}{\beta V^2} \int d\mathbf{r} d\mathbf{r}' k_d(|\mathbf{r} - \mathbf{r}'|)U_\kappa(|\mathbf{r} - \mathbf{r}'|) \sim \Delta \tag{49}$$

for $r_s \sim 1$ (the coefficient depends on r_s, see [18]). Therefore, filling a state $\psi_{i\uparrow}$ pushes up the level $\psi_{i\downarrow}$ (with respect to other eigenstates) by an amount of order of Δ. This removes a bimodal structure of the distribution of peak spacings and slightly modifies the value of the coefficient a.

Basing on the above analysis, we can make the following general statement. Imaging that we fix $r_s \sim 1$ (i.e. fix the electron density and thus the Fermi wave length) and the system geometry, and then start to increase the linear dimension L of the system. Then, while the average value of the peak spacing S_N scales as $\langle S_N \rangle \approx e^2/C \propto 1/L$, its fluctuations will scale differently: r.m.s.$(S_N) \sim \Delta \propto 1/L^2$. This result is not at all trivial, since in an analogous problem for classical particles [49, 50] the fluctuations are proportional to the mean value $\langle S_N \rangle$. The physical reason for smaller fluctuations in the quantum case is in the delocalized nature of the electronic wave functions, which are spread roughly uniformly over the system.

The above prediction was confirmed by a recent experiment [46], where a thorough study of the peak spacing spacing statistics was carried out. It was found that the low-temperature value of r.m.s.(S_N), as well the typical temperature scale for its change are approximately given by the mean level spacing Δ (while in units of E_c the magnitude of fluctuations was very small, typically 2–4%). We note also that in recent numerical simulations [48] fluctuations of the addition energies were found to be approximately 0.7Δ, in agreement with our results.

ACKNOWLEDGMENTS

It is a pleasure to thank my collaborators Ya. M. Blanter, Y. V. Fyodorov, and B. A. Muzykantskii. Useful discussions with O. Agam, S. Fishman, Y. Gefen, V. E. Kravtsov, and C. Marcus are gratefully acknowledged. I also thank S. R. Patel for a useful remark. This research was supported by SFB195 der Deutshen Forschungsgemeinschaft.

REFERENCES

1. R. A. Jalabert, A. D. Stone, and Y. Alhassid, Phys. Rev. Lett. **68**, 3468 (1992).
2. V. N. Prigodin, K. B. Efetov, and S. Iida, Phys. Rev. Lett. **71**, 1230 (1993).
3. E. R. Mucciolo, V. N. Prigodin, and B. L. Altshuler, Phys. Rev. B **51**, 1714 (1995).
4. A. M. Chang, H. U. Baranger, L. N. Pfeiffer, K. W. West, and T. Y. Chang, Phys. Rev. Lett. **76**, 1695 (1996).

5. J. A. Folk, S. R. Patel, S. F. Godijn, A. G. Huibers, S. M. Cronenwett, C. M. Marcus, K. Campman, and A. C. Gossard, Phys. Rev. Lett. **76**, 1699 (1996).

6. H. J. Stöckmann and J. Stein, Phys. Rev. Lett. **64**, 2215 (1990); J. Stein and H. J. Stöckmann, Phys. Rev. Lett. **68**, 2867 (1992).

7. S. Sridhar, Phys. Rev. Lett. **67**, 785 (1991); A. Kudrolli, V. Kidambi, and S. Sridhar, Phys. Rev. Lett. **75**, 822 (1995).

8. A. D. Mirlin and Y. V. Fyodorov, J. Phys. A: Math. Gen. **26**, L551 (1993); Y. V. Fyodorov and A. D. Mirlin, Int. Journ. Mod. Phys. B **8**, 3795 (1994).

9. Y. V. Fyodorov and A. D. Mirlin, Phys. Rev. B **51**, 13403 (1995).

10. V. I. Fal'ko and K. B. Efetov, Phys. Rev. B **52**, 17413 (1995).

11. A. D. Mirlin, J. Math. Phys. **38**, 1888 (1997).

12. M. V. Berry, J. Phys. A **10**, 2083 (1977).

13. V. N. Prigodin, Phys. Rev. Lett. **74**, 1566 (1995); V. N. Prigodin, N. Taniguchi, A. Kudrolli, V. Kidambi, and S. Sridhar, Phys. Rev. Lett. **75**, 2392 (1995).

14. M. Srednicki, Phys. Rev. E **54**, 954 (1996); M. Srednicki and F. Stiernelof, J. Phys. A **29**, 5817 (1996).

15. Ya. M. Blanter and A. D. Mirlin, Phys. Rev. E **55**, 6514 (1997).

16. Ya. M. Blanter and A. D. Mirlin, Phys. Rev. B **57**, 4566 (1998).

17. Y. V. Fyodorov and A. D. Mirlin, Phys. Rev. B **55**, R16001 (1997).

18. Ya. M. Blanter, A. D. Mirlin, and B. A. Muzykantskii, Phys. Rev. Lett. **78**, 2449 (1997).

19. Ya. M. Blanter, A. D. Mirlin, and B. A. Muzykantskii, Phys. Rev. Lett. **80**, 4161 (1998).

20. B. L. Altshuler and B. I. Shklovskii, Zh. Eksp. Teor. Fiz. **91**, 220 (1986) [Sov. Phys. JETP **64** (1986), 127].

21. K. B. Efetov, Adv. Phys. **32**, 53 (1983).

22. V. E. Kravtsov and A. D. Mirlin, Pis'ma Zh. Eksp. Teor. Fiz. **60**, 645 (1994) [JETP Lett. **60**, 656 (1994)].

23. B. A. Muzykantskii and D. E. Khmelnitskii, Pis'ma Zh. Éksp. Teor. Fiz. **62**, 68 (1995) [JETP Lett. **62**, 76 (1995)]; cond-mat/9601045 (unpublished).

24. A. V. Andreev, O. Agam, B. Simons, and B. L. Altshuler, Phys. Rev. Lett. **76**, 3947 (1996); A. V. Andreev, B. Simons, O. Agam, and B. L. Altshuler, Nucl. Phys. B **482**, 536 (1996).

25. B. Eckhardt, S. Fishman, J. Keating, O. Agam, J. Main, and K. Müller, Phys. Rev. E **52**, 5893 (1995).

26. S. Hortikar and M. Srednicki, Phys. Rev. Lett. **80**, 1646 (1998); Phys. Rev. E **57**, 7313 (1998).

27. E. Heller, in *Chaos and Quantum Physics*, M.-J. Giannoni, A. Voros, and J. Zinn-Justin, eds. (North-Holland, 1991), p.547, and references therein.

28. O. Agam and S. Fishman, J. Phys. A **26**, 2113 (1993).

29. F. Wegner, Z. Phys. B **36**, 209 (1980).

30. C. Castellani, C. di Castro, and L. Peliti, J. Phys. A: Math. Gen. **19**, L1099 (1986);

31. J. T. Chalker and G. J. Daniell, Phys. Rev. Lett. **61**, 593 (1988).

32. B. Huckestein and L. Schweitzer, Phys. Rev. Lett. **72**, 713 (1994); T. Brandes, B. Huckestein, L. Schweitzer, Ann. Physik **5**, 633 (1996).

33. B. L. Altshuler, I. K. Zharekeshev, S. A. Kotochigova, and B. I. Shklovskii, Zh. Eksp. Teor. Fiz. **94**, 343 (1988) [Sov. Phys. JETP **67**, 625 (1988)].

34. B. I. Shklovskii, B. Shapiro, B. R. Sears, P. Lambrianides, and H. B. Shore, Phys. Rev. B **47**, 11487 (1993).

35. S. N. Evangelou, Phys. Rev. B **49**, 16805 (1994).

36. D. Braun and G. Montambaux, Phys. Rev. B **52**, 13903 (1995).

37. I. Kh. Zharekeshev and B. Kramer, Phys. Rev. B **51**, 17356 (1995).

38. A. D. Mirlin and Y. V. Fyodorov, J. Phys. A: Math. Gen. **24**, 2273 (1991); Y. V. Fyodorov and A. D. Mirlin, Phys. Rev. Lett. **67**, 2052 (1991).

39. Y. V. Fyodorov, A. D. Mirlin, and H.-J. Sommers, J. Phys. I France **2**, 1571 (1992).

40. A. D. Mirlin and Y. V. Fyodorov, Phys. Rev. Lett. **72**, 526 (1994); J. Phys. I France **4**, 655 (1994).

41. K. Pracz, M. Janssen, P. Freche, J. Phys. Cond. Matter **8**, 7147 (1996).

42. L. P. Kouwenhoven, C. M. Marcus, P. L. McEuen, S. Tarucha, R. M. Westerwelt, and N. S. Wingreen, in *Mesoscopic Electron Transport*, ed. by L. L. Sohn, L. P. Kouwenhoven, and G. Schön (Kluwer, 1997), p.105.

43. S. Tarucha, D. G. Austing, T. Honda, R. J. van der Hage, and L. P. Kouwenhoven, Phys. Rev. Lett. **77**, 3613 (1996).

44. U. Sivan, R. Berkovits, Y. Aloni, O. Prus, A. Auerbach, and G. Ben-Joseph, Phys. Rev. Lett. **77**, 1123 (1996).

45. F. Simmel, T. Heinzel, and D. A. Wharam, Europhys. Lett. **38**, 123 (1997).

46. S. R. Patel, S. M. Cronenwett, D. R. Stewart, A. G. Huibers, C. M. Marcus, C. I. Duruöz, J. S. Harris, K. Campman, and A. C. Gossard, Phys. Rev. Lett. **80**, 4522 (1998).

47. O. Prus, A. Auerbach, Y. Aloni, U. Sivan, and R. Berkovits, Phys. Rev. B **54**, R14289 (1996).

48. M. Stopa, preprint cond-mat/9709119.

49. J. R. Morris, D. M. Deaven, and K. M. Ho, Phys. Rev. B **53**, R1740 (1996).

50. A. A. Koulakov, F. G. Pikus, and B. I. Shklovskii, Phys. Rev. B **55**, 9223 (1997).

SPATIAL CORRELATIONS IN CHAOTIC EIGENFUNCTIONS

Mark Srednicki

Department of Physics
University of California
Santa Barbara, CA 93106 USA

My focus in this talk will be on the statistical properties of the quantum energy eigenfunctions in classically chaotic systems. Twenty years ago, Berry[1] conjectured that these eigenfunctions could be treated as gaussian random variables, and that the spatial correlations would be those following from taking the expected value of the Wigner density for each eigenfunction to be microcanonical. More specifically, the probability that the actual eigenfunction $\psi_\alpha(\mathbf{q})$ for energy $E = E_\alpha$ (the corresponding energy eigenvalue) is between $\psi(\mathbf{q})$ and $\psi(\mathbf{q}) + d\psi(\mathbf{q})$ for all coordinate points $\mathbf{q}$ is given by

$$P(\psi|E) \propto \exp\left[-\frac{\beta}{2} \int d^f q \int d^f q' \; \psi^*(\mathbf{q}) K(\mathbf{q}, \mathbf{q}'|E)\psi(\mathbf{q}')\right] \mathcal{D}\psi . \tag{1}$$

If the system is time-reversal invariant, the eigenfunctions are real, $\beta = 1$, and the measure is $\mathcal{D}\psi = \prod_{\mathbf{q}} d\psi(\mathbf{q})$; if it is not, the eigenfunctions are complex, $\beta = 2$, and the measure is $\mathcal{D}\psi = \prod_{\mathbf{q}} d\operatorname{Re}\psi(\mathbf{q}) \, d\operatorname{Im}\psi(\mathbf{q})$. In either case, the kernel $K(\mathbf{q}, \mathbf{q}'|E)$ is the functional inverse of the two-point correlation function

$$C(\mathbf{q}', \mathbf{q}|E) \equiv \langle \psi(\mathbf{q}')\psi^*(\mathbf{q})\rangle , \tag{2}$$

where the angle brackets denote averaging over $P(\psi|E)$. Taking the Wigner density to be microcanonical results in

$$C(\mathbf{q}', \mathbf{q}|E) = \frac{1}{\bar{\rho}(E)} \int \frac{d^f p}{(2\pi\hbar)^f} \; e^{i\mathbf{p}\cdot(\mathbf{q}'-\mathbf{q})/\hbar}\delta(E - H_W(\mathbf{p}, \bar{\mathbf{q}})) , \tag{3}$$

where $\bar{\mathbf{q}} = \frac{1}{2}(\mathbf{q}+\mathbf{q}')$ is the midpoint, $H_W(\mathbf{p}, \mathbf{q})$ is the classical hamiltonian (more specifically, it is the Weyl symbol of the hamiltonian operator), and $\bar{\rho}(E)$ is the semiclassical density of states,

$$\bar{\rho}(E) = \int \frac{d^f p \, d^f q}{(2\pi\hbar)^f} \; \delta(E - H_W(\mathbf{p}, \mathbf{q})) . \tag{4}$$

Berry's conjecture for $C(\mathbf{q}', \mathbf{q}|E)$ can be tested numerically in two-dimensional billiard systems; in this case, $H = \mathbf{p}^2/2m$, and we have $C(\mathbf{q}', \mathbf{q}|E) = J_0(k|\mathbf{q}'-\mathbf{q}|)$, where

Supersymmetry and Trace Formulae: Chaos and Disorder
Edited by Lerner *et al.*, Kluwer Academic / Plenum Publishers, New York, 1999

$E = \hbar^2 k^2/2m$ and $J_0(x)$ is an ordinary Bessel function. An "experimental" correlation function $C_{\exp}(\mathbf{s}, D|E)$ can be defined in terms of a particular energy eigenfunction $\psi_\alpha(\mathbf{q})$ via

$$C_{\exp}(\mathbf{s}, D|E) \equiv \frac{A_B}{A_D} \int_D d^2q \, \psi_\alpha^*(\mathbf{q} + \tfrac{1}{2}\mathbf{s})\psi_\alpha(\mathbf{q} - \tfrac{1}{2}\mathbf{s}) \,, \tag{5}$$

where $E = E_\alpha$, A_B is the area of the billiard, and A_D is the area of a domain D over which the central point $\mathbf{q}$ is averaged. Early computations[2, 3] of $C_{\exp}(\mathbf{s}, D|E)$ found that it typically resembled its predicted value $J_0(k|\mathbf{s}|)$, but with large fluctuations. Later it was realized[4] that these fluctuations were in fact entirely consistent with the expected gaussian fluctuations[5] of $\psi_\alpha(\mathbf{q})$. Recent detailed computations[6, 7] leave little doubt that (at asymptotically high energies) the eigenfunctions of chaotic billiards are very well described by Berry's conjecture.

Berry's conjecture also follows from other approaches to quantum chaos. The random-matrix analogy[8, 9] suggests that the energy eigenstates have a Porter-Thomas distribution of amplitudes in a suitably chosen basis. For billiards, the natural choice is the momentum basis, and from this starting point one can derive eqs. (1) and (3)[10]. If one adds a white-noise, spatially uncorrelated random potential to the system, and takes the limit where the mean-free path is much larger than the system size, then eq. (1) can be derived via the supersymmetric sigma-model technique[11, 12, 13, 14, 15, 16].

However, we cannot expect eq. (3) to be valid when the separation between $\mathbf{q}$ and $\mathbf{q}'$ becomes large compared to distance scales which are important classically. This is because the hamiltonian evaluated at just the midpoint $\bar{\mathbf{q}}$ does not contain enough information about the classical landscape between $\mathbf{q}$ and $\mathbf{q}'$. To find the correct formula for $C(\mathbf{q}', \mathbf{q}|E)$ when $|\mathbf{q}'-\mathbf{q}|$ is large[17], we first consider the theory with an added random potential. In this case an expression for $C(\mathbf{q}', \mathbf{q}|E)$ can be derived explicitly[12]:

$$C(\mathbf{q}', \mathbf{q}|E) = \frac{1}{2\pi i \bar{\rho}(E)} \left[\overline{G}(\mathbf{q}, \mathbf{q}'|E)^* - \overline{G}(\mathbf{q}', \mathbf{q}|E) \right] \,. \tag{6}$$

Here $G(\mathbf{q}', \mathbf{q}|E)$ is the energy Green's function,

$$G(\mathbf{q}', \mathbf{q}|E) = \sum_\alpha \frac{\psi_\alpha(\mathbf{q}')\psi_\alpha^*(\mathbf{q})}{E - E_\alpha + i0^+} \,, \tag{7}$$

$\rho(E)$ is the density of states,

$$\rho(E) = \frac{1}{2\pi i} \int d^f q \, [G(\mathbf{q}, \mathbf{q}|E)^* - G(\mathbf{q}, \mathbf{q}|E)] \,, \tag{8}$$

and the bar stands for averaging over the random potential. We now make the assumption that, for a chaotic system, eq. (6) continues to hold if we first take the limit of small $\hbar$, and then disregard the random potential.

There are two different semiclassical approximations which can be used to compute the Green's function in the small-$\hbar$ limit[18, 19]. The first applies when the distance between $\mathbf{q}$ and $\mathbf{q}'$ is small, in the sense that the classical path of least action with energy E which connects $\mathbf{q}$ to $\mathbf{q}'$ is well approximated by a linear function of time (in any coordinate system). This path then dominates the Green's function, and one finds

$$\overline{G}(\mathbf{q}', \mathbf{q}|E) = \int \frac{d^f p}{(2\pi\hbar)^f} \, e^{i\mathbf{p}\cdot(\mathbf{q}'-\mathbf{q})/\hbar} \frac{1}{E - H_W(\mathbf{p}, \bar{\mathbf{q}}) + i0^+} \,, \tag{9}$$

which immediately yields eq. (3) for $C(\mathbf{q}', \mathbf{q}|E)$. If, however, the classical path of least action is not well approximated by a linear function of time, and the value of this action is much larger than $\hbar$, then we have instead

$$\overline{G}(\mathbf{q}',\mathbf{q}|E) = \frac{1}{i\hbar(2\pi i\hbar)^{(f-1)/2}} \sum_{\text{paths}} |D_{\text{p}}|^{1/2} e^{iS_{\text{p}}/\hbar - i\nu_{\text{p}}\pi/2} \ . \tag{10}$$

Here the sum is over all classical paths connecting $\mathbf{q}$ to $\mathbf{q}'$ with energy E, action

$$S_{\text{p}} = \int_{\mathbf{q}}^{\mathbf{q}'} \mathbf{p} \cdot d\mathbf{q} \ , \tag{11}$$

focal point number ν_{p}, and fluctuation determinant

$$D_{\text{p}} = \det \begin{pmatrix} \frac{\partial^2 S_{\text{p}}}{\partial \mathbf{q}_2 \partial \mathbf{q}_1} & \frac{\partial^2 S_{\text{p}}}{\partial E \partial \mathbf{q}_1} \\[2mm] \frac{\partial^2 S_{\text{p}}}{\partial \mathbf{q}_2 \partial E} & \frac{\partial^2 S_{\text{p}}}{\partial E^2} \end{pmatrix} \ . \tag{12}$$

Eqs. (6) and (10) give the correct formula for $C(\mathbf{q}',\mathbf{q}|E)$ when $|\mathbf{q}' - \mathbf{q}|$ is large[17].

This formula is most useful under circumstances where the sum over paths is dominated by a single path of least action. An interesting example is a two-dimensional billiard in a perpendicular magnetic field $\mathbf{B}$. In the billiard interior the hamiltonian is $H = (\mathbf{p} - e\mathbf{A})^2/2m$, and we will work in the gauge in which the vector potential is $\mathbf{A} = \frac{1}{2}\mathbf{B} \times \mathbf{q}$. Assuming that it is not blocked, the classical path of least action is a circular arc with length ℓ, related to the separation $L = |\mathbf{q}' - \mathbf{q}|$ and classical cyclotron radius $R = (2mE)^{1/2}/|eB|$ via $\ell = 2R\sin^{-1}(L/2R)$. The action for this path can be divided into a geometric part and a gauge-dependent part, $S_{\text{p}} = S_{\text{geom}} + S_{\text{gauge}}$. The geometric part is

$$\begin{aligned} S_{\text{geom}} &= \hbar k \left(\ell - \frac{\mathcal{A}}{R}\right) \\[2mm] &= \hbar k L \left(1 - \frac{L^2}{24R^2} + \dots\right) \ , \end{aligned} \tag{13}$$

where $\hbar k = (2mE)^{1/2}$, and $\mathcal{A} = \frac{1}{2}R\ell - \frac{1}{2}R^2\sin(\ell/R)$ is the area enclosed by the circular arc and the straight line connecting $\mathbf{q}$ to $\mathbf{q}'$. For our gauge choice, the gauge-dependent part is

$$S_{\text{gauge}} = \frac{1}{2}e\mathbf{B} \cdot (\mathbf{q} \times \mathbf{q}') \ . \tag{14}$$

In any gauge, the determinant $|D_{\text{p}}|$ is given by

$$|D_{\text{p}}| = \frac{m^2}{\hbar k L} \left(1 - \frac{L^2}{4R^2}\right)^{-1/2} \ . \tag{15}$$

Keeping only the contribution of this one path, and recalling that $\bar{\rho}(E) = mA_B/2\pi\hbar^2$ for a two-dimensional billiard with area A_B, we find from eqs. (6) and (10) that

$$C(\mathbf{q}',\mathbf{q}|E) = A_B^{-1} \exp(iS_{\text{gauge}}/\hbar) \frac{\cos(S_{\text{geom}}/\hbar - \pi/4)}{(\pi k L/2)^{1/2}(1 - L^2/4R^2)^{1/4}} \ . \tag{16}$$

We need $kL \gg 1$ for this formula to hold. Also, to avoid an integrable region of phase space, a circle of radius R must not be able to fit inside the billiard; this implies $L < 2R$.

We can compare eq. (16) with the result of using Berry's formula, eq. (3); with our gauge choice we find

$$C(\mathbf{q}',\mathbf{q}|E) = A_B^{-1} \exp(iS_{\text{gauge}}/\hbar) J_0(kL) \ . \tag{17}$$

For large kL, we can use the asymptotic form of the Bessel function to get

$$C(\mathbf{q}', \mathbf{q}|E) = A_B^{-1} \exp(iS_{\text{gauge}}/\hbar) \frac{\cos(kL - \pi/4)}{(\pi kL/2)^{1/2}} \,. \tag{18}$$

We see that Berry's formula misses the corrections due to the finite cyclotron radius R which are present in eq. (16).

If we make a gauge transformation

$$\mathbf{A}(\mathbf{q}) \to \mathbf{A}(\mathbf{q}) + \nabla\Phi(\mathbf{q}) \,, \tag{19}$$

where $\Phi(\mathbf{q})$ is any smooth function, then

$$S_{\text{gauge}} \to S_{\text{gauge}} + e\Phi(\mathbf{q}') - e\Phi(\mathbf{q}) \,, \tag{20}$$

and so eq. (16) implies

$$C(\mathbf{q}', \mathbf{q}|E) \to e^{+ie[\Phi(\mathbf{q}') - \Phi(\mathbf{q})]/\hbar} C(\mathbf{q}', \mathbf{q}|E) \,. \tag{21}$$

That this is correct can be seen by recalling that a wave function $\psi(\mathbf{q})$ transforms as

$$\psi(\mathbf{q}) \to e^{+ie\Phi(\mathbf{q})/\hbar} \psi(\mathbf{q}) \tag{22}$$

under (19), and that $C(\mathbf{q}', \mathbf{q}|E_\alpha)$ is the expected value of $\psi_\alpha(\mathbf{q}')\psi_\alpha^*(\mathbf{q})$. On the other hand, eq. (3) implies instead that

$$C(\mathbf{q}', \mathbf{q}|E) \to e^{+ie(\mathbf{q}' - \mathbf{q}) \cdot \nabla\Phi(\bar{\mathbf{q}})/\hbar} C(\mathbf{q}', \mathbf{q}|E) \,, \tag{23}$$

which again illustrates the fact that Berry's formula is valid only when $|\mathbf{q}' - \mathbf{q}|$ is sufficiently small.

I now turn to another issue, the statistical properties of transition matrix elements $A_{\alpha\beta} \equiv \langle\alpha|A|\beta\rangle$ in chaotic systems. Here the operator A is a smooth, $\hbar$-independent function of the coordinates and momenta, and $|\alpha\rangle$ and $|\beta\rangle$ are different energy eigenstates. We will be interested in the average value of $|A_{\alpha\beta}|^2$ when the energies E_α and E_β are each varied over a range which is small classically, but encompasses many quantum energy levels. For later convenience let us define

$$\bar{E} = \tfrac{1}{2}(E_\alpha + E_\beta) \qquad \text{and} \qquad \hbar\omega = E_\beta - E_\alpha \,. \tag{24}$$

We will do our computations in the limit $\hbar \to 0$ with $\bar{E}$ and ω held fixed. The mean level spacing near energy $\bar{E}$ is $1/\bar{\rho}(\bar{E}) \sim \hbar^f$, so our condition on the ranges of E_α and E_β is easy to fulfill.

One approach[20, 21, 22] to this problem is to relate the average value $\langle|A_{\alpha\beta}|^2\rangle$ to the classical time-correlation function of the observable A via

$$\langle|A_{\alpha\beta}|^2\rangle = \frac{1}{\tau_{\mathrm{H}}} \int_{-\tau_{\mathrm{H}}}^{\tau_{\mathrm{H}}} dt\, e^{i\omega t} \int d\mu_{\bar{E}}\, A_W(\mathbf{p}_t, \mathbf{q}_t) A_W(\mathbf{p}, \mathbf{q}) \,. \tag{25}$$

Here $\tau_{\mathrm{H}} = 2\pi\hbar\bar{\rho}(\bar{E})$ is the Heisenberg time, $(\mathbf{p}_t, \mathbf{q}_t)$ is the point in phase space which is reached classically at time t when starting from $(\mathbf{p}, \mathbf{q})$ at time zero, $A_W(\mathbf{p}, \mathbf{q})$ is the Weyl symbol of the operator A, and $d\mu_E$ is the Liouville measure on the surface in phase space with energy E, given by

$$d\mu_E = \frac{1}{\bar{\rho}(E)} \frac{d^f p\, d^f q}{(2\pi\hbar)^f} \delta(E - H_W(\mathbf{p}, \mathbf{q})) \,. \tag{26}$$

Note that the notation, though useful later, is somewhat misleading: $d\mu_E$ is a purely classical, $\hbar$-independent object. We will not discuss the derivation[20, 21, 22] of eq. (25) here; a brief and non-rigorous review is given elsewhere[23].

Another approach[4, 24, 7] to computing $\langle|A_{\alpha\beta}|^2\rangle$ is to write $|A_{\alpha\beta}|^2$ in terms of the energy eigenfunctions $\psi_\alpha(\mathbf{q})$ and $\psi_\beta(\mathbf{q})$, and then average $|A_{\alpha\beta}|^2$ over the eigenfunction probability distribution of eq. (1). This calculation is simplified if A is a function of $\mathbf{q}$ only (rather than both $\mathbf{q}$ and $\mathbf{p}$), and so we specialize to this case. We get

$$\langle|A_{\alpha\beta}|^2\rangle = \int d^f q'\, d^f q\, C(\mathbf{q},\mathbf{q}'|E_\alpha) A_W(\mathbf{q}') C(\mathbf{q}',\mathbf{q}|E_\beta) A_W(\mathbf{q}) \ . \tag{27}$$

It is not at all clear that eq. (27) gives the same result for $\langle|A_{\alpha\beta}|^2\rangle$ as eq. (25), an issue which was raised (in the context of the diagonal matrix elements) by Austin and Wilkinson[25]. If Berry's formula is used for $C(\mathbf{q}',\mathbf{q}|E)$, then eqs. (25) and (27) do not necessarily agree, as can be seen by working out some simple examples. In this case, it is eq. (27) which is wrong. The reason is that $\mathbf{q}$ and $\mathbf{q}'$ are independently integrated in eq. (27), and so $|\mathbf{q}' - \mathbf{q}|$ is generically large. Therefore we should use eqs. (6) and (10) for $C(\mathbf{q}',\mathbf{q}|E)$, rather than eq. (3). If we do, then eqs. (25) and (27) give the same result for $\langle|A_{\alpha\beta}|^2\rangle$.

To see this still requires some work. A detailed exposition is given elsewhere[23], and here we will just highlight the key elements. One is the diagonal approximation[26]; after substituting eqs. (6) and (10) into eq. (27), the double sum over paths is collapsed to a single sum, since the off-diagonal terms will have rapidly oscillating phases. In different but related contexts[26, 24], this requires restricting the single sum to paths whose elapsed times are less than the Heisenberg time, and we will assume the same is true here. The action difference in the single sum is

$$
\begin{aligned}
S_{\mathrm{p}\beta} - S_{\mathrm{p}\alpha} &= (E_\beta - E_\alpha)\frac{\partial S_\mathrm{p}}{\partial E}\bigg|_{E=\bar{E}} + \dots \\
&= \hbar\omega\tau_\mathrm{p} + O(\hbar^2) \ ,
\end{aligned}
\tag{28}
$$

where τ_p is the elapsed time along the path. We assume that $\nu_{\mathrm{p}\beta} = \nu_{\mathrm{p}\alpha}$, since ν_p is a topological quantity which in general will not change when the energy of the path is varied slightly. Finally, we note that the determinant D_p can be written as

$$D_\mathrm{p} = \det\begin{pmatrix} -\dfrac{\partial \mathbf{p}}{\partial \mathbf{q}'} & \dfrac{\partial \tau}{\partial \mathbf{q}'} \\[2mm] -\dfrac{\partial \mathbf{p}}{\partial E} & \dfrac{\partial \tau}{\partial E} \end{pmatrix} \ . \tag{29}$$

Here $\mathbf{p} = -\partial S_\mathrm{p}/\partial \mathbf{q}$ is the momentum at the beginning of the path, and $\tau = \tau_\mathrm{p}$ is the elapsed time along the path. Eq. (29) shows us that $|D_\mathrm{p}|$ can be thought of[18] as a jacobian for a change of variables from the final position $\mathbf{q}'$ and total energy $\bar{E}$ to the initial momentum $\mathbf{p}$ and elapsed time τ. Eq. (27) already has an integral over $\mathbf{q}'$, and to get one over $\bar{E}$ we insert $1 = \int d\bar{E}\, \delta(\bar{E} - H_W(\mathbf{p},\mathbf{q}))$. Now we can make the change of integration variables suggested by eq. (29), and we find

$$\langle|A_{\alpha\beta}|^2\rangle = \frac{1}{\pi\hbar\bar{\rho}(\bar{E})^2} \int_0^{\tau_\mathrm{H}} d\tau \int \frac{d^f p\, d^f q}{(2\pi\hbar)^f} \sum_{\text{paths}} \delta(\bar{E} - H_W(\mathbf{p},\mathbf{q})) \cos(\omega\tau) A_W(\mathbf{q}') A_W(\mathbf{q}) \ . \tag{30}$$

The sum is now over all paths which begin at $(\mathbf{p},\mathbf{q})$ and have elapsed time τ. However, there is only one such path, and so the sum over paths may be dropped. Also, $\mathbf{q}'$ is the position at time τ, and it is now more properly denoted $\mathbf{q}_\tau$. Using eq. (26), the fact that time-translation invariance implies that $\int d\mu_{\bar{E}}\, A_W(\mathbf{q}_\tau) A_W(\mathbf{q})$ is an even function of τ (even if the system is not time-reversal invariant), and $\tau_H = 2\pi\hbar\bar\rho(\bar E)$, we see that eq. (30) can be rewritten as

$$\langle |A_{\alpha\beta}|^2 \rangle = \frac{1}{\tau_H} \int_{-\tau_H}^{+\tau_H} d\tau\, e^{i\omega\tau} \int d\mu_{\bar E}\, A_W(\mathbf{q}_\tau) A_W(\mathbf{q}) \,, \tag{31}$$

which is equivalent to eq. (25).

Another quantity of interest is the size of the fluctuations in the diagonal matrix elements $A_{\alpha\alpha}$. If we first shift A (if necessary) so that $\langle A_{\alpha\alpha}\rangle = \int d\mu_{E_\alpha} A_W(\mathbf{q}) = 0$, then the object we wish to evaluate is $\langle |A_{\alpha\alpha}|^2\rangle$. This has been done previously[27, 28, 24] by making use of the trace formula[18, 29, 30] and properties of periodic orbits. Here we will instead compute $\langle |A_{\alpha\alpha}|^2\rangle$ by averaging over the probability distribution for energy eigenfunctions[4, 24, 7, 23]. In the case of a system which is not invariant under time reversal, the energy eigenfunctions are generically complex, and the relevant formula is[15]

$$\langle \psi_1^* \psi_2 \psi_3^* \psi_4\rangle = \langle \psi_1^* \psi_2\rangle\langle \psi_3^* \psi_4\rangle + \langle \psi_1^* \psi_4\rangle\langle \psi_3^* \psi_2\rangle \,, \tag{32}$$

where $\psi_i = \psi_\alpha(\mathbf{q}_i)$. If the system is invariant under time reversal, the energy eigenfunctions are real, and we have instead[15]

$$\langle \psi_1 \psi_2 \psi_3 \psi_4\rangle = \langle \psi_1 \psi_2\rangle\langle \psi_3 \psi_4\rangle + \langle \psi_1 \psi_4\rangle\langle \psi_2 \psi_3\rangle + \langle \psi_1 \psi_3\rangle\langle \psi_2 \psi_4\rangle \,. \tag{33}$$

Combined with the previous results for $\langle |A_{\alpha\beta}|^2\rangle$, we find that

$$\langle |A_{\alpha\alpha}|^2\rangle = \frac{2/\beta}{\tau_H} \int_{-\tau_H}^{\tau_H} d\tau \int d\mu_{E_\alpha}\, A_W(\mathbf{q}_\tau) A_W(\mathbf{q}) \,. \tag{34}$$

Here $\beta = 1$ for a system which is invariant under time reversal, and $\beta = 2$ for a system which is not. Eq. (34) is in agreement with the earlier results[27, 28, 24].

Acknowledgments

This work was supported in part by NSF Grant PHY–97–22022.

REFERENCES

1. M. V. Berry, J. Phys. A **10**, 2083 (1977).
2. S. W. MacDonald and A. N. Kaufman, Phys. Rev. A **37**, 3067 (1988).
3. R. Aurich and F. Steiner, Physica D **64**, 185 (1993).
4. M. Srednicki, cond-mat/9403051, Phys. Rev. E **50**, 888 (1994).
5. P. O'Connor, J. Gehlen, and E. J. Heller, Phys. Rev. Lett. **58**, 1296 (1987).
6. B. Li and M. Robnik, chao-dyn/9501003, J. Phys. A **27**, 5509 (1994).
7. M. Srednicki and F. Stiernelof, chao-dyn/9603012, J. Phys. A **29**, 5817 (1996).
8. O. Bohigas, M.-J. Giannoni, and C. Schmit, Phys. Rev. Lett. **52**, 1 (1984).
9. O. Bohigas, in *Les Houches LII, Chaos and Quantum Physics*, edited by M.-J. Giannoni, A. Voros, and J. Zinn-Justin (North–Holland, Amsterdam, 1991).
10. Y. Alhassid and C. H. Lewenkopf, cond-mat/9510023, Phys. Rev. Lett. **75**, 3922 (1995).
11. V. N. Prigodin, K. B. Efetov, and S. Idia, Phys. Rev. Lett. **71**, 1230 (1993).
12. V. N. Prigodin, B. L. Altshuler, K. B. Efetov, and S. Idia, Phys. Rev. Lett. **72**, 546 (1994).
13. V. N. Prigodin, Phys. Rev. Lett. **74**, 1566 (1995).
14. V. N. Prigodin, N. Taniguchi, A. Kudrolli, V. Kidambi, cond-mat/9504089, and S. Sridhar, Phys. Rev. Lett. **75**, 2392 (1995).
15. M. Srednicki, cond-mat/9512115, Phys. Rev. E **54**, 954 (1996).
16. Ya. M. Blanter and A. D. Mirlin, cond-mat/9604139, Phys. Rev. E **55**, 6514 (1997).
17. S. Hortikar and M. Srednicki, chao-dyn/9710025.
18. M. C. Gutzwiller, J. Math. Phys. **8**, 1979 (1967).

19. M. V. Berry and K. E. Mount, Rep. Prog. Phys. **35**, 315 (1972).
20. M. Feingold and A. Peres, Phys. Rev. A **34**, 591 (1986).
21. M. Wilkinson, J. Phys. A **20**, 2415 (1987).
22. T. Prosen, Ann. Phys. **235**, 115 (1994).
23. S. Hortikar and M. Srednicki, chao-dyn/9711020.
24. B. Eckhardt, S. Fishman, J. Keating, O. Agam, J. Main, and K. Müller, chao-dyn/9509017, Phys. Rev. E **52**, 5893 (1995).
25. E. Austin and M. Wilkinson, Europhys. Lett. **20**, 589 (1992).
26. M. V. Berry, Proc. R. Soc. London A **400**, 229 (1985).
27. M. Wilkinson, J. Phys. A **21**, 1173 (1988);
28. B. Eckhardt and J. Main, chao-dyn/9508006, Phys. Rev. Lett. **75**, 2300 (1995).
29. M. C. Gutzwiller, J. Math. Phys. **10**, 1004 (1969); **11**, 1791 (1970); **12**, 343 (1971).
30. B. Eckhardt, S. Fishman, K. Müller, and D. Wintgen, Phys. Rev. A **45**, 3531 (1992).

LEVEL CURVATURE DISTRIBUTION BEYOND RANDOM MATRIX THEORY

V.E.Kravtsov[1,2,3], I.V.Yurkevich[4], C.M.Canali[5]

[1]International Centre for Theoretical Physics,
 P.O. Box 586, 34100 Trieste, Italy.
[2] Landau Institute for Theoretical Physics,
 2 Kosygina str., 117940 Moscow, Russia.
[3] Newton Institute for Mathematics,
 20 Clarkson Rd, Cambridge, CB3 0EH, UK.
[4] School of Physics and Astronomy,
 University of Birmingham,
 Edgbaston, Birmingham B15 2TT, UK.
[5] Department of Theoretical Physics I,
 University of Lund,
 Sölvegatan 14A, S - 223 62 LUND, Sweden.

INTRODUCTION

As first suggested by Edwards and Thouless [1], the sensitivity of spectrum $\{E_n\}$ of disordered conductors to a small twist of phase ϕ in the boundary conditions $\Psi(x = 0, \rho) = e^{i\phi}\Psi(x = L, \rho)$ is widely considered as a powerful tool to probe the space structure of eigenfunctions and to distinguish between the extended and the localized states. More precisely, the quantity K_n was introduced in Ref.[1] in order to describe this sensitivity quantitatively:

$$K_n = \frac{1}{\Delta} \left. \frac{\partial^2 E_n(\phi)}{\partial \phi^2} \right|_{\phi=0}, \tag{1}$$

where the mean level spacing $\Delta = (\nu L^d)^{-1}$ is related with the mean density of states $\nu = \langle \nu(E) \rangle$ and the size of the d-dimensional sample L. The quantity K_n in Eq.(1) is referred now as "level curvature", since it is proportional to the *curvature* of the function $E_n(\phi)$ at $\phi = 0$.

In complex or disordered quantum systems, the quantity K_n fluctuates over the ensemble of energy levels $\{E_n\}$ or, for a given level, over the ensemble of realizations of disorder. The typical width of the distribution of level curvatures $P(K)$ is of the order of the dimensionless conductance $g = D/(L^2\Delta)$, where D is the diffusion coefficient. Therefore by studying this distribution one can study the Anderson transition from metal to insulator with increasing of disorder [see Fig.1].

Supersymmetry and Trace Formulae: Chaos and Disorder
Edited by Lerner *et al.*, Kluwer Academic / Plenum Publishers, New York, 1999

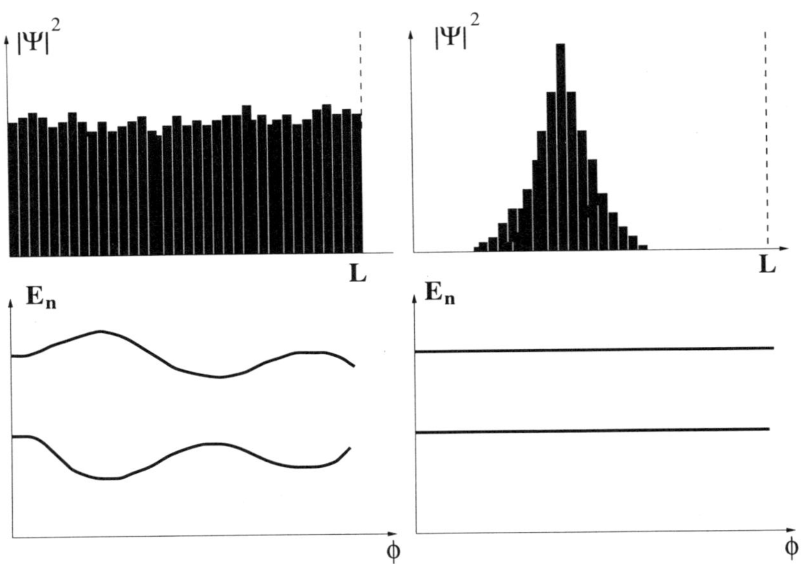

Figure 1. Typical dependences of an energy level E_n on ϕ for extended and localized eigenstates.

In the recent works [2] it has been shown that the distribution of the fluctuating quantity K_n is a particular example of the *parametric level statistics*, i.e. statistics of spectral responses of the system to a perturbation proportional to some parameter ϕ.

A remarkable property of the parametric level statistics [2] similar to that of the usual level statistics [3,4], is that in a certain limit they are universal for all classically chaotic and disordered systems and can be described by the random matrix theory (RMT) of Wigner and Dyson [3,4]. For disordered systems considered here, this limit coincides [5] with $g \to \infty$. For chaotic systems the same role is played [6] by the ratio $g = \gamma_1/\Delta$, where γ_1 is the first non-zero mode in the spectrum of the Perron-Frobenius operator that describes the chaotic behavior of the corresponding classical system. In particular, for the time-reversal-invariant system without spin-dependent interactions (orthogonal ensemble) the distribution of level curvatures, Eq.(1), was found [7,8,9] in this limit to have the form:

$$P_{WD}(k) = \frac{1}{2\left(1 + k^2\right)^{\frac{3}{2}}}, \qquad k = \frac{K}{\langle |K| \rangle}, \tag{2}$$

where in the RMT limit the mean absolute value of the level curvature is proportional to the mean conductance:

$$\langle |K| \rangle_{RMT} = 2g. \tag{3}$$

This proportionality is known as the Thouless relationship.

Further study [9] showed that the form, Eq.(2), remains the same if weak localization is taken into account, only the dimensionless conductance in the expression for $\langle |K| \rangle$ is changed appropriately.

The form, Eq.(2), is universal. It does not depend, e.g. on details of the system and the perturbation. The only important point is that the system is T-invariant with $T^2 = 1$ and the perturbation breaks this invariance. The underlying physics behind this universality is that of basis-invariance of the RMT which is equivalent to the structureless eigenfunctions.

Anderson localization apparently breaks down the basis invariance. Thus at a sufficiently small value of g one may expect the universality of the spectral statistics to break down as well. In the strong localization limit one would expect [10] the logarithmically normal rather than the power-law tails of the Edwards-Thouless curvature distribution function. This is because the fluctuations of level curvature $K \propto e^{-L/\xi}$ can be considered in this case as a consequence of the Gaussian fluctuations of the localization radius ξ for the exponentially localized wave functions. This picture is qualitatively confirmed by the recent analytical [11] and numerical [12,10] calculations.

In the present Chapter we address a question of the deviation $\delta P(k) = P(k) - P_{WD}(k)$ of the distribution of level curvatures from the result obtained in the framework of the random matrix theory. This chapter is mostly based on the results published recently in [10,14,13] as well as on the result of an unpublished work [38].

It turns out that there are two completely different contributions that amount to $\delta P(k)$. One of them $\delta P_{reg}(k)$ is regular in the small parameter g^{-1} and can be obtained by the perturbative treatment [15] of the non-zero spatial modes of the nonlinear supersymmetric sigma-model [5]. The main result of this treatment [14] is that *sign* of the correction $\delta P_{reg}(k)$ depends on the topological nature of perturbation. It is different for the "global" Edwards-Thouless curvature where the perturbation is represented by the global twist of phase in the boundary conditions *, and for the "local" T-breaking perturbation such as the magnetic impurities or random magnetic fluxes. Below we present a numerical evidence of that fact.

On top of the regular correction, there is [13] also a non-perturbative in g^{-1} correction $\delta P_s(k)$ which is proportional to $\exp(-1/g^{-1})$. The latter correction is due to the so called pre-localized states [17-20,21], i.e. eigenstates with anomalously high peak(s) in the probability density $|\Psi(\mathbf{r})|^2$.

There are reasons to consider the highly irregular, multifractal eigenstates [22] in the critical region near the Anderson transition as a result of proliferation of such pre-localized states. This point of view is supported, in part, by the observation that the weakly localized states in the lower critical dimensionality $d = 2$ exhibit also a (weak) multifractality which can be demonstrated using the very same methods (renormalization group [23,17] or space inhomogeneous saddle-point approximation [18]) as those applied to discover the pre-localized states responsible for slow current relaxation [23,20].

This idea enables to extend the results for $\delta P_s(k)$ obtained by the novel saddle-point approximation [20,14] for 2D metals to the critical state at the Anderson transition in $2 + \epsilon$ dimensions. Thus we can explain the branching non-analyticity at $k = 0$ found numerically in [10] for the 3D critical curvature distribution function $P_c(k)$. Furthermore, we suggest the relationship between the branching power and the exponent d_2 describing critical states with weak multifractality [22]. This relationship fits well the numerical results even at the 3D Anderson transition and provides the link between the spectral statistics and statistics of critical eigenstates.

FUNCTIONAL REPRESENTATION FOR $P(K)$.

For disordered systems in the metallic regime $g \gg 1$, the proper machinery to go beyond the RMT approximation is provided by the field-theoretic approach based on the supersymmetric nonlinear sigma-model [5]. This model has been derived starting from the model of non-interacting particles in a random Gaussian potential and thus it contains all the information about the behavior of the system at scales larger than the

---

*or by the magnetic flux in the problem of persistent current [16]

elastic scattering length l. In the limit $g \to \infty$ the so called zero-mode approximation suffices which is known [26] to be equivalent to the random matrix theory. However, also at a finite g more sophisticated methods allow to extract from the sigma-model both regular and non-analytical in $1/g$ corrections to the RMT result.

Let us consider disordered mesoscopic d-dimensional system with the random white-noise impurity potential $V(\mathbf{r})$ perturbed by a small vector-potential $\vec{\phi}/L$. It is described by the microscopic Hamiltonian of the form:

$$H = \frac{1}{2m}\left(\frac{1}{i}\nabla - \frac{\vec{\phi}}{L}\right)^2 + V(\mathbf{r}), \tag{4}$$

In the **case I** we assume the sample to be closed to a ring geometry and pierced by a static magnetic flux Φ. Then the problem is equivalent to the twist of phase $\phi = 2\pi\Phi/\Phi_0$ ($\Phi_0 = hc/e$ is the flux quantum) in the boundary condition and $\vec{\phi}/L = (\phi/L)\mathbf{n} = (\phi/L)\{1, 0, ..0\}$ is a constant.

In the "local" **case II** we consider $\vec{\phi}(\mathbf{r})$ to be a random δ-correlated vector-potential:

$$\langle \phi_\alpha(\mathbf{r})\,\phi_\beta(\mathbf{r})\rangle = v_\tau \phi^2\,\delta(\mathbf{r}-\mathbf{r}')\,\delta_{\alpha\beta}, \quad v_\tau = \frac{D}{2\pi\nu v_F^2}. \tag{5}$$

The parameter ϕ is introduced in this way in order to keep $\langle |K|\rangle_{RMT} = 2g$ the same as for the "global" curvature (case I).

The curvature distribution function

$$P(K) = \Delta\left\langle \sum_n \delta(K - K_n)\,\delta(E - E_n)\right\rangle \tag{6}$$

can be expressed through the two-level parametric correlation function $R(\omega, \phi)$

$$R(\omega, \phi) = \nu^{-2}\langle \nu(E + \omega, \phi)\,\nu(E, \phi = 0)\rangle \tag{7}$$

in the form similar to that derived in Ref.[25] for the distribution of level velocities:

$$P(K) = \lim_{\phi \to 0}\frac{\phi^2}{2}R\left(\omega = \frac{1}{2}K\Delta\phi^2, \phi\right). \tag{8}$$

Indeed, using an exact expression for the fluctuating density of states $\nu(E, \phi) = L^{-d}\sum_n \delta(E - E_n(\phi))$ we have:

$$R(\omega, \phi) = \Delta^2 \sum_{m,n}\langle\delta(\omega + E_n(0) - E_m(\phi))\,\delta(E - E_n(0))\rangle. \tag{9}$$

Because of the level repulsion, in the limit $\omega \to 0$ and $\phi \to 0$ only terms with $n = m$ contribute to the sum Eq.(9). On the other hand, with the perturbation which is odd under time reversal, the T-invariant level energy $E_n(\phi) \approx E_n(0) + \frac{1}{2}K_n\Delta\,\phi^2$ must be [†] even in ϕ. Then choosing $\omega = \frac{1}{2}K\Delta\,\phi^2$ we immediately arrive at Eq.(8).

It is worth noting that the limiting procedure in Eq.(8) is extremely important. In Fig.2 it is illustrated how the entire curvature distribution function arises from the broadened self-correlation peak in $R(\omega, \phi)$ at $\omega = 0$. This peak is blown up and reduced by the limiting procedure that simultaneously kills all the regular terms in

---

[†]this statement is not true [25] in the case where there is a degeneracy of spectrum at $\phi = 0$ (which is lifted for $\phi \neq 0$) for different values of T-odd quantities such as spin (Kramers degeneracy) or angular momentum (in perfect systems).

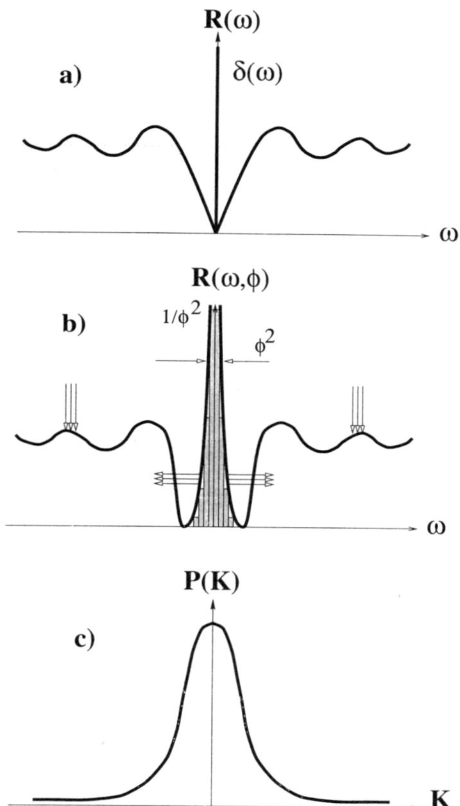

Figure 2. Limiting procedure in Eq.(8): a). typical form of $R(\omega, 0)$; b). broadening of the self-correlation peak for $\phi \neq 0$; c). typical form of the curvature distribution function.

$R(\omega) = R(\omega, 0)$. Since the entire curvature distribution $P(K)$ arises from the feature in $R(\omega, \phi)$ at very small $|\omega| < \phi\Delta$, we conclude that $P(K)$ is totally beyond the semi-classical approximation that works for $|\omega| \gg \Delta$.

The two-level correlation function $R(\omega, \phi)$ can be represented in the form of a functional integral using the Efetov's supersymmetry approach. A straightforward application of the results of Ref.[5] and Eq.(8) leads to:

$$P\left(K\right) = -\frac{1}{\pi^2} \frac{\partial^2 Z}{\partial J_1 \partial J_2}\bigg|_{J_1=J_2=0}, \qquad Z = \lim_{\phi \to 0}\left\{\phi^2 \Re \int DQ \exp\left(-F\left[Q\right]\right)\right\}, \qquad (10)$$

where for the **case I** the functional $F[Q]$ takes the form:

$$F\left[Q\right] = -\frac{\pi}{8} \int \frac{d\mathbf{r}}{V} Str\left\{g\left(L\nabla Q + i\left[\widehat{\phi}, Q\right]\right)^2 + iK\phi^2\Lambda Q + JQ\right\}, \qquad (11)$$

A similar representation for $P(K)$ has been used in Ref.[9].

In Eq.(11) we introduce notations:

$$\widehat{\phi} = \overrightarrow{\phi} P_+ \tau, \qquad J = \widehat{k}\left[J_1 P_+ + J_2 P_-\right],$$

and

$$P_+ = \frac{1}{2}\left(1 + \Lambda\right), \qquad P_- = \frac{1}{2}\left(1 - \Lambda\right).$$

The coordinate dependent 8×8 supermatrices $Q\left(\mathbf{r}\right)$ are parametrized as $Q = T^{-1}\Lambda T$, where T belongs to a graded coset space $UOSP\left(2, 2|4\right) / UOSP\left(2|2\right) \otimes UOSP\left(2|2\right)$ [26].

Other matrices are specified as follows:

$$\Lambda = diag\left(I_2, I_2, -I_2, -I_2\right)_{R-A}, \qquad \tau = diag\left(\tau_3, \tau_3, 0, 0\right)_{R-A},$$
$$\widehat{k} = diag\left(I_2, -I_2, I_2, -I_2\right)_{R-A}, \qquad \tau_3 = diag\left(1, -1\right), \quad I_2 = diag\left(1, 1\right).$$

Here we implied the following hierarchy of blocks of supermatrices: retarded-advanced $(R - A)$ blocks, boson-fermion $(B - F)$ blocks, and blocks corresponding to time reversal.

In the **case II** the linear in ϕ term in Eq.(11) is absent but otherwise the functional $F[Q]$ is the same provided that ϕ is introduced as in Eq.(5). A similar functional $F[Q]$ appears [5] if one considers a small concentration of magnetic impurities as perturbation. In both cases the structure of the "covariant derivative" $\mathcal{D}Q = \nabla Q + \frac{i}{L}\left[\widehat{\phi}, Q\right]$ which implies a sort of global gauge invariance, is broken down.

It is important in deriving the functional $F[Q]$ for the case II that the correlation radius of the random vector-potential is much less than the elastic scattering length. In this case the averaging over $\overrightarrow{\phi}\left(\mathbf{r}\right)$ should be done *prior* to switching to the slowly varying in space Q-variables. In the opposite limit of large correlation radius, one can average $e^{-F[Q]}$ over $\overrightarrow{\phi}\left(\mathbf{r}\right)$ and arrive at a much more complicated functional.

REGULAR CORRECTIONS TO $P(K)$.

A general approach to calculate such corrections using the nonlinear supersymmetric sigma-model [5] has been suggested in [15] and applied to distributions of different quantities [9,15,24]. It is based on the perturbative consideration of the non-zero diffusion modes which are integrated out to produce corrections to the zero-mode supersymmetric sigma-model [5]. The latter must be then integrated exactly.

The representation, Eqs.(10)-(11), in terms of the field $Q(\mathbf{r})$ contains all the spatial diffusion modes $\gamma_q = (D/L^2)\mathbf{q}^2$. However, in doing the limit $\phi \to 0$ in Eq.(10) the main role is played by the zero mode which corresponds to $\mathbf{q} = 0$. At $\phi = 0$ this mode does not cost any energy no matter how large are the components of the field Q in the non-compact boson-boson sector [5]. It is an arbitrary large amplitude of the zero mode components of the field Q that compensates the infinitesimal parameter ϕ in Eqs.(10)-(11) and leads to a finite result for $P(K)$. Thus the space independent zero mode Q_0 must be considered non-perturbatively.

In the limit $g \to \infty$ all the non-zero modes can be neglected [5], and one arrives [9] at the RMT result, Eq.(2). For finite $1/g$ the non-zero modes should be also taken into account. However, all the non-zero modes can be treated perturbatively for $g \gg 1$ to lead to some corrections to the zero-mode action. In order to obtain these corrections we have to separate zero modes from all other modes and then integrate over all the non-zero modes using a certain perturbative scheme.

Following the method suggested in [15] we decompose matrices $Q(\mathbf{r})$ as follows:

$$Q(\mathbf{r}) = T_0^{-1} \tilde{Q}(\mathbf{r}) T_0, \quad \tilde{Q} = \Lambda \frac{1 + \widetilde{W}/2}{1 - \widetilde{W}/2} \tag{12}$$

where T_0 describes the zero mode and $\tilde{W}(\mathbf{r}) = \sum_{\mathbf{q}\neq 0} \tilde{W}_{\mathbf{q}} e^{i\mathbf{q}\mathbf{r}}$ does not contain the zero mode at all.

As has been already noticed, the main contribution to the functional integral is done by the zero mode. The zero-mode approximation, $Q = Q_0 = T_0^{-1}\Lambda T_0$ is known to be equivalent to the random matrix theory [26]. To go beyond the RMT we integrate perturbatively over $\tilde{W}(\mathbf{r})$ to obtain the effective zero-mode action $F^{eff}[Q_0]$ as follows:

$$F^{eff}[Q_0] = -\ln \int D\tilde{Q} \cdot J[\tilde{Q}] \exp\left\{-F[Q_0, \tilde{Q}]\right\},$$

where $F[Q_0, \tilde{Q}]$ is obtained from $F[Q]$ by substituting the decomposition, Eq.(12), and $J[\tilde{Q}]$ is the Jacobian of the corresponding nonlinear transformation.

This scheme is implemented in Ref.[14,38]. The remarkable result of these calculations is that in the limit of interest the new zero-mode functional $F^{eff}[Q_0]$ is a polynomial in the vertices $STr\left(\hat{\phi}Q_0\right)^2$, $STr(\Lambda Q_0)$, and $STr[JQ_0]$ contained in the initial zero-mode functional $F[Q_0]$ that leads to the RMT result. This means that the RMT curvature distribution function, Eq.(2) is a *generating function* for the regular correction $\delta P_{reg}(K)$:

$$\delta P_{reg}(k) \propto \left.\frac{\partial^2}{\partial a^2}\left(a^{-1} P_{WD}(k/a)\right)\right|_{a=1}. \tag{13}$$

This simplifies the problem dramatically. For the case of T-breaking perturbations over the orthogonal ensemble we finally obtain [14,38]:

$$\delta P_{reg}(k) = C_d \frac{2 - 11k^2 + 2k^4}{2\left(1 + k^2\right)^{7/2}}, \quad k = \frac{K}{\langle |K| \rangle} \ll g, \tag{14}$$

where

$$C_d = \frac{1}{(\pi g)^2} \sum_{\mathbf{q}\neq 0} \frac{1}{(\mathbf{q}^2)^2} \times \begin{cases} \left(\frac{4}{d} - 1\right), & \text{case I} \\ -1, & \text{case II} \end{cases}. \tag{15}$$

Here $\mathbf{q} = \{q_1, ... q_d\}$, where $q_i = 2\pi n_i$, $(n_i = 0, \pm1, \pm2...)$ in the case of the periodic boundary conditions (for an unperturbed system) considered in this paper.

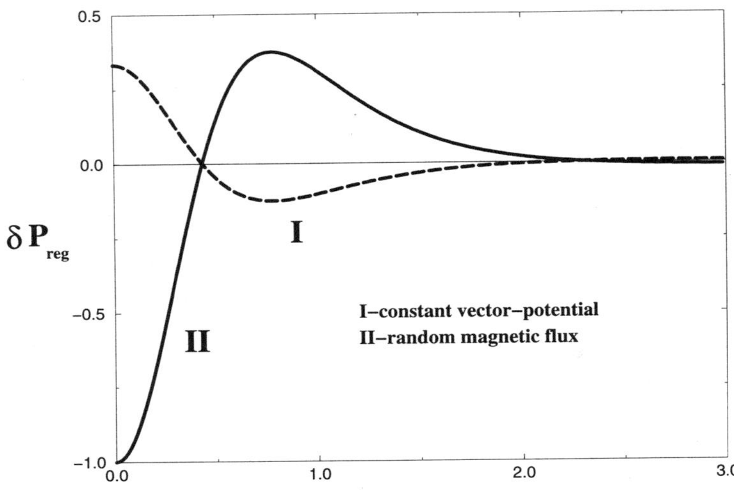

Figure 3. Regular in $1/g$ corrections to $P(k)$ (arbitrary units) in the case of the global (I) and random (II) magnetic flux.

A remarkable feature of Eqs.(14),(15) for $d < 4$ is that the *sign* of the correction is different for global (**case I**) and local (**case II**) T-breaking perturbations [see Fig.2].

The positive sign at small k in **case I** reflects the tendency towards the weaker spectral response to the boundary conditions with decreasing g.

Another principal result of these calculations is that the Thouless relationship, Eq.(3), breaks down beyond the random matrix theory. The ratio $r(g) = \langle|K|\rangle/2g$ *increases* above its RMT value $r = 1$, the correction being equal to [14,38]:

$$\delta r(g) = \frac{\delta\langle|K|\rangle}{2g} = \frac{1}{(\pi g)^2}\sum_{\mathbf{q}\neq 0}\frac{1}{(\mathbf{q}^2)^2} \times \left\{ \begin{array}{ll} \left(\frac{9}{2} - \frac{16}{d} + \frac{36}{d(d+2)}\right), & \text{case I} \\ \frac{9}{2}, & \text{case II} \end{array} \right. . \tag{16}$$

SIGNATURE OF PRE-LOCALIZED STATES IN THE LEVEL CURVATURE DISTRIBUTION

It has been known for quite a while that the relaxation of current and the local density of states in disordered conductors exhibit an anomaly even in the weak-localization regime. Namely, it has been shown in [17] that with a small (but not exponentially small) probability one can find a current relaxation time or a local DOS that is much larger than the corresponding mean values. These anomalies have been attributed to the quasi-localized (or pre-localized) states [18-20], i.e. the states with an anomalously large peak in $|\Psi(\mathbf{r})|^2$ at some point $\mathbf{r} = \mathbf{r}_0$.

Very recently the problem of current relaxation in disordered conductors has been reconsidered [20,27] by an elegant instanton approximation [20] applied to supersymmetric version of the nonlinear sigma-model [5]. In these works the main result of the previous

276

consideration [17] has been confirmed for 2D systems. However, the new method is able to describe some unknown regimes of current relaxation and to set correct limits of validity for the regimes found earlier. Later the same idea [20] has been applied [18] to find directly the distribution of $|\Psi(\mathbf{r})|^2$ and the distribution of local densities of states [28]. Thus the existence of the pre-localized states has been proven and the corresponding configuration of the random impurity potential has been found [20,21].

The main idea of Edwards and Thouless [1] is that it is possible to distinguish between the localized and extended states studying the sensitivity of spectrum to the twist of phase in the boundary conditions. This sensitivity is significant only for states with the localization radius bigger than the sample size L and negligible for the strongly localized states. It is clear that the existence of the pre-localized states should lead to an enhancement in $P(K)$ at small K. In low dimensional systems $d = 1, 2$ where the pre-localized states correspond to the localized states with an anomalously small localization radius, one may expect the singularity of $P(K)$ at $K = 0$. In 3D metal the typical pre-localized state looks like a sharp peak in $|\Psi(\mathbf{r})|^2$ on top of the extended background $|\Psi(\mathbf{r})|^2 \sim L^{-d} = const$. The level curvature that corresponds to such a state does not vanish but only slightly decreases. Thus the pre-localized states in 3D metal should have much weaker effect on the level curvature distribution.

Instanton Approximation.

In order to check these predictions we consider, instead of $P(K)$ at small K, its Fourier-transform $\tilde{P}(\lambda) = \int dK\, P(K)\, e^{-iK\lambda}$ at $\lambda \gg 1$.

It is easy to see that for both the RMT result Eq.(2) and the regular correction Eq.(14) the function $\tilde{P}(\lambda)$ vanishes exponentially for $\lambda \gg 1$. In what follows we will look for the slowly-decreasing contributions to the characteristic function. A formal reason why such contribution may exist is clear from the functional representation of $P(K)$, Eq.(10), Eq.(11). Indeed one can easily see that the level curvature k in Eq.(11) plays the same role as the frequency in the problem of current relaxation [20], thus λ being similar to time. Therefore, one may expect the long, non-exponential tails in $\tilde{P}(\lambda)$ by analogy with those in the current relaxation function $I(t)$. However, it is far from being clear that two problems are equivalent, since the boundary conditions are different and there are additional terms that describe the T-breaking perturbation in the nonlinear sigma-model Eq.(11).

The main idea of Ref.[20] which we will exploit here is that at large λ the configurations of the field $Q(\mathbf{r})$ that are $\mathbf{r}$–independent or slowly varying in space, are too expensive. It appears to be much less expensive to consider essentially space-dependent configurations in the vicinity of the classical (instanton) solution $Q_{ins}(\mathbf{r})$ that minimize the action $F[Q]$. At large g the fluctuations around this solution are expected to be small and one arrives at:

$$\tilde{P}(\lambda) \equiv A e^{-S(\lambda)} = \lim_{\phi \to 0} \Re \int \mathcal{D}Q \int_{-\infty}^{+\infty} dK\, A[Q_{ins}; \phi]\, e^{-\{F[Q_{ins}]+iK\lambda\}}. \tag{17}$$

where $A[Q_{ins}; \phi]$ is a pre-exponential factor.

The Grassmann variables in the action, Eq.(11) can lead only to a renormalization of the pre-exponential factor A in Eq.(17), since the integration over these variables is equivalent to a differentiation. So, with the exponential accuracy we neglect all the Grassmann variables in the Efetov's parametrization for $Q(\mathbf{r})$. Next, the finite contribution to $S(\lambda)$ in the limit $\phi \to 0$ comes only from the infinitely large boson-boson components of the field $Q(\mathbf{r})$. Therefore we consider only the leading terms in the non-compact angles θ_1 and θ_2 in the Efetov's parametrization [5] for the orthogonal ensemble.

By varying the functional $F[Q_{ins}]+iK\lambda$ over the relevant variables $\theta=(\theta_1+\theta_2)/2$, and φ parametrizing $Q_{ins}(\mathbf{r})$, and the curvature K we find [13,38]:

$$\partial^2\theta + \phi^2[\kappa - (\partial v - \mathbf{n})^2]\sinh\theta = 0, \tag{18}$$

$$\partial\left[(\partial v - \mathbf{n})(\cosh\theta - 1)\right] = 0, \tag{19}$$

and

$$\frac{\pi}{4}\phi^2\int(\cosh\theta + 1)\,d^d\rho = \lambda, \tag{20}$$

where $\kappa = iK/2g$, $d^d\rho = \frac{d^d\mathbf{r}}{L^d}$ and $\varphi = \phi v$.

Eqs.(18)-(20) correspond to the global **case I**. As usual, in the local **case II** the linear in $\mathbf{n} = \{1, 0, 0, ...\}$ terms are absent.

The limit $\phi \to 0$ is done simply by absorbing ϕ^2 into θ. We introduce $\tilde\theta = \theta + \ln\phi^2$. Then in the limit $\phi \to 0$ we have $\sinh\theta \approx \cosh\theta = \frac{1}{2}e^{\tilde\theta}\phi^{-2}$ and Eqs.(18),(19),(20) take the form:

$$\partial^2\tilde\theta + \frac{1}{2}[\kappa - (\partial v - \mathbf{n})^2]e^{\tilde\theta} = 0, \tag{21}$$

$$\partial\left[(\partial v - \mathbf{n})e^{\tilde\theta}\right] = 0, \tag{22}$$

$$\frac{\pi}{8}\int e^{\tilde\theta}\,d^d\rho = \lambda. \tag{23}$$

where $\tilde\theta, v \in [-\infty, +\infty]$ obey the periodic boundary conditions.

The effective action $S(\lambda)$ in Eq.(17) is expressed in terms of the saddle-point solutions $\tilde\theta(\mathbf{r})$ and κ as follows:

$$S(\lambda) = \frac{\pi}{4}g\int(\partial\tilde\theta)^2\,d^d\rho + 2g\kappa\lambda. \tag{24}$$

One can solve Eq.(22):

$$(\partial v - \mathbf{n}) = [\nabla \times \mathbf{A}]\,e^{-\tilde\theta}, \tag{25}$$

where $[\nabla \times \mathbf{A}] = const$ in 1D and is a *curl* of an arbitrary vector function $\mathbf{A}(\mathbf{r})$ in higher dimensions. Below we consider only the simplest solution that corresponds to $[\nabla \times \mathbf{A}] \equiv -\mathbf{n}/N = const$.

Let us consider first the local **case II**. Doing the space integration of Eq.(25) which does not contain the term proportional to $\mathbf{n}$ in this case, and using the periodic boundary conditions for $v(\mathbf{r})$ one immediately arrives at $[\nabla \times \mathbf{A}] = \partial v = 0$. Then the same procedure with Eq.(21) leads to the conclusion that the only solution $\tilde\theta = const$ that obeys the periodic boundary conditions, exists only for $\kappa = \mathbf{n}^2 = 1$, and the corresponding action is $S(\lambda) = 2g\lambda$. Thus the instanton approximation in the local **case II** gives only an exponentially small tail $\tilde P(\lambda) \propto e^{-2g\lambda}$ that has been already obtained by the perturbative approach. In this case an analogy with the problem of current relaxation appears to be wrong.

Now consider the global **case I**. Integrating Eq.(25) over space and using the periodicity of $v(\mathbf{r})$ we have:

$$N = \int e^{-\tilde\theta}\,d^d\rho. \tag{26}$$

Substituting Eqs.(25),(26) into Eq.(21) we finally arrive at:

$$\partial^2\tilde\theta + \frac{\partial U}{\partial\tilde\theta} \equiv \partial^2\tilde\theta + \frac{\kappa}{2}e^{\tilde\theta} - \frac{1}{2N^2}e^{-\tilde\theta} = 0. \tag{27}$$

It appears that the global nature of perturbation and the corresponding linear in $\mathbf{n}$ term in Eq.(25) leads to the term proportional to $e^{-\tilde{\theta}}$ in Eq.(27) that builds a second "wall" in the effective "potential" $U(\tilde{\theta})$ and makes it possible for the periodic solutions ("oscillations") to exist.

Eq.(27) takes a more symmetric form if we make a shift $\tilde{\theta} = u - \zeta$, where:

$$\cosh\zeta = \left(\kappa + \frac{1}{N^2}\right)\frac{N}{2\sqrt{\kappa}}, \quad \sinh\zeta = \left(\kappa - \frac{1}{N^2}\right)\frac{N}{2\sqrt{\kappa}}. \tag{28}$$

Finally we have the system of equations [13,38]:

$$\partial^2 u + \gamma^2 \sinh u = 0, \tag{29}$$

$$\frac{1}{N} = \gamma^2 \int e^{-u}\, d^d\rho, \tag{30}$$

$$\lambda = \frac{\pi}{8\gamma^2 N^2} \int e^{u}\, d^d\rho, \tag{31}$$

where $\gamma^2 = \sqrt{\kappa}/N$.

Solving these equations with the periodic boundary conditions for a hyper-cubic sample $-1/2 < \rho_i < 1/2$ one finds $u(\mathbf{r}, \lambda)$, $N(\lambda)$ and $\gamma(\lambda)$ which enter the instanton action $S(\lambda)$:

$$S(\lambda) = \frac{\pi}{4} g \int (\partial u)^2\, d^d\rho + 2g\gamma^4 N^2 \lambda. \tag{32}$$

Non-Exponential Tails of $\tilde{P}(\lambda)$ in Low-Dimensional Systems.

We will see below that for large λ the parameter γ is small. For $\gamma \ll 1$ the term $\gamma^2 \sinh u$ is very small unless $\sinh u$ is exponentially large. This means that we can approximate $\gamma^2 \sinh u \approx \frac{\gamma^2}{2} e^{|u|}\, sign(u)$. Thus we come to the Liouville equation instead of Eq.(29):

$$\partial^2 u + \frac{\gamma^2}{2} e^{|u|}\, sign(u) = 0. \tag{33}$$

Quasi-1D Case. The generic solution to the 1D Liouville equation reads:

$$e^{|u|} = \frac{4k^2}{\gamma^2 \cosh^2(kx + b)}, \tag{34}$$

where k and b are real constants.

The solution on a ring $-\frac{1}{2} < x < \frac{1}{2}$ is constructed by a reflection of a positive solution with $b = 0$ anti-symmetrically about the points $x = \pm\frac{1}{4}$. The second constant k is found from the condition of continuity of $u(x)$ (together with the first derivative) $u(\pm 1/4) = 0$:

$$4k^2 = \gamma^2 \cosh^2(k/4), \quad k \approx \ln(1/\gamma^4) \gg 1. \tag{35}$$

Indeed, the so constructed solution $u(x)$ describes motion of a classical particle between two almost rigid walls [see Fig.4.].

The anti-symmetric nature of the solution immediately leads to an identity:

$$I = \int e^{u}\, dx = \int e^{-u}\, dx. \tag{36}$$

Then from the self-consistency Eqs.(30),(31) we obtain in the limit $\lambda \gg 1$:

$$N = \frac{1}{8k}, \quad \gamma^4 = \frac{64\pi k^3}{\lambda}. \tag{37}$$

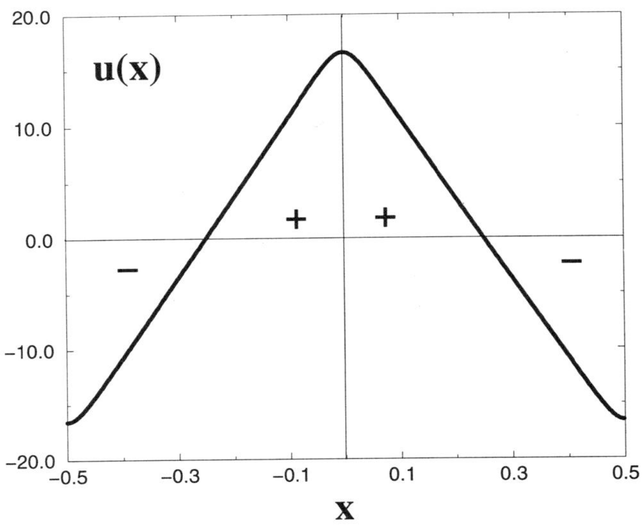

Figure 4. Solution to the 1D Liouville equation.

Finally using Eqs.(32),(34) we arrive at:

$$S(\lambda) = \pi g k^2 \approx \pi g \ln^2 \lambda. \tag{38}$$

Thus the characteristic function in a quasi-1D systems is *logarithmically-normal*:

$$\tilde{P}(\lambda) = A \exp\left[-\frac{g_1}{2}\ln^2 \lambda\right] \quad g_1 = 2\pi g, \tag{39}$$

The validity of the above result is restricted by the validity of the nonlinear sigma-model Eq.(11). This model and hence the saddle-point equations, work only for slowly-varying fields $Q(\mathbf{r})$, namely $|\partial u| < L/l$, where l is an elastic scattering length. It follows immediately from Eq.(34),(35) that the above result is valid for $1 \ll \lambda \ll exp(L/l)$.

The logarithmically-normal tail in $\tilde{P}(\lambda)$ described by Eq.(39) is exactly of the same functional form as the current relaxation function $I(t)$ in Ref.[20] for the orthogonal ensemble.

2D Case. In full analogy with the quasi-1D case, we construct the double-periodic solution to the Liouville equation on a torus $-\frac{1}{2} < x, y < \frac{1}{2}$ by a reflection. We consider a positive solution $u(z)$ inside the square Ω with the vertices at $z = \pm 1/2, \pm i/2$ and then continue it anti-symmetrically about a side of the square in any quater of the sample $|\Re z| < 1/2$, $|\Im z| < 1/2$. By construction, the symmetry relationship Eq.(36) is valid for such a 2D solution too [see Fig.5.].

The procedure of finding the solution is described in detail in Ref.[38]. We note that for our purposes we need only the solution for $|z| = r \ll 1$. It is rotationally-invariant

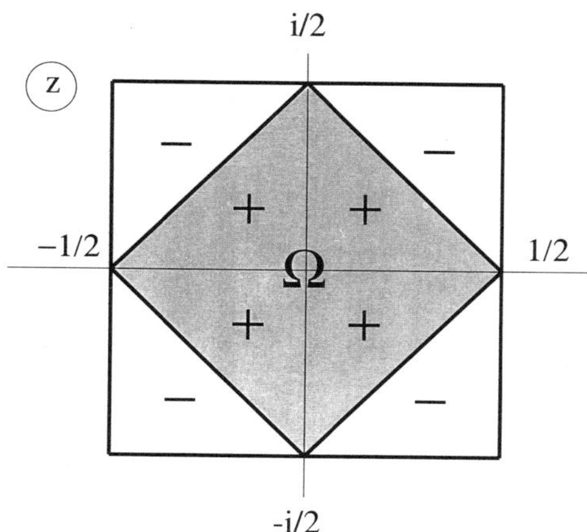

Figure 5. The symmetry of the double-periodic solution to the Liouville equation: the regions of positive and negative $u(z)$ are shown by $+$ and $-$.

and has the form:

$$e^{u(r)} = \frac{16b(k-1)^2\, r^{2k-4}}{(\gamma^2 + b\, r^{2k-2})^2}, \tag{40}$$

where

$$b = 16\left(\frac{\pi^2}{2}\right)^k (k-1)^2. \tag{41}$$

Certainly, the solution, Eq.(40) can be immediately obtained from the radial Liouville equation, with k and b being two constants of integration. The requirement of periodicity of $u(z)$ helps to establish a connection, Eq.(41), between these constants.

The remaining constant k is found in a standard way from the requirement of convergence of $\int (\partial u)^2 dx dy$ in the action $S(\lambda)$. ¿From this condition we immediately find:

$$k = 2, \quad b = 4\pi^4. \tag{42}$$

Because of the symmetry of $u(z)$, the integral $\int_\Omega (\partial u)^2\, d^2\rho$ over the square Ω is exactly one-half of the total integral over the period (over the sample) $\int (\partial u)^2\, d^2\rho$. It diverges logarithmically at $r \gg \gamma$, and we arrive at:

$$\int (\partial u)^2\, d^2\rho = 32\pi \ln\left(\frac{Cb}{\gamma^2}\right), \tag{43}$$

where $C = \frac{2}{\pi^2 e}$ can be found from an exact solution in the region $|z| \sim 1$.

The result is almost independent of b at small γ and is essentially determined by the logarithmic solution to the Poisson equation that follows from Eq.(29) at $\gamma = 0$.

The constants $\gamma(\lambda)$ and N are found from the self-consistency Eqs.(30),(31) and require the full solution Eq.(40) of the radial Liouville equation:

$$N = \frac{1}{16\pi}, \quad \gamma^4 = \frac{8^3 \pi^4}{\lambda}. \tag{44}$$

The final expression for the instanton action in 2D reads:

$$S(\lambda) = 4\pi^2 g \left[\ln\left(\frac{\lambda}{8}\right) - 1 \right]. \tag{45}$$

Accordingly, the characteristic function $F(\lambda)$ turns out to have a power-law asymptotic behavior at large $\lambda \gg 1$:

$$F(\lambda) = A \left(\frac{c}{\lambda}\right)^{2g_2}, \qquad g_2 = 2\pi^2 g, \tag{46}$$

where $c = 8e$.

Few notes should be made on the validity of the result Eq.(46). Firstly, the above instanton approximation with the logarithmic in λ action $S(\lambda)$ is only justified at $g \gg 1$, since the pre-exponent A could also be a power-law function of λ but with the exponent of order 1. Secondly, the nonlinear sigma-model and hence the saddle-point equations work only for $|\partial u| < L/l$, where l is an elastic scattering length. It follows immediately from Eq.(40),(44) that the above result is valid for $1 \ll \lambda \ll (L/l)^4$.

Non-analyticity of the Level Curvature Distribution.

In this section we show that the slowly decreasing tails in the characteristic function $\tilde{P}(\lambda)$ at $\lambda \gg 1$ given by Eqs.(39), (46) result in a non-analytical behavior of $P(K)$ at $K = 0$. As usual, the true non-analyticity arises only in the thermodynamic limit $L/l \to \infty$ [‡], since only in this limit the tails are extended to infinity. For any finite L/l the function $P(K)$ is still analytical at $K = 0$ but the region of the regular behavior of $P(K)$ shrinks to zero with increasing L/l. Below we will assume the limit $L/l \to \infty$ to be done.

Let us consider the **quasi-1D** case first. In this case all derivatives of $P(K)$ are finite at $K = 0$:

$$P^{(2n)}(0) = \int_{-\infty}^{+\infty} (-1)^n \lambda^{2n} \, \tilde{P}(\lambda) \frac{d\lambda}{2\pi} \propto \exp[(2n+1)^2/2g_1]. \tag{47}$$

Yet the function $P(K)$ is *non-analytical* at K=0, since the Taylor series $P(K) = \sum_n \frac{P^{(2n)}(0)}{(2n)!} K^{2n}$ has zero radius of convergence because of the very fast growth of $P^{(2n)}(0)$ with n.

The non-analyticity at $K = 0$ is much stronger in **2D case**. In this case all derivatives $P^{(2n)}(0)$ with $2n + 1 > 2g_2$ are proportional to $(L/l)^{4(2n+1-2g_2)}$ and diverge in the thermodynamic limit. Let us define m as an integer obeying the inequality of $|g_2 - m| \leq \frac{1}{2}$. Then the expansion of $P(K)$ at small K has the form:

$$P(K) = c_0 + c_1 K^2 + \ldots c_{m-1} K^{2(m-1)} + c_m K^{2m-\alpha_m} + o(K^{2m}), \tag{48}$$

where the non-trivial exponent $0 < \alpha_m < 2$ that describes branching in Eq.(48), is given by:

$$\alpha_n = (2n+1) - 2g_2. \tag{49}$$

In the **3D** metal case we failed to find a solution to the saddle-point problem that would lead to the finite action $S(\lambda)$ in the thermodynamic limit. This means that the characteristic function $\tilde{P}(\lambda)$ has only regular corrections at $g \gg 1$ and thus decays exponentially for $\lambda \gg 1$.

[‡]it is always possible by a proper choice of parameters (e.g. the cross-section in quasi-1d case) to implement this limit while keeping the conductance g fixed.

NON-ANALYTICITY OF $P(K)$ AT $K = 0$ AND MULTIFRACTALITY OF EIGENFUNCTIONS.

We see that the strength of non-analyticity of $P(K)$ at $K = 0$ depends on dimensionality in a non-monotonous way. In a quasi-1D metal it is very weak, in a 2D metal where $P(K)$ has a branching non-analyticity, it reaches maximum and a in 3D metal the level curvature distribution is analytical.

Such a behavior is related with the fact that $d = 2$ is the lower critical dimension for the Anderson transition, and the wavefunctions in the 2D weak-localization regime share some features of the critical wavefunctions at the Anderson transition in higher dimensions.

$P(K)$ at the Anderson Transition in $2 + \epsilon$ Dimensions.

A usual way to describe the critical state near the Anderson transition is the $(d - 2) = \epsilon$-expansion. To this end one considers the quantity of interest in a 2D system with $g_2 \gg 1$ and then replaces g_2 by the critical conductance $g_d^* = 1/(d - 2)$ which is the fixed point [§] of the scaling equation [29]:

$$\frac{d \ln g_d}{d \ln L} = (d - 2) - \frac{1}{g_d} + o\left(\frac{1}{g_d^2}\right). \tag{50}$$

So, for the orthogonal ensemble in $d = 2 + \epsilon$ dimensions we find to the leading order in $\epsilon \ll 1$:

$$\tilde{P}_c(\lambda) \propto \left(\frac{1}{\lambda}\right)^{\mu}, \qquad \mu = \frac{2}{\epsilon} + o(1). \tag{51}$$

Note that in the critical point, the conductance g_d^* is *exactly* size-independent, and one can consider the thermodynamic limit $L \to \infty$ without tuning other parameters in order to keep g_d fixed. Therefore Eq.(51) defines a true critical exponent μ that describes the power-law behavior of $\tilde{P}(\lambda)$ in the entire region $\lambda \gg 1$.

By setting $\epsilon = 1$ in Eq.(51) we find $\mu \approx 2$ for the 3D Anderson transition. Then it follows from Eq.(48) that already the second derivative of $P(K)$ at $K = 0$ is divergent and we arrive at [13,38]:

$$P_c(K) = c_0 - c_1 |K|^{2-\alpha}, \qquad \alpha = 3 - \mu. \tag{52}$$

Exponent μ and Multifractality of Critical Eigenstates.

Unfortunately it is known that the accuracy of the $d - 2 = \epsilon$- expansion is quite poor and insufficient for a precise determination of critical exponents. In this situation one can try to find relationships between different critical exponents rather than to evaluate them using the ϵ-expansion. This certainly requires some *assumptions* about underlying physics.

As has been mentioned in the Introduction, a unique property of the critical states is multifractality. This property is characterized by the nontrivial *power-law* scaling of averaged powers of eigenfunction amplitudes $|\Psi_E(\mathbf{r})|$. [30,22]:

$$\sum_{\mathbf{r},n} \langle |\Psi_n(\mathbf{r})|^{2q} \delta(E - E_n) \rangle \propto L^{-d_q(q-1)}, \tag{53}$$

where $d_q < d$ is a fractal dimension that depends on q (*"multifractality"*).

[§]In this equation $g_1 = 2\pi g$, $g_2 = 2\pi^2 g$, and $g_3 = 4\pi^2 g$, where $g = D/(L^2 \Delta)$.

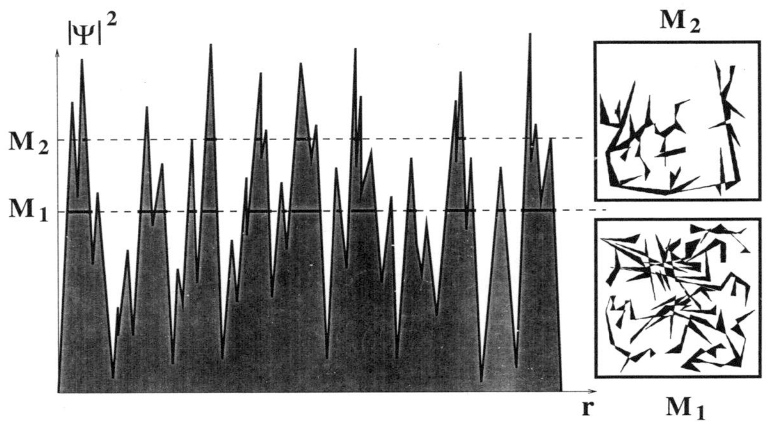

Figure 6. Multifractality of a critical eigenfunction: the black patterns in the inserts are fractals with the fractal dimensions depending on the cut-off level M.

In order to visualize this property of critical eigenfunctions one can imagine [see Fig.6] a generic irregular function $|\Psi(\mathbf{r})|^2$ that is cut at some level $|\Psi(\mathbf{r})|^2 = M$. Let us show in black all the space regions where $|\Psi(\mathbf{r})|^2 > M$. Then the multifractality implies the black pattern to be a fractal with the fractal dimension depending on the cut-off level.

It is remarkable that Eq.(53) can be proven [18] in 2D metals by the instanton approximation similar to the one we used in this paper. The spectrum of fractal dimensions d_q obtained in this approximation turns out to be linear:

$$d_q = d - \frac{\eta}{2}\, q, \qquad \eta = d - d_2 = \frac{2}{\beta g_2}, \tag{54}$$

where $\beta = 1, 2, 4$ for the orthogonal, unitary and symplectic ensembles.

It is reasonable to assume that the power-law tail in $\tilde{P}(\lambda)$ is another signature of multifractality [10]. Then one may hope that the expression for μ in terms of the structural constant of multifractality $\eta = d - d_2$ provides a better approximation for μ than the ϵ-expansion. By using Eq.(54) and the relationship $\mu = 2\beta g_2$, we obtain [13,38]:

$$\mu = \frac{4}{\eta}. \tag{55}$$

The conjecture Eq.(55) is based on two assumptions: i). the exponent μ is determined by the spectrum of multifractality d_q and ii). this spectrum is linear (for $q \ll 1/\eta$). Since for any critical state with weak multifractality the spectrum of d_q is expected to be linear up to very large values of q, we believe that Eq.(55) is valid for *any* critical state with weak multifractality. In contrast to Eq.(51), the relationship between μ and η, Eq.(55), is independent of dimensionality and the symmetry parameter β and should apply to 2D critical states in the Quantum Hall regime and for systems with the spin-orbit interaction [30,22].

NUMERICAL RESULTS

In this section we present the results [10,38] of numerical simulations on the Anderson model with diagonal disorder. For our numerical analysis we consider a tight-bibding model on a square lattice of L^d sites. The one-particle Hamiltonian is:

$$H = \sum_i \epsilon_i c_i^\dagger c_i + \sum_{\langle\langle ij\rangle\rangle} (e^{i\theta_{ij}} c_i^\dagger c_j + e^{-i\theta_{ij}} c_j^\dagger c_i). \tag{56}$$

The site energies ϵ_i are randomly distributed with uniform probability between $-W/2$ and $W/2$. The parameter W controls the amount of disorder in the system. The phase shifts θ_{ij} in the hopping term connecting nearest neighbors represent the effect of an external perturbation that breaks the T-invariance of the system. As for the analytical calculations we consider two types of such perturbations. The first one (**case I**) is the usual Aharonov-Bohm flux $\Phi = (\phi/2\pi)\Phi_0$ that pierces the system closed to a ring geometry giving rise to a *global* shift of the boundary conditions in one direction. We will choose a gauge such that each hop in x-direction picks up a phase $\theta_{ij} = \phi/L$, so that total twist of the boundary condition is ϕ. The second one (**case II**) is a random magnetic flux. In this case the gauge is such that the phase θ_{ij} relative to a hop in the x-direction is Gaussian distributed with zero average and variance equal to $\langle\theta_{ij}^2\rangle = (\phi/L)^2$. For this gauge, the vector-potential $\mathbf{A}(i) \propto \{\theta_{i,i+1}, 0, ...\}$ is defined on a dual lattice with sites in the middle of bonds in x-direction. This is a random vector-potential model with a short-range correlations $\langle A_\alpha(i)A_\beta(j)\rangle = (\Phi/L)^2 \delta_{\alpha,x}\delta_{\beta,x}\,\delta_{ij}$ of the type given in the continuous approximation by Eq.(5). The only difference is that the correlations is *anisotropic*. This difference is not important, since it leads only to the constant factor $1/d$ in v_τ that can be absorbed in the parameter ϕ. This kind of perturbation is qualitatively different from **case I**, since it acts *locally*.

Corrections to $P(K)$ Beyond RMT in 3D Metal.

In this section we present the numerical results for the finite g corrections to the shape of the curvature distribution in the metallic regime, comparing the numerical results with the regular corrections, Eq.(14),(15).

Global Vector Potential In Fig.7 we show the numerical results for $\delta P(k) = P(k) - P_{WD}(k)$ for the 3D metallic regime in **case I**. The calculations are performed for system size $L = 8$ and disorder $W = 12$. The number of disorder realizations is 1500. The deviation of $P(k)$ from the RMT result is very small, less than one percent. The magnitude of the statistical noise present after averaging is done, appears to be only a little smaller than the signal itself. Nevertheless the general trend of the curve agrees with the analytical prediction Eq.(14): $P(k)$ is *above* the RMT result at small k. We have used the coefficient C_d in Eq.(14) as a free parameter in the least square fitting of the numerical results. The value $C_3 = 0.0044$ found from such a fitting is probably a reliable estimate of the magnitude of the correction $\delta P(k)$ in the above case.

Random Magnetic Flux. The same correction $\delta P(k)$ for the case of a random magnetic flux is displayed in Fig.8. The values of the parameters of the Hamiltonian are the same as for the previous case. Despite the statistical fluctuations are still rather strong, the numerical results are quite significant. We see that again the expression Eq.(14) provides a rather good one-parameter fitting function for the numerical results. However, in this case the coefficient $C_d = -0.014$ is *negative* in full agreement with the analytical prediction. Moreover, the numerical results are consistent with the analytical

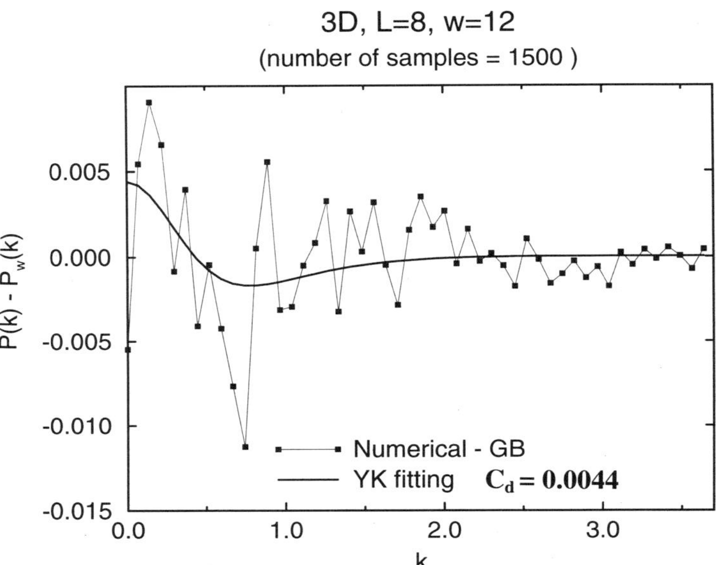

Figure 7. The correction to the curvature distribution for a 3D system in the metal regime for the system size L=8 and disorder W=12. The T-breaking perturbation is caused by the constant vector-potential. The number of disorder realizations is 1500. The solid line is the one-parameter fitting using Eq.(14).

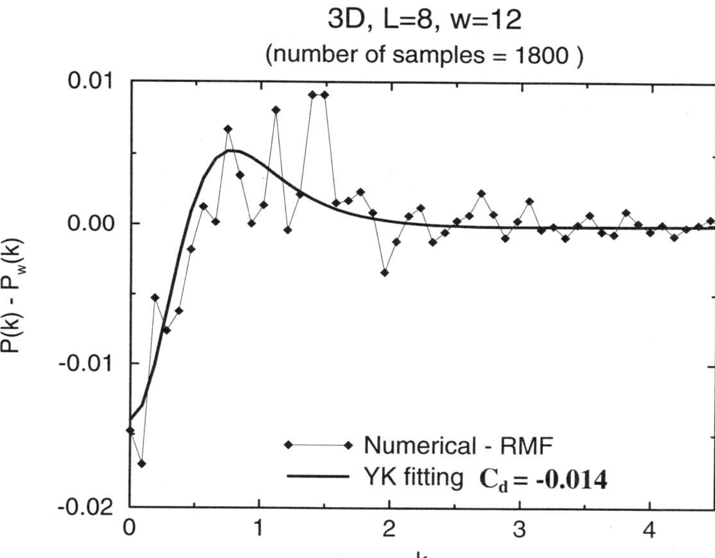

Figure 8. The correction to the curvature distribution for a 3D system in the metal regime for the system size L=8 and disorder W=12. The T-breaking perturbation is caused by a random magnetic flux. The number of disorder realizations is 1800. The solid line is the one-parameter fitting using Eq.(14).

prediction even *quantitatively*. It follows from Eq.(15) that there is a magic relationship for the ratio of amplitudes of the correction in **case I** and **case II**:

$$R = \frac{C_3^{(I)}}{C_3^{(II)}} = -\frac{1}{3},$$

(57)

Our calculations give a result $R = -0.32$ which is in an amazingly good agreement with Eq.(57).

Curvature Distribution at the Mobility Edge in 3D.

A numerical investigation of the distribution $P(k)$ at the Anderson transition critical point has been carried out in Ref.[10]. The main finding of the numerical simulation is that the distribution function at the mobility edge is remarkably well fitted by the formula:

$$P_\alpha(k) = \frac{A_\alpha}{\left(1 + |k|^{(2-\alpha)}\right)^{\frac{3}{2-\alpha}}}.$$

(58)

with $\alpha \approx 0.4$. Eq.(58) defines a function that has a branching point of the type Eq.(52) at $k = 0$ and the asymptotic behavior $P_\alpha(k) \propto |k|^{-3}$ for $|k| \gg 1$ that is expected in all cases where there is a level repulsion $R(\omega, 0) \propto |\omega|$ at $\omega \ll 1$.

In Fig.9 we plot the results for the difference $\delta P(k) = P(k) - P_{WD}(k)$ for the critical disorder $W = 16.5$ and the system size $L = 12$ as compared to two one-parameter fitting curves provided by Eq.(14) and Eq.(58). It is clearly seen that despite the analytical function $\delta P_{reg}(k)$ given by Eq.(14) reproduces a correct qualitative behavior of $\delta P(k)$, there is a sharp feature at small $|k|$ that is captured better by the non-analytical fitting function, Eq.(58).

Using the numerical results for the branching power α in the curvature distribution function one can calculate the fractal dimension $d_2 = 3 - \eta$ from the relationship, Eq.(55), between η and $\mu = 3 - \alpha$. For $\alpha \approx 0.4$ we have $\mu \approx 2.6$ and $d_2 \approx 1.5$. This value is in a good agreement with direct evaluation [32] of d_2 from Eqs.(53).

CONCLUSION.

The main goal of our study is to establish a relationship between the statistics of eigenfunctions in disordered conductors and the corresponding spectral statistics. The level curvature distribution has been chosen as a target of investigation, since it is the simplest example of the parametric spectral statistics that has been suggested long ago as a spectral probe of the structure of eigenfunctions.

The main results of this contribution are formulated by Eqs.(14),(15) and Eqs.(52),(55). The first two equations describe the regular corrections beyond RMT to the curvature distribution in metals. These corrections stem from the long-range correlations in the wave functions with the typical scale of the order of the sample size. The latter two equations summarize the effect of the local irregularities (sharp peaks) in the structure of eigenfunctions in its most developed form (multifractality) in the critical region near the mobility edge.

Moreover, Eq.(55) suggests an explicit relationship between the fractal dimension $d_2 = d - \eta$ of a critical eigenfunction and the exponent μ in the power-law tail Eq.(51) of the characteristic function $P_c(\lambda)$.

Note that Eq.(51) is more general than Eq.(52). The latter one requires rather strong multifractality $\eta > \frac{4}{3}$, while the former one applies to a generic critical state.

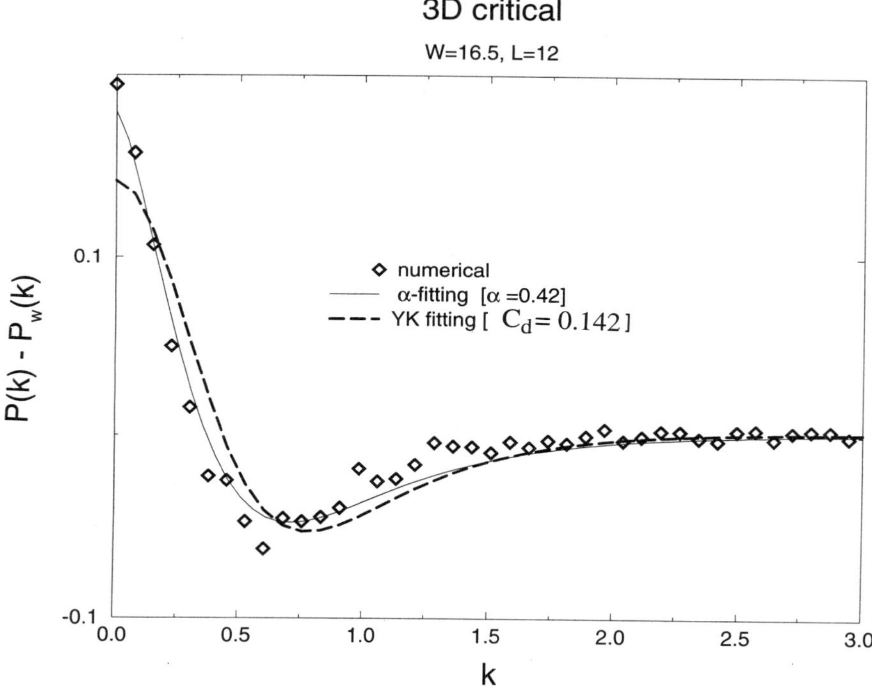

Figure 9. The correction to the curvature distribution for a 3D system in the critical regime for disorder W=16.5 and size L=12. The T-breaking perturbation is caused by a constant vector-potential. Two different one-parameter fitting functions are used: the dashed line is the regular fitting function, Eq.(14), the solid line is the non-analytical fitting function, Eq.(58)

For instance, it would be interesting to check its validity for the critical state in the Quantum Hall effect, where $\eta \approx 0.5$ and we predict $\mu \approx 8$. The recent progress [33] in numerical simulations on the Chalker-Coddington network [34] seems to make the task attainable.

As far as the regular corrections are concerned, there is an interesting question of what happens to them for $d > 4$. The sum in Eq.(15) is an independent of parameters number only for $d < 4$ when it converges. For $d > 4$ the sum is divergent and requires a cut-off at large $|\mathbf{q}|$. Thus for the correct evaluation of this sum it is necessary to go beyond the diffusion approximation and the approximation of slow spatial variations of the field $Q(\mathbf{r})$ in the nonlinear sigma-model. The divergent sum in Eq.(15) implies that the correction $\delta P(k)$ to the level curvature distribution is dominated for $d > 4$ by the *short-range* spatial correlations of eigenfunctions, in contrast to the situation for $d < 4$ when it is dominated by the *long-range* correlations. It is probably the short-range nature of eigenfunction correlations at $d > 4$ that makes the correction $\delta P(k)$ in the "global" case I to change sign (as compared to $d < 4$) and to show qualitatively the same behavior as for "local" perturbations.

One of the most important results Eq.(16) of our calculations is that the ratio $r(g)$ of the mean level curvature $\langle |K| \rangle$ and the mean Drude conductance $2g$ is not a constant and is always above the RMT level $r = 1$.

This is in a qualitative agreement with the result [35,36] that $\langle |K| \rangle \propto \sqrt{g}$ in a strictly one-dimensional case where $g \ll 1$ and effects of localization are strong. Indeed, let us assume that the square-root dependence is typical for strongly localized states in any dimensions. Then the function $r(g)$ should behave like $r(g) \propto g^{-1/2}$ at small $g \ll 1$ and $r(g) \to 1$ for $g \to \infty$. If in addition we make a natural assumption that $r(g)$ is a monotonous function, we arrive at a conclusion that $\delta r(g) > 0$ everywhere in agreement with Eq.(16).

A similar deviation from the proportionality $\langle |K| \rangle \propto g$ has been observed in numerical simulations [37].

Acknowledgements We thank B.L.Altshuler, E.Akkermans, V.I.Fal'ko, Y.V.Fyodorov, I.V.Lerner and A.D.Mirlin for stimulated discussions. V.E.K. is grateful for the hospitality extended to him at the Newton Institute (Cambridge, UK) where a final part of this work has been done. Support from grants RFBR/INTAS No.95-675 and CRDF No.RP1-209 (V.E.K.) is also gratefully acknowledged.

REFERENCES

1. J. T.Edwards and D. J. Thouless, J. Phys. C. , 802, (1972); D. J. Thouless, Phys. Rep. **13**, 93 (1974).
2. B. D. Simons and B. L. Altshuler, Phys. Rev. B. **48**, 5422 (1993); A. Szafer and B. L. Altshuler, Phys. Rev. Lett. **70**, 587 (1993); B. D. Simons, P. A. Lee and B. L. Altshuler, Phys. Rev. Lett. **70**, 4122 (1993).
3. E. P. Wigner, Proc. Cambridge Philos. Soc. **47**, 790 (1951); F. J. Dyson, J. Math. Phys. **3**, 140 (1962).
4. M. L. Mehta, *Random matrices* (Academic Press, Boston, 1991).
5. K. B. Efetov, Adv.Phys. **32**, 53 (1983).
6. A. V. Andreev, O. Agam, B. D. Simons, and B. L. Altshuler, Phys. Rev. Lett. **76**, 3947 (1996).
7. J. Zakrewski and D. Delande, Phys. Rev. E. **47**, 1650 (1993).
8. F. von Oppen, Phys. Rev. E. **51**, 2647 (1995);
9. Y. V. Fyodorov and H.- J. Sommers, Phys. Rev. E. **51**, R2719 (1995).
10. C. M. Canali, C. Basu, W. Stephan and V. E. Kravtsov, Phys. Rev. B. **54**, 1431 (1996).
11. M. Titov, D.Braun and Y. V. Fyodorov, cond-mat/9611235.
12. K. Zyczkowski, L. Molinari, and F. M. Izrailev, J. Phys. I France. **4**, 1469 (1994).

13. V. E.Kravtsov and I. V.Yurkevich, Phys. Rev. Lett. **78**, 3354 (1997).

14. I. V. Yurkevich and V. E. Kravtsov, Phys. Rev. Lett. , **78**, 701 (1997).

15. V. E. Kravtsov and A. D. Mirlin, Pis'ma Zh. Exp. Teor. Fiz. **60**, 645 (1994) [JETP Lett. **60**, 656 (1994)].

16. M. Buttiker, Y. Imry, and R. Landauer, Phys. Lett. A. **96**, 365 (1983).

17. B. L. Altshuler, V. E. Kravtsov and I. V. Lerner in *Mesoscopic Phenomena in Solids*, ed. B.L.Altshuler et al., (Elsevier, Amsterdam 1991), p.449 and references therein.

18. V. I. Falko, K. B. Efetov, Phys. Rev. B. **52**, 17413 (1995).

19. V. E. Kravtsov, Habilitationsschrift, Heidelberg University 1992 (unpublished).

20. B. A. Muzykantskii and D. E. Khmelnitskii, Phys. Rev. B. **51**, 5480 (1995).

21. I. E. Smolyarenko and B. L. Altshuler, cond.-math/9606181 (unpublished).

22. C. Castellani and L. Peliti, J. Phys. A **19**, L429 (1986); W. Pook and M. Janssen, Z. Phys. B. **82**, 295 (1991).

23. F. Wegner, Z. Phys. B **36**, 209 (1980).

24. Y. V.Fyodorov and A. D. Mirlin, Phys. Rev. B **51**, 13403 (1995).

25. V. E. Kravtsov and M. R. Zirnbauer, Phys. Rev. B. **46**, 4332 (1992).

26. J. J. Verbaarschot, H. A. Weidenmuller and M. R. Zirnbauer, Phys. Rep. **129**, 367 (1985).

27. A. D. Mirlin, Pis'ma Zh. Exp. Teor. Fiz. **62**, 583 (1995) [JETP-Lett. **62**, 603 (1995)].

28. A. D. Mirlin, Phys. Rev. B. **53**, 1186 (1996).

29. E. Abrahams, P. W. Anderson, D. C Licciardello, and T. V. Ramakrishnan, Phys. Rev. Lett. **42**, 673 (1979).

30. J. T. Chalker, Physica A **167**, 253 (1990); B. Huckenstein and L. Schweitzer, Phys. Rev. Lett. **72**, 713 (1994).

31. B. L. Altshuler, V. E. Kravtsov, and I. V. Lerner, Zh. Eksp. Teor. Fiz. **91**, 2276 (1986) [Sov. Phys. JETP **64**, 1352 (1986)].

32. T Brandes, B. Huckestein and L. Schweitzer, cond-math/9605062;
T. Ohtsuki and T. Kawarabayashi, cond-math/9701013 and references therein.

33. R.Klesse and M.Metzler, Phys. Rev. Lett. **79**, 721 (1997).

34. J. T.Chalker and P. D.Coddington, J. Phys. C. **21**, 2665 (1988).

35. P. W.Anderson and P. A.Lee, Prog. Theor. Phys. **69**, 212 (1980).

36. E.Akkermans, J. Math. Phys. **38** 1781 (1997).

37. D.Braun, E.Hoffstetter, G.Montambaux and A. MacKinnon, cond-mat/9611059.

38. C.Basu, C.M.Canali, V.E.Kravtsov and I.V.Yurkevich (unpublished).

ALMOST-HERMITIAN RANDOM MATRICES: APPLICATIONS TO THE THEORY OF QUANTUM CHAOTIC SCATTERING AND BEYOND

Yan V. Fyodorov*

Fachbereich Physik, Universität-GH Essen, D-45117 Essen, Germany

Introduction

As is well-known, statistics of highly excited bound states of *closed* quantum chaotic systems of quite different microscopic nature is universal. Namely, it turns out to be independent of the microscopic details when sampled on the energy intervals large in comparison with the mean level separation, but smaller than the energy scale related by the Heisenberg uncertainty principle to the relaxation time necessary for the classically chaotic system to reach equilibrium in the phase space [1]. Moreover, the spectral correlation functions turn out to be exactly those which are provided by the theory of large random matrices on the *local* scale determined by the typical separation $\Delta = \langle X_i - X_{i-1} \rangle$ between neighboring eigenvalues situated around a point X, with brackets standing for the statistical averaging [2]. Microscopic justifications of the use of random matrices for describing the universal properties of quantum chaotic systems have been provided recently by several groups, based both on traditional semiclassical periodic orbit expansions [3, 4] and on advanced field-theoretical methods [5, 6]. These facts make the theory of random Hermitian matrices a powerful and versatile tool of research in different branches of modern theoretical physics, see e.g.[2, 7].

Very recently complex eigenvalues of non-Hermitian random matrices have also attracted much research interest due to their relevance to several branches of theoretical physics. Most obvious motivation comes from the quantum description of *open* systems [8, 9, 10] whose fragments can escape, at a given energy, to infinity or come from infinity. For systems of this kind the notion of discrete energy levels loses its validity. Actually, chaotic scattering manifests itself in terms of a high density of poles of the scattering matrix placed irregularly in the complex energy plane. Each of these poles, or *resonances*, $E_k = \mathcal{E}_k - \frac{i}{2}\Gamma_k$, is characterized not only by energy $\mathcal{E}_k$ but also by a finite width Γ_k defined as the imaginary part of the corresponding complex energy and reflecting the finite lifetime of the states in the open system. Recently, the progress in numerical techniques and computational facilities made available high accuracy patterns of resonance poles for realistic atomic and molecular systems in the regime of quantum chaos, see e.g. [12, 13, 14].

---

*on leave from: Petersburg Nuclear Physics Institute, Gatchina 188350, Russia

Supersymmetry and Trace Formulae: Chaos and Disorder
Edited by Lerner *et al.*, Kluwer Academic / Plenum Publishers, New York, 1999

Due to the presence of these resonances, elements of the scattering matrix show irregular fluctuations as functions of the energy of incoming waves, see [15] and references therein. The main goal of the theory of quantum chaotic scattering is to provide an adequate statistical description of such a behavior.

Whereas the issue of energy level statistics in closed chaotic systems was addressed in an enormous amount of papers statistical characteristics of resonances are much less studied and attracted significant attention only recently, see [8, 9, 10, 11] and references in [16].

One possible way of doing this is to address resonances in the so-called "Heidelberg approach" suggested in the pioneering paper [17] and described in much detail in [16]. The approach turns out to be the most natural framework for incorporating random matrix description of the chaotic scattering.

The starting point of this approach is a division of the Hilbert space of the scattering system into two parts: the "interaction region" and the "channel region". The channel region is supposed to describe a situation of two fragments being apart far enough to neglect any interaction between them. Under these conditions their motion along the collision coordinate is described by a superposition of incoming and outgoing plane waves with wavevectors depending on the internal quantum states of the fragments. We assume that at given energy E exactly M different quantum states of the fragments are allowed, defining M "scattering channels" numbered by the index a.

At the same time, the second part of the Hilbert space is to describe the situation when fragments are close to one another and interact strongly. Correspondingly, any wavefunction of the system $|\Phi(E)\rangle$ can be represented as two-component vector: $|\Phi(E)\rangle = \begin{pmatrix} \mathbf{u} \\ \psi \end{pmatrix}$, with $\mathbf{u}$ and ψ describing the components of the wave function inside the interaction (respectively, channel) region.

Using the standard methods of the scattering theory exposed in detail in [16] one can relate two parts of the wavefunction to one another and finally arrive at the following representation of the energy-dependent scattering matrix $\hat{S}$ in terms of an effective non-Hermitian Hamiltonian $\mathcal{H}_{ef} = \hat{H} - i\hat{\Gamma}$:

$$S_{ab}(E) = \delta_{ab} - 2i\pi \sum_{ij} W_{ai}[E - \mathcal{H}_{ef}]^{-1}_{ij} W_{jb} \tag{1}$$

with the Hermitian Hamiltonian $\hat{H}$ describing the *closed* counterpart of the open system (i.e. interaction region decoupled from the channel one) and the anti-Hermitian part $\hat{\Gamma}$ arising due to a coupling to open scattering channels. In this expression the Hamiltonian $\hat{H}$ is written in some arbitrary basis of states $|i\rangle$, such that $H_{ij} = \langle i|\hat{H}|j\rangle$. The amplitudes $W_{ai}, \quad a = 1, 2, ..., M$ are matrix elements coupling the internal motion in an "internal" state $|i\rangle$ to one out of M open channels a. One also has to choose the anti-Hermitian part to be $\hat{\Gamma} = \pi \sum_a W_{ia} W_{ja}^*$ in order to ensure the unitarity of the $M \times M$ scattering matrix $\hat{S}(E)$ [18, 19, 20].

It is natural to expect, that universal properties of open chaotic systems are inherited from the corresponding universality of levels of their closed counterparts. Of course, one can expect a relation of this kind only when incoming particles stay inside the interaction region long enough to be able to experience the chaoticity of internal dynamics. Going from the time domain to the energy domain, this fact suggests that only scattering characteristics on a scale shorter than inverse classical relaxation time on the energy shell are expected to be universal. Another characteristic energy scale in this domain is a typical level spacing Δ of the closed counterpart of our quantum open system. Thus, we expect the scattering characteristics (in particular, the statistics of resonances) to be universal on the scale comparable with Δ. In contrast, smooth energy dependence of $S-$matrix elements on a much larger energy scale must be system-specific.

The next step is to incorporate the random matrix description of quantum chaotic systems by replacing the Hamiltonian $\hat{H}$ by a random matrix of appropriate symmetry. Namely, chaotic systems with preserved time-reversal invariance (TRI) should be described by matrices H_{ij} which are real symmetric. Such matrices form the Gaussian Orthogonal Ensemble, whereas for systems with broken TRI one uses complex Hermitian matrices from the Gaussian Unitary Ensemble [2, 7].

The third essential ingredient of the Heidelberg approach is performing the ensemble averaging non-perturbatively in the framework of the so-called supersymmetry technique. It was invented initially by Efetov in the context of theory of disordered metals and the Anderson localization [21, 22] and adjusted for the description of quantum chaotic scattering by Verbaarschot, Weidenmüller and Zirnbauer [17].

The Heidelberg approach turns out to be a very powerful tool for extracting universal properties of open chaotic systems. In the first part of the paper we outline the derivation of the mean resonance density in the complex plane following Fyodorov and Sommers [9]. Another recent example of the utility of the non-Hermitian effective Hamiltonian $\hat{H} - i\hat{\Gamma}$ is that its resolvent is related to the probability for an excited system to decay via open channels. For this reason it can be used to calculate the statistics of such quantities as e.g. photodissociation cross-section in the regime of quantum chaos. We present the formula for cross-section autocorrelation function derived recently by Fyodorov and Alhassid [23].

The fact that non-selfadjoint operators appear quite generally when one considers open systems of various types is known for a long time [20]. It is therefore not surprising that open quantum systems were the first examples of applications of non-Hermitian random matrices[8, 9, 16], see also recent papers [10, 24, 25]. Other early applications included also studies of dissipative quantum maps [26, 27, 28] and chaotic dynamics of asymmetric neural networks [29].

Recently, however, random matrices (more generally, random linear operators) with complex eigenvalues emerged in many other physical contexts. Let us briefly mention the most interesting examples.

- The effective Hamiltonian describing a thermal motion of an isolated vortex in disordered type-II superconductors with columnar defects has a form of that for a quantum particle in an *imaginary* vector potential $\mathbf{A}$:

$$\hat{H} = \frac{1}{2m} \left(-i\hbar\nabla + i\mathbf{A} \right)^2 + V(\mathbf{r}), \tag{2}$$

  with $V(\mathbf{r})$ being a random potential generated by defects. The imaginary vector potential makes the Hamiltonian to be a non-Hermitian one. This fact pointed out by Hatano and Nelson [30] gave a boost to several interesting studies [31, 32].

- A classical diffusing particle advected by a stationary random velocity field $\mathbf{v}$ is described by a non-Hermitian Fokker-Plank random operator $\mathcal{L}_{FP}$ [33]:

$$\frac{\partial}{\partial t} n(\mathbf{r}, t) = \mathcal{L}_{FP} n(\mathbf{r}, t) = \left(D\nabla^2 - \nabla\mathbf{v} \right) n(\mathbf{r}, t) \tag{3}$$

  where $n(\mathbf{r}, t)$ is the density of particles and D is the diffusion constant.

- Recent attempts to understand the universal features of chiral symmetry breaking in Quantum Chromodynamics required to consider quarks in a finite chemical potential μ interacting with the Yang-Mills gauge field. The corresponding partition function is given by:

$$\mathcal{Z}(m,\mu) = \left\langle \prod_{f=1}^{N_f} \det\left(m_f + \mu\hat{\gamma}_0 + \hat{\mathcal{D}}\right) \right\rangle_{\mathcal{A}} \tag{4}$$

  where $\hat{\mathcal{D}} = \hat{\gamma}_\mu \partial_\mu + i\hat{\gamma}_\mu \mathcal{A}_\mu$ is the Euclidean Dirac operator in the gauge vector potential $\mathcal{A}$, and $\langle \ldots \rangle_{\mathcal{A}} = \int D\mathcal{A}(\ldots) \exp - \int d^4 x F_{\mu\nu}^2$ stands for the averaging over all configurations of the gauge field $F_{\mu\nu} = \partial_\mu \mathcal{A}_\nu - \partial_\nu \mathcal{A}_\mu + i[\mathcal{A}_\mu, \mathcal{A}_\nu]$

  A finite chemical potential μ makes the corresponding operator to be a non-selfadjoint one with complex eigenvalues. This fact makes a problem of numerical evaluation of the partition function by lattice simulations to be a very difficult one. Recently it was suggested, that some universal features of the model can be correctly recovered if one replaces the true gauge-field averaging by averaging over random gauge-field configurations. As a result, one comes to a class of non-Hermitian random matrix problems of a particular type [34].

- Finally, let us mention that there exist several interesting links between complex eigenvalues of non-Hermitian random matrices and systems of interacting particles in one and two spatial dimensions [35, 36].

At the same time, our knowledge of statistical properties of random non-selfadjoint matrices is quite scarce and incomplete. This fact recently stimulated efforts of different groups to improve our understanding in this direction [37–42].

Traditional mathematical treatment of random matrices with no symmetry conditions imposed goes back to the pioneering work by Ginibre [43] who determined all the correlation functions of the eigenvalues in an ensemble of complex matrices with independent Gaussian entries. The progress in the field was rather slow but steady, see [44, 45, 29, 46, 47, 48, 40, 49].

Surprisingly, all these studies completely disregarded the existence of a nontrivial regime of *weak non-Hermiticity* recognized in the work by Fyodorov, Khoruzhenko and Sommers [38], see more detailed discussion in [50]. The guiding idea to realize the existence of such a regime comes from the experience with resonances [9]. Guided by that example one guesses that a new regime occurs when the imaginary part of typical eigenvalues is comparable with a mean *separation* between neighboring eigenvalues along the real axis.

One can again employ the same supersymmetry approach which was used to study resonances and obtain the mean density of complex eigenvalues in the regime of weak non-Hermiticity for matrices with independent elements [38, 50]. The density turned out to be described by a formula containing as two opposite limiting cases both the Wigner semicircular density of real eigenvalues typical for Hermitian random matrices and the uniform density of complex eigenvalues discovered for strongly non-Hermitian random matrices already by Ginibre [43], in much details addressed by Girko [45] and studied for different cases by other authors [29, 48, 49].

Very recently, Efetov [32] showed the relevance of almost-Hermitian random matrices to the very interesting problem of motion of flux lines in superconductors with columnar defects [30]. He also managed to derive the density of complex eigenvalues for a related, but different set of almost-symmetric real random matrices. This development clearly shows that, apart from being a rich and largely unexplored mathematical object, almost-Hermitian random matrices enjoy direct physical applications and deserve to be studied in more detail.

Actually, the non-Hermitian matrices considered in [38] and [32] are just two limiting cases of a general three-parameter family of non-Hermitian ensembles [50]. In second section of the paper we outline the derivation of this fact and present the resulting expression in terms of a non-linear σ-model integral.

Although giving an important insight into the problem, the supersymmetry nonlinear $\sigma-$ model technique suffers from at least two deficiencies. The most essential one is that the present state of art in the application of the supersymmetry technique gives little hope of access to quantities describing *correlations* between different eigenvalues in the complex plane due to insurmountable technical difficulties. At the same time, conventional theory of random Hermitian matrices suggests that these *universal* correlations are the most interesting features. The second drawback is less important for a physicist, but a crucial one for the mathematicians: at the moment the supersymmetry technique can not be considered as a rigorous mathematical tool and has the status of a powerful heuristic method.

Fortunately, for the simplest case of almost-Hermitian Gaussian random matrices one can develop the rigorous mathematical theory based on the method of orthogonal polynomials. Such a method is free from the above mentioned problem and allows one to study correlation properties of complex spectra to the same degree as is typical for earlier studied classes of random matrices [39]. We briefly discuss the obtained results in the end of the paper. The detailed exposition of the method and the derivation of the results can be found in [50]. Unfortunately, the paper [39] contained a number of misleading misprints. For this reason we indicate those misprints in the present text by using footnotes.

Non-Hermitian random matrices in the theory of chaotic quantum scattering

To calculate the density of resonance poles in the complex energy plane we notice that they are just eigenvalues of the effective non-Hermitian Hamiltonian introduced in Eq.(1).

Without any loss of generality coupling amplitudes W_{aj} can be chosen in a way ensuring that the average $S-$matrix is diagonal in the channel basis: $\overline{S_{ab}} = \delta_{ab}\overline{S_{aa}}$. Then one finds the following expression [17]:

$$\overline{S_{aa}} = \frac{1 - \gamma_a g(X)}{1 + \gamma_a g(X)}; \quad \gamma_a = \pi \sum_i W_{ia}^* W_{ia} \tag{5}$$

where $g(X) = iX/2 + \pi\overline{\nu(X)}$ and $\pi\overline{\nu(X)} = (1 - X^2/4)^{1/2}$ is the semicircular level density. The strength of coupling to continua is convenient to characterize via the transmission coefficients $T_a = 1 - |\overline{S_{aa}}|^2$. These quantities measure a part of the incoming flux in a given channel that spends a substantial part of the time in the interaction region [17, 51]. The case $T_a \ll 1$ corresponds to almost closed channel a, whereas the opposite limiting case $T_a = 1$ corresponds to the perfect coupling between the interaction region and the channel a. It is easy to see that both limits $\gamma_a \to 0$ and $\gamma_a \to \infty$ equally correspond to the weak effective coupling regime whereas the strongest coupling (at fixed energy X) corresponds to $\gamma_a = 1$.

In the case of weak effective coupling to continua individual resonances do not overlap: $\langle \Gamma \rangle \ll \Delta$, with $\Delta = (\nu(X)N)^{-1}$ standing for the mean level spacing of the "closed" system and $\langle \Gamma \rangle$ standing for the mean resonance width. Under these conditions one can use a simple first order perturbation theory to calculate resonance widths in terms of eigenfunctions of the closed system. Since different components of eigenvectors of large random matrices are decorrelated and Gaussian-distributed, one finds in such a procedure that the scaled widths $y_s = \frac{\Gamma}{\langle \Gamma \rangle}$ are distributed according to the so-called χ^2-distribution:

$$\rho(y_s) = \frac{(\nu/2)^{\nu/2}}{\Gamma(\nu/2)} y_s^{\nu/2-1} e^{-\frac{\nu}{2} y_s} \tag{6}$$

where $\Gamma(z)$ stands for the Gamma function and the parameter $\nu = M$ $(\nu = 2M)$ for systems with preserved (broken) time reversal invariance, M being the number of open scattering channels. The case $\nu = 1$ is known as Porter-Thomas distribution and was shown to be in agreement with experimental data (see some references in [52]).

Experimentally, one quite frequently encounter the case of only $M \sim 1$ open channels and $\langle \Gamma \rangle \sim \Delta$, see e.g. [53, 54, 55]. Under this situation resonances overlap considerably and one can not use perturbation theory any longer. The problem of determining the statistical characteristics of the chaotic scattering in the regime of (partly or completely) overlapping resonances is of essentially non-perturbative nature. As a result, one has to use some non-perturbative methods allowing to evaluate the ensemble averaging exactly for arbitrary ratio $\langle \Gamma \rangle / \Delta$.

Fortunately, one can study very efficiently various universal statistical features of chaotic quantum scattering by performing the ensemble averaging with the use of the supersymmetry method [17].

One can recover the spectral density

$$\rho(Z) = \sum_{k=1}^{N} \delta^{(2)}(Z - Z_k) = \sum_{k=1}^{N} \delta(X - X_k)\delta(Y - Y_k) \equiv \rho(X, Y) \tag{7}$$

of complex eigenvalues $Z_k = X_k + iY_k$, $k = 1, 2, ..., N$ if one knows the "potential" [29]:

$$\Phi(X, Y, \kappa) = \frac{1}{2\pi} \ln \overline{Det[(Z - \mathcal{H}_{ef})(Z - \mathcal{H}_{ef})^\dagger + \kappa^2]}$$

in view of the relation:$\rho(X, Y) = \lim_{\kappa \to 0} \partial^2 \Phi(X, Y, \kappa)$, where ∂^2 stands for the two-dimensional Laplacian. Technically, it is convenient to introduce the generating function (cf.[16])

$$\mathcal{Z} = \frac{Det\left[(Z - \mathcal{H}_{ef})(Z - \mathcal{H}_{ef})^\dagger + \kappa^2\right]}{Det\left[(Z_b - \mathcal{H}_{ef})(Z_b - \mathcal{H}_{ef})^\dagger + \kappa^2\right]} \tag{8}$$

in terms of which

$$\rho(Z) = -\frac{1}{\pi} \lim_{\kappa \to 0} \frac{\partial}{\partial Z^*} \lim_{Z_b \to Z} \frac{\partial}{\partial Z_b} \mathcal{Z} \tag{9}$$

To facilitate the ensemble averaging we follow the standard route and represent the ratio of the two determinants in Eq.(8) in terms of a Gaussian superintegral over eight-component supervectors $\Phi_i = \begin{pmatrix} \Psi_i(+) \\ \Psi_i(-) \end{pmatrix}$ where $\Psi_i(\pm) = \begin{pmatrix} \vec{R}_i(\pm) \\ \vec{\eta}_i(\pm) \end{pmatrix}$ and

$$\vec{R}_i(\pm) = \begin{pmatrix} r_i(\pm) \\ r_i^*(\pm) \end{pmatrix} \quad ; \quad \vec{\eta}_i(\pm) = \begin{pmatrix} \chi_i(\pm) \\ \chi_i^*(\pm) \end{pmatrix} \quad ; \quad \vec{\eta}_i^\dagger(\pm) = (\chi_i^*(\pm); -\chi_i(\pm))$$

with components $r_i(+), r_i(-);$ $i = 1, 2, ..., N$ being complex commuting variables and $\chi_i(+), \chi_i(-)$ forming the corresponding Grassmann parts of the supervectors $\Psi_i(\pm)$.

Further evaluation goes along the lines discussed in [16] in more detail. After a set of standard manipulations one arrives at the following expression for the density $\rho_X(y) \equiv \frac{1}{\pi}\rho(X,Y)\Delta^2(X)$ of scaled resonance widths $y = \frac{\pi\Gamma}{\Delta}$ (measured in units of the local mean level spacing $\Delta(X)$ of the closed system) for the resonances whose positions are within a narrow window around the point X of the spectrum:

$$\langle \rho_X(y) \rangle = \tag{10}$$

$$\frac{1}{16} \int d\mu(\hat{Q}) \mathrm{Str}\left(\hat{\sigma}_\tau^{(F)}\hat{Q}\right) \mathrm{Str}\left(\hat{\sigma}_\tau \hat{Q}\right) \exp \frac{i}{4} y \mathrm{Str}\left(\hat{\sigma}_\tau \hat{Q}\right) \prod_{a=1}^{M} \mathrm{Sdet}^{-1/4}\left[1 - \frac{i}{2g_a}\left\{\hat{Q},\hat{\sigma}_\tau\right\}\right]$$

Here the integration goes over the set of 8×8 supermatrices $\hat{Q}$ satisfying the constraint $\hat{Q}^2 = -1$ and $\left\{\hat{Q},\hat{\sigma}_\tau\right\} = \hat{Q}\hat{\sigma}_\tau + \hat{\sigma}_\tau\hat{Q}$ stands for the anticommutator. Properties of these matrices and the integration measure $d\mu(\hat{Q})$ can be found in [21]. Other 8×8 supermatrices entering the expression Eq.(10) are as follows:

$$\hat{\sigma}_\tau^{(F)} = \begin{pmatrix} \hat{0}_4 & \hat{\tau}_3^{(F)} \\ \hat{\tau}_3^{(F)} & \hat{0}_4 \end{pmatrix}; \quad \hat{\sigma}_\tau = \begin{pmatrix} 0 & \hat{\tau}_3 \\ \hat{\tau}_3 & 0 \end{pmatrix}$$

and $\hat{\tau}_3;$, $\hat{\tau}_3^{(F)}$ are 4×4 diagonal supermatrices: $\hat{\tau}_3 = \mathrm{diag}\{\hat{\tau},\hat{\tau}\}$; $\hat{\tau}_3^{(F)} = \mathrm{diag}\{\hat{0}_2,\hat{\tau}\}$. with $\hat{\tau} = \mathrm{diag}(1,-1)$. We also introduced quantities $g_a = \frac{1}{2\pi\bar{\nu}(X)}(\gamma_a + \gamma_a^{-1})$ related to the transmission coefficients as $g_a = 2/T_a - 1$ and used the symbols $\mathrm{Str},\mathrm{Sdet}$ for the graded trace and the graded determinant, correspondingly.

The expression above is valid for chaotic systems with preserved as well as with broken time-reversal invariance. To extract the explicit form of the distribution function one still has to perform the integration over the manifold of the supermatrices Q which is different for two cases. In general it is a rather difficult calculation due to a cumbersome parametrisation of that manifold. At the moment the result is known for the simplest case of the systems with broken time-reversal invariance [9, 16]. For the sake of simplicity we present this distribution for the case of equivalent channels $a = 1, ..., M$ with equal transmission coefficients $T_a = T$.

First of all, it turns out that the mean resonance width is related to the transmission coefficient T as:

$$\langle \Gamma \rangle = -\Delta \frac{M}{2\pi} \ln (1 - T) \equiv \Delta \frac{M}{2\pi} \ln \frac{g+1}{g-1} \tag{11}$$

The formula Eq.(11) is well known in nuclear physics as Moldauer-Simonius relation [56].

It is convenient to use the parameter $\kappa = -\frac{M}{2} \ln (1 - T)$ as a measure of the resonance overlap. It is related to the mean widths as $\kappa = \pi \frac{\langle \Gamma \rangle}{\Delta}$ and therefore gives a typical number of neighboring resonances that overlap substantially.

Measuring the resonance widths in units of the mean widths $\langle \Gamma \rangle$ one finds the following distribution function.

$$\rho\left(y_s = \frac{\Gamma}{\langle \Gamma \rangle}\right) = \frac{1}{2\Gamma(M)\kappa y^2} \int_{y_s b_-}^{y_s b_+} dt\, t^M e^{-t} \tag{12}$$

$$= (-1)^M \frac{y_s^{M-1}}{\Gamma(M)} \frac{d^M}{dy_s^M}\left(\exp -[\kappa \coth (\kappa/M)y_s]\frac{\sinh \kappa y_s}{\kappa y_s}\right).$$

where we used the notations $b_\pm = \kappa e^{\pm\kappa/M}/\sinh (\kappa/M)$ and $\Gamma(M) = (M - 1)!$ for the Euler gamma-function.

Properties of this distribution are discussed in much details in [16], also for the case of non-equivalent channels. Let us only briefly mention the most interesting features.

Quick inspection of eq.(12) shows that it is indeed reduced to the χ^2 distribution, eq.(6) when the effective coupling to continua is weak: $\kappa \ll 1$. Under this condition resonances are typically too narrow to overlap with others: $\Gamma \ll \Delta$. However, as long as the effective coupling becomes stronger: $T \to 1$, hence $g \gg 1$ the parameter κ grows large. Under these conditions another domain of resonance widths becomes more and more important:

$$e^{-\kappa/M}\frac{\sinh\left(\kappa/M\right)}{(\kappa/M)} < y_s < e^{\kappa/M}\frac{\sinh\left(\kappa/M\right)}{(\kappa/M)},$$

where the distribution eq.(12) shows the powerlaw decrease: $\rho(y_s) \approx \frac{1}{2\kappa}My_s^{-2}$. The most drastic difference from eq.(6) occurs for the maximal effective coupling $g = 1$ (i.e $\kappa = \infty$). In this regime the powerlaw tail extends up to infinity, making all positive moments (starting from the first one) to be apparently divergent. One can argue that the powerlaw tail My_s^{-2} turns out to be dictated by classical processes of exponential escape typical for fully chaotic systems [16]. The rate of this escape in the semiclassical limit $M \gg 1$ is provided by the value of the gap in the distribution of resonance width, see [8] for a more detailed discussion.

The best candidates for checking the applicability of eq.(12) to real physical systems are realistic models of ballistic mesoscopic devices subject to an applied magnetic field that serves to break the TRI [53]. It is however quite clear that all the basic qualitative features of the distribution eq.(12) (in particular, the powerlaw behavior $\rho(y_s) \propto My_s^{-2}$ for the overlapping resonance regime) should be valid for the systems with preserved TRI as well. Recent numerical data [57] support the validity of this conjecture.

We have seen, that the non-Hermitian random matrix Hamitonian $\hat{H}_{ef}$ appeared naturally in the scattering matrix description of open quantum systems. Actually, such a Hamiltonian is the most adequate tool to describe the quantum relaxation processes such as escape of the particle from the interaction region in the regime of quantum chaos. Some aspects of such a relaxation were studied some time ago in [58] and reconsidered in more details recently by Savin and Sokolov [24] who used insights provided by the resonance widths distribution Eq.(12)

It is therefore quite natural that the resolvent of the non-Hermitian effective Hamiltonian $\hat{H}_{ef} = \hat{H} - i\hat{\Gamma}$ is related to the probability for an excited system to decay via one of open channels. For this reason it can be used to calculate the statistics of such quantities as the Wigner *time delay* [59] which is a measure of mean time spent by a scattered particle inside the interaction region.

One more example of the utility of the resolvent of the non-Hermitian effective Hamiltonian $\hat{H}_{ef}$ is that it can be related to the total photodissociation cross section in the regime of quantum chaos. In this way fluctuations of the cross section with energy were studied recently by Fyodorov and Alhassid [23]. Below we outline the derivation and present the explicit expression for the cross section autocorrelation function.

The total energy-dependent cross section $\sigma(E)$ is defined as a probability to be excited from a ground state $|g\rangle$ and to dissociate at a given energy E per unit time and per unit incoming photon flux density. In the dipole approximation it is given by (see, e.g. detailed discussion in [60]) :

$$\sigma(E) = \sigma_0 \sum_a \left|\left\langle g|\hat{\mu}|\Phi_a^{out}(E)\right\rangle\right|^2 \tag{13}$$

Here $\hat{\mu}$ is the dipole operator $\hat{\mu} = e\mathbf{r}$ of the system, σ_0 is a constant proportional to the excitation energy E and $|\Phi_a^{out}(E)\rangle$ is the exact wave function of the system at energy E subject to *outgoing* boundary conditions in one of the open channels $a = 1, 2, ..., M$.

It turns out, that for many systems of interest (e.g. molecules HO_2[54] and H_3^+[14]), the cross section patterns look like irregular fluctuating signals consisting of many randomly positioned (partly) overlapping resonance peaks. Such a behavior (typical also for the resonance scattering in atomic systems [12, 55]) has its origin in the underlying pattern of resonances in the complex plane.

We already mentioned that one can relate the "internal" and "external" parts of the wavefunctions by using the resolvent of the non-Hermitian effective Hamiltonian $\hat{H}_{ef}$. In the present case such a relation can be writen as (see e.g. [61]):

$$\mathbf{u} = \left(E - \hat{H}_{ef}^\dagger\right)^{-1} \hat{W}\mathbf{B} \tag{14}$$

where M-component vector $\mathbf{B}$ contains amplitudes of outgoing waves in each of the open channels.

The ground state wavefunction describes a bound state and as such has no components outside the interaction region. Using this fact and Eq.(14) one finds after some algebraic manipulations that the cross section Eq.(13) can be rewritten in the following convenient form:

$$\sigma(E) \propto Im \left\langle g \left| \mu \frac{1}{E - \mathcal{H}_{ef}} \mu \right| g \right\rangle \tag{15}$$

which is just a particular case of the optical theorem. One also can arrive at the expression Eq.(15) by resumming the perturbation theory, see [55] for more details and relevant references.

The advantage of the form Eq.(15) is that it expresses the photodissociation cross section in terms of the resolvent of an effective non-Hermitian operator $\mathcal{H}_{ef} = H_{in} - i\pi WW^\dagger$ which is known to describe open chaotic systems in the random matrix formalism. It allows to apply very well developed methods of evaluating averages of products of resolvents based on the Efetov supermatrix formalism. Measuring energy separations in units of the mean level spacing of the closed system Δ one finds in such a calculation the cross section autocorelation function [23]:

$$S\left(\omega = \pi\Omega/\Delta\right) = \frac{\langle\sigma(E - \Omega/2)\sigma(E + \Omega/2)\rangle}{\langle\sigma(E)\rangle^2} - 1,$$

to be a sum of two terms $S(\omega) = S_1(\omega) + S_2(\omega)$ given by the following expressions:

$$S_{1,2}(\omega) = \int_{-1}^1 d\lambda \int_1^\infty d\lambda_1 \int_1^\infty d\lambda_2 \frac{\cos\left[\omega(\lambda_1\lambda_2 - \lambda)\right](1 - \lambda^2)}{[\lambda_1^2 + \lambda_2^2 + \lambda^2 - 2\lambda_1\lambda_2\lambda - 1]^2}$$

$$\times f_{1,2}(\lambda, \lambda_1, \lambda_2) \prod_{c=1}^M \frac{(g_a + \lambda)}{[(g_a + \lambda_1\lambda_2)^2 - (\lambda_1^2 - 1)(\lambda_2^2 - 1)]^{1/2}} \tag{16}$$

where

$$f_1(\lambda, \lambda_1, \lambda_2) = (\lambda_1\lambda_2 - \lambda)^2; \quad f_2(\lambda, \lambda_1, \lambda_2) = 2\lambda_1^2\lambda_2^2 - \lambda_1^2 - \lambda_2^2 - \lambda^2 + 1$$

The parameters g_a were introduced before and related to the transmission coefficients as $g_a = 2/T_a - 1$.

It is worth mentioning that each of the contributions $S(\omega)_{1,2}$ represent an interesting object by itself. Namely, $S_1(\omega)$ coincides with the autocorrelation function of the so-called Wigner time delays studied in some detail in [59], whereas $S_2(\omega)$ is related by the Fourier-transform to the so-called "norm leakage" out the interaction region. The latter quantity was introduced recently by Savin and Sokolov as a characteristic of the process of quantum relaxation in chaotic systems and studied in detail for the simplest case of broken time-reversal invariance [24].

Actually, the starting Fermi golden rule formula Eq.(13) is valid for an arbitrary excitation of the system with a weak perturbation $\hat{\mu}$. For this reason the autocorrelation function of crossections presented above is also of a general applicability.

Finally, it is necessary to mention that in the limit $T_a = 0$ for all a (corresponding to a closed system with purely bound spectra and no possibility for photodissociation) the expression Eq.(16) reduces to the "oscillator strength" correlation function found by Taniguchi et al. [62].

Non-Hermitian matrices with independent elements: Universal properties in the regime of weak non-Hermiticity

To begin with, any $N \times N$ matrix $\hat{J}$ can be decomposed into a sum of its Hermitian and skew-Hermitian parts: $\hat{J} = \hat{H}_1 + i\hat{H}_2$, where $\hat{H}_1 = (\hat{J} + \hat{J}^\dagger)/2$ and $\hat{H}_2 = (\hat{J} - \hat{J}^\dagger)/2i$. Following this, we consinder an ensemble of random $N \times N$ complex matrices $\hat{J} = \hat{H}_1 + iv\hat{H}_2$ where $\hat{H}_p$; $p = 1, 2$ are both Hermitian: $\hat{H}_p^\dagger = \hat{H}_p$. The parameter v is used to control the degree of non-Hermiticity.

In turn, complex Hermitian matrices $\hat{H}_p$ can always be represented as $\hat{H}_1 = \hat{S}_1 + iu\hat{A}_1$ and $\hat{H}_2 = \hat{S}_2 + iw\hat{A}_2$, where $\hat{S}_p = \hat{S}_p^T$ is a real symmetric matrix, and $\hat{A}_p = -\hat{A}_p^T$ is a real antisymmetric one. From this point of view the parameters u, w control the degree of being non-symmetric.

Throughout the paper we consider the matrices $\hat{S}_1, \hat{S}_2, \hat{A}_1, \hat{A}_2$ to be mutually statistically independent, with i.i.d. entries normalized in such a way that:

$$\lim_{N \to \infty} \frac{1}{N} Tr \hat{S}_p^2 = \lim_{N \to \infty} \frac{1}{N} Tr \hat{A}_p \hat{A}_p^T = 1 \tag{17}$$

As is well-known [2], this normalisation ensures that for any value of the parameter $u \neq 0$, such that $u = O(1)$ when $N \to \infty$ statistics of real eigenvalues of the Hermitian matrix of the form $\hat{H} = \hat{S} + iu\hat{A}$ is identical (up to a trivial rescaling) to that of $u = 1$, the latter case known as the Gaussian Unitary Ensemble (GUE). On the other hand, for $u \equiv 0$ real eigenvalues of the real symmetric matrix $\hat{S}$ follow another pattern of the so-called Gaussian Orthogonal Ensemble (GOE).

The non-trivial crossover between GUE and GOE types of statistical behavior happens on a scale $u \propto 1/N^{1/2}$ [63]. This scaling can be easily understood by purely perturbative arguments [64]. Namely, for $u \propto 1/N^{1/2}$ a typical shift $\delta\lambda$ of eigenvalues of the symmetric matrix S due to the antisymmetric perturbation $iu\hat{A}$ is of the same order as the mean spacing Δ between unperturbed eigenvalues : $\delta\lambda \sim \Delta \sim 1/N$.

Similar perturbative arguments show [38], that the most interesting behavior of *complex* eigenvalues of non-Hermitian matrices should be expected for the parameter v being scaled in a similar way: $v \propto 1/N^{1/2}$. It is just the regime when the *imaginary* part ImZ_k of a typical eigenvalue Z_k due to non-Hermitian perturbation is of the same order as the mean spacing Δ between unperturbed real eigenvalues : $ImZ_k \sim \Delta \sim 1/N$. Under these conditions a non-Hermitian matrix J still "remembers" the statistics of its Hermitian part $\hat{H}_1$. As will be clear afterwards, the parameter w should be kept of the order of unity in order to influence the statistics of the complex eigenvalues.

It is just this regime of *weak non-Hermiticity* which we are interested in. Correspondingly, we scale the parameters as [†]:

$$v = \frac{\alpha}{2\sqrt{N}}; \quad u = \frac{\phi}{2\sqrt{N}} \tag{18}$$

and consider α, ϕ, w fixed of the order O(1) when $N \to \infty$.

To be specific, we consider the real symmetric matrix $\hat{S}_1$ to be taken from the ensemble of *sparse* random matrices [65, 66] characterized by the following probability density of a given entry S_{ij} :

$$\mathcal{P}(S_{ij}) = (1 - \frac{p}{N})\delta(S_{ij}) + \frac{p}{N}h(S_{ij}) \tag{19}$$

where $h(s) = h(-s)$ is an arbitrary even distribution function satisfying the conditions: $h(0) < \infty$; $\int h(s)s^2 ds < \infty$ and p stands for the mean value of non-zero matrix elements per column. Actually, this ensemble is the most general one among those with independent elements, and statistics of its eigenvalues was proved to be completely universal [65, 67], up to a rescaling by *ensemble-dependent* mean eigenvalue density $\nu(X)$. Statistics of the matrix elements of all other matrices $S_2, A_{1,2}$ is immaterial as long as their elements are statistically independent as well.

The calculation of the mean density of complex eigenvalues follows essentially the same route as that outlined in the previous section for the resonances. The method used [50] is a generalization of the Efetov's technique to the case of sparse random matrices suggested in [65] (see some details also in [66]). As the result, one arrives at the following expression [50]:

$$\langle \rho(X,y) \rangle = \frac{N\nu(X)}{16} \int d\mu(\hat{Q})\mathrm{Str}\left(\hat{\sigma}_\tau^{(F)}\hat{Q}\right)\mathrm{Str}\left(\hat{\sigma}_\tau\hat{Q}\right)\exp{-S(\hat{Q})} \tag{20}$$

$$S(\hat{Q}) = -\frac{i}{2}y\mathrm{Str}\left(\hat{\sigma}_\tau\hat{Q}\right) - \frac{a^2}{16}\mathrm{Str}\left(\hat{\sigma}_\tau\hat{Q}\right)^2 + \frac{b^2}{16}\mathrm{Str}\left(\hat{\tau}_2\hat{Q}\right)^2 - \frac{c^2}{16}\mathrm{Str}\left(\hat{\sigma}\hat{Q}\right)^2 \tag{21}$$

where we introduced the scaled imaginary parts $y = \pi\nu(X)NY$ and used the notations: $a^2 = (\pi\nu(X)\alpha)^2$, $b^2 = (\pi\nu(X)\phi)^2$, $c^2 = (\pi\nu(X)\alpha w)^2$. The supermatrices $\hat{\tau}_2$ and $\hat{\sigma}$ entering this expressions are as follows:

$$\hat{\tau}_2 = \mathrm{diag}\{\hat{\tau}_3, \hat{\tau}_3\}; \quad \sigma = \begin{pmatrix} 0 & \hat{I}_4 \\ \hat{I}_4 & 0 \end{pmatrix}$$

and the supermatrices $\hat{\sigma}_\tau$ and $\hat{\tau}_3$ were defined after Eq.(10). The expression (20) is just the universal $\sigma-$ model representation of the mean density of complex eigenvalues in the regime of weak non-Hermiticity we were looking for. The universality is clearly manifest: all the particular details about the ensembles entered only in the form of mean density of real eigenvalues $\nu(X)$. The density of complex eigenvalues turns out to be dependent on three parameters: a, b and c, controlling the degree of non-Hermiticity (a), and symmetry properties of the Hermitian part (b) and non-Hermitian part (c).

The following comment is appropriate here. The derivation above can be done not only for ensembles with i.i.d. entries but also for any "rotationaly invariant" ensemble of real symmetric matrices $\hat{S}_1$. To do so one can employ the procedure invented by Hackenbroich and Weidenmüller [68] allowing one to map the correlation functions of the invariant ensembles (plus perturbations) to that of Efetov's $\sigma-$model.

[†]In the Letter 39 there is a misprint in the definition of the parameter α.

Still, in order to get an explicit expression for the density of complex eigenvalues one has to evaluate the integral over the set of supermatrices $\hat{Q}$. In general, it is an elaborate task due to complexity of that manifold.

At the present moment such an evaluation was successfully performed for two important cases: those of almost-Hermitian matrices and real almost-symmetric matrices. The first case (which is technically the simplest one) corresponds to $\phi \to \infty$, that is $b \to \infty$. Under this condition only that part of the matrix $\hat{Q}$ which commutes with $\hat{\tau}_2$ provides a nonvanishing contribution. As the result, $\mathrm{Str}\left(\hat{\sigma}\hat{Q}\right)^2 = \mathrm{Str}\left(\hat{\sigma}_\tau\hat{Q}\right)^2$ so that second and fourth term in Eq.(20) can be combined together. Evaluating the resulting integral, and introducing the notation $\tilde{a}^2 = a^2 + c^2$ one finds [38]:

$$\rho_X(y) = \sqrt{\frac{2}{\pi}}\frac{1}{\tilde{a}}\exp\left(-\frac{2y^2}{\tilde{a}^2}\right)\int_0^1 dt\,\cosh(2ty)\exp\left(-\tilde{a}^2t^2/2\right), \qquad (22)$$

where $\rho_X(y)$ is the density of the scaled imaginary parts y for those eigenvalues, whose real parts are situated around the point X of the spectrum (cf. Eq.(10)).

It is easy to see, that when $\tilde{a}$ is large one can effectively put the upper boundary of integration in Eq.(22) to be infinity due to the Gaussian cut-off of the integrand. This immediately results in the uniform density $\rho_X(y) = (\tilde{a}^2)^{-1}$ inside the interval $|y| < \tilde{a}^2/2$ and zero otherwise. Translating this result to the two-dimensional density of the original variables X, Y, we get:

$$\rho(X,Y) = \begin{cases} \frac{N}{4\pi v^2(1+w^2)} & \text{for } |Y| \leq 2\pi\nu(X)v^2(1+w^2) \\ 0 & \text{otherwise} \end{cases} \qquad (23)$$

This result is a natural generalization of the so-called "ellipic law" known for strongly non-Hermitian random matrices [43, 45, 29]. Indeed, the curve encircling the domain of the uniform eigenvalue density is an ellipse: $\frac{Y^2}{2v^2(1+w^2)} + \frac{X^2}{4} = 1$ as long as the mean eigenvalue density of the Hermitian counterpart is given by the semicircular law. The semicircular density is known to be shared by ensembles with i.i.d. entries, provided the mean number p of non-zero elements per row grows with the matrix size as $p \propto N^\alpha$; $\alpha > 0$, see [65]. In the general case of sparse or "rotationally invariant" ensembles the function $\nu(X)$ might be quite different from the semicircular law. Under these conditions Eq.(23) still provides us with the corresponding density of complex eigenvalues.

The second nontrivial case for which the result is known explicitly is due to Efetov [32]. It is the limit of slightly asymmetric real matrices corresponding in the present notations to: $\phi \to 0; w \to \infty$ in such a way that the product $\phi w = \tilde{c}$ is kept fixed. The density of complex eigenvalues turns out to be given by:

$$\rho_X(y) = \delta(y)\int_0^1 dt\,\exp\left(-\tilde{c}^2t^2/2\right) \qquad (24)$$

$$+2\sqrt{\frac{2}{\pi}}\frac{|y|}{\tilde{c}}\int_1^\infty du\,\exp\left(-\frac{2y^2u^2}{\tilde{c}^2}\right)\int_0^1 dt\,t\,\sinh(2t|y|)\exp\left(-\tilde{c}^2t^2/2\right), \qquad (25)$$

The first term in this expression shows that everywhere in the regime of "weak asymmetry" $\tilde{c} < \infty$ a finite fraction of eigenvalues remains on the real axis.

Such a behavior is qualitatively different from that typical for the case of "weak non-Hermiticity" $\tilde{a} < \infty$, where eigenvalues acquire a nonzero imaginary part with probability one.

In the limit $\tilde{c} >> 1$ the portion of real eigenvalues behaves like $\tilde{c}^{-1}$. Remembering the normalization of the parameter v, Eq.(17), it is easy to see that for the case of

$v = O(1)$ the number of real eigenvalues should scale as $\sqrt{N}$. The fact that of the order of $N^{1/2}$ eigenvalues of strongly asymmetric real matrices stays real was first found numerically by Sommers et al. [29, 46], and proved by Edelman [48].

Gaussian almost-Hermitian matrices: from Wigner-Dyson to Ginibre eigenvalue statistics.

In the present section we concentrate on the particular case of almost-Hermitian random matrices with i.i.d. entries $\hat{J} = \hat{H}_1 + iv\hat{H}_2$, where $\hat{H}_1$ and $\hat{H}_2$ are taken independently from the Gaussian Unitary Ensemble (GUE) of *Hermitian* matrices with the probability density $\mathcal{P}(\hat{X}) = Q_N^{-1} \exp\left(-N/2J_0^2 \ \mathrm{Tr}\ \hat{X}^2\right)$, $\hat{X} = \hat{X}^\dagger$.

Let us now introduce a new parameter $\tau = (1 - v^2)/(1 + v^2)$ and choose the scale constant J_0^2 to be equal to $(1 + \tau)/2$, for the sake of convenience. The parameter τ controls the magnitude of correlation between J_{jk} and J_{kj}: $\langle J_{jk} J_{kj} \rangle = \tau/N$, hence the degree of non-Hermiticity. This is easily seen from the probability density function for our ensemble of the random matrices $\hat{J}$:

$$\mathcal{P}(\hat{J}) = C_N^{-1} \exp\left[-\frac{N}{(1 - \tau^2)} \ \mathrm{Tr}(\hat{J}\hat{J}^\dagger - \tau \ \mathrm{Re}\ \hat{J}^2)\right], \tag{26}$$

where $C_N = [\pi^2(1 - \tau^2)/N^2]^{N^2/2}$. All the J_{jk} have zero mean and variance $\langle |J_{jk}|^2 \rangle = 1/N$ and only J_{jk} and J_{kj} are pairwise correlated. If $\tau = 0$ all the J_{jk} are mutually independent and we have maximum non-Hermiticity. When τ approaches unity, J_{jk} and J_{kj}^* are related via $J_{jk} = J_{kj}^*$ and we are back to an ensemble of Hermitian matrices.

Our first goal is to determine the n-eigenvalue correlation functions in the ensemble of random matrices specified by Eq. (26). The density of the joint distribution of eigenvalues in the ensemble is given by

$$\mathcal{P}_N(Z_1, \ldots, Z_N) = \frac{N^{N(N+1)/2}}{\pi^N 1! \cdots N! (1 - \tau^2)^{N/2}} \times \tag{27}$$

$$\exp\left\{\frac{-N}{1 - \tau^2} \sum_{j=1}^{N} \left[|Z_j|^2 - \frac{\tau}{2}(Z_j^2 + Z_j^{*2})\right]\right\} \prod_{j<k} |Z_j - Z_k|^2.$$

To derive Eq. (27) we integrate $\mathcal{P}(\hat{J})$ from Eq. (26) over the surface of all complex matrices whose eigenvalues are $Z_1, \ldots Z_N$. Following Dyson ([44], p.501, see also [48]) we decompose every complex matrix with distinct eigenvalues as $\hat{J} = \hat{U}(\hat{Z} + \hat{R})\hat{U}^\dagger$, where $\hat{Z} = \mathrm{diag}\{Z_1, \ldots Z_N\}$, $\hat{U}$ is a unitary matrix, and $\hat{R}$ is a strictly upper-triangular one. If we label the eigenvalues and require the first non-zero element in each column of $\hat{U}$ to be positive, then the decomposition is unique. The Jacobian of the transformation $\hat{J} \to \{\hat{Z}, \hat{R}, \hat{U}\}$ depends only on $\hat{Z}$ and is given by the squared modulus of the Vandermonde determinant. So, integrating out $\hat{R}$ and $\hat{U}$ is straightforward and the resulting expression is Eq. (27).

The form of the distribution Eq. (27) allows one to employ the powerful method of orthogonal polynomials [44]. Let $H_n(z)$ denote the n-th Hermite polynomial,

$$H_n(z) = \frac{(\pm i)^n}{\sqrt{2\pi}} \exp\left(\frac{z^2}{2}\right) \int_{-\infty}^{\infty} dt \ t^n \exp\left(-\frac{t^2}{2} \mp izt\right). \tag{28}$$

The crucial observation borrowed from the paper [69] (see also the related paper [70]) is that the polynomials

$$p_n(Z) = \frac{\tau^{n/2}\sqrt{N}}{\sqrt{\pi}\sqrt{n!}(1 - \tau^2)^{1/4}} H_n\left(\sqrt{\frac{N}{\tau}}Z\right), \tag{29}$$

$n = 0, 1, 2, \ldots$, are orthogonal in the *complex plane* $Z = X + iY$ with the weight function

$$w^2(Z) = \exp\left\{ -\frac{N}{(1-\tau^2)}\left[|Z|^2 - \frac{\tau}{2}(Z^2 + Z^{*2}) \right] \right\},$$

i. e. $\int d^2 Z p_n(Z) p_m(Z^*) w^2(Z) = \delta_{nm}$, where $d^2 Z = dX dY$. The standard machinery of the method of orthogonal polynomials [44] yields the n-eigenvalue correlation functions

$$R_n(Z_1, ..., Z_n) = \frac{N!}{(N-n)!} \int d^2 Z_{n+1} ... d^2 Z_N \mathcal{P}_N\{Z\} \tag{30}$$

in the form

$$R_n(Z_1, ..., Z_n) = \det\left[K_N(Z_j, Z_k^*) \right]_{j,k=1}^n,$$

where the kernel $K_N(Z_1, Z_2^*)$ is given by

$$K_N(Z_1, Z_2^*) = w(Z_1) w(Z_2^*) \sum_{n=0}^{N-1} p_n(Z_1) p_n(Z_2^*). \tag{31}$$

With Eqs. (29)–(31) in hand, let us first examine the regime of strong non-Hermiticity, i.e. the case when $\lim_{N\to\infty}(1-\tau) > 0$. In this regime the averaged density of eigenvalues $N^{-1}R_1(Z)$ is asymptotically zero outside the ellipse $[\mathrm{Re}z/(1+\tau)]^2 + [\mathrm{Im}z/(1-\tau)]^2 \le 1$. Inside the ellipse $\lim_{N\to\infty} N^{-1}R_1(Z) = [\pi(1-\tau^2)]^{-1}$. This sets a microscopic scale on which the averaged number of eigenvalues in any domain of unit area remains finite when $N \to \infty$. Remarkably, the τ-dependence is essentially trivial on this scale: the statistical properties of eigenvalues are described by $\tilde{R}_n(z_1, \ldots, z_n) \equiv N^{-n} R_n(\sqrt{N}Z_1, \ldots, \sqrt{N}Z_n)$ and

$$\lim_{N\to\infty} \tilde{R}_n(z_1, \ldots, z_n) = \tag{32}$$
$$\left[\frac{1}{\pi(1-\tau^2)} \right]^n e^{-\frac{1}{1-\tau^2}\sum_{j=1}^n |z_j|^2} \det\left[e^{\frac{1}{1-\tau^2} z_j z_k^*} \right]_{j,k=1}^n.$$

This limiting relation can be inferred [50] from Mehler's formula for the Hermite polynomials [71]. After the trivial additional rescaling $z \to z\sqrt{1-\tau^2}$ the expression on the right-hand side in Eq. (32) becomes identical to that found by Ginibre [43].

Now we move on to the regime of weak non-Hermiticity. We know that in this regime new non-trivial correlations occur on the scale: $\mathrm{Im}Z_{1,2} = O(1/N)$, $\mathrm{Re}Z_1 - \mathrm{Re}Z_2 = O(1/N)$. Correspondingly, we introduce new variables x, y_1, y_2, ω in such a way that: $x = \mathrm{Re}(Z_1 + Z_2)/2$, $y_{1,2} = N\mathrm{Im}(Z_{1,2})$, $\omega = N\mathrm{Re}(Z_1 - Z_2)$, and consider them finite when performing the limit $N \to \infty$.

Substituting Eq.(28) into Eq.(31) and using the above definitions we can explicitly perform the limit $N \to \infty$, taking into account that $\lim_{N\to\infty} N(1-\tau) = \alpha^2/2$. The details of the procedure are given elsewhere [50]. In this regime

$$\lim_{N\to\infty} \frac{1}{N^2} K_N\left(x + \frac{\omega/2 + iy_1}{N}, x - \frac{\omega/2 + iy_2}{N} \right) = \tag{33}$$
$$\frac{1}{\pi\alpha} \exp\left\{ -\frac{y_1^2 + y_2^2}{\alpha^2} + \frac{ix(y_1 - y_2)}{2} \right\} \int_{-\pi\nu_{sc}(x)}^{\pi\nu_{sc}(x)} \frac{du}{\sqrt{2\pi}} \exp\left[-\frac{\alpha^2 u^2}{2} - u(y_1 + y_2) + 2i\omega u \right],$$

with $\nu_{sc}(X) = \frac{1}{2\pi}\sqrt{4 - X^2}$ standing for the Wigner semicircular density of real eigenvalues of the Hermitian part $\hat{H}$ of the matrices $\hat{J}$.

Equation (33) constitutes the most important result of the present section. The kernel K_N given by Eq. (33) determines all the properties of complex eigenvalues in the regime of weak non-Hermiticity. For instance, the mean value of the density $\rho(Z) = \sum_{i=1}^{N} \delta^{(2)}(Z - Z_i)$ of complex eigenvalues $Z = X + iY$ is given by $\langle \rho(Z) \rangle = K_N(Z, Z^*)$. Putting $y_1 = y_2$ and $\omega = 0$ in Eqs (33) we immediately recover the density Eq.(22) found by the supersymmetry approach [‡].

One of the most informative statistical measures of the spectral correlations is the 'connected' part of the two-point correlation function of eigenvalue densities:

$$\langle \rho(Z_1)\rho(Z_2) \rangle_c = \langle \rho(Z_1) \rangle \, \delta^{(2)}(Z_1 - Z_2) - \mathcal{Y}_2(Z_1, Z_2), \tag{34}$$

In particular, it determines the variance $\Sigma^2(D) = \langle n(D)^2 \rangle - \langle n(D) \rangle^2$ of the number $n = \int_D d^2 Z \rho(Z)$ of complex eigenvalues in any domain D in the complex plane as:

$$\Sigma_2(D) = \int_D d^2 Z_1 \int_D d^2 Z_2 [\langle \rho(Z_1)\rho(Z_2) \rangle - \langle \rho(Z_1) \rangle \langle \rho(Z_2) \rangle] = \tag{35}$$

$$\int_D d^2 Z \langle \rho(Z) \rangle - \int_D d^2 Z_1 \int_D d^2 Z_2 \mathcal{Y}_2(Z_1, Z_2)$$

Comparing Eq.(34) with the definition Eqs. (29)–(31) we see that the *cluster function* $\mathcal{Y}_2(Z_1, Z_2)$ is expressed in terms of the kernel K_N as $\mathcal{Y}_2(Z_1, Z_2) = |K_N(Z_1, Z_2^*)|^2$.

It is evident that in the limit of weak non-Hermiticity the kernel K_N depends on X only via the semicircular density $\nu_{sc}(X)$. Thus, it does not change with X on the local scale comparable with the mean spacing along the real axis $\Delta \sim 1/N$.

The cluster function is given by the following explicit expression:

$$\mathcal{Y}(\omega, y_1, y_2) = \frac{N^4}{\pi^2 \alpha^2} e^{-2\frac{y_1^2 + y_2^2}{\alpha^2}} \left| \int_{-\pi\nu(X)}^{\pi\nu(X)} \frac{du}{(2\pi)^{1/2}} \exp\left[-\frac{\alpha^2 u^2}{2} - u(y_1 + y_2) + iu\omega \right] \right|^2 \tag{36}$$

The parameter $a = \pi\nu(X)\alpha$ controls the deviation from Hermiticity [§]. When $a \to 0$ the cluster function tends to GUE form $\mathcal{Y}_2(\omega, y_1, y_2) = \frac{N^4}{\pi^2} \delta(y_1)\delta(y_2) \frac{\sin^2 \pi\nu(X)\omega}{\omega^2}$. In the opposite case $a \gg 1$ the limits of integration in Eq.(36) can be effectively put to $\pm\infty$ due to the Gaussian cutoff of the integrand. The corresponding Gaussian integration is trivially performed yielding in the original variables Z_1, Z_2 the expression equivalent (up to a trivial rescaling) to that found by Ginibre [43]: $\mathcal{Y}_2(Z_1, Z_2) = (N^2/\pi\alpha^2)^2 \exp\{-N^2|Z_1 - Z_2|^2/\alpha^2\}$.

The operation of calculating the Fourier transform of the cluster function over its arguments ω, y_1, y_2 amounts to simple Gaussian and exponential integrations. Performing them one finds the following expression for the *spectral form-factor*:

$$b(q_1, q_2, k) = \int_{-\infty}^{\infty} d\omega \int_{-\infty}^{\infty} dy_1 \int_{-\infty}^{\infty} dy_2 \mathcal{Y}_2(\omega, y_1, y_2) \exp\{2\pi i(\omega k + y_1 q_1 + y_2 q_2)\} \tag{37}$$

$$= N^4 \exp\{-\frac{\alpha^2}{2}\left(q_1^2 + q_2^2 + 2k^2\right)\} \frac{\sin\left[\pi^2\alpha^2(q_1 + q_2)(\nu(X) - |k|)\right]}{\pi^2\alpha^2(q_1 + q_2)} \theta(\nu(X) - |k|)$$

where $\theta(u) = 1$ for $u > 0$ and zero otherwise.

We see, that everywhere in the regime of weak non-Hermiticity $0 < \alpha < \infty$ the formfactor shows a kink-like behavior at $|k| = \nu(X)$. This feature is inherited from the

---

[‡] In the present section we normalized $\hat{H}_2$ in such a way that for weak non-Hermiticity regime we have $\lim_{N \to \infty} \mathrm{Tr} \hat{H}_2^2 = N$, whereas the nomalization Eq.(17) gives $\lim_{N \to \infty} \mathrm{Tr} \hat{H}_2^2 = N(1 + w^2)$. It is just because of this difference the parameter $\tilde{a}$ entering Eq.(22) contains an extra factor $1 + w^2$ as compared to the present case.

[§] In our earlier Letter [39] we used the definition of the parameter a different by a factor of 2 from the present one.

corresponding Hermitian counterpart-the Gaussian Unitary Ensemble. It reflects the oscillations of the cluster function with ω which is a manifestation of the long-ranged order in eigenvalue positions along the real axis [2]. When non-Hermiticity increases the oscillations become more and more damped.

As we already discussed above the knowledge of the formfactor allows one to determine the variance Σ_2 of a number of eigenvalues in any domain D of the complex plane. Small Σ_2 is a signature of a tendency for levels to form a cristal-like structure with long correlations. In contrast, increase in the number variance signals about growing decorrelations of eigenvalues.

In a general case this expression is not very transparent, however. For this reason we restrict ourselves to the simplest case, choosing the domain D to be the infinite strip of width L_x (in units of mean spacing along the real axis $\Delta = (\nu_{sc}(0)N)^{-1}$) oriented perpendicular to the real axis: $0 < \mathrm{Re}Z < L_x\Delta; \quad -\infty < \mathrm{Im}Z < \infty$. Such a choice means that we look only at real parts of complex eigenvalues irrespective of their imaginary parts. It is motivated, in particular, by the reasons of comparison with the GUE case, for which the function $\Sigma(L_x)$ behaves at large L_x logarithmically: $\Sigma(L_x) \propto \ln L_x$ [2].

After simple calculations one finds ¶

$$\Sigma_2(L_x) = L_x \left[1 - \frac{2}{\pi^2} \int_0^{L_x} \frac{dk}{k^2} (1 - \frac{k}{L_x}) \sin^2(\pi k) e^{-(\frac{ak}{L_x})^2} \right] \tag{38}$$

First of all, it is evident that Σ_2 grows systematically with increase in the degree of non-Hermiticity $a = \pi\nu(0)\alpha$. This fact signals on the gradual decorrelation of the *real* parts $\mathrm{Re}Z_i$ of complex eigenvalues. It can be easily understood because of increasing possibility for eigenvalues to avoid one another along the $Y = \mathrm{Im}Z$ direction, making their projections on the real axis X to be more independent.

In order to study the difference from the Hermitian case in more detail let us consider again the large L_x behavior. In that case the upper limit of the integral in Eq.(38) can be set to infinity. Then it is evident, that the number variance is only slightly modified by non-Hermiticity as long as $a \ll L_x$. We therefore consider the case $a \gg 1$ when we expect essential differences from the Hermitian case.

In a large domain $1 \ll L_x \sim a$ the second term in the integrand of Eq.(38) can be neglected and the number variance grows like $\Sigma(L_x) = L_x f(L_x/a)$. We find it more transparent to rewrite the function $f(u)$ in an equivalent form:

$$f(u) = 1 + \frac{2}{\sqrt{\pi}} \left\{ \frac{1}{2\pi u} \left(1 - e^{-\pi^2 u^2} \right) - \int_0^{\pi u} dt\, e^{-t^2} \right\}.$$

which can be obtained from Eq.(38) after a simple transformation.

For $u = L_x/a \ll 1$ we have simply $f \approx 1$ and hence a linear growth of the number variance. For $u \gg 1$ we have $f \approx (\pi^{3/2} u)^{-1}$. Thus, $\Sigma_2(L_x)$ slows down: $\Sigma_2(L_x) \approx \frac{a}{(\pi)^{3/2}}$.

Only for exponentially large L_x such that $\ln(L_x/a) \sim a$ second term in Eq.(38) contributes significantly. Calculating its contribution explicitly and remembering that $\Sigma_2^{(1)}\big|_{L_x >> a} \approx a/(\pi^{3/2})$ we finally find:

$$\Sigma_2(L_x \gg a) = \frac{a}{\pi^{3/2}} + \frac{1}{\pi^2} \left(\ln\left(\frac{L_x}{a}\right) - \frac{\gamma}{2} \right)$$

¶In our earlier Letter [39] the expression Eq.(38) and formulae derived from it erroneously contained πa instead of a.

where γ is Euler's constant. This logarithmic growth of the number variance is reminiscent of that typical for real eigenvalues of the Hermitian matrices.

Another important spectral characteristics which can be simply expressed in terms of the cluster function is the small-distance behavior of the nearest neighbor distance distribution [44, 2, 40].

We define the quantity $p(Z_0, S)$ as the probability density of the following event: i) There is exactly one eigenvalue at the point $Z = Z_0$ of the complex plane. ii) Simultaneosly, there is exactly one eigenvalue on the circumference of the circle $|Z - Z_0| = S$ iii) All other eigenvalues Z_i are *out* of that circle: $|Z_i - Z_0| > S$.

As a consequence, the normalization condition is: $\int d^2 Z_0 \int_0^\infty dS\, p(Z_0, S) = 1$. In particular, for Hermitian matrices with real eigenvalues one has the relation: $p(Z_0, S) = \delta(\mathrm{Im} Z_0)\tilde{p}_X(S)$, with $\tilde{p}_X(S)$ being the conventional "nearest neighbor spacing" distribution at the point X of the real axis [44].

We are interested in finding the leading small-S behavior for the function $p(Z_0, S)$. It turns out to be given by the following expression [50]:

$$p(Z_0, S) \approx \frac{S}{N} \int_0^{2\pi} d\theta \left[\langle \rho(Z_0) \rangle \langle \rho\left(Z_0 + Se^{i\theta}\right) \rangle - \mathcal{Y}_2\left(Z_0, Z_0 + Se^{i\theta}\right) \right] \tag{39}$$

where we used the definition of the cluster function, Eq.(34).

In the regime of weak non-Hermiticity this formula is valid as long as the parameter S is small in comparison with a typical separation between real eigenvalues of the Hermitian counterpart: $S \ll \Delta \sim 1/N$.

Substituting the expression Eqs.(22,36) for the mean density and the cluster function into Eq.(39) one arrives after a simple algebra to the probability density to have one eigenvalue at the point $Z_0 = X + iy_0\Delta$ and its closest neighbor at the distance $|z_1 - z_0| = s\Delta$, $\Delta = (\nu(X)N)^{-1}$, such that $s \ll 1$:

$$p_\alpha(X + iy_0\Delta, s\Delta)|_{s\ll 1} = \frac{\pi \nu^2(X)}{2} \left[g_a(y_0)\frac{\partial^2}{\partial y_0^2} g_a(y_0) - \left(\frac{\partial}{\partial y_0} g_a(y_0) \right)^2 \right]$$

$$\times e^{-4\frac{y_0^2}{a^2}} \frac{s^3}{a^2} \int_0^\pi d\theta \exp\left[-\frac{2}{a^2}(s^2 \cos^2\theta - 2y_0 s \cos\theta) \right] \tag{40}$$

where

$$g_a(y) = \int_{-1}^1 \frac{du}{(2\pi)^{1/2}} \exp\{-\frac{a^2 u^2}{2} - 2uy\}$$

First of all it is easy to see that in the limit $a \gg 1$ one has: $p_{a\gg 1}(Z_0, s \ll 1) = \frac{2}{\pi}(s/a^2)^3$ in agreement with the cubic repulsion generic for strongly non-Hermitian random matrices [43, 26, 40]. On the other hand one can satisfy oneself that in the limit $a \to 0$ we are back to the familiar GUE quadratic level repulsion: $p_{a\to 0}(Z_0, s \ll 1) \propto \delta(y_0)s^2$. In general, the expression Eq.(40) describes a smooth crossover between the two regimes, although for any $a \neq 0$ the repulsion is always cubic for $s \to 0$.

To this end, an interesting situation may occur when deviations from the Hermiticity are very weak: $a \ll \sqrt{2}$ and 'observation points' Z_0 are situated sufficiently far from the real axis: $2|y_0|/a \gg 2^{-1/2}$.

Under this condition the following three regions for the parameter s should be distinguished: i) $\frac{s}{a} \ll \frac{a}{4|y_0|}$ ii) $\frac{a}{4|y_0|} \ll \frac{s}{a} \ll 2\frac{|y_0|}{a}$ and finally iii) $2^{-1/2} \ll 2\frac{|y_0|}{a} \ll \frac{s}{a} \ll a^{-1}$.

In the regimes i) and ii) the term linear in $\cos\theta$ in the exponent of Eq.(40) dominates yielding the result of integration to be the modified Bessel function $\pi I_0\left(\frac{4y_0 s}{a^2}\right)$. In the regime iii) the term quadratic in $\cos\theta$ dominates producing $2\pi e^{-(s/a)^2} I_0\left[(s/a)^2\right] \approx$

$(2\pi a/s)^{1/2}$. As the result, the distribution $p(Z_0, s)$ displays the following behavior:

$$p_\alpha(Z_0, s) = \frac{\pi^2 \nu^2}{2} \left[g_0(y_0) \frac{\partial^2}{\partial y_0^2} g_0(y_0) - \left(\frac{\partial}{\partial y_0} g_0(y_0) \right)^2 \right]$$

$$\times e^{-4\frac{y_0^2}{a^2}} \begin{cases} \frac{s^3}{a^2} & \text{for} \quad \frac{s}{a} \ll \frac{a}{4|y_0|} \\ \frac{s^{5/2}}{2a\sqrt{2\pi|y_0|}} & \text{for} \quad \frac{a}{4|y_0|} \ll \frac{s}{a} \ll 2\frac{|y_0|}{a} \\ \sqrt{\frac{2}{\pi}} \frac{s^2}{a} & \text{for} \quad 2\frac{|y_0|}{a} \ll \frac{s}{a} \ll a^{-1} \end{cases} \qquad (41)$$

with $g_0(y) \equiv g_a(y)|_{a=0}$.

Unfortunately, it might be a very difficult task to detect numerically the unusual power law $p(s) \propto s^{5/2}$ because of the low density of complex eigenvalues in the observation points reflected by the presence of the Gaussian factor in the expression Eq.(41).

Conclusion

In the present paper we addressed the issue of eigenvalue statistics of large weakly non-Hermitian matrices.

Our original motivation came from the field of resonance chaotic scattering. The resonances, which are complex poles of the scattering matrix enter the theory as complex eigenvalues of a non-Hermitian effective Hamiltonian of a particular type: $\mathcal{H}_{ef} = \hat{H} - i\hat{\Gamma}$. We demonstrated that one can extract mean density of such poles employing a mapping onto the supermatrix non-linear σ−model. We also have shown how the resolvent of the non-Hermitian Hamiltonian $\mathcal{H}_{ef}$ can be used to describe the process of chaotic photodissociation and presented the crossection autocorrelation function.

Guided by our experience with the resonances, we found a regime of weak non-Hermiticity for other types of non-selfadjoint random matrices. The regime can be defined as that for which the imaginary part $\mathrm{Im}Z$ of a typical complex eigenvalue is of the same order as the mean eigenvalue separation Δ for the corresponding Hermitian counterpart.

Exploiting a mapping to the non-linear σ−model we are able to show that there are three different "pure" symmetry classes of weakly non-Hermitian matrices: i) almost Hermitian with complex entries ii) almost symmetric with real entries and iii) complex symmetric ones. Within each of these classes the eigenvalue statistics is *universal* in a sense that it is the same irrespective of the particular distribution of matrix entries up to an appropriate rescaling. There are also crossover regimes between all three classes.

Our demonstration of universality was done explicitly for the density of complex eigenvalues of matrices with independent entries. Within the non-linear σ−model formalism one can easily provide a heuristic proof of such a universality for higher correlation functions as well as for "rotationally invariant" matrix ensembles, see [68]. The above feature is a great advantage of the supersymmetry technique.

A weak point of that method is a very complicated representation of the ensuing quantities. It seems, that the explicit evaluation of the higher correlation functions is beyond our reach at the moment, and even a calculation of the mean density requires a lot of effort, see [38, 32]. As a result, at present time the mean density is known explicitly only for the cases i) and ii). (For the case iii) see Ref. [74].)

Fortunately, because of the mentioned universality another strategy can be pursued. Namely, one can concentrate on the particular case of matrices with independent, Gaussian distributed entries for which alternative analytical techniques might be available. Such a strategy turned out to be a success for the simplest case of complex almost-Hermitian matrices, where we found the problem to be an exactly soluble one by the method of orthogonal polynomials. This fact allowed us to extract all the correlation functions in a mathematically rigorous way [39, 50].

One might hope that combining the supersymmetric method and the method of orthogonal polynomials one will be able to elevate our understanding of properties of almost-Hermitian random matrices to the level typical for their Hermitian counterparts.

¿From this point of view a detailed numerical investigation of different types of almost-Hermitian random matrices is highly desirable. Recently, an interesting work in this direction appeared motivated again by the theory of chaotic scattering [10]. Unfortunately, matrices $\mathcal{H}_{ef}$ emerging in that theory are different from the Gaussian matrices because of the specific form of the antihermitean perturbation $i\hat{\Gamma}$ necessary to ensure the unitarity of the scattering matrix. This fact makes impossible a direct quantitative comparison of our results with those obtained in [10]. The qualitative fact of increase in number variance with increase in non-Hermiticity agrees well with our findings. At the same time it turns out that the knowledge of the *time-delay* correlations [16] allows one to make a plausible conjecture about the form of the number variance for the scattering systems with broken time-reversal symmetry. These results, as well as the proof of the conjecture, will be published elsewhere [73],[75].

The author is much obliged to H.-J. Sommers, B.A.Khoruzhenko and Y.Alhassid for collaboration on different aspects of resonance scattering and non-Hermitian random matrices considered in the present paper and to D.Savin, J.Main, J.J.M.Verbaarschot and especially to V. Sokolov for useful discussions.

The author is grateful to the organizers of the program "Disordered Systems and Quantum Chaos" for the financial support of his stay at the Newton Institute, Cambridge.

The work was supported by SFB 237 "Unordnung und grosse Fluktuationen" and EPRSC Research Grant GR/L31913. The warm hospitality of the School of Mathematical Sciences, Queen Mary& Westfield College, University of London, of the Sloane Physics Laboratory, Yale University where different parts of the work were done is acknowledged with thanks.

REFERENCES

1. B.L.Altshuler and B.D.Simon in: *Mesoscopic Quantum Physics* ed. by E.Akkermans et al, Les Houches Summer School, Session LXI 1994, edited E.Akkermans et al., Elsever Science.
2. O.Bohigas, in *Chaos and Quantum Physics. Proceedings of the Les-Houches Summer School. Session LII*, ed. by M.J.Giannoni et.al (North Holland, Amsterdam, 1991), p.91
3. M.Berry, *Proc.R.Soc.London*, Ser.A **400**, 229 (1985);
4. E.Bogomolny and J.Keating, *Phys.Rev.Lett* **77**, 1472 (1996)
5. B.A.Muzykantsky and D.E.Khmelnisky *JETP Lett.* **62** , 76 (1995);
6. A.Andreev, O. Agam, B. Altshuler and B.Simons Phys.Rev.Lett., **76**, 3947 (1996)
7. T.Guhr, A.Müller-Groelling and H.A.Weidenmüller, to appear in Rev.Mod.Phys.
8. V.V.Sokolov and V.G.Zelevinsky Phys.Lett.B **202**, 10 (1988); Nucl.Phys.A **504**, 562 (1989); F.Haake, F.Izrailev, N.Lehmann, D.Saher, and H.-J.Sommers, Z.Phys.B **88**, 359 (1992); N.Lehmann, D.Saher, V.V.Sokolov, and H.-J.Sommers, Nucl.Phys.A **582**, 223 (1995); M. Müller, F.-M.Dittes, W.Iskra, and I. Rotter, Phys.Rev.E **52**, 5961 (1995).
9. Fyodorov Y V and Sommers H-J *Pis'ma ZhETF* v.63,970 (1996); [*JETP Letters* v.63 ,1026 (1996)]
10. T.Gorin, F.-M.Dittes, M.Müller, I.Rotter and T.H.Seligman , *Phys.Rev.E*, v.56 , 2481 (1997)
11. G.Hackenbroich and J.Nöckel *Europh.Lett.* v.39, 371 (1997)
12. J.Main and G.Wunner *J.Phys.B: At.Mol.*, **27**, 1994 (1994); B.Gremaud and D.Delande *Europh.Lett.* v.40 , p.363 (1997)
13. R.Blumel, *Phys.Rev.E*, **54**, 5420 (1996);

14. V.A.Mandelshtam and H.S.Taylor, *J.Chem.Soc.Faraday.Trans.* , **93** ,847 (1997) and *Phys.Rev.Lett.*, **78**, 3274 (1997)

15. U.Smilansky in *Chaos and Quantum Physics. Proceedings of the Les-Houches Summer School. Session LII*, ed. by M.J.Giannoni et.al (North Holland, Amsterdam, 1991),p.372

16. Y.V.Fyodorov and H.-J.Sommers , *J.Math. Phys.*, **38** , 1918 (1997)

17. J.J.M.Verbaarschot, H.A.Weidenmüller, M.R.Zirnbauer , *Phys.Rep.* v.129 , 367 (1985)

18. C.Mahaux and H.A.Weidenmüller, *Shell Model Approach in Nuclear Reactions* (North Holland, Amsterdam), 1969

19. I.Yu.Kobzarev, N.N.Nikolaev and L.B.Okun, Yad.Phys. **10**, 864 (1969) [in Russian]

20. M.S.Livsic *Operators, Oscillations, Waves: Open Systems*, Amer.Math.Soc.Trans. v.34 (Am.Math.Soc.,Providence,RI,1973)

21. K.B. Efetov *Supersymmetry in Disorder and Chaos* (Cambridge University Press,1996)

22. Y.V. Fyodorov in "Mesosocopic Quantum Physics", Les Houches Summer School, Session LXI,1994, edited E.Akkermans et al., Elsever Science, p.493

23. Y.V.Fyodorov and Y.Alhassid , *Phys.Rev.A*, **58**, 3375 (1998)

24. D.V.Savin and V.V.Sokolov, *Phys.Rev.E* v.56, R4911 (1997)

25. E.Gudowska-Nowak, G.Papp and J.Brickmann *Chem.Physics* v.220, 125 (1997)

26. R.Grobe, F.Haake, and H.-J.Sommers, Phys.Rev.Lett. **61**, 1899 (1988)

27. F. Haake *Quantum Signature of Chaos* (Berlin, Springer, 1991)

28. L.E.Reichl, Z.Y.Chen and M.Millonas *Phys.Rev.Lett.* v.63, 2013 (1989)

29. H.-J.Sommers, A.Crisanti, H.Sompolinsky, and Y.Stein, Phys.Rev.Lett. **60**, 1895 (1988); H. Sompolinsky , A. Crisanti and H.-J. Sommers *Phys. Rev. Lett.* **61** 259, 1988; B. Doyon ,B. Cessac , M. Quoy and M. Samuelidis *Int. J. Bifurc. Chaos* **3** 279 (1993);

30. N.Hatano and D.R.Nelson, Phys.Rev.Lett. **77**, 570 (1996); *Phys.Rev.B* v.56 (1997), 8651

31. P.W.Brouwer, P.G.Silvestrov and C.W.J. Beenakker *Phys.Rev.B* v.56 (1997), 4333; R.A.Janik et al., e-preprint cond-mat/9705098; B.A. Khoruzhenko and I.Goldscheid *Phys.Rev.Lett.* **80**, 2897 (1998)

32. K.B.Efetov, *Phys.Rev.Lett.* **79**, 491 (1997)

33. J. Miller and J. Wang, *Phys. Rev. Lett.* **76**, 1461 (1996); J. Chalker and J. Wang,*Phys.Rev.Lett.* **79** , 1797 (1997)

34. M.A. Stephanov, *Phys. Rev. Lett.* **76**, 4472 (1996); R.A.Janik et al., *Phys. Rev. Lett.* **77**, 4876 (1996); M.A. Halasz, A.D. Jackson and J.J.M. Verbaarschot , *Phys.Rev.D*, v.56 , 5140 (1997); M.A. Halasz, J.C.Osborn and J.J.M. Verbaarschot *Phys. Rev. D*, **56**, 7059 (1997).

35. T. Akuzawa and M. Wadati, *J. Phys. Soc. Jpn.* **65** , 1583 (1996).

36. M. V. Feigelman and M. A. Skvortsov, *Nucl. Phys.B* **506**, 665 (1997); A. Khare and K. Ray, *Phys. Lett. A* **230**, 139 (1997).

37. B.A.Khoruzhenko, J.Phys.A **29**, L165 (1996).

38. Y.V. Fyodorov, B. Khoruzhenko and H.-J. Sommers, *Physics Letters A* **226**, 46 (1997);

39. Y.V. Fyodorov, B. Khoruzhenko and H.-J. Sommers, *Phys.Rev.Lett.* v. 79, 557 (1997)

40. G. Oas, *Phys. Rev. E* **55**, 205 (1997)

41. R.A. Janik, M.Nowak,G.Papp and I.Zahed Nucl.Phys.B **501**, 603 (1997); J. Feinberg and A. Zee, Nucl.Phys.B. **501**,643(1997) and Nucl.Phys.B. **504**, 579 (1997)

42. M. Kus, F. Haake, D. Zaitsev and A.Huckleberry, *J. Phys. A: Math. Gen.* **30**, 8635 (1997).

43. J. Ginibre, *J. Math. Phys.* **6**, 440 (1965).

44. M.L.Mehta, *Random Matrices* (Academic Press Inc., N.Y., 1990)

45. V. Girko, *Theor. Prob. Appl.* **30**, 677 (1986).

46. N.Lehmann and H.-J.Sommers, Phys.Rev.Lett. **67**, 941 (1991).

47. P.J. Forrester, *Phys. Lett. A* **169**,21 (1992); *J. Phys. A:Math. Gen.* **26**, 1179 (1993).

48. A. Edelman, *J. Multivariate Anal.* **60**, 203 (1997); A.Edelman, E.Kostlan and M.Shub *J.Am.Math.Soc.* v.7 , 247 (1994)

49. Z. D. Bai, *Ann. Prob.* **25**, 494 (1997).

50. Y.V.Fyodorov, B.A. Khoruzhenko and H.-J.Sommers *Ann.Inst.Henri Poincare: Physique Theorique*, **68**, 449 (1998)

51. C.H.Lewenkopf and H.A.Weidenmüller, *Ann.Phys.* v.212 ,53 (1991)

52. P.Gaspard in "Quantum Chaos", *Proceedings of E.Fermi Summer School,1991* ed. by G.Casati et al. (North Holland, Amsterdam,1991), p.307

53. D.Stone in *Mesoscopic Quantum Physics*, see [1]

54. R.Schinke, H.-M.Keller, M.Stumpf and A.J.Dobbyn , *J.Phys.B: At.Mol.*, **28**, 2928 (1995)

55. V.V.Flambaum, A.A.Gribakina and G.F.Gribakin, *Phys.Rev.A* v.54, 2066,(1996)

56. P.A.Moldauer, *Phys.Rev.* v.157,907 (1967); M.Simonius *Phys.Lett.* v.52B, 279 (1974)

57. S.Albeverio,F.Haake,P.Kurasov,M.Kus and P.Seba , *J.Math.Phys.* v.37, 4888 (1996)

58. H.L.Harney, F.M.Dittes and A. Müller *Ann.Phys.* v.220, 159 (1992)

59. N.Lehmann , D.Savin, V.V. Sokolov and H.-J. Sommers *Physica D* **86** 572 (1995); Y.V. Fyodorov, D.Savin and H.-J.Sommers , *Phys.Rev.E*, **55**,4857 (1997)

60. R.Schinke *Photodissociation Dynamics*, (Cambridge University, 1993);

61. V.V.Sokolov and V.G.Zelevinsky *Phys.Rev.C* **56** ,311 (1997);

62. N. Taniguchi,A.V. Andreev and B.L. Altshuler , *Europh.Lett.* , **29**, 515 (1995)

63. A.Pandey and M.L.Mehta, *Commun.Math.Phys.* v.87 , 449 (1983)

64. A.Altland, K.B.Efetov, S.Iida, *J.Phys.A:Math.Gen* v.26 ,2545 (1993)

65. A.D.Mirlin, Y.V.Fyodorov, *J.Phys.A* v.24 , 2273 (1991);
Y.V.Fyodorov, A.D.Mirlin, *Phys.Rev.Lett.* v.67 , 2049 (1991)

66. Y.V.Fyodorov and H.-J.Sommers,*Z.Phys.B* v.99 ,123 (1995)

67. Strictly speaking, the form of the correlation function of eigenvalue densities for sparse matrices was shown to be identical to that known for the corresponding Gaussian ensemble provided p exceeds some critical value $p = p_l$. The "threshold" value p_l is nonuniversal and depends on the form of the distribution $\mathcal{P}(\hat{H})$ [65]. However, direct numerical simulations, see S.Evangelou *J.Stat.Phys.* v.69 (1992), 361 show that actual value is $1 < p_l < 2$. Thus, even existence of two nonvanishing elements per row already ensure, that the corresponding statistics belongs to the Gaussian universality class. In the present paper we assume that $p > p_l$.

68. G. Hackenbroich and H.A. Weidenmüller *Phys.Rev.Lett.* **74** 4118 (1995)

69. F. Di Francesco, M.Gaudin, C.Itzykson, and F.Lesage Int.J.Mod.Phys.A **9**, 4257 (1994).

70. P.J.Forrester and B.Jancovici, Int.J.Mod.Phys.A **11**,941 (1997)

71. G.Szegö, *Orthogonal polynomials*, 4th ed. (AMS, Providence, 1975), p. 380.

72. I.S.Gradshteyn, I.M.Ryzhik "Table of Integrals, Series, and Products" (Academic Press, N.Y. 1980).

73. Y.V.Fyodorov, M.Titov and H.-J.Sommers, *Phys.Rev.E*, **58**, 1195 (1998)

74. H.-J.Sommers, Y.V.Fyodorov and M.Titov, e-preprint cond-mat/9807015

75. Y.V.Fyodorov and B.A.Khoruzhenko, under preparation

TOPOLOGICAL FEATURES OF THE MAGNETIC RESPONSE IN INHOMOGENEOUS MAGNETIC FIELDS

E. Akkermans[1] and R. Narevich[1,2]

[1]Department of Physics, Technion
32000 Haifa, Israel
[2]Department of Physics, University of Maryland
College Park, MD 20742, USA

Abstract

We present topological features of the magnetic response (orbital and spin) of a two-dimensional non interacting electron gas due to inhomogeneous applied magnetic fields. These issues are analysed from the point of view of the Index theory with a special emphasis on the non perturbative aspects of this response. The limiting case of a Aharonov-Bohm magnetic flux line is studied in details and the results are extended to more general situations.

INTRODUCTION

The aim of this paper is to discuss some features of the magnetic response of a degenerate two-dimensional electron gas in the non interacting limit. Although there is a vast literature devoted to that subject, we would like to present a different point of view which may help to bring new results in some of the left open issues. In those systems, it is quite common to discuss separately the two components (orbital and spin) of the magnetic response. Very often, there is indeed a clear dichotomy between orbital and spin effects which might be due to some physical constraint (e.g. full spin polarization in a strong magnetic field) or to the independence of the two components of the response, for instance for a homogeneous magnetic field where (neglecting the spin-orbit coupling) both the orbital and Zeeman parts in the Hamiltonian do commute. Recently, it was noticed by a number of authors that more complicated situations for instance Dirac fermions in a random field[1, 2] may lead to new and unexpected effects: transition between localized and extended states, multifractal structure, etc. We do not want to discuss here the richness of some specific model but instead to present some features of the total magnetic response in inhomogeneous magnetic fields shared by most of these models in the general case where orbital and spin effects cannot be simply disentangled. These features as we shall see are essentially non perturbative and do require for their study some new tools imported from the Index theory of elliptic operators[3].

Supersymmetry and Trace Formulae: Chaos and Disorder
Edited by Lerner *et al.*, Kluwer Academic / Plenum Publishers, New York, 1999

The outline of this paper is as follows. In the remaining part of the Introduction, we shall set up a general form for the Hamiltonian we aim to study. Then, we shall discuss its factorizability and define the associated Index. In part 2, we present a detailed study of the magnetic response to a Aharonov-Bohm flux line as a limiting case of inhomogeneous field. This will be the opportunity to discuss the physical meaning of the Index and its relation to the magnetic response. In part 3, some properties of the associated Heat Kernel are outlined. Then in part 4, those results are extended to other systems, and a relation between the spin magnetization and the Index is given.

We start writing a general expression for the magnetic Schrödinger Hamiltonian for a single electron in the two-dimensional plane submitted to a perpendicular magnetic field of strength $B(r)$. We choose the gauge $\mathrm{div}\vec{A} = 0$ such that the vector potential $\vec{A}(x,y)$ obeys the two equations:

$$\begin{cases} \partial_x A_x + \partial_y A_y = 0 \\ \partial_x A_y - \partial_y A_x = B(r) \end{cases} \tag{1}$$

A solution of (1) is $(A_x, A_y) = \frac{\Phi(r)}{2\pi r^2}(-y, x)$, and the function $\Phi(r)$ is related to the magnetic field by $B(r) = \frac{1}{2\pi r}\partial_r \Phi(r)$ such that the Schrödinger Hamiltonian is

$$\frac{2m}{\hbar^2} H = -\frac{1}{r}\partial_r(r\partial_r) + (\frac{i}{r}\partial_\theta + \frac{\phi(r)}{r})^2 \tag{2}$$

where $\phi(r) \equiv \frac{e}{hc}\Phi(r) = \frac{\Phi(r)}{\Phi_0}$. This Hamiltonian can be expressed as well in terms of the two (formally) self-adjoint first order differential operators D and $D^\dagger$ as

$$\frac{2m}{\hbar^2} H = DD^\dagger - \frac{2\pi}{\Phi_0} B(r) \tag{3}$$

where $D = e^{i\theta}(\partial_r + \hat{J})$ and $D^\dagger = e^{-i\theta}(-\partial_r + \hat{J})$ such that $[D, D^\dagger] = 4\pi B(r)$. The operator $\hat{J} = \frac{i}{r}\partial_\theta + \frac{\phi(r)}{r}$ describes the azimuthal current.

The Zeeman term describing the coupling of the electron spin to the magnetic field is given by $H_s = -\frac{1}{2}g\mu_B\sigma_z B(r)$, where $\mu_B = \frac{e\hbar}{2mc}$ is the Bohr magneton, g is the gyromagnetic ratio we shall take equal to 2 and σ_z is the Pauli matrix. The total Pauli Hamiltonian (in appropriate units) is

$$\frac{2m}{\hbar^2} H_P = DD^\dagger - 2\pi B(r)(1 + \sigma_z) \tag{4}$$

It has the well known and interesting property to be exactly factorizable regardless of the shape of the magnetic field profile $B(r)$, i.e. it may be rewritten as the product of two first order differential operators (formally self-adjoint) Q and $Q^\dagger$ such that $H = QQ^\dagger$. This is a special example of a $N = 2$ supersymmetric Hamiltonian[4]. This feature is at the origin of the peculiar (non perturbative) topological properties of those systems. For instance, Aharonov and Casher[5] did show explicitly that if the magnetic field is of finite flux, then the ground state degeneracy is $N - 1$ where N is the closest integer to the total flux (in units of Φ_0). This is an example of the Atiyah Singer Index theorem[6] which, for that case states that

$$\mathrm{Index}Q = \dim\mathrm{Ker}Q - \dim\mathrm{Ker}Q^\dagger = N - 1 \tag{5}$$

where $\dim\mathrm{Ker}Q$ (resp. $\dim\mathrm{Ker}Q^\dagger$) is the (finite) number of zero modes of Q (resp. $Q^\dagger$) i.e. the number of solutions of the first order differential equation $Q\Psi = 0$ (resp. $Q^\dagger\Psi = 0$). The index is an integer and as such a topological invariant in that sense that it remains unchanged under any classically permissible gauge transformation where for instance we may change randomly the profile of the magnetic field $B(r)$ but keeping unchanged the total magnetic flux. This results from the factorizability of the Pauli Hamiltonian a property which in general is not met by Schrödinger Hamiltonians unless either we impose some special boundary conditions or we consider a uniform magnetic field B where the corresponding Hamiltonian (Landau) is factorizable (see (3)) in terms of the operators D and $D^\dagger$ up to a constant $-\frac{2\pi}{\Phi_0}B$, which sets the ground state energy (i.e. the lowest Landau level). In this latter case, it is again possible (at least formally) to define an Index. It turns out to be infinite (and ill defined) which still corresponds to the infinite degeneracy of the lowest Landau level. Since the extensive (i.e. proportional to the surface) degeneracy of the ground state is an important ingredient for the Hall quantization in those systems, it might be tempting to preserve and extend the Index theorem to finite geometries. But then, it can be shown that any local choice of boundary conditions (e.g. Dirichlet or Neumann) destroys the factorizability property of the Schrödinger Hamiltonian and therefore the condition of applicability of the Index theorem with this consequence that the corresponding ground state is always non degenerate. It was shown recently[7] that the proper degeneracy as given by the Index can be restored using a special kind of non local boundary conditions.

To go further and relate these topological features to the magnetic response, we shall first focus on a specific example namely the case of a Aharonov-Bohm magnetic flux line.

THE MAGNETIC RESPONSE OF AHARONOV-BOHM SYSTEMS

We consider now the limiting case of a localized magnetic field of finite flux which corresponds to a Aharonov-Bohm flux line i.e. to a delta function magnetic field $\vec{B}(r) = \Phi\delta(\vec{r})\hat{e}_z = \frac{hc}{e}\phi\delta(\vec{r})\hat{e}_z$ and to a vector potential $\vec{A}(\vec{r}) = \phi\frac{\hat{e}_z\times\vec{r}}{r^2}$, where $\hat{e}_z$ is the unit vector perpendicular to the plane. The corresponding Schrödinger equation is obtained from Eq.(2) using $\phi(r) = \phi$. The angular momentum is a good quantum number and then the equation is separable. In each sector $m \in \mathbb{Z}$, a single valued solution of the radial equation is

$$\Psi_m(kr) = aJ_{|m+\phi|}(kr) + bJ_{-|m+\phi|}(kr) \tag{6}$$

where $E = \frac{\hbar^2 k^2}{2m}$ is the energy, a and b are constants and $J_\nu(kr)$ are Bessel functions. To describe an impenetrable flux line, we impose the boundary condition $\Psi_m(0) = 0$. Since $J_{-|\nu|}(kr)\sim_{r\to 0}(kr)^{-|\nu|}$ this amounts to take $b = 0$ in order to have square integrable solutions at the origin. This choice is not as innocuous as it seems and we shall comment on it later. By choosing conveniently the normalization of the wavefunction we obtain $a = 1$. Finally, in order to define completely our Hilbert space, we demand the two operators D and $D^\dagger$ to be self adjoint such that for any states $f(r,\theta)$ and $g(r,\theta)$, $\langle f|Dg\rangle = \langle D^\dagger f|g\rangle$. The part $\frac{\phi}{r}$ does not make any problem and by evaluating the radial integral on a disk of radius R (eventually $R \to \infty$), we obtain

$$R\int_0^{2\pi} d\theta e^{i\theta} f^* g|_{r=R} = 0 \tag{7}$$

which is not fulfilled in the general case (there is an exception using Dirichlet boundary conditions). A way to solve this problem is to translate the angular momentum of the eigenfunctions f on the domain of the operator $D^\dagger$ by half a unit and therefore to consider the set $J_{|m+\frac{1}{2}+\phi|}(kr)e^{i(m+\frac{1}{2})\theta}$. Then, on the domain of the operator D, we consider functions g of angular momentum decreased by half a unit. This amounts to consider a spinor like wavefunction, i.e. an effective Pauli Hamiltonian but where we keep only one of the two spin components. We shall come later to this using another point of view.

To characterize the magnetic response of the system, we calculate the (so called) persistent current I and the magnetization $\vec{M}$. Both are obtained from the local current density

$$\vec{j}(r) = \frac{\hbar}{m}\mathrm{Im}(\Psi^*(\vec{\nabla} - \frac{ie}{\hbar c}\vec{A})\Psi) \tag{8}$$

where at zero temperature, we have to sum over all the occupied states up to the Fermi energy. Due to symmetry, only the azimuthal component j_θ is non zero such that $I = \int_0^\infty dr j_\theta(r)$ and the total magnetic moment is $\vec{M} = \frac{\pi}{c}\int_0^\infty drr^2 j_\theta(r)\hat{e}_z$. ¿From (8) we obtain

$$j_\theta = \int_0^{k_f} \frac{kdk}{2\pi} \sum_{m=-\infty}^{\infty} \frac{m+\phi+\frac{1}{2}}{r} J^2_{|m+\frac{1}{2}+\phi|}(kr) \tag{9}$$

such that

$$I = \frac{k_f^2}{8\pi} \sum_{m=-\infty}^{\infty} \mathrm{sgn}(m + \frac{1}{2} + \phi) \tag{10}$$

We first have to give a meaning to the divergent series $\eta \equiv \sum_m \mathrm{sgn}(m + \frac{1}{2} + \phi)$. To that purpose, we first derive a relation between η and the eigenvalues of the azimuthal current operator $\hat{J} = \frac{i}{r}\partial_\theta + \frac{\phi+\frac{1}{2}}{r}$. Consider the projection of $\hat{J}$ on a circle of radius $r = R$ (the exact value of R is irrelevant and taken to be one). The spectrum of the projected operator $\hat{J}_P$ is $\lambda = m + \frac{1}{2} + \phi$. Defining after Atiyah, Patodi and Singer[8] the quantity $\eta(\hat{J}_P) = \sum_{\lambda<0} 1 - \sum_{\lambda\geq0} 1$, we can rewrite it

$$\eta(\hat{J}_P) = \frac{1}{\sqrt{\pi}} \int_0^\infty \frac{dt}{\sqrt{t}} Tr(\hat{J}_P e^{-t\hat{J}_P^2})$$

$$= \sum_\lambda \frac{1}{\sqrt{\pi}} \mathrm{sgn}(\lambda) \int_0^\infty \frac{dt}{\sqrt{t}} |\lambda| e^{-t|\lambda|^2}$$

which gives $\eta(\hat{J}_P) = \eta$. It is of interest to rewrite $\eta(\hat{J}_P)$ under the form

$$\eta(\hat{J}_P) = -\frac{1}{2\sqrt{\pi}}\partial_\phi \int_0^\infty \frac{dt}{t^{\frac{3}{2}}} Z(t,\phi) \tag{11}$$

where $Z(t,\phi) = \sum_m e^{-t(m+\frac{1}{2}+\phi)^2} = Tr(e^{-t\hat{J}_P^2})$ is the partition function (or the Heat Kernel) at temperature $\frac{1}{t}$ of an electron constrained to move on a one-dimensional ring pierced by a Aharonov-Bohm flux. Before calculating η explicitly we notice that it was already considered in different contexts. To study the statistical properties of anyons, it was calculated[9, 10] using a Feynman-Kac integral. In the context of the Index theory of elliptic operators on manifolds with boundary, it was calculated[11] using a zeta function regularization i.e. by writing $\eta = \lim_{s\to0}\sum_{\lambda\neq0} \mathrm{sgn}(\lambda)|\lambda|^{-s}$. Here we shall evaluate it using the Poisson summation formula for the partition function $Z(t,\phi)$

$$Z(t,\phi) = \sum_{n=-\infty}^{\infty} \int_{-\infty}^{\infty} dx e^{2i\pi nx} e^{-t(x+\frac{1}{2}+\phi)^2}$$

$$= \sqrt{\frac{\pi}{t}} + 2\sqrt{\pi} \sum_{n\neq0} \frac{1}{\sqrt{t}} e^{\frac{-\pi^2 n^2}{t}} e^{2i\pi n(\phi+\frac{1}{2})}$$

Inserting this expression in (11), we obtain

$$\eta(\hat{J}_P) = -\frac{1}{2\pi^2}\partial_\phi \sum_{n=1}^{\infty} \frac{\sin^2 \pi n(\phi + \frac{1}{2})}{n^2}$$

$$= 2\{\phi + \frac{1}{2}\} - 1$$

where $\{...\}$ represents the fractional part. The persistent current is then given by

$$I = \frac{E_f}{2}\eta(\hat{J}_P) \tag{12}$$

The amplitude of the total current is proportional to the Fermi energy E_f and is therefore very large. This might be surprising since it is sometimes claimed that in the limit of an infinite system the normal persistent currents should vanish. This is indeed true for a one-dimensional ring of radius $R \to \infty$ but is incorrect in general. This point was discussed[12] and we shall come back to it later when calculating the magnetic moment. We would like now to discuss the topological features of the current and relate it to the Index associated with the Aharonov-Bohm problem. We saw previously that for a factorizable Hamiltonian the Index which counts the zero modes is defined by (5). Unfortunately, the Aharonov-Bohm Hamiltonian is at first sight non factorizable due to the additional factor $-2\pi B(r)$. However, by demanding self-adjointness of D and $D^\dagger$, we ended up with a (factorizable) Pauli Hamiltonian. This result was obtained in a different (and perhaps more physical) way, noticing that the energy spectrum for the conveniently regularized problem[9], is a non analytic function of the reduced flux ϕ. This has its origin in the behaviour of the wavefunction for the angular momentum $m = 0$. There, the unperturbed Hilbert space contains functions which do not vanish at the origin, while for $\phi \neq 0$, they do vanish like $r^{|\phi|}$. This gives rise to singularities in perturbation theory which can be dealt with by adding a repulsive contact term in the Hamiltonian[9]. To calculate the Index, we notice that since there is only one spin component (depending on the sign of the magnetic field), only one of the two operators D and $D^\dagger$ may have zero modes. The solutions of $Df = 0$ are (for large r) of the form $f \propto r^{m-\phi-\frac{1}{2}}$ and their square integrability at infinity requires $m < \phi - \frac{1}{2}$. To obtain well behaved solutions near the origin, we consider instead of the singular flux line a finite cylinder of radius ϵ in a uniform magnetic field B such that the magnetic flux is $\Phi = B\pi\epsilon^2$. The zero modes inside this disk are solutions of $(\partial_r + \frac{i}{r}\partial_\theta + \frac{B}{2}r)f = 0$ which in the sector m are of the form $f(r) \propto r^m e^{-\frac{Br^2}{4}}$. It is straightforward to check that these solutions do match for $r = \epsilon$ where both inside and outside logarithmic derivatives are given by $\partial_r f|_{r=\epsilon} = \frac{1}{\epsilon}(m - \phi - \frac{1}{2})f(\epsilon)$. Near $r = 0$, the zero modes behave like r^m so that square integrability requires $m \geq 0$. Finally, the number of zero modes i.e. the Index is given by Index $= [\phi + \frac{1}{2}]$, where $[...]$ represents the integer part. This corresponds to the degeneracy of the Aharonov-Bohm Hamiltonian (with its boundary condition).

There is another way to obtain the Index which may shed some more light on its physical interpretation. To that purpose, we consider the scattering description of the Aharonov-Bohm effect[13]. Berry et al.[14] proposed in that context to study the phase $\chi(r)$ of the scattered part of the wavefunction and in particular the dislocations of the wavefronts defined as points where the modulus of the wavefunction vanishes[15]. The circulation $\frac{1}{2\pi}\oint_c d\chi = \frac{1}{2\pi}\oint_c \vec{\nabla}\chi \cdot d\vec{r}$ of this phase over a close contour encircling the dislocation (i.e. the flux line) is an integer equals to $[\phi + \frac{1}{2}]$ i.e. to the Index. This result is not fortuitous and was clearly recognised in the mathematical literature[16] where the equivalence between the winding number around a singular point as introduced by

Poincare[17], the degree of a continuous map from the circle to the punctured complex plane and the Index of a conveniently defined elliptic operator was discussed. Each of those different points of view depends on the way we look at this problem. This way is geometric when considering instead of the initial map, say f, the mapping $\frac{f}{|f|}$ from the circle to the circle and count algebraically the number of intersections of the path with an arbitrary ray emanating from the origin. It is combinatorial when approximating the initial mapping by a piece-wise linear path and use combinatorial methods. It is differential in the way Berry et al.[14] considered it and analytic when studied from the point of view of the Index of an elliptic operator. It is the equivalence of those descriptions which is part of the richness of this problem. Although the differential point of view may appear at first sight more physical, the approach using the Index of an operator is more systematic which is useful in such non perturbative issues. Finally, comparing the different points of view, we arrive to the result

$$\text{Index} = [\phi + \frac{1}{2}] = \frac{1}{2\pi} \oint_c \vec{\nabla}\chi \cdot d\vec{r} \tag{13}$$

Here, the first equality is obtained by calculating the zero modes of D, and the second comes from the definition of the azimuthal current density which tells us that the integral over a closed contour can be evaluated either over a circle of large radius thus using the scattering form of the wavefunction[14] or on a circle of radius $r \to 0$ thus retaining in the series of Bessel functions (9) only the lowest indices which corresponds to a calculation of the zero modes. The Index as a function of the flux ϕ shows plateaus and jumps at half integers. As explained by Berry et al.[14], these jumps correspond to a long range reorganisation of the wavefronts in contrast to the behaviour at integers which describes local changes of the wavefronts around the flux line. It might be interesting at this stage to compare the previous results with those obtained in the context of superfluids or superconductors[18]. For superfluids, the gradient of the phase of the macroscopic Onsager-Feynman wavefunction measures the superfluid velocity and then the total current. Being a gradient, the latter describes an irrotational flow such that the circulation of the velocity on a closed curve is quantized as observed experimentally by Vinen[19]. As an outcome, the force exerted on a body immersed in the fluid vanishes (d'Alembert paradox). For a superconductor, the bulk current density vanishes (Meissner effect), such that from the circulation on a closed curve we obtain the quantization of the magnetic flux. But this is different from the Index theorem which states that the circulation on a closed curve of the gradient of the phase which must be an integer, depends on the flux as given by (13).

Considering the sum of $\eta(\hat{J}_P)$ and of the Index, we obtain the relation

$$\text{Index} + \eta(\hat{J}_P) = \phi \tag{14}$$

which in other words gives a sum rule between the radial and azimuthal integrals of the current density j_θ. This relation is the expression of a general result[8] which generalises to non compact spaces the Atiyah Singer Index theorem.

It might be also of interest to rewrite (14) using for $\frac{I}{E_f}$ its expression in terms of scattering phase shifts[12], $\frac{I}{E_f} = \frac{1}{\pi}\partial_\phi \sum_m \delta_m(\phi)$ where the phase shifts for a spin one half are given [20] by $\delta_m(\phi) = \frac{\pi}{2}(|m + \phi + \frac{1}{2}| - |m|)$. Then,

$$\text{Index} + \frac{1}{\pi}\partial_\phi \sum_m \delta_m(\phi) = \phi \tag{15}$$

Under this form, it is easy to see that the phase shift term represents the boundary contribution to the Index theorem for non compact manifolds.

The relation between the persistent currents and the scattering phase shifts suggests an interesting analogy between this problem and the screening of an electric charge as given by the Friedel sum rule[21]. In the latter case, inserting an external charge Z in a metal the electrical neutrality is expressed self-consistently by the sum rule $\pi Z = \delta(E_f)$ where $\delta(E_f)$ is the total scattering phase shift calculated at the Fermi energy describing the scattering of electrons by the external charge. For the Aharonov-Bohm case, we can interpret the persistent currents in a similar way saying that the magnetic flux line is screened by current loops of electrons sitting at infinity. Using (15), the current can be expressed in terms of a topological invariant, namely the Index. The Friedel sum rule may be understood similarly. The total phase shift which in principle depends on the microscopic details of the potential created by the external charge is in fact a function of Z only, irrespective of the way this charge is distributed.

Finally, we need to evaluate the total magnetic moment. Substituting (9) into the Biot-Savart law gives

$$
M = \mu_B \int_0^R r^2 dr \int_0^{k_f} k\,dk \sum_{m=-\infty}^{\infty} \frac{m + \phi + \frac{1}{2}}{r} J^2_{|m+\frac{1}{2}+\phi|}(kr). \tag{16}
$$

Using the Euler-Maclaurin summation formula[22] we obtain an asymptotic expansion in terms of the large parameter $k_f R$ such that the magnetic moment can be written as a series $\sum_{m=-\infty}^{\infty} F(m + \phi + \frac{1}{2})$, where the function F is

$$
F(x) = x \mu_B \int_0^R r\,dr \int_0^{k_f} k\,dk\, J^2_{|x|}(kr).
$$

A naive application of the Euler-Maclaurin summation formula would give zero due to the vanishing of the function $F(x)$ at plus and minus infinity, for all finite k_f and R. However this function is singular at $x = 0$, where its second derivative is discontinuous with a jump equal to $-2\mu_B$ (this result has small corrections of the order of $\mu_B/k_f R$). Then, the usual derivation of the Euler-Maclaurin summation formula[23] should be revised, taking into account the singularity at $m + \phi + \frac{1}{2} = 0$. Finding an integer m_1 such that $m_1 + \phi + \frac{1}{2} < 0 < m_1 + \phi + \frac{3}{2}$, it turns out that the important parameter (which determines the position of the singularity) in the derivation of the modified Euler-Maclaurin summation formula is $m_1 + \phi + \frac{3}{2}$, which is equal to the fractional part of the total flux ϕ. One can prove then that the correction itself is proportional to the jump of the second derivative times the Bernoulli polynomial of third order, evaluated at $\{\phi\}$. More precisely we obtain

$$
M \sim \frac{\mu_B}{3} B_3(\{\phi\}) + O(\mu_B/k_f R), \tag{17}
$$

where $B_3(x) = x^3 - 3x^2/2 + x/2$ is a Bernoulli polynomial. The magnetization of the system, which is according to the definition the magnetic moment per unit area, vanishes in the thermodynamic limit. Indeed, (17) corresponds to the finite magnetic moment of an infinite system. Therefore, even though the current is large, the magnetic response is experimentally inaccessible.

HEAT KERNEL AND PARTITION FUNCTION FOR THE AHARONOV-BOHM PROBLEM

¿From (11), we obtained an expression for the persistent current I in terms of an integral over the partition function $Z(t, \phi)$ of an electron moving on a one-dimensional ring pierced by a Aharonov-Bohm flux at an effective temperature $\frac{1}{t}$. On the other hand, since the persistent current is a thermodynamic quantity, it is possible in principle to express it in terms of the Heat Kernel (or the partition function) $P(\beta, \phi) = Tr(e^{-\beta H(\phi)})$ where $H(\phi)$ is the Hamiltonian defined by (2) for a constant ϕ and where β is the inverse temperature. To obtain a relation between these two quantities, we start writing the thermodynamic grand potential $\Omega(\beta, \phi) = \int_0^\infty dE f(E, \mu, \beta) N(E)$ where f is the Fermi-Dirac distribution function, μ the chemical potential and $N(E)$ the integrated density of states (the counting function). The current is then given by $I = -\frac{\partial \Omega}{\partial \phi}$ in the appropriate units. We then define after Sondheimer and Wilson[24] the function $z(E, \phi)$ by

$$\frac{P(\beta, \phi)}{\beta^2} = \int_0^\infty dE e^{-\beta E} z(E, \phi) \tag{18}$$

The grand potential as a function of $z(E, \phi)$ rewrites $\Omega(\beta, \phi) = \int_0^\infty dE \frac{\partial f}{\partial E} z(E, \phi)$. In the asymptotic limit $\beta\mu \gg 1$, i.e. for very low temperatures we obtain $I = \frac{\partial z(\mu, \phi)}{\partial \phi}$ from which we deduce

$$\frac{1}{\beta^2} \frac{\partial P(\beta, \phi)}{\partial \phi} = \int_0^\infty dE e^{-\beta E} I(E, \phi) \tag{19}$$

By inserting (11) for I, we obtain

$$\frac{\partial P(\beta, \phi)}{\partial \phi} = -\frac{1}{4\sqrt{\pi}} \partial_\phi \int_0^\infty \frac{dt}{t^{\frac{3}{2}}} Z(t, \phi) \tag{20}$$

which relates the partition function $P(\beta, \phi)$ of the two-dimensional problem to the partition function $Z(t, \phi)$ of an effective one-dimensional Aharonov-Bohm problem.

BEYOND THE AHARONOV-BOHM CASE

We would like now to discuss the extension of the previous results to more general magnetic field configurations. We noticed that the Pauli Hamiltonian (4) is always factorizable irrespective of the field configuration $B(r)$. Using this property, the ground state degeneracy is given by the Index. In addition to that result we obtained for the case of a flux line a relation (14) between the Index and $\eta(\hat{J}_P)$. This relation holds for any factorizable Hamiltonian. Is it possible to interpret it in terms of the magnetic response of the system? To that purpose we consider once again the simpler case of a uniform magnetic field. There, the orbital and spin parts of the magnetization can be separated. In the radial gauge $\vec{A} = \frac{B}{2}(-y, x)$, the Pauli Hamiltonian is $H = \frac{1}{2m}(\vec{p} - \frac{e}{c}\vec{A})^2 - \mu_B B \sigma_z \equiv H_0 - \mu_B B \sigma_z$. This can be written in a matricial form

$$H = \begin{pmatrix} H_+ & 0 \\ 0 & H_- \end{pmatrix}$$

where $H_\pm = H_0 \pm \mu_B B$. Defining the supercharge $Q = \frac{1}{\sqrt{4m}} \vec{\Pi}.\vec{\sigma} = \frac{1}{\sqrt{4m}}(\Pi_x \sigma_x + \Pi_y \sigma_y)$ where $\Pi_x = p_x - \frac{e}{c} A_x$ and $\Pi_y = p_y - \frac{e}{c} A_y$, we can write the Hamiltonian under the supersymmetric form $H = 2Q^2$. The associated creation and annihilation operators A and $A^\dagger$ are given by $A = \frac{1}{\sqrt{2m}}(\Pi_x - i\Pi_y)$ and $A^\dagger = \frac{1}{\sqrt{2m}}(\Pi_x + i\Pi_y)$, such that $H_+ = AA^\dagger$ and $H_- = A^\dagger A$. According to (5), the Index is then given[25] by Index = dimKer$A$ − dimKer$A^\dagger$. To go further and connect it with the thermodynamics of the system, we notice that Ker$A$ = Ker$A^\dagger A$ = KerH_- and Ker$A^\dagger$ = Ker$AA^\dagger$ = KerH_+.

322

Since the spectra of H_+ and H_- are identical except perhaps for the zero modes, it is possible to rewrite the Index using the regularization

$$\text{Index} = \lim_{t \to 0} \text{Tr}(e^{-tH_+} - e^{-tH_-}) = \lim_{t \to 0} \text{Tr}(\sigma_z e^{-tH}) \tag{21}$$

The spectrum of the Pauli Hamiltonian is given as for the Landau case by a set of infinitely degenerate Landau levels. The supersymmetric pairing of all the excited states means that they do not participate to the spin magnetization since there are no states with respectively spin up and spin down which are unpaired on an excited Landau level. Then, a finite spin magnetization stems only from possible unpaired states in the lowest Landau level. For a non interacting electron gas, the Fermi energy E_f is determined by fixing the total number N of electrons through $N = \text{Tr}(\Theta(E_f - H))$. The spin magnetization can be written $M_S = \mu_B(N_+ - N_-)$ where N_+ and N_- are the number of electrons with spin up and spin down respectively. Using a regularization equivalent to (21) we obtain the result

$$M_S = \mu_B \text{Tr}(\sigma_z \Theta(E_f - H)) = \mu_B \text{Index} \tag{22}$$

In order to calculate explicitly the Index, we write the traces in (21)

$$\text{Index} = \lim_{t \to 0} \int d^2 x \{ \langle x | e^{-tH_+} | x \rangle - \langle x | e^{-tH_-} | x \rangle \} \tag{23}$$

The Heat Kernels for respectively spin up and spin down which do appear in (23) are given for the case of a uniform magnetic field by

$$\int d^2 x \langle x | e^{-tH_\pm} | x \rangle = \frac{1}{2} \frac{\Phi}{\Phi_0} e^{\pm \frac{\hbar \omega_c t}{2}} \frac{1}{\sinh(\frac{\hbar \omega_c t}{2})} \tag{24}$$

where $\omega_c = \frac{eB}{mc}$ is the cyclotron frequency. The Index is effectively t-independent and given by the total magnetic flux (in units of the flux quantum Φ_0) and is therefore infinite.

The outcome of this is the relation between the Index and the spin magnetization given by (22). As anticipated from the role of the Zeeman term for the Pauli Hamiltonian to be factorizable, the Index appears as a measure of the spin magnetization. Nevertheless, this result is very peculiar and applies only for the case of a uniform magnetic field, where the orbital and Zeeman parts in the Hamiltonian do commute, such that the respective magnetic responses are independent. In the general case of a inhomogeneous magnetic field, it is not anymore possible to disentangle those two parts. Another peculiarity of the uniform field case is that the Index is infinite and formally ill-defined. For a magnetic field of finite flux, it is not anymore the case and the total magnetic response is still given by (14) as a consequence of the Atiyah, Patodi, Singer[8] theorem.

CONCLUSION

We have presented a study of the magnetic response of a degenerate non interacting electron gas in a inhomogeneous magnetic field from the point of view of the Index Theory. For the limiting case of a Aharonov-Bohm flux line, it allowed us to find a relation between the persistent current and a topological invariant, namely the Index of an operator and to calculate it. We obtained a relation between this Index and the winding number defined geometrically from the behaviour of the wavefunction. To provide another physical interpretation of the Index, we showed that for the case of a uniform magnetic field, it is proportional to the spin magnetization and we presented a general regularization scheme to calculate it from the partition function (or the Heat Kernel). This connection to the magnetic response can be generalised to other systems as an application of the Atiyah, Patodi, Singer Theorem. The extension of this point of view to study transport coefficients in the Quantum Hall regime is appealing. It allows[26] to relate the Hall conductance to a suitably defined Index for systems with a boundary. This generalises other approaches using topological numbers[27, 28]. Among other problems where the point of view presented here may be relevant, is the puzzle of mesoscopic persistent currents in small (but multichannel) rings. There, as discussed by Leggett[29], the existence of nodal lines where the wavefunction vanishes would involve more sophisticated topological tools in order to make some progress.

Acknowledgements

This work is supported in part by a grant from the Israel Academy of Sciences and by the fund for promotion of Research at the Technion. It is a pleasure to thank M. Atiyah, J. E. Avron, M. Berry and K. Mallick for very useful discussions. R. N. also acknowledges the support of the NSF grant DMR 9625549 and thanks R. E. Prange for enlightening discussions.

REFERENCES

1. A. W. W. Ludwig, M. P. A. Fisher, R. Shankar and G. Grinstein, *Phys. Rev.* **B50**:7526 (1994).
2. C. Chamon, C. Mudry, X. G. Wen *Phys. Rev. Lett.* **77**:4194 (1996); C. Chamon, C. Mudry, X. G. Wen *Phys. Rev.* **B53**:R7638 (1996); J. S. Caux, I. I. Kogan and A. M. Tsvelik *Nucl. Phys.* **B466**:444 (1996).
3. P. Gilkey (1995). "Invariance Theory, The Heat Equation and the Atiyah Singer Index Theorem," CRC Press (1995).
4. For a recent review of Supersymmetric Quantum Mechanics see A. Comtet and C. Texier *Cond-Mat/9707313* (1997).
5. Y. Aharonov and A. Casher *Phys. Rev.* **A19**:2461 (1979).
6. M. F. Atiyah and I. M. Singer *Bull. Am. Math. Soc.* **69**:422 (1963).
7. E. Akkermans, J. E. Avron, R. Narevich and R. Seiler *Eur. Phys. Jour.* **B1**:1 (1998).
8. M. F. Atiyah, V. K. Patodi and I. M. Singer *Math. Proc. Camb. Phil. Soc.* **77**:43 (1975).
9. S. Ouvry *Phys. Rev.* **D50**:5296 (1994); A. Comtet, S. Mashkevich and S. Ouvry *Phys. Rev.* **D52**:5294 (1995).
10. A. Comtet, Y. Georgelin and S. Ouvry *J. Phys. A: Math. Gen.* **22**:3917 (1989).
11. M. F. Atiyah, V. K. Patodi and I. M. Singer *Math. Proc. Camb. Phil. Soc.* **78**:405 (1975).
12. E. Akkermans, A. Auerbach, J. E. Avron and B. Shapiro *Phys. Rev. Lett.* **66**:76 (1991).
13. Y. Aharonov and D. Bohm *Phys. Rev.* **115**:485 (1959).
14. M. V. Berry, R. G. Chambers, M. D. Large, C. Upstill and J. C. Walmsley *Eur. J. Phys.* **1**:154 (1980).
15. J. F. Nye and M. V. Berry *Proc. Roy. Soc.* **A336**:165 (1974).
16. M. F. Atiyah *Comm. Pure Appl. Math.* **XX**:237 (1967).
17. S. Lefschetz. "Differential Equations: Geometric theory," Dover Publications, N.Y. (1977).
18. P. G. de Gennes *in:* "Many Body Physics," Les Houches 1967, C. de Witt and R. Balian, eds., Gordon and Breach, (1968).
19. W. F. Vinen *Prog. in Low Temp. Physics* Vol. III:25 (1961).
20. C. R. Hagen *Phys. Rev. Lett.* **64**:503 (1990).
21. J. Friedel *Nuovo Cimento* **7**:287 (1958).
22. Personal communication with R. E. Prange.
23. T. J. I. Bromwich. "An Introduction to the Theory of Infinite Series," MacMillan and Co., London (1931).
24. E. H. Sondheimer and A. H. Wilson *Proc. Roy. Soc.* **A 210**:173 (1952).

25.   to be rigorous, the quantity defined here is the Fredholm Index. Nevertheless, for the case considered here, it coincides with (5). For a further discussion of this point, see H. L. Cycon, R. G. Froese, W. Kirsch and B. Simon. "Schrödinger operators," Chap.6, Springer Verlag (1987).

26.   E. Akkermans and R. Narevich (1996), unpublished results.

27.   J. E. Avron and R. Seiler *Phys. Rev. Lett.* **54**:259 (1985).

28.   D. J. Thouless, M. Kohmoto, P. Nightingale and M. den Nijs *Phys. Rev. Lett.* **49**:40 (1982).

29.   A. J. Leggett, *in*: "Granular Nanoelectronics," D. K. Ferry, ed., Plenum Press, N.Y. (1991).

FROM CLASSICAL TO QUANTUM KINETICS

D.E.Khmelnitskii [1,2] and B.A.Muzykantskii,[3]

[1]Cavendish Laboratory, University of Cambridge,
Madingley Road, Cambridge CB3 0HE, UK
[2]L.D.Landau Institute for Theoretical Physics,
Moscow, Russia
[3]Department of Physics, University of Warwick,
Coventry CV4 7AL, UK

We discuss interconnections between quantum dynamics and classical kinetics on examples from our research practice.
The time dispersion of the averaged conductance $G(t)$ of a mesoscopic sample is calculated in the long time limit when t is much larger than the diffusion traveling time t_D. In this case the functional integral in the effective supersymmetric field theory is determined by the saddle point contribution. If t is shorter than the inverse level spacing Δ ($\Delta t/\hbar \ll 1$), then $G(t)$ decays as $\exp[-t/t_D]$. In the ultra-long time limit ($\Delta t/\hbar \gg 1$) the conductance $G(t)$ is determined by the electron states that are poorly connected with the outside leads. The probability to find such a state decreases more slowly than any exponential function as t tends to infinity. It is worth mentioning, that the saddle point equation looks very similar to the well known Usadel equation in the theory of dirty superconductors.
An effective field theory for disordered conductors is suggested to describe quantum kinetics of ballistically propagating electrons. This theory contains non-linear σ-model [1] as its long wave limit.
A new configuration of the random potential and the optimal fluctuation method are used for the evaluation of the long time delay in Ohmic response of a finite piece of a disordered one-dimensional chain. The result agrees, to exponential accuracy, with that, obtained earlier [2]*.

0. These lecture aims to show the delicate relations between quantum dynamics of electron in the random potential and classical kinetics. The mentioned interconnections of these two subjects seem to us possessing an heuristic power. The lecture consists a presentation of previously published results [3,4,5,6], which appear here together, accompanied by comments.

1.1. We begin with consideration of the long time relaxation phenomena in a disordered conductor that is attached to ideal leads. For simplicity we assume that

---

*This Lecture is based on joint papers and was presented by DEK

the electrons in this conductor do not interact with each other, the temperature is zero ($T = 0$) and there are no inelastic processes. The total current $I(t)$ at time t depends upon the voltage according to the Ohm law:

$$I(t) = \int_{-\infty}^{t} dt' G(t - t') V(t').$$ (1)

We are interested in the asymptotic form of the conductance $G(t)$ as $t \to \infty$.

The same problem has been considered earlier by Altshuler, Kravtsov and Lerner (AKL) [7], who effectively analyzed divergences of the series for $G(t)$ in $1/t$ in the long limit for dimension $d = 2$. We are going to suggest a method of a more direct calculation, which enables to obtain also all intermediate asymptotes as well as brings a new heuristic idea, which will lead to the quantum action for the ballistics (see discussion below).

There are three time scales in the problem:

1. The mean free time $\tau = l/v_F$, where v_F is the Fermi velocity and l is the mean free path. This time scale determines the dispersion of the Drude conductivity $\sigma_0 \sim e^{-t/\tau}$

2. The time of diffusion through the sample $t_D = L^2/D$, where $D = l^2/3\tau$ is the diffusion coefficient, and L is the sample size.

3. The inverse mean level spacing $\hbar/\Delta = \hbar \nu V$, where ν is the density of states and V is the volume of the sample.

In a macroscopic sample the inequality $\tau \ll t_D \ll \hbar/\Delta$ is valid, provided that $L \gg l$ and the disorder is weak. Indeed, the product Δt_D is connected with the dimensionless conductance of the sample $g = 2\pi\hbar/(t_D\Delta)$, which is large for a weak disorder. The times t_D and $\hbar/\Delta$ enter into time dispersion only due to quantum corrections to conductivity.

At times $t \ll \hbar/\Delta$ an electron can be considered a wave packet of many superimposed states propagating semi-classically. Therefore, it is natural to assume that the conductance $G(t)$ is proportional to the probability of finding a Brownian trajectory that remains in the sample for the time t. For $t \gg t_D$ such a probability decays as $\exp\left[-t/t_D\right]$. Our calculations confirm this result.

In the opposite limit, for $t \gg \hbar/\Delta$, the conductance $G(t)$ is proportional to the probability of finding an electron state with the life time t. In order to trap an electron for a long time the state must be poorly connected with the leads (nearly localized). We show that the probability of finding such a state decays non-exponentially with time. Namely, $G(t) \sim \exp\left[-g\log^2(t\Delta)\right]$ for $d = 1$ and $G(t) \sim (t\Delta)^{-g}$ for $d = 2$. These results are not valid in the very long time limit. We discuss this later together with the question of dimensional crossover.

Instead of calculating the conductance as a function of time, we could have worked in the frequency representation. In that way we would have found a singularity in $G(\omega)$ as $\omega \to 0$. This singularity, however, does not affect the value of the d.c. conductance and therefore has an obscure physical meaning, while the time domain results have the direct interpretation.

1.2. Since the long time asymptote corresponds to the rare events when the electron is nearly trapped in the sample, it is natural to use the saddle-point approximation. We employ the field theory [1] with the partition function

$$F = \frac{\pi\nu}{8} \int dr \mathrm{str}[D(\nabla Q)^2 + 2i\omega\Lambda Q], \qquad Z = \int_{Q^2=1} \mathcal{D}Q e^{-F}$$ (2)

and carry out the following program:

1. Express the averaged conductance[†] as a functional integral over supermatrices[‡] (see [1,8] for review):

$$G(t) = G_0 e^{-t/\tau} +$$
$$\int \frac{d\omega}{2\pi} e^{-i\omega t} \int \mathcal{D}Q(r) P\{Q\} \exp\left[-A\right],$$
$$A = \frac{\pi\nu}{8} \int dr \, \mathrm{Str}\{D(\nabla Q)^2 + 2i\omega\Lambda Q\}, \tag{3}$$

2. Vary the action A with respect to Q, taking into account the constraint $Q^2 = 1$, and obtain the saddle-point condition which recalls the diffusion limit of the Eilenberger equation (the Usadel equation) [9] :

$$2D\nabla(Q\nabla Q) + i\omega\left[\Lambda, Q\right] = 0 \tag{4}$$

3. Derive the condition at the boundary with the lead

$$Q|_{\mathrm{lead}} = \Lambda. \tag{5}$$

4. Perform the integration over ω in Eq. (3) and obtain the self consistency condition

$$\int \frac{dr}{V} \, \mathrm{Str}(\Lambda Q) = -\frac{4t\Delta}{\pi\hbar} \tag{6}$$

which allows us to exclude ω from Eq. (4).

5. Substitute the solution of Eq. (4) with boundary conditions (5) in Eq. (3) and obtain the results with exponential accuracy.

1.3. The 8×8 supermatrix Q has commutative and anticommutative matrix elements. Since $Q^2 = 1$ it can be chosen in the form[§] :

$$Q = V^{-1}HV, \quad V = \begin{pmatrix} u & 0 \\ 0 & v \end{pmatrix},$$

$$\Lambda = \begin{pmatrix} 1 & 0 \\ 0 & -1 \end{pmatrix}, \quad H = \begin{pmatrix} \cos\hat\theta & i\sin\hat\theta \\ -i\sin\hat\theta & -\cos\hat\theta \end{pmatrix}, \tag{7}$$

$$\hat\theta = \begin{pmatrix} \theta_1 & 0 & 0 & 0 \\ 0 & \theta_1 & 0 & 0 \\ 0 & 0 & i\theta & 0 \\ 0 & 0 & 0 & i\theta \end{pmatrix}$$

This decomposition allows us to present the action A in the form

$$A = \frac{\pi\nu}{8} \int dr \, \mathrm{Str}\{D(\nabla H)^2 + DM^2 + 2i\omega\Lambda H\}, \tag{8}$$

[†]Our results represent the time-dependent conductance $G(t)$ of a large set of mesoscopic junctions with identical macroscopic parameters. For a single junction the fluctuations of the conductance $G(t)$ are small compared to its averaged value, if $t\Delta \ll 1$, and are not small in the opposite limit.
[‡]The explicit form of prefactor $P\{Q\}$ is irrelevant within the exponential accuracy.
[§]To be specific, we consider the unitary ensemble. As it follows from the derivation below, to the exponential accuracy, the results do not depend on the ensemble.

where $M = [V^{-1}\nabla V, H]$. The minimum action is reached for $V = \text{const}$, and Eq. (8) may be expressed in terms of θ-variables only:

$$A = \frac{\pi\nu}{2}\int dr\{[D(\nabla\theta)^2 - 2i\omega\cosh\theta] \qquad (7)$$
$$+[D(\nabla\theta_1)^2 + 2i\omega\cos\theta_1].$$

Consequently, Eq. (2) has the form:

$$D\nabla^2\theta + i\omega\sinh\theta = 0, \qquad (9)$$
$$D\nabla^2\theta_1 + i\omega\sin\theta_1 = 0 \qquad (10)$$

The boundary condition (5) follows from the fact that Q does not fluctuate in the bulk electrodes, $Q = \Lambda$. Hence, at the boundary with the ideal lead $\theta = \theta_1 = 0$ [10]. The time decay of the conductance $G(t) \sim \exp(-i\omega t)$ corresponds to real and positive values of $i\omega$. The permitted values of frequency ω in Eq. (10) are bounded from below by the value $\omega_1 \sim 1/t_D$, which corresponds to the linearized form of Eq. (10). For smaller frequencies $\omega < \omega_1$, which will turn out to be the only relevant ones, Eq. (10) has only trivial solutions $\theta_1 = 0$. Thus, the self-consistency equation (6) has the form:

$$\int \frac{dr}{V}\{\cosh\theta - 1\} = \frac{t\Delta}{\pi\hbar}. \qquad (11)$$

The solutions of Eq. (9) depend on the sample geometry. We start by considering a one dimensional wire of length L, attached to ideal leads at $x = \pm L/2$. If $t\Delta \ll 1$, then, to satisfy the self-consistency condition (11) we must choose $\theta \ll 1$. Therefore, Eq. (9) can be linearized. The solutions that satisfy the boundary conditions is

$$\theta = C\cos(\pi n x/L), \qquad \omega_n = \frac{-i\pi^2 n^2}{t_D}, \qquad (12)$$

where n is an arbitrary integer. The above formula for the frequency implies that in the discussed regime

$$G(t) \sim e^{-i\omega_1 t} = \exp\left(-\frac{\pi^2 t}{t_D}\right). \qquad (13)$$

To obtain this result we determine the amplitude C from the linearized self-consistency equation, and then substitute (12) into the action A.

1.4. For arbitrary times Eqs. (9) and (11) in dimensionless coordinates have the form:

$$\frac{d^2\theta}{dz^2} + \frac{\gamma^2}{2}\sinh\theta = 0, \qquad z = \frac{x}{L}, \qquad (14)$$
$$\int_{-1/2}^{1/2} dz[\cosh\theta - 1] = \frac{\Delta t}{\pi\hbar}, \qquad \gamma^2 = 2i\omega t_D. \qquad (15)$$

The solution of (14) is symmetric $\theta(z) = \theta(-z)$, and in the region $z > 0$ is given by the quadrature:

$$z = \frac{1}{\gamma}\int_{\theta(z)}^{\theta_0} \frac{d\theta'}{\sqrt{\cosh\theta_0 - \cosh\theta'}} \qquad (16)$$
$$\theta_0 = \theta(0) = 2\log\frac{1}{\gamma} + 2\log\log\frac{1}{\gamma}, \quad \text{for } \gamma \ll 1 \qquad (17)$$

The function $\theta(z)$ is almost linear $\theta = \theta_0(1 - 2|z|)$ everywhere except in the region $|z| < 1/\log(1/\gamma) \ll 1$. Substituting Eq. (16) into Eqs. (11) and (3) we get

$$i\omega = \frac{2g}{t}\log\frac{t\Delta}{\hbar}, \quad G(t) \sim \exp\left[-g\log^2\frac{t\Delta}{\hbar}\right] \tag{18}$$

As mentioned earlier, the contribution from the individual nearly localized states dominates in $G(t)$ whenever $t\Delta/\hbar \gg 1$. The square modulus of the wave function for such a state $|\Psi|^2$ equals $\cosh\theta$. As we can see, this value decays exponentially towards the leads, where $|\Psi(x = \pm L/2)|^2 = \cosh\theta(\pm 1/2) = 1$. Because of the latter condition, the current through the wire is equal to unity. Therefore, the escape time t is proportional to the normalization integral. This is exactly what is stated in the self-consistency condition (15) for $\theta \gg 1$. To summarize, the wave function is localized in the region $|x| \ll \xi \ll L$ with the localization length $\xi = L/\log(t\Delta/\hbar)$ and the probability to find such a state is given by Eq.(18). For very long times, when ξ becomes less than the transverse size of the sample, the one-dimensional regime crosses over to a two- or three-dimensional one.

1.5. In the two-dimensional case we consider a mesoscopic disk of radius R surrounded by a well conducting electrode. The Laplacian operator in Eq. () is now two-dimensional and the boundary condition is $\theta(R) = 0$ at the circumference of the disk. It is natural to assume that the minimal action corresponds to θ that depends on the radius only and, therefore, obeys the equation:

$$\theta'' + \theta'/z + i\omega t_D \sinh\theta = 0, \quad \theta(1) = 0 \tag{19}$$

where $z = r/R$ and $t_D = R^2/D$.

For $t \ll \hbar/\Delta$, Eq. (19) can be linearized. Its solution is the Bessel function

$$\theta = CJ_0(\gamma z), \; \gamma = \sqrt{i\omega t_D} = \mu_n, \tag{20}$$

where μ_n denotes the n-th zero of the Bessel function. The conductance is

$$G(t) \sim \exp\left(-\frac{\mu_1^2 t}{t_D}\right), \quad t_D \ll t \ll \hbar/\Delta. \tag{21}$$

For a long time tail $t \gg \hbar/\Delta$, the non-linear term in Eq. (19) is large near the origin and can be neglected elsewhere. As a result,

$$\theta(z) = C\log\frac{1}{z} \tag{22}$$

for all but very small z. On the other hand, for $z \ll 1$, the parameter θ is large and $\sinh\theta = e^\theta/2$. The substitutions $z = e^{-\eta}$ and $\theta = \psi(\eta) + 2\eta$ transform Eq. (19) into the form $\psi'' + \gamma^2/4 e^\psi = 0$ which has the first integral. With this approximation the solution of Eq. (19) can be found having the asymptote

$$\theta(z) = -\theta(0) + 6\log 2 + \log\frac{4}{\gamma^2} + 4\log\frac{1}{z}, \tag{23}$$

for $\gamma \ll z \le 1$. Comparing with Eq. (22), we have $\theta(0) = 6\log 2 + \log(4/\gamma^2)$ and $C = 4$. To calculate the integral in the self-consistency equation

$$\frac{\Delta t}{2\pi\hbar} = \int_0^1 \{\cosh\theta(z) - 1\}z\,dz \tag{24}$$

we multiply Eq. (19) by z, integrate in the limits 0 and 1, and obtain

$$z\frac{d\theta}{dz}\bigg|_0^1 + i\omega t_D \int_0^1 \sinh\theta(z)z\,dz = 0. \tag{25}$$

Since $\theta(0) \gg 1$, we neglect the difference between the integrals in Eqs. (24) and (25), and with asymptote (22) finally get $i\omega = 4g/t$. The action A is dominated by the contribution of the tail (22):

$$A = 4g\log\frac{t\Delta}{2\pi\hbar}, \qquad G \sim \left(\frac{\hbar}{\Delta t}\right)^{4g}. \tag{26}$$

The characteristic size of the averaged 2D wave function is $\xi = \gamma R = R(\hbar/t\Delta)^{1/2}$. The crossover to a 3D case occurs when ξ becomes comparable with the film thickness.

1.6. The consideration of the 3D case makes relevant the question of the validity of the diffusion approximation. As before, we consider a disordered drop of radius R surrounded by a well conducting lead. Analogously to what has been done in the 2D case, the function θ depends on the radius r only and obeys Eq. (9), where the Laplace operator is substituted by its 3D radial component. The boundary condition is $\theta(r = R) = 0$. The analysis of the linear regime is similar to that for 1D and 2D cases and gives for $t_D = R^2/D \ll t \ll \hbar/\Delta$:

$$A = \pi^2 t/t_D, \qquad G(t) \sim \exp(-\pi^2 t/t_D) \tag{27}$$

The nonlinear in θ regime leads to the equation

$$\frac{d^2\theta}{dz^2} + \frac{2}{z}\frac{d\theta}{dz} + i\omega t_D \sinh\theta = 0, \quad z = \frac{r}{R} \tag{28}$$

The analysis of this equation shows that the permitted values of ω are larger than a certain value $\omega_0 > 0$, and that the integral in Eq. (11) remains finite even for the solutions of Eq. (28) with $\theta(r = 0) \to \infty$. As a result, the self-consistency equation cannot be satisfied for sufficiently long time $t \geq \hbar/\Delta$. Thus, for $|\omega t_D| \ll 1$, all non-trivial solutions of Eq. (28) satisfying the condition $\theta(1) = 0$ are singular at $z \to 0$. Therefore, the derivative $d\theta/dr$ becomes comparable with the inverse mean-free path $1/l$ for a certain radius r_*. The diffusion approximation inevitably breaks down for smaller distances, where non-local corrections become important.

It is sensible now to analyze whether the diffusion treatment is valid in the 1D and 2D cases. Using the solutions of Eqs. (14) and (19) we find the value of t_* such that for $t < t_*$ the derivative $l\,d\theta/dr$ is less than unity. This gives:

$$t_* = \frac{\hbar}{\Delta}\begin{cases} \exp\{L/l\}, & d = 1 \\ (R/l)^2, & d = 2 \end{cases} \tag{29}$$

Therefore, we can expect that the asymptotes (18) and (26) are valid for $t < t_*$. At longer times $t > t_*$ for all dimensions $d = 1, 2, 3$ the asymptote cannot be found within the diffusion approximation. A detailed kinetic analysis of this problem was done in a our paper [5]. In the following an estimate is presented first putted forward in our paper [3] and later improved by Mirlin [18]. We do not try to solve the kinetic problem but assume that the mentioned non-locality smoothes out the singularity at the origin. We also assume that, similarly to what has happened in the 1D and 2D cases in the diffusion approximation, the nonlinear term in Eqs. (28, 19) can be neglected at $r > r_*$ and is important for $r \sim r_*$. Thus

$$\theta(r) \sim C_3\left(\frac{R}{r} - 1\right), \quad r > r_*$$

$$1 = l\left(\frac{d\theta}{dr}\right)_{r=r_*} = \frac{C_3 l R}{r_*^2},$$

and $r_* = (C_3 l R)^{1/2}$. Then $\theta_* = \theta(r_*) = (C_3 R/l)^{1/2}$ and, finally,

$$\omega t_D \exp(\theta_*) \sim \frac{\theta_* R^2}{r_*^2} \sim \frac{1}{\theta_*}\left(\frac{R}{l}\right)^2,$$

which gives

$$\theta_* = \log\left[\frac{1}{\omega t_D}\left(\frac{R}{l}\right)^2\right], \quad C_3 = \frac{l}{R}\log^2\left[\frac{1}{\omega t_D}\left(\frac{R}{l}\right)^2\right].$$

Using the self-consistency condition we express the frequency ω through the time t and obtain a rough estimate for the action

$$A \sim \left(\frac{p_F l}{\hbar}\right)^2 \log^3\left(\frac{t}{\tau}\right), \quad G(t) \sim \exp\left[-\kappa\left(\frac{p_F l}{\hbar}\right)^2 \log^3\left(\frac{t}{\tau}\right)\right], \tag{30}$$

where the coefficient κ in the exponent can only be determined from the solution of the kinetic problem (see [5]).

For two-dimensional case the very same estimates give:

$$\theta(r) \sim C_2 \log\left(\frac{R}{r}\right), \quad r > r_*$$

$$1 = l\left(\frac{d\theta}{dr}\right)_{r=r_*} = \frac{C_2 l}{r},$$

$$r_* = C_2 l, \quad \theta_* = \theta(r_*) = C_2 \ln\left(\frac{R}{C_2 l}\right),$$

and, finally,

$$\omega t_D \exp \theta_* \sim \frac{\theta_* R^2}{r_*^2} \sim \frac{\theta_*}{C_2 l^2},$$

which gives

$$\theta_* = 2\log\left[\frac{1}{\omega t_D}\left(\frac{R\ln(R/l)}{l}\right)^2\right];$$

$$C_2 = 2 + \frac{2\ln\ln(R/l) - \ln \omega t_D}{\ln(R/l)} \rightarrow -\frac{\ln \omega t_D}{\ln(R/l)}.$$

Using the self-consistency condition we express the frequency ω through the time t, obtain a rough estimate for the action A and for the conductance $G(t)$. One can see that at times $\hbar/\Delta \ll t \ll (R/l)\hbar/\Delta$ the power law asymptote (26) recovers. In the long-time limit $t \gg (R/l)\hbar/\Delta$ the asymptote of the conductance $G(t)$ has the form:

$$G(t) \sim \exp\left[-\frac{\pi g}{2}\frac{\ln^2(t/g\tau)}{\ln R/l}\right], \tag{31}$$

which coincides with a minor variation with the AKL result. This seems natural, because the authors studied the coefficient's growth rate in a power expansion of $G(\omega)$ in

$\omega\tau$. Since the power expansion has an asymptotic character, it is determined by a non-analytic contribution. A new insight, which comes from the presented calculation of a non-analitic saddle-point contribution, is that the AKL long-time asymptote originates in a saddle-point solution of a ballistic problem.

In the case of a tunnel barrier at the sample-lead interface , the time dispersion of the conductance can be considered in the same way with the usage of the generalized boundary condition For an arbitrary transparency of the sample-lead interface T the boundary condition has the form [10] $lQ\nabla Q + 3/4(T/(1-T))[\Lambda, Q] = 0$.

1.7. Everything discussed above was obtained by the means of non-linear σ-model, i.e. under assumption that the random potential $U(\mathbf{r})$ remains small compare to the energy of electrons (the Fermi energy E_F)[¶]. This very assumption had been quite correctly criticized by Shklovskii and by Smolyarenko and Altshuler (SSA) [19] as being too restrictive. For the case of the Gaussian distribution of the random potential these authors managed to find an optimal fluctuation of the random potential which traps electron in small volume and isolates it from the rest of conductor by a high barrier, so, the electron must tunnel through this barrier. In 3D case the contribution of small traps of this kind to time dependent conductance G(t) is

$$G(t) \sim \exp\left[-\kappa\frac{p_F l}{\hbar}\ln^3\frac{t\Delta}{\hbar}\right], \qquad \kappa \sim 1. \tag{32}$$

The right hand side of Eq (32) consists of a parametrically smaller exponential factor, than that of Eq (30). This makes small traps statistically preferable in the long time limit. Nevertheless, this does not explain whether small traps of (SSA) exhibit the only minimum of the action or this minimum exists in parallel with one, we found by the means of non-linear σ-model. In the latter case the contributions of both minima exist and win over one another depending on parameters. Study of this problem touches a very sensitive point of whole theory of electron localization and its resolution could bring a significant progress to our understanding of all related problems.

2.1. As we have seen, the non-linear σ-model is proven to be a useful tool in the description of various properties of disordered conductors. Any property, such as conductivity, averaged over different realizations of the random potential can be presented in this model as a statistical average with the free energy

$$Z = \int_{Q^2=1}\mathcal{D}Qe^{-F}, \qquad F = \frac{\pi\nu}{8}\int d\mathbf{r}\,\mathrm{str}[D(\nabla Q)^2 + 2i\omega\Lambda Q] \tag{33}$$

The functional integral is taken over the 8×8 super-matrix $Q(\mathbf{r})$ which is subjected to the constraint $Q^2 = 1$. Here and below we use the super-matrix version [1,8] of the nonlinear σ model.

This discription is valid under the following two conditions:

1. The Fermi wave length $\lambda_F = \hbar/p_F$ is much smaller than the mean free path l, i.e. $p_F l/\hbar \gg 1$.

2. The typical wave vector q of the super-matrix fluctuations is smaller than $1/l$, i.e. $ql \ll 1$.

These conditions mean that (i) the semi-classical description is applicable to the electrons with the Fermi energy, and (ii) their motion is described by the diffusion

---

[¶]This means that localization, which leads to the long-time trapping arises due to over-barrier scattering, similar to that, which is discussed in section **3.1** of this lecture

equation. There are physical situations when the condition (i) is fulfilled, while the condition (ii) is not and electrons propagate ballistically. This happens, for example in a metallic grain with a diffusive boundary scattering if the bulk mean free path l is much larger than the grain size L, i.e. $l \gg L$. We are going to suggest a generalized version of the model (33) whose validity is no longer restricted by condition (ii). The generalized partition function correctly accounts for the fluctuations with wave vectors $q \sim 1/l$ and therefore can be used for the description of systems with ballistic electron motion.

We begin with a general expression for the free energy which is obtained after averaging over the random potential, the Hubbard-Stratonovich decomposition of the quartic form and integration over the electron degrees of freedom (see [1] for details and notations).

$$F = -\frac{1}{2}\operatorname{str}\ln[-i\hat{K}] + \frac{\pi\nu}{8\tau}\int \operatorname{str}Q^2(\mathbf{r})d\mathbf{r}, \quad Z = \int \mathcal{D}Qe^{-F}, \tag{34}$$

$$\hat{K} = E - \hat{H}_0 + \frac{\omega}{2}\Lambda + \frac{i}{2\tau}Q, \quad \hat{H}_0 = \frac{(-i\hbar\nabla)^2}{2m} \tag{35}$$

This expression appears at a preliminary stage in the derivation of Eq. (33) and the supermatrix Q is not yet restricted by the constraint $Q^2 = 1$.

Equation (34), in principle, could have served as a required generalisation of the free energy (33). However, it is too detailed being valid for the super-matrices Q fluctuating with arbitrary wave vectors $\mathbf{q}$. It will be simplified in order to describe the small $\mathbf{q}$ fluctuations only ($q \ll p_F/\hbar$). The first step in the simplification is the same as in the derivation of the quantum kinetic equation in the Keldysh approach (see, for example, [9]).

2.2. The Green function $G(\mathbf{r}, \mathbf{r}'|Q)$ of the operator $\hat{K}$ obeys the equations

$$\left[E - \hat{H}_0(\mathbf{r}) + \frac{\omega}{2}\Lambda + \frac{i}{2\tau}Q(\mathbf{r})\right]G(\mathbf{r}, \mathbf{r}'|Q) = i\delta(\mathbf{r} - \mathbf{r}') \tag{36}$$

$$\left[E - \hat{H}_0(\mathbf{r}')\right]G(\mathbf{r}, \mathbf{r}'|Q) + G(\mathbf{r}, \mathbf{r}'|Q)\left[\frac{\omega}{2}\Lambda + \frac{i}{2\tau}Q(\mathbf{r}')\right] = i\delta(\mathbf{r} - \mathbf{r}') \tag{37}$$

Subtracting Eq. (37) from Eq. (36) and going to the Wigner representation

$$G(\mathbf{r}, \mathbf{r}') = \int (d\mathbf{p})\,\tilde{G}(\frac{\mathbf{r} - \mathbf{r}'}{2}, \mathbf{p})\,e^{i\mathbf{p}(\mathbf{r}-\mathbf{r}')} \tag{38}$$

we can find after the integration over the modulus of the momentum $\mathbf{p}$ an equation for

$$g_\mathbf{n}(\mathbf{r}) = \frac{1}{\pi}\int d\xi\tilde{G}(\mathbf{r}, \mathbf{n}\frac{\xi}{v_F}), \qquad \mathbf{n}^2 = 1. \tag{39}$$

This equation can be presented in the form

$$2v_F\mathbf{n}\frac{\partial g_\mathbf{n}(\mathbf{r})}{\partial \mathbf{r}} = \left[i\omega\Lambda - \frac{Q}{\tau}, g_\mathbf{n}\right], \tag{40}$$

which resembles the quantum kinetic equation in the Eilenberger form [9]. The matrix $g_\mathbf{n}(\mathbf{r})$ in this equation has the meaning of distribution function at a coordinate $\mathbf{r}$ and momentum $\mathbf{p} = \mathbf{n} \cdot p_F$.

Being linear, Eq. (40) does not define $g_\mathbf{n}$ uniquely and must be supplied with the normalisation condition [9]

$$g_\mathbf{n}^2 = 1; \qquad \operatorname{tr}g_\mathbf{n} = 0. \tag{41}$$

The matrix $Q(\mathbf{r})$ is invariant with respect to the charge conjugation
$$\bar{Q} \equiv CQ^T C^T = Q, \tag{42}$$
where $\hat{C}$ is a certain matrix (see [1]), $C^T C = 1$. Taking the charge conjugate of Eq (36) and using Eq (42), we see that $\bar{G}(\mathbf{r}, \mathbf{r}')$ obeys Eq (37). Therefore
$$\bar{G}(\mathbf{r}, \mathbf{r}') = G(\mathbf{r}', \mathbf{r}), \quad \bar{\tilde{G}}(\mathbf{r}, \mathbf{p}) = G(\mathbf{r}, -\mathbf{p}), \quad \bar{g}_{\mathbf{n}}(\mathbf{r}) = g_{-\mathbf{n}}(\mathbf{r}). \tag{43}$$
Thus, Eq. (40) with the normalisation condition (41) and the symmetries (43) is a long wave limit of Eqs. (36,37). Our goal is to perform analogous simplification of the free energy (34).

2.3. An intermediate step is finding a functional Φ, which reaches its extrema for solutions of Eq. (40). This equation resembles the equation of motion of a magnetic moment $\mathbf{M}$ in external magnetic field $\mathbf{B}$:
$$\frac{\partial \mathbf{M}}{\partial t} = [\mathbf{M} \times \mathbf{B}], \qquad \mathbf{M}^2 = 1. \tag{44}$$
The action for this problem has the form (see, for instance, [12]
$$\mathcal{A} = \int_0^t dt' \mathbf{B} \mathbf{M}(t') + \int_0^t dt' \int_0^1 du \tilde{\mathbf{M}} \cdot \left[\frac{\partial \tilde{\mathbf{M}}}{\partial t} \times \frac{\partial \tilde{\mathbf{M}}}{\partial u} \right], \tag{45}$$
where the function $\tilde{\mathbf{M}}(t, u)$ is introduced as
$$\tilde{\mathbf{M}}(t, 0) = \mathbf{M}_0; \qquad \tilde{\mathbf{M}}(t, 1) = \mathbf{M}(t). \tag{46}$$
The second term in Eq (45) does not depend upon the choice of $\mathbf{M}_0$ and values of $\tilde{\mathbf{M}}(t, u)$ for $0 < u < 1$, provided $\mathbf{M}(0) = \mathbf{M}(t)$.

Following this analogy we present Φ in the form
$$\Phi = \int d\mathbf{r} \operatorname{str} \left[(\frac{1}{\tau} Q(\mathbf{r}) + i\omega\Lambda)\langle g(\mathbf{r}) \rangle \right] + \frac{v_F}{2} \mathcal{W}\{g_{\mathbf{n}}\}, \tag{47}$$
$$\langle g(\mathbf{r}) \rangle = \int \frac{d\Omega_{\mathbf{n}}}{4\pi} g_{\mathbf{n}}(\mathbf{r}), \tag{48}$$
$$\mathcal{W}\{g_{\mathbf{n}}\} = \int d\mathbf{r} \int \frac{d\Omega_{\mathbf{n}}}{4\pi} \int_0^1 du \operatorname{str} \tilde{g}_{\mathbf{n}}(\mathbf{r}, u) \left[\frac{\partial \tilde{g}_{\mathbf{n}}}{\partial u}, \mathbf{n} \frac{\partial \tilde{g}_{\mathbf{n}}}{\partial \mathbf{r}} \right], \tag{49}$$
$$\tilde{g}_{\mathbf{n}}(\mathbf{r}, 0) = \Lambda; \qquad \tilde{g}_{\mathbf{n}}(\mathbf{r}, 1) = g_{\mathbf{n}}(\mathbf{r}). \tag{50}$$
The functional derivative $\delta\Phi/\delta g_{\mathbf{n}}$ must be taken with constraint (41) which guaranties that $g_{\mathbf{n}} \delta g_{\mathbf{n}} + \delta g_{\mathbf{n}} g_{\mathbf{n}} = 0$ and an arbitrary variation $\delta g_{\mathbf{n}}$ has the form $\delta g_{\mathbf{n}} = [g_{\mathbf{n}}, a_{\mathbf{n}}]$. As a result
$$\delta\Phi = \int d\mathbf{r} \int \frac{d\Omega_{\mathbf{n}}}{4\pi} \operatorname{str} \left(\left[\frac{1}{\tau} Q(\mathbf{r}) - i\omega\Lambda, g_{\mathbf{n}} \right] a_{\mathbf{n}} \right) + \frac{v_F}{2} \delta\mathcal{W}, \tag{51}$$
where
$$\delta\mathcal{W} = 4 \int d\mathbf{r} \int \frac{d\Omega_{\mathbf{n}}}{4\pi} \operatorname{str} \left(\mathbf{n} \frac{\partial g_{\mathbf{n}}}{\partial \mathbf{r}} a_{\mathbf{n}} \right). \tag{52}$$
Thus, Eq. (47) gives the required functional.

One can show that in the limit $l \gg \lambda_F$ the partition function (34) reduces to the form
$$Z = \int_{g_{\mathbf{n}}^2=1} \mathcal{D} g_{\mathbf{n}}(\mathbf{r}) e^{-F}, \tag{53}$$
$$F = \frac{\pi\nu}{4} \left[\int d\mathbf{r} \operatorname{str} \left\{ i\omega\Lambda \langle g(\mathbf{r}) \rangle - \frac{\langle g(\mathbf{r}) \rangle^2}{2\tau} \right\} - \frac{v_F}{2} \cdot \mathcal{W}\{g_{\mathbf{n}}\} \right], \tag{54}$$
$$\mathcal{W}\{g_{\mathbf{n}}\} = \int d\mathbf{r} \int \frac{d\Omega_{\mathbf{n}}}{4\pi} \int_0^1 du \operatorname{str} \tilde{g}_{\mathbf{n}}(\mathbf{r}, u) \left[\frac{\partial \tilde{g}_{\mathbf{n}}}{\partial u}, \mathbf{n} \frac{\partial \tilde{g}_{\mathbf{n}}}{\partial \mathbf{r}} \right]. \tag{55}$$

2.4. For small gradients, the free energy (53, 54, 55) reduces to the standard σ-model (33). To show this we expand the matrix $g_{\mathbf{n}}$ into the sum over sperical functions $Y_{L,M}(\mathbf{n})$

$$g_{\mathbf{n}}(\mathbf{r}) = \sum_{L=0}^{\infty} \sum_{M=-L}^{L} g_{L,M}(\mathbf{r}) \cdot Y_{L,M}(\mathbf{n})$$

and note that only zero and first harmonics contribute to the functional integral (53, 54, 55):

$$g_{\mathbf{n}} = Q(\mathbf{r}) + \mathbf{J}(\mathbf{r}) \cdot \mathbf{n} - \frac{Q\mathbf{J}^2}{6}. \tag{56}$$

The constraint $g^2 = 1$ now reads

$$Q^2 = 1, \qquad Q\mathbf{J} + \mathbf{J}Q = 0. \tag{57}$$

Substituting the Eq. (56) into Eqs. (53, 54, 55) and using conditions (57) we obtain the partition function in the form

$$Z = \int \mathcal{D}Q \int \mathcal{D}\mathbf{J} e^{-F(Q,\mathbf{J})},$$

$$F(Q,\mathbf{J}) = \frac{\pi\nu}{4} \int d\mathbf{r}\,\mathrm{str}\{i\omega\Lambda Q + \frac{\mathbf{J}^2}{6\tau} - \frac{v_F}{3}(\nabla Q)Q\mathbf{J}\} \tag{58}$$

After the Gaussian integration over $\mathbf{J}$ in Eq (58) we arrive, finally, at Eq (33).

2.5. Equations (53, 54, 55) can be generalized in order to describe the ballistic motion in the presence of external fields. In a general case the electron is described by the classical Hamiltonians $H(p_i, x_i)$ and the kinetic equation (40) has the form (see [9]):

$$\{H(x,p), g(x,p)\} = \left[\left(\frac{i\omega\Lambda}{2} - \frac{Q}{2\tau} \right), g(p,x) \right] \tag{59}$$

where $\{H, g\}$ denotes the Poisson brackets

$$\{H(x,p), g(x,p)\} = \frac{\partial H}{\partial p_i} \frac{\partial g}{\partial x_i} - \frac{\partial H}{\partial x_i} \frac{\partial g}{\partial p_i}$$

Equation (59) is still the first order differential equation and the generalization of expression (54) for the free energy has the form

$$F = \frac{\pi}{4} \int dx_i dp_i \delta(E - H(p,x))\mathrm{str}\,\{i\omega\Lambda g -$$

$$\frac{g\langle g\rangle}{2\tau} - \frac{1}{2} \int_0^1 du \tilde{g}(x,p,u) \left[\frac{\partial \tilde{g}}{\partial u}, \{H, \tilde{g}\} \right] \} \tag{60}$$

where

$$\langle g(x)\rangle = \frac{1}{\nu} \int dp_i' \delta(E - H(p',x))g(x,p').$$

2.6. As an application of Eq. (60), let us consider the derivation of the Pruisken action [13]. for a two-dimensional electron gas in a perpendicular magnetic field B. To simplify the treatment, we consider only the case of classically weak field

$$\Omega_c \tau \ll 1; \qquad \Omega_c = \frac{eB}{mc}, \tag{61}$$

when there is no Landau quantization and the density of states ν is a constant. Nevertheless, we take into account that in the presence of magnetic field the symmetry of

g-matrix is reduced, and g belongs to the unitary ensemble. The Poisson brackets in magnetic field are

$$\{H, g\} = v_F \mathbf{n} \frac{\partial g_{\mathbf{n}}}{\partial \mathbf{r}} + \Omega_c \left[\mathbf{n} \times \frac{\partial g_{\mathbf{n}}}{\partial \mathbf{n}}\right] \tag{62}$$

and the free energy (60) has the following form

$$F = \frac{\pi \nu}{4} \int d\mathbf{r} \, \mathrm{str} \left\{ i\omega \Lambda \langle g \rangle - \frac{\langle g \rangle^2}{2\tau} - \right.$$
$$\left. -\frac{1}{2} \int_0^1 du \, \langle \tilde{g}(x, p, u) \left[\frac{\partial \tilde{g}}{\partial u}, \, v_F \mathbf{n} \frac{\partial \tilde{g}}{\partial \mathbf{r}} + \Omega_c \left(\mathbf{n} \times \frac{\partial \tilde{g}}{\partial \mathbf{n}}\right)\right]\rangle \right\} \tag{63}$$

In the diffusive limit the expansion (56) can be used, which leads to the following expression for the free energy as a functional of Q and $\mathbf{J}$;

$$Z = \int \mathcal{D}Q \int \mathcal{D}\mathbf{J} e^{-F(Q, \mathbf{J})},$$
$$F(Q, \mathbf{J}) = \frac{\pi \nu}{4} \int d\mathbf{r} \, \mathrm{str} \left\{ i\omega \Lambda Q + \frac{\mathbf{J}^2}{4\tau} - \frac{v_F}{2}(\nabla Q)Q\mathbf{J} - \frac{\Omega_c}{2} Q \left[\mathbf{J} \times \mathbf{J}\right] \right\} \tag{64}$$

The last term in the free energy (64) does not vanish because the components of the matrix $\mathbf{J}$ do not commute. Under the conditions (61), the Gaussian integration over $\mathbf{J}$ may be performed, with the vector product in Eq. (64) as a perturbation, to yeild, finally, the free energy in the form

$$F = \frac{\pi}{8e^2} \int d\mathbf{r} \mathrm{str} \left(\sigma_{xx}(\nabla Q)^2 + 2\sigma_{xy}Q[\nabla_x Q, \nabla_y Q]\right) \tag{65}$$

where

$$\sigma_{xx} = e^2 \nu D, \qquad \sigma_{xy} = \sigma_{xx} \cdot \Omega_c \tau \tag{66}$$

2.7. There is a topological question, related to the $\mathcal{W}$-term in the free energy (53, 54, 55): is it always possible to construct the functional $\mathcal{W}\{g\}$, whose variation is given by Eq. (52)? The prescription (49) gives the $\mathcal{W}$-term for the functions $g(\mathbf{r})$, which are close to $g_0(\mathbf{r}) \equiv \Lambda$. The question is whether such a functional can be defined globally.

The answer depends upon the topology of the constant energy surface $H(\mathbf{r}, \mathbf{p}) = E$ in the phase space $\{x_i, p_i\}$. For the cases of billiards and space dimension $d > 1$ the functional $\mathcal{W}$ does exist.

For a one-dimensional system $\mathcal{W}$ can only be found as a multivalued functional, just as the action (45). This causes no trouble, provided $\pi \hbar \nu v_F$ is an integer. This integer exactly equal to the wave-guide channel number in the wire.

2.8. So far, we have considered only the systems with finite amount of disorder. One can see, however, that the expression (60) remains meaningful even as $\tau \to \infty$. Therefore, we expect that the free energy $F_\infty = F(\tau \to \infty)$ describes a clean system with the Hamiltonian H. As a consequence, the partition function $Z_\infty = \int \mathcal{D}g \exp(-F_\infty)$ with the proper source terms gives the level statistics.

In the low-frequency limit ($\omega \to 0$) only the zero-mode $g^0(r, p)$ such that $\{H, g^0\} = 0$ contributes to Z_∞. There are two possibilities:

1. The hamiltonian system under consideration is integrable and there exists a set of integrals of motion $\{I_1, \ldots I_n\}$, $\{H, I_k\} = 0$. Under this condition the energy levels are characterized by the eigenvalues of $\{I_1, \ldots I_n\}$ and do not repel each other. Therefore the level statistics is Poissonian.

2. The classical dynamics is chaotic and the only integral of motion is energy. In this case the zero-mode is constant in the phase space and Z_∞ is reduced to the form

$$Z_\infty = \int_{g^2=1} \mathcal{D}g \exp\left(-\frac{\pi\nu\omega}{4}\text{str}(\Lambda g)\right) \tag{67}$$

which leads to the Wigner-Dyson (WD) level statistics [1]

In the chaotic case deviations form the WD statistics occur for the frequencies larger than the inverse time of flight through the system. These deviations are described by the small fluctuations of g about several stationary points Λ_i, similar to what has been recently shown by Andreev and Altshuler (AA) for diffusive systems [15]. In complete agreement with a general AA-conjecture, the deviation from the WD statistics is described by the determinant of some operator. It follows from our consideration that this is the Liovillean operator

$$\hat{L} = \frac{\partial H}{\partial p} \cdot \frac{\partial}{\partial x} - \frac{\partial H}{\partial x} \cdot \frac{\partial}{\partial p}$$

2.9. In conclusion, we would like to emphasize that the theory presented here contains the diffusive σ-model as a limiting case and supplies it with the physically motivated regularization of the infinities at short distances.

2.10. Part **2.8** is presented here in exactly that form as it had been published in the paper [4]. After the paper [4] had been published, an alternative step towards the same target has been made [16] for chaotic systems with no disorder. The authors of Ref [16] used averaging over large number of eigenstates in the interval $E \gg \Delta$, what should fix the modulus of the super-matrix. The whole approach should lead to the action (53, 54, 55) with $\tau = \infty$. At this stage these authors struggled with singularities at real frequencies ω and emphasized necessity of regularization. Recently, some difficulty had been discovered on this way (see [17]).
To our understanding, statistical properties of chaotic system could be calculated if the same properties for a disordered system with a long scattering time τ are studied and then a transition to the limit $\tau \to \infty$ is made. We anticipated that on this way no uncertainty will occur. Of course, all this remains a conjecture so far.

3.1. Altshuler and Prigodin [2] have studied the long time asymptote of the averaged conductance $G(t)$ in a disordered one-dimensional chain

$$G(t) = \int_{-\infty}^{+\infty} \frac{d\omega}{2\pi} G(\omega) e^{-i\omega t} \tag{68}$$

and found that as $t \to \infty$

$$G(t) \sim \exp\left(-\frac{l}{L}\ln^2 t\Delta\right), \tag{69}$$

where L is the length of the chain, l is the mean free path, $\Delta = 1/L\nu$ is the mean level spacing and ν is the density of states.

Formula (69) can be understood as a probability of an optimal potential fluctuation that traps an electron at Fermi energy E_F for time t. In a weak potential $U(x) \ll E_F$ the wave function can be presented in the form

$$\Psi(x) = \phi_+(x)e^{ip_F x} + \phi_-(x)e^{-ip_F x} \tag{70}$$

with the amplitudes $\phi_\pm(x)$ changing slowly: $\nabla\phi_\pm \ll p_F\phi$. Let us consider a quasi-stationary state obeying the open boundary conditions

$$\phi_+(0) = \phi_-(L) = 0, \tag{71}$$

which correspond to the outward flow of current through the ends of the wire. The life time of such a state is inversely proportional to the outward current

$$t = \frac{\int dx |\Psi|^2}{v_F \left(|\phi_-(0)|^2 + |\phi_+(L)|^2\right)}.$$ (72)

The maximum delay time is achieved when the currents through both ends are equal $(\phi_-(-L/2) = \phi_+(L/2))$. Fixing normalization by

$$|\phi_-(0)| = |\phi_+(L)| = 1$$ (73)

we reduce Eq. (72) to the from

$$\frac{t\Delta}{\pi\hbar} = \int \frac{dx}{L} |\Psi|^2$$ (74)

which resembles the self-consistency condition from the paper [3], obtained there for arbitrary dimensions.

To achieve life times $t \gg \hbar/\Delta$ the wave function must grow towards the middle of the wire; assuming that the growth is exponential, $\Psi \sim \exp[(L/2 - |x|)/\xi]$, we obtain for the localization length of the quasi-stationary state

$$\xi = \frac{L}{\ln t\Delta}.$$ (75)

A typical random potential $\tilde{U}$ causes one-dimensional wave functions to be localized with $\xi \sim l$. The shorter localization length $\xi \ll l$ corresponds to life times longer than $\hbar\Delta^{-1}\exp(L/l)$ and can be achieved in the potential

$$U(x) = \tilde{U} + U_0 \cos(2p_F x).$$ (76)

with the additional $2p_F$-Fourier component having the amplitude

$$U_0 = \frac{2\hbar v_F}{\xi}.$$ (77)

The probability of this potential realization in given by the Gaussian distribution

$$\exp\left(-\frac{\pi\nu\tau}{2}\int U(x)^2 dx\right) \sim \exp\left(-\frac{l}{L}\ln^2 t\Delta\right)$$ (78)

and coincides with (69) with the correct numerical factor in the exponent. We therefore conclude that the electron states in one-dimensional chain with long life times are locked by the Bragg reflection and can be found with the probability proportional to that of potential fluctuation with the Bragg mirror of appropriate strength.

The authors are grateful to K. B. Efetov and A. D. Mirlin for constructive criticism on a part, regarding quantum traps and B. I. Shklovskii for discussions and criticism of the part, related to qualitative picture of trapping.

REFERENCES

1 K. B. Efetov, Adv. Phys.**32**,53 (1983). K. B. Efetov, Supersymmetry in Disorder in Chaos, Cambridge University Press, Cambridge 1997.

2 B. L. Altshuler and V. N. Prigodin, JETP Letters **41**, 43 (1988).

3 B. A. Muzykantskii and D. E. Khmelnitskii, Phys. Rev. **B 51**, 5480 (1995).

4 B. A. Muzykantskii and D. E. Khmelnitskii, JETP letters **62**, 76 (1995), Pis'ma Zh. Eksp Teor Fiz. vol. 62 p. 68-74 (1995). cond-mat 9506093.

5 B. A. Muzykantskii and D. E. Khmelnitskii cond-mat 9601045.

6 B. A. Muzykantskii and D. E. Khmelnitskii Phys. Reports **288**, 259 (1997)

7 B. L. Altshuler, V. E. Kravtsov, I. V. Lerner, Sov.Phys. - JETP Lett. **45**,199 (1987) ; Sov.Phys. - JETP **67**,799 (1988) ; *in* Mesoscopic Fhenomena in Solids ed. by B. L. Altshuler, P. A. Lee and R. A. Webb, North Holland p.449 (1991)

8 J. J. M. Verbaarschot, H. A. Weidenmuller and M. Zirnbauer, Phys.Rep.**129**, 367 (1985).

9 G. Eilenberger Z.Phys.,**44**,1288 (1968); K. D. Usadell, Phys. Rev. Lett. **25**, 507 (1970); A. I. Larkin and Yu. N. Ovchinnikov, Sov.Phys. - JETP,**46**, 155 (1977); Schmid A., *in* Nonequilibrium Superconductivity, Phonons and Kapitza Boundaries (Proceedings NATO ASI) ed. by K. E. Gray, Plenum (1981), p.423.

10 M. Yu. Kupriyanov and V. F. Lukichev, Sov. Phys. JETP **67**, 1163 (1988)

11 G. Eilenberger, Z.Phys.,**44**,1288 (1968);

12 E. Fradkin, Field Theories of Condensed Matter Systems, Addison-Wesley. 1991

13 A. M. M. Pruisken, *Nucl. Phys.* ,**B235**, 277,(1984)

14 B. L. Altshuler and B. I. Shklovskii, Zh.Exp.Theor.Phys., **91**, 220 (1986) [Soviet Phys. JETP **64**, 127 (1986)].

15 A. V. Andreev and B. L. Altshuler, Phys. Rev. Lett. **75**, 902, (1995).

16 A. V. Andreev, B. D. Simons, O. Agam and B. L. Altshuler, Phys. Rev. Lett. **75**, 902, (1996); Nucl. Phys. **75**, 902, (1996))

17 B. D. Simons, (see this book).

18 A. D. Mirlin, JETP Lett. **62**, 603 (1995)

19 B. I. Shklovskii, Private letter on 13 January 1996; I. E. Smolyarenko and B. L. Altshuler, Phys.Rev. **B 55**, 10451 (1997)

STOCHASTIC SCATTERING

H. A. Weidenmüller

Max–Planck–Institut für Kernphysik, Heidelberg, Germany

This is a summary of two lectures given at the NATO Advanced Study Institute. The description of stochastic scattering is based on two assumptions: (i) the scattering is dominated by resonances; (ii) the resonances have stochastic features. The formalism is laid out. Properties of the stochastic scattering matrix are displayed. Some results and applications are presented.

INTRODUCTION

Stochastic scattering is of central importance in several fields of Physics. Several years ago, I had the opportunity to explain and justify this claim in a series of lectures[1]. There is no point in reiterating these statements here. Suffice it to say that generically resonances occur in the scattering of nucleons by nuclei, in light scattering by molecules, in the passage of light through a medium with a randomly varying index of refraction, and in the passage of electrons through a disordered or chaotic mesoscopic sample. In all these cases, the resonances share stochastic properties. It is the aim of my lectures to describe a theoretical framework in which such stochastic resonance phenomena can be handled on a common footing. This framework is based on random matrix theory. Another important approach to some of the physical problems mentioned above uses the semiclassical approximation. This approach was outlined by other lecturers at the Advanced Study Institute and is not treated here.

The resonance formalism for the scattering matrix is introduced in the following section. Several important properties of the scattering matrix are displayed. Then, stochasticity is implemented into this formalism. The implementation makes use of concepts of random matrix theory. These concepts are not explained in any detail. I refer to a recent review[2]. In what follows, some aspects of averaging the intensity or calculating higher moments are mentioned. This is usually done using Efetov's supersymmetry technique[3]. The technique has been presented in some detail by other lecturers and is not reviewed here again. In the last Section, some applications and results of the S–matrix approach to stochastic scattering are mentioned.

These lectures comprise in condensed form results obtained by many people in the course of many years. Giving due credit to all authors in all relevant places would have changed the character of these notes, and would have resulted in my writing a review. Therefore, I mention here only some main contributors. These are the groups in
– Cambridge, Mass. (Feshbach, Kerman, Koonin);

Supersymmetry and Trace Formulae: Chaos and Disorder
Edited by Lerner *et al.*, Kluwer Academic / Plenum Publishers, New York, 1999

- Essen and Karlsruhe (Fyodorov, Haake, Lehmann, Mirlin, Saher, Sommers);
- Heidelberg (Agassi, Dittes, Elattari, Engelbrecht, Gossiaux, Guhr, Hackenbroich, Harney, Hartmann, Hofmann, Iida, Kagalovsky, Lewenkopf, Mantzouranis, Müller, Müller–Groeling, Nishioka, Pluhar, Richert, Verbaarschot, Weidenmüller, Yoshida, Zirnbauer, Zuk);
- Leiden (Beenakker, Brouwer, Frahm);
- Novosibirsk (Israilev, Savin, Sokolov, Zelevinsky);
- Mexico and Saclay (Mello, Pereira, Seligman, Pichard).

References to the works of these authors are to be found in the review papers and books cited at the end of this paper. I apologize to all authors not mentioned explicitly.

FORMULATION OF SCATTERING THEORY

Resonances are generically modelled in terms of the Breit–Wigner formula. Here, we need a generalization which allows for overlapping resonances (resonance spacing smaller than total resonance width) and, in some cases, for a "direct" coupling between channels. Formulas allowing for this generalization have been derived, in different form, by several authors, most of them working in atomic or nuclear reaction theory. I mention the names of Fano, Wigner, and Feshbach. In what follows, I will use the formulation given in Ref. 4 because it is most suitable to our purpose and has found wide application in the present context. The central idea goes back to Dirac's book on Quantum Mechanics. For simplicity, I consider time–reversal invariant systems throughout. In applications to atomic, molecular, or nuclear physics, all states considered carry the same quantum numbers (spin, parity).

Resonance Mechanisms

I know of two mechanisms which lead to the occurrence of resonances: Single–particle resonances caused by barrier penetration, and auto–ionizing states, a typical many–body phenomenon. Both must be studied in order to arrive at a general description of resonance phenomena.

(i) Single–particle resonances.

Such resonances occur when a particle is trapped within a potential well with a barrier separating it from the outside world. The particle may tunnel through the barrier. The width of the resonance is proportional to the tunneling probability. Single–particle states within a quantum dot or neutron states within the nuclear mean–field potential with an angular–momentum barrier serve as examples. The Hamiltonian may generically be modeled in the form proposed by Anderson:

$$H = \sum_{c,k} \epsilon_c(k) a_c^\dagger(k) a_c(k) + \sum_{\mu}^{N} E_\mu d_\mu^\dagger d_\mu + \sum_{ck\mu} \left[W_{c\mu}(k) a_c^\dagger(k) d_\mu + \text{h.c.} \right]. \tag{1}$$

Here, the N states within the barrier are labelled μ, with energies E_μ and creation and annihilation operators $d_\mu^\dagger$ and d_μ, respectively. The states beyond the barrier form a continuum, with wave number k, energy $\epsilon(k)$, and creation and annihilation operators $a^\dagger(k)$ and $a(k)$, respectively. The resonance may decay into one of several channels labelled c. An example is the coupling of a quantum dot to two leads, each supporting several transverse modes at the Fermi energy. The matrix elements $W_{c\mu}(k)$ are the tunneling amplitudes connecting the states μ with the channel states labelled (c, k). It is straihtforward to calculate the scattering matrix corresponding to H. In the case of a quantum dot, it may sometimes be necessary to add to H terms which account for the

344

Coulomb interaction of electrons within the dot. Then, the problem turns into a Kondo problem and can no longer be treated as a single–particle scattering problem unless a mean–field approximation is used. This is the limitation of the present approach.

(ii) Auto–Ionizing States.

We consider a mean–field approximation to a many–body problem in atomic, molecular or nuclear physics. The many–body wave functions are Slater determinants of single–particle states. We suppose that the mean field allows for both bound states and scattering states. For simplicity, we consider only Slater determinants where all particles or all particles but one occupy bound states. We label the first set $|\Phi_\mu >$, with Hartree-Fock energies E_μ, and the second set $|\chi_c(E) >$, where the energy E is a continuous variable, and where c labels the channels. These states are taken to be orthonormal, with a delta–function normalization in energy for the continuum states. We consider a bound state with energy E_μ larger than the threshold energy E_c in channel c. The residual two–body interaction which is not accounted for by the mean field, couples $|\Phi_\mu >$ and $|\chi_c(E) >$. This causes the state $|\Phi_\mu >$ to become instable against particle decay into channel c. The state $|\Phi_\mu >$ turns into a resonance. In atomic physics, such states are referred to as auto-ionizing states. In Ref. 4, the term bound states embedded in the continuum was used. I supress complications due to thresholds, see Ref. 4). Then, the Hamiltonian has the form

$$H = \sum_c \int_{E_c}^\infty dE\ E|\chi_c(E) >< \chi_c(E)| + \sum_\mu^N E_\mu\ |\Phi_\mu >< \Phi_\mu|$$
$$+ \sum_{c,\mu} \left[\int_{E_c}^\infty W_{c\mu}(E)|\chi_c(E) >< \Phi_\mu| + \text{h.c.} \right]. \qquad (2)$$

Here, W is the matrix element of the residual interaction. We note the close similarity of the Hamiltonians in Eqs. (1,2). Because of the normalization of the continuum wave functions, in both cases W has the dimension $E^{1/2}$.

In applications, it is useful to consider a further generalization. It is obtained by taking into account possible interactions between the N bound–state configurations $|\Phi_\mu >$. Then, with a self–explanatory notation, the Hamiltonian reads

$$H = \sum_c \int_{E_c}^\infty dE\ E|\chi_c(E) >< \chi_c(E)| + \sum_{\mu\nu}^N H_{\mu\nu}\ |\Phi_\mu >< \Phi_\nu|$$
$$+ \sum_{c,\mu} \left[\int_{E_c}^\infty W_{c\mu}(E)|\chi_c(E) >< \Phi_\mu| + \text{h.c.} \right]. \qquad (3)$$

Canonical form of the Scattering Matrix

Both models (1,3) yield identical forms of the scattering matrix. Before I give the general result, it is instructive to consider the case of a single channel and of a single resonance. Putting $E_c = 0$, $\mu = 0$ and omitting the label c, we have for the scattering function

$$S(E) = \exp(2i\delta) \left[1 - 2i\pi \frac{W_0^2(E)}{E - E_0 - F(E)} \right] \qquad (4)$$

where

$$F(E) = \mathcal{P} \int_0^\infty dE'\ \frac{W_0^2(E')}{E - E'} - i\pi W_0^2(E) \qquad (5)$$

and where $\mathcal{P}$ denotes the principal–value integral. The background phase shift is denoted by δ. Obviously, unitarity holds, $|S(E)|^2 = 1$. Moreover, Eq. (4) has the form of a Breit–Wigner resonance. In most applications, it is realistic to assume that $W_0(E)$ is smooth over the width of the resonance. Then,

$$S(E) = \exp(2i\delta)\left[1 - 2i\pi\frac{W_0^2}{E - E_0 + i\pi W_0^2}\right]. \tag{6}$$

The formula for the resonance width, $\Gamma = 2\pi W_0^2$, looks like the golden rule but actually is a non–perturbative result.

In the general case, we deal with the scattering matrix, a matrix of dimension Λ given by the number of open channels. It has the form

$$S_{ab}(E) = \exp(i\delta_a)\left[\delta_{ab} - 2i\pi\sum_{\mu\nu} W_{a\mu}\left(D^{-1}\right)_{\mu\nu} W_{\nu b}\right]\exp(i\delta_b). \tag{7}$$

The quantity D is a matrix in the space of the N bound states $|\mu>$ and is given by

$$D_{\mu\nu}(E) = E\delta_{\mu\nu} - H_{\mu\nu} + i\pi\sum_c W_{\mu c}W_{c\nu}, \tag{8}$$

with $H_{\mu\nu}$ introduced in Eq. (3). We have again suppressed the energy dependence of the matrix elements W. It is straightforward to check that the matrix S is unitary. Thus, Eqs. (7,8) constitute the unitary extension of the Breit–Wigner formula to N resonances. They apply both for isolated and for overlapping resonances. We note that as the energy increases and passes a threshold with energy E_c, one or several channels open, and the dimension Λ of the S–matrix increases.

For later purposes, we assume that the matrix elements W obey the relations

$$\sum_\mu W_{a\mu}(E)W_{\mu b}(E) = Nv_a^2\delta_{ab}. \tag{9}$$

This assumption is not as restrictive as it may seem, cf. the end of the present section.

It is sometimes necessary to consider a further generalization of the model in Eq. (3). In the model it is assumed that there is no dynamical coupling between the states $|\chi_c(E)>$ pertaining to different channels. This may not be realistic. The generalization consists in allowing for such "direct" reactions by replacing the term $\sum_c \int_{E_c}^\infty dE\, E|\chi_c(E)><\chi_c(E)|$ in Eq. (3) by the term $\sum_{cc'}\int_{E_c}^\infty dE\int_{E_{c'}}^\infty dE'|\chi_c(E)> V_{cc'}(E,E')<\chi_{c'}(E')|$. The S–matrix for this case is obtained by the following sequence of steps. (i) We disregard the matrix elements W and consider the formal solutions $|\Psi_c^\pm(E)>$ of the resulting channel–scattering problem. We denote the corresponding scattering matrix by $S_{ab}^{(0)}$. (ii) We take account of the presence of the matrix elements $W_{\mu c}(E) = <\mu|W|\chi_c(E)>$ and define the new elements $W_{\mu c}^{(0)}(E) = <\mu|W|\Psi_c^+(E)>$. Then, Eqs. (7,8) with W replaced by $W^{(0)}$, with δ_{ab} replaced by $S_{ab}^{(0)}$, and with $\delta_c = 0$, give the scattering matrix of the generalized model.

The resulting expression for S is very general but quite complex. However, a transformation exists which reduces this scattering matrix to the form given in Eqs. (7,8,9). The steps are the following. (i) Find the orthogonal transformation O in channel space which brings the symmetric matrix $S^{(0)}$ to diagonal form, $OS^{(0)}O^T = \exp(2i\delta)$. The symbol δ denotes a diagonal matrix the elements of which are the (real) eigenphases of $S^{(0)}$. Define $\overline{W} = OW^{(0)}$. (ii) Find a second orthogonal transformation $O^{(1)}$ in channel space which diagonalizes the symmetric bilinear form $\sum_\mu \overline{W}_{a\mu}\overline{W}_{\mu b}$ so that $\sum_{ab} O_{ca}^{(1)}\sum_\mu \overline{W}_{a\mu}\overline{W}_{\mu b}O_{bd}^{(1)} = Nv_c^2\delta_{cd}$. Define $W = O^{(1)}\overline{W}$. Define the unitary matrix $U = O^{(1)}\exp(-i\delta)O$ and write $S^{new} = USU^T$. The resulting S–matrix S^{new} has the desired form and all the properties used in Eqs. (7,8,9). This is why we refer to it as to the canonical form.

Whenever the matrix elements W in the model (3) do not obey the condition (9), step (ii) of this construction can be used to attain it. Hence, the condition (9) can always be imposed without loss of generality.

Properties of the Scattering Matrix

By construction, the S–matrix depends on the energy E and on the wave numbers $k_c = \sqrt{2m_c(E - E_c)}/\hbar$ in all the channels. Here, m_c is the reduced mass of the scattered particle in channel c. The dependence on k_c arises because the states $|\chi_c>$ depend on the k_c's, and so do the matrix elements $W_{\mu c}$. As a function of the complex variable E, the S–matrix therefore has branch points on the real E–axis located at the energies E_c. Thus, sections of the real E–axis separated by a branch point connect to different Riemann sheets. Poles of S are given by zeros of $\det(D)$. Causality requires these poles to occur below the real physical E–axis. But these poles have different locations on different sheets. The canonical simplification used in all applications of this formalism to stochastic scattering consists in omitting all channels with threshold energies in the energy interval of interest. Without this simplification, all methods of averaging fail. The simplification is justified if the omitted channels are weakly coupled to the system.

With this simplification, the S–matrix has N poles in the lower E–plane. For the model (7,8), it takes the form (I omit the background phase shifts)

$$S_{ab} = \delta_{ab} - 2i\pi \sum_{\mu}^{N} \frac{g_{a\mu} g_{\mu b}}{E - \mathcal{E}_\mu}, \tag{10}$$

where $\mathrm{Im}(\mathcal{E}_\mu) \leq 0$. It is tempting to use the form (10) as the starting point for further analytical work since it displays explicitly all N resonances. This, however, is not easy because of the constraints imposed by unitarity on the resonance parameters $g_{a\mu}$ and $\mathcal{E}_\nu$. For isolated resonances (resonance spacing large compared to resonance width) unitarity yields only the relation -2 $\mathrm{Im}(\mathcal{E}_\mu) = \Gamma_\mu = \sum_a |g_{\mu a}|^2$: The total width equals the sum of the partial widths over all open channels. But whenever the resonances overlap, the constraints imposed by untarity lead to a set of $\Lambda(\Lambda - 1)/2$ equations which connect all partial width amplitudes $g_{\mu a}$ with all resonance energies $\mathcal{E}_\nu$. This is why it is preferable in general to use the expressions (7,8) as starting point since these obey unitarity automatically.

These statements have a straightforward physical interpretation. An isolated pole does signify an isolated resonance visible as a local enhancement versus energy of the cross–section (or of a related intensity). Poles with spacings smaller than their distance from the real axis describe overlapping resonances. Such resonances jointly contribute to a perhaps very complicated behavior of the cross–section. In this case it is not possible to establish a one–to–one correspondence between a specific feature of the cross section and one of the poles of S. Therefore, it is not possible to attach physical meaning to any one of these poles individually.

A related point occurs when the coupling to the channels described by the matrix elements W becomes very large. In this case, it is convenient to bring the symmetric level matrix $\sum_c W_{\mu c} W_{c\nu}$ appearing in Eq. (8) to diagonal form. This can be accomplished by an orthogonal transformation denoted by $\overline{O}$. We denote the eigenvalues by w_μ^2 with $\mu = 1 \ldots N$. We note that for all μ, we have $w_\mu^2 \geq 0$. The form of the matrix $\sum_c W_{\mu c} W_{c\nu}$ implies that only Λ of its eigenvalues differ from zero. Typically, the number Λ of open channels is much smaller than the number N of levels. The transformed matrix $\overline{O}D\overline{O}^T$ has the form $E\delta_{\mu\nu} + i\pi w_\mu^2 \delta_{\mu\nu} - [\overline{O}H\overline{O}^T]_{\mu\nu}$. If the non–vanishing eigenvalues w_μ^2 are much bigger than the non–diagonal elements of $[\overline{O}H\overline{O}^T]_{\mu\nu}$, Λ poles of the S–matrix have a distance from the real axis which is much larger than that of the remaining $(N - \Lambda)$ ones. These far–away poles only change the overall phase of the scattering matrix but do not cause the same rapid energy–dependence as the $(N - \Lambda)$ close–lying ones. This makes it difficult to assign the same dynamical significance to these far–away poles as to the close–lying ones. Example: For a single open channel, it is straightforward to show that a far–away pole changes the background term from unity into (-1), corresponding to a shift of the background phase δ by π. But it is anyway difficult to pinpoint the dynamical significance of δ, let alone a change of δ by π. Hence, there is no ready dynamical interpretation of the far–away pole.

This discussion also suggests that the quantity $2\pi w_\mu^2$ is not a suitable measure for the strength of the coupling between channels and resonances. Indeed, the $(N - \Lambda)$ close–lying poles may cause very sharp resonances in spite of the non–vanishing eigenvalues w_μ^2 being very large. In the context of our stochastic model, a more suitable measure of the strength of the coupling between resonances and channels is given by the transmission coefficients. These are introduced in Eq. (12) below.

The eigenvalues of the symmetric and unitary scattering matrix have the form $\exp(i\delta)$. For a single channel and an isolated resonance, the phase shift δ increases by π as the energy increases over the width of the resonance. The extension of this statement to the many–channel case is: The sum of the Λ eigenphases increases by π over the width of an (isolated) resonance. The eigenphases obey the von Neumann–Wigner non–crossing theorem. Therefore, at an isolated resonance, each eigenphase increases on average only by π/Λ.

IMPLEMENTATION OF STOCHASTICITY

Two ways exist of implementing stochastic features into the scattering problem described by Eqs. (7,8). The first one will be referred to as the Random Hamiltonian approach. It stipulates that the Hamiltonian H appearing in Eq. (8) is replaced by a suitable ensemble of random Hermitean matrices. The second one takes the S–matrix itself as a member of an ensemble of random matrices, without using the detour of implementing stochasticity into the Hamiltonian. This is referred to as the Random S–Matrix approach. I will describe the advantages and weaknesses of both approaches. Whenever I refer to moments of S higher than the first, I mainly have in mind averages involving (powers of) both S and its complex conjugate $S^\star$, without saying so explicitly. It is these averages which are physically relevant.

Random Hamiltonian Approach

In the simplest case, the ensemble is taken to be the Gaussian Orthogonal Ensemble (GOE). Because of time–reversal invaiance, $H_{\mu\nu}$ can be chosen real and symmetric. The matrix elements $H_{\mu\nu}$ with $\mu \geq \nu$ are uncorrelated Gaussian random variables with mean value zero. The second moments, indicated by a bar, are given by $\overline{H_{\mu\nu}H_{\mu'\nu'}} = (\lambda^2/N)(\delta_{\mu\mu'}\delta_{\nu\nu'} + \delta_{\mu\nu'}\delta_{\nu\mu'})$. The measure in matrix space is given by the product of the differentials of the independent matrix elements. The parameter λ has the dimension of energy and determines the local mean level spacing $d \sim \lambda/N$. The fluctuations about the mean are predicted in a parameter–free fashion. This ensemble is invariant under orthogonal transformations of the basis of states $|\mu >, \mu = 1, \ldots, N$, hence its name.

The dimension N of the matrices is considered finite, but eventually the limit $N \to \infty$ is taken. The mean level density has the form of Wigner's semicircle and extends from -2λ to $+2\lambda$. This form is unrealistic. We are interested only in local fluctuations, however, which occur on the scale of d. On this scale, the global form of the spectrum is irrelevant. On the scale of d, it is also irrelevant whether we use the GOE or other, non–Gaussian ensembles. This is true with the proviso that we admit only ensembles for which the mean level density is confined to a finite interval. Such ensembles are obtained by replacing the Gaussian weight factor $\exp(-[\lambda^2/N]\mathrm{tr}(H^2))$ of the GOE by the more general form $\exp(-[\lambda^2/N]\mathrm{tr}(V(H)))$, with V a polynomial in H. For analytical work, the GOE plays a preferred role because it is the simplest of the lot.

Because of the orthogonal invariance of the GOE, all states $|\mu>$ are treated on the same footing. This assumption is realistic only if the time scale τ_{eq} for intrinsic equilibration among the states $|\mu>$ is small compared to the decay time τ_{dec} of the system due to coupling to the channels. In many–body systems, τ_{eq} is given by the strength of the residual interaction. In disordered systems, τ_{eq} is given by the inverse of the Thouless energy E_c. In classically chaotic systems, τ_{eq} is given by the period of the shortest periodic orbit. In each of these three cases, the condition $\tau_{\mathrm{eq}} \ll \tau_{\mathrm{dec}}$ is not always met. If it fails, the stochastic model for $H_{\mu\nu}$ must be altered. The typical alteration consists in using for $H_{\mu\nu}$ a random band matrix 2. By relinquishing orthogonal invariance, one enables the model to describe diffusive systems, or many-boy systems with a physically relevant time scale for equilibration.

Needless to say, the stochastic approach can never reproduce specific features of a given system. Rather, the ensemble average $\overline{F(E)}$ of an observable $F(E)$ is calculated. The result is compared with the "running" average $< F(E) >$ obtained by averaging a set of data points for $F(E)$ over some finite energy interval comprising a number N_0 of resonances in the system. It is important to establish the conditions under which $\overline{F(E)} =< F(E) >$. In analogy to statistical mechanics, this is referred to as the ergodic problem 5. While the equality $\overline{F(E)} =< F(E) >$ cannot be ascertained analytically for a given system, one asks whether it holds for almost all members of the ensemble. A necessary condition is

$$\overline{\left(\overline{F(E)} - < F(E) >\right)^2} = 0. \tag{11}$$

Under the assumptions that $\overline{F(E)}$ is independent of E and that the correlation function $C(E_1, E_2) = \overline{F(E_1)F(E_2)} - \overline{F(E_1)}\ \overline{F(E_2)}$ depends only on $(E_1 - E_2)$ so that $C(E_1, E_2) = C(E_1 - E_2)$, one easily finds that condition (11) reduces to $\lim_{|E|\to\infty}(EC(E)) = 0$. In several cases, this last condition has been shown to hold. If we take the necessary condition (11) to be also sufficient and consider the limit $N_0 \to \infty$, the equality of ensemble average and running average is then guaranteed for almost all members of the ensemble.

It may appear that the stochastic model defined by substituting for H the GOE and using Eqs. (7,8,9), is ill–defined. Indeed, the number of parameters (the $W_{\mu a}$'s and λ) is $\Lambda \times N + 1$ and diverges as $N \to \infty$. However, because of the orthogonal invariance of the GOE, all ensemble averages can depend only on orthogonal invariants constructed from these parameters, i.e. on the quantities $\sum_\mu W_{a\mu}W_{\mu b}$ and λ. Eq. (9) reduces this set to λ and Nv_a^2, $a = 1\ldots\Lambda$. The S–matrix is dimensionless. Hence, only the dimensionless parameters $x_a = \pi Nv_a^2/\lambda$, $a = 1\ldots\Lambda$ can play a role. The input for the stochastic model consists in the values of the average S–matrix elements $\overline{S}_{ab}$. Because of Eq. (9), $\overline{S}_{ab} = \delta_{ab}\overline{S}_{aa}$. This shows that the number Λ of parameters x_a of the model equals the number of input variables $\overline{S}_{aa}$: The stochastic model predicts S–matrix fluctuations uniquely in terms of average S–matrix elements. Actually and except for overall phase factors, the higher moments of S do not depend on the variables $\overline{S}_{aa}$, but only on the "sticking probabilities" or "transmission coefficients" T_a defined by

$$T_a = 1 - |\overline{S}_{aa}|^2. \tag{12}$$

Each T_a depends non–linearly on the parameter x_a and vanishes for both $x_a \to 0$ and $x_a \to \infty$. This statement clarifies the point raised in the second paragraph on p.348 and shows that T_a vanishes both for very weak and, perhaps surprisingly, also for very strong coupling of the channel a to the levels. This point is briefly taken up again below. The form of T_a has a simple physical interpretation: By ergodicity, we have $\overline{S} = < S >$. According to the uncertainty principle, the energy–averaged S–matrix $< S >$ describes the fast part of the reaction. The coefficients T_a measure the unitarity deficit of this part of S. In other words, the T_a's measure that part of the incident flux which is not scattered instantaneously but populates the long–lived resonant states. Fluctuation properties and correlation functions of S depend only on this part.

For non–invariant Hamiltonian ensembles, the stochastic model requires at least one additional parameter. It is physically equivalent to the equilibration time or, in quasi one–dimensional disordered systems, to the diffusion constant.

It is a strength of the Random Hamiltonian approach that it is capapble of predicting correlation functions, both versus energy and, in systems with charged particles, versus magnetic field strength. The reason is that in Eqs. (7,8), both energy E and Hamiltonian H appear explicitly. The dependence on magnetic field strength is easily incorporated into the latter. For instance, in the case of broken time–reversal invariance due to an external magnetic field, the Hamiltonian ensemble has the form proposed by Mehta and Pandey,

$$H = H^{\mathrm{GOE}} + i\sqrt{t/N}H^A. \tag{13}$$

Here, H^{GOE} stands for the GOE and H^A for the ensemble of Gaussian real antisymmetric matrices. The parameter t can be related to the strength of the magnetic field in a given system. For $t \to 0$, we have the GOE and, for $t \to \infty$, the GUE, i.e. the ensemble of Gaussian unitary matrices which corresponds to systems with broken time–reversal invariance.

The weakness of the Random Hamiltonian approach is that the actual calculation of moments of S higher than the second and of intensity correlation functions, is very difficult. This is because no way has yet been found to apply Mehta's method of orthogonal polynomials to this problem, and because the technical difficulties in using Efetov's supersymmetry technique grow with the number of S–matrix elements appearing in the observable over which the average is to be taken.

Random S–Matrix Approach

The earliest example is provided by Dyson's circular ensembles. (Dyson's papers and other interesting early work on random matrices can be found in Porter's book[6]). For time–reversal invariant systems with symmetric S–matrix, the Circular Orthogonal Ensemble (COE) is defined by writing $S = U\,U^T$, and by defining a measure (the Haar measure) for the ensemble of unitary matrices U. For this ensemble, $\overline{S} = 0$. An extension to the case where $\overline{S} \neq 0$ uses the maximum entropy approach with given values for $\overline{S}$ and yields the probability density $P(S, \overline{S})$. It has been shown that the COE is completely equivalent to the Random Hamiltonian appraoch with $\overline{S} = 0$ or $T_a = 1$ for all a. A similar equivalence has not been established yet for the case where $\overline{S} \neq 0$ although its existence is very likely.

The Random S–Matrix approach has found an important application to properties of quasi one–dimensional disordered conductors. There, one considers a division (transverse to the direction of the current) of the conductor into bins. In each bin, the transfer matrix connects the values of the wave function and its derivative on one side of the bin, with those of the other. The dimension of the transfer matrix is determined by the number M of transverse modes at the Fermi surface. Stochasticity is injected by using a maximum entropy principle for the transfer matrix. Letting the bin size go to zero, the number of bins go to infinity, and keeping the length of the sample fixed, one finds an equation of Fokker–Planck type for the probability density $P(\lambda, x)$. Here, $\lambda = (\lambda_1 \ldots \lambda_M)$ are real parameters related to the eigenvalues of the transfer matrix, and x denotes the length of the sample. Also in this case, the equivalence between the random transfer matrix approach and the Random Hamiltonian approach using a random band matrix, has been shown.

The strength of the Random S–Matrix approach lies in the fact that it deals directly with the quantity of interest, i.e. the scattering matrix, and avoids introducing the Hamiltonian. It yields an expression for $P(S, \overline{S})$ from which all moments of S can be obtained, although the practical calculation may be very hard.

The weakness of the Random S–Matrix approach is its inability so far to allow for the calculation of correlation functions, except for select cases. It is not clear how a dependence of energy and/or magnetic field can be incorporated into this approach in a physically correct fashion.

AVERAGES AND SUPERSYMMETRY

The only non–perturbative method presently available for the calculation of higher moments and correlation functions of the scattering matrix in the Random Hamiltonian approach is Efetov's supersymmetry method 3. For spectral correlation functions, this method yields for the effective action the well–known non–linear sigma model expression,

$$\mathcal{L} = \frac{\pi \nu}{8} \int \mathrm{d}^f r \ \mathrm{trg} \ [\mathcal{D} \ (\mathrm{grad} Q)^2 + 2i\omega L Q], \tag{14}$$

where we follow the notation of Ref. 2, and where $\mathcal{D}$ is the diffusion constant, ν the density of states per unit volume, ω the energy difference, and Q a graded matrix with $Q^2 = 1$. The space integration extends over f dimensions. In the application of this method to scattering problems, an additional term arises in the effective action. It is due to the coupling to the channels and has the form

$$\mathrm{tr}_\mu \mathrm{trg} \ln \left(1 + QL\frac{\pi}{\lambda} \sum_a W_{\mu a} W_{a\nu} \right). \tag{15}$$

Expanding the logarithm, using Eq. (9), and resumming, one obtains $\sum_a \mathrm{trg} \ln(1 + QLx_a)$, with x_a defined in subsection . Because of $Q^2 = 1$, this last expression is for any a invariant under the substitution $x_a \to 1/x_a$ and depends only on the transmission coefficients introduced in Eq. (12). Needless to say that in addition to this term, also the source terms change. They, too, depend on the transmission coefficients.

SELECTED RESULTS

It is impossible to give here a complete account of the results obtained in the framework of the stochastic approach to scattering. Suffice it to list a number of topics studied, and to describe some select results in more detail.

Topics studied are
- Compound–Nucleus Scattering.
- Precompound Reactions.
- Resonance Fluorescence in Methylglyoxal.
- Wigner Time Delay in Reactions.
- Pole Distribution of the Scattering Matrix.
- Conductance Properties of Chaotic Mesoscopic Billiards.
- Conductance Properties of Quasi One–Dimensional Disordered Mesoscopic Devices.
- Resonance Enhancement of Diffusive Light Scattering.

Most of this work is reviewed in Ref. 2. I turn to some select results.

Compound–Nucleus Scattering and Wigner Time Delay

Here, the stochastic model used is either that of the Random S–Matrix approach or that of the Random Hamiltonian approach with the GOE as input, as the case may be. With d the mean resonance spacing, and Γ the average total resonance width, we distinguish two regimes, defined by the ratio Γ/d. For isolated resonances with $d \gg \Gamma$, it is found that the positions of the resonances and their partial widths are uncorrelated random variables. The resonance positions obey Wigner–Dyson statistics, and the partial widths follow the Porter–Thomas distribution, a χ–square distribution with one degree of freedom. For strongly overlapping resonances where $\Gamma \gg d$, the elements of the scattering matrix S have a Gaussian distribution with mean value zero. More generally, S–matrix elements at different energies form a Gaussian random process with mean value zero. The only non–vanishing parts of the S–matrix autocorrelation function are given by (for simplicity, I consider only non–identical channels $a \neq b$)

$$\overline{S_{ab}(E)S^{\star}_{ab}(E+\omega)} = \frac{T_a T_b}{\sum_c T_c + i\pi\omega/d}. \tag{16}$$

In this regime which is usually referred to as the Ericson regime, the distribution of S–matrix elements is known completely, and so is therefore the cross–section autocorrelation function. Eq. (16) shows that this and all other correlation functions have Lorentzian form, with width $\Gamma = (d/(2\pi)) \sum_c T_c$. In the intermediate case $\Gamma \sim d$, only the S–matrix autocorrelation function (but not the higher moments of $S, S^{\star}$) are known. It is given in terms of a threefold integral involving Efetov's eigenvalues.

To study the Wigner–Smith time delay, the quantity $\hbar \, \mathrm{tr} \, \overline{(S^{\star}(E)(\mathrm{d}/\mathrm{d}E)S(E))}$ is calculated. It is found that for $\Gamma \gg d$, the average decay of compound–nucleus resonances is nearly exponential in time. Deviations from the exponential law occur for large times.

Resonance Enhancement of Diffusive Light Scattering

The scattering of light in a medium with a random index of refraction is usually described in terms of the scalar wave equation

$$[\Delta + \epsilon(\vec{r})k^2]\Phi = 0, \tag{17}$$

where k is the wave number. This avoids using the vector Maxwell's equations. The index ϵ of refraction varies randomly with position $\vec{r}$. When ϵ is written as the sum of an average part ϵ_0 and a fluctuating part $\delta\epsilon(\vec{r})$, the scalar wave equation (17) resembles the Schrödinger equation except that the analogue of the potential, i.e. the term $\delta\epsilon(\vec{r}) \, k^2$, depends on k. This dependence causes the two equations to generate different Ward identities. In spite of this difference, the non–linear sigma model derived for the scalar wave equation is identical to that for the Schrödinger equation. The entire difference resides in the source terms. This result is of interest because systems described by very different wave equations show identical spectral fluctuation properties, all given by random matrix theory. It is conceivable that all these wave equations lead to the same non–linear sigma model.

Diffusive scattering of light is often investigated experimentally by shining a Laser beam onto a cavity filled with a liquid into which some powder with a different index of refraction has been stirred. An interesting modification of this setup uses powder grains of such size that a Mie resonance occurs close to the frequency of the Laser light. This causes strong deviations from the standard relation $\mathcal{D} = (1/3)\, v\, l$ connecting the diffusion constant $\mathcal{D}$, the energy transport velocity v, and the elastic mean free path l. Typically, $\mathcal{D}$ is much smaller than predicted by this relation for quasi one–dimensional samples. The intuitive explanation is simple: The light waves are captured by the Mie resonances and spend more time passing through the sample. While the total transmitted intensity is not affected, all correlation functions are.

The observed effect has been understood in several ways. In the framework of stochastic scattering, it is accounted for as follows. Because of the Mie resonances in all the powder grains, the density of states for light waves in the cavity acquires a peak at the resonance energy. In the effective Lagrangean of the non–linear sigma model (14), this results in a modification of the density of states factor ν. However, the product $\nu\mathcal{D}$ remains unchanged. This guarantees that the transmitted intensity is unchanged. As a result, the frequency dependence of the rhs of Eq. (14) is altered. This modification accounts for the experimental findings.

The results described in this last Section are not contained in any of the references given below. They have been obtained in collaboration with B. Elattari and V. Kagalovsky. The work has been submitted to Europhysics Letters, and to Physical Review E.

ACKNOWLEDGMENTS

This work was done while I was visiting the Isaac Newton Institute for the Mathematical Sciences, Cambridge, UK. I am grateful to the Institute for partial support.

REFERENCES

1.   H. A. Weidenmüller, in: Lecture Notes in Physics **411** Springer–Verlag (1992) 121.
2.   T. Guhr, A. Müller–Groeling, and H. A. Weidenmüller, Physics Reports, **249** (1998) 189.
3.   K. B. Efetov, *Supersymmetry in Disorder and Chaos*, Cambridge University Press (1997).
4.   C. Mahaux and H. A. Weidenmüller, *Shell–Model Approach to Nuclear Reactions*, North–Holland Publishing Company, Amsterdam (1969).
5.   T. A. Brody, J. Flores, J. B. French, P. A. Mello, A. Pandey, and S. S. M. Wong, Rev. Mod. Phys **53** (1981) 385.
6.   C. E. Porter, *Statistical Theories of Spectra: Fluctuations*, Academic Press, New York (1965).

$H=xp$ AND THE RIEMANN ZEROS

M.V. Berry[1] and J. P. Keating[2]

[1]H. H.Wills Physics Laboratory,
Tyndall Avenue, Bristol BS8 1TL, U.K.
[2]School of Mathematics, University Walk,
Bristol BS8 1TW, U.K.,
and Basic Research Institute in the Mathematical Sciences,
Hewlett-Packard Laboratories Bristol, Filton Road,
Stoke Gifford, Bristol BS12 6QZ, U.K.

1. INTRODUCTION

The Riemann hypothesis [1,2] states that the complex zeros of $\zeta(s)$ lie on the critical line Re $s=1/2$; that is, the nonimaginary solutions E_n of

$$\zeta\left(\tfrac{1}{2}+iE_n\right) = 0 \tag{1}$$

are all real. Here we will present some evidence that the E_n are energy levels, that is eigenvalues of a hermitian quantum operator (the 'Riemann operator'), associated with the classical hamiltonian

$$H_{\text{cl}}(x,p) = xp \tag{2}$$

where x is the (one-dimensional) position coordinate and p the conjugate momentum. This is frankly speculative, because large gaps remain that are not merely technical.

We were prompted to write this paper by Connes[3] (see also [4]) who has devised a hermitian operator whose eigenvalues are the Riemann zeros that lie on the line. His operator is the transfer (Perron-Frobenius) operator of a *classical* transformation. Such classical operators (Liouville operators times i in the case of flows) formally resemble quantum hamiltonians, but usually have very complicated non-discrete spectra and singular eigenfunctions. Connes gets a discrete spectrum by making the operator act on an abstract space where the primes appearing in the Euler product for $\zeta(s)$ are built in; the space is constructed from collections of p-adic numbers (adeles) and the associated units (ideles). The proof of the Riemann hypothesis is thus reduced to the proof of a certain classical trace formula. His construction succeeds in overcoming certain difficulties [5] associated with the quantum analogy. Nevertheless, our hope for some time has been that a simpler characterisation of the Riemann operator can be found along the lines we explore here; perhaps it will be equivalent to that of Connes.

Supersymmetry and Trace Formulae: Chaos and Disorder
Edited by Lerner *et al.*, Kluwer Academic / Plenum Publishers, New York, 1999

We start by listing and briefly commenting on the properties of the Riemann operator that are suggested by the quantum analogy (see also [5-7]). We will call the operator H.

a. H has a classical counterpart (the 'Riemann dynamics'), corresponding to a hamiltonian flow, or a symplectic transformation, on a phase space. This is based on a formal resemblance between the von Mangoldt expansion [2] for the logarithm of the Euler product for $\zeta(1/2+iE)$ and the semiclassical expansion [8, 9] of quantum traces as sums over classical periodic orbits, and also on statistical evidence (see property b below).

b. The Riemann dynamics is chaotic, that is unstable and bounded. This is based on the observation that the local statistics of the E_n are those of the eigenvalues of random matrices [10-14], and the connection of random-matrix statistics with the quantum mechanics of classically chaotic motion [6, 15-17]. Long-range correlations, between distant E_n, differ from those predicted by random-matrix theory [17, 18], and the differences are characteristic of quantum systems that have classical counterparts.

c. The Riemann dynamics does not have time-reversal symmetry. This is because the statistics of the E_n are locally those of the gaussian unitary ensemble of complex hermitian random matrices [19, 20], rather than the gaussian orthogonal ensemble of real matrices (which corresponds to systems with time-reversal symmetry). Related to this is the recent discovery [21, 22] of modified statistics of the low zeros for the ensemble of Dirichlet L-functions, associated with a symplectic structure.

d. The Riemann dynamics is homogenously unstable. This is suggested by the fact that the instability (Lyapunov) exponents of the periodic orbits are all unity, which follows from the exponential decay of the terms in the von Mangoldt formula: $q^{-m/2}=\exp(-T_{m,q}/2)$, where $T_{m,q}$ is the orbit period defined in (3).

e. The classical periodic orbits of the Riemann dynamics have periods that are independent of energy E, and given by multiples of logarithms of prime numbers, that is

$$T_{m,q} = m\log q \quad (m = 1,2,\ldots; q \text{ prime}) \tag{3}$$

and the associated actions are

$$S_{m,q} = Em\log q \tag{4}$$

This follows from the form of the oscillatory terms in the analogy with the semiclassical trace formula. In terms of symbolic dynamics, the Riemann dynamics is peculiar, and resembles Chinese: each primitive orbit is labelled by its own symbol (the prime q) in contrast to the usual situation where periodic orbits can be represented as words made of letters in a finite alphabet.

f. The Maslov phases associated with the orbits are also peculiar: they are all π. This follows [5] from the negative signs of the terms in the von Mangoldt formula. The result appears paradoxical in view of the relation between these phases and the winding numbers of the stable and unstable manifolds associated with periodic orbits [23], but finds an explanation in the scheme of Connes[3].

g. The Riemann dynamics possesses complex periodic orbits (instantons) whose periods are

$$T_{\text{complex},m} = im\pi \tag{5}$$

This is suggested by the small exponentials arising in the large-E asymptotics of $\zeta(1/2+iE)$, associated with the high orders of the Riemann-Siegel expansion [24] and the high orders of the Stirling series for the gamma functions representing the smooth part of the counting function for the zeros [25].

h. For the Riemann operator, leading-order semiclassical mechanics is exact: $\zeta(1/2+iE)$ is a product over classical periodic orbits, without corrections (as in the case of the Selberg trace formula [26] for geodesic motion on surfaces of constant negative curvature).

i. The Riemann dynamics is quasi-one-dimensional. There are two indications of this. First, the number of zeros less than E increases as $E\log E$ (see (9) below); for a d-dimensional scaling system, with energy parameter $\alpha(E)$ proportional to $1/\hbar$, the number of energy levels increases as $\alpha(E)^d$. Second, the presence of the factor $q^{-m/2}$ in the von Mangoldt formula, rather than the determinant in the more general Gutzwiller formula, suggests that there is a single expanding direction and no contracting direction.

We note immediately that the system (2) represents the simplest form of instability, because it has a hyperbolic point at $x=0$, $p=0$. Hamilton's equations, and their solutions, are

$$\dot{x} = x, \quad \text{i.e. } x(t) = x(0)\exp(t); \quad \dot{p} = -p, \quad \text{i.e. } p(t) = p(0)\exp(-t) \tag{6}$$

Thus classical evolution is simply dilation in x (that is, multiplication) and contraction in p, and the stretching exponent is unity, so that the instability is indeed homogeneous as required. In addition, xp does not possess time-reversal symmetry, because it is not invariant under $p \rightarrow -p$; more fundamentally, reversal of velocity $\dot{x}$ for fixed x does not lead to retracing of the orbit, for the simple reason that $\dot{x}$ is tied to x and so cannot be reversed independently. Furthermore, dynamics generated by xp is semiclassically exact.

2. SEMICLASSICAL LEVEL COUNTING

For any classically bound hamiltonian $H_{cl}(x, p)$ in one dimension, the number of quantum levels with energy less than E, the counting function, is

$$N(E) = A(E)/h + \ldots \tag{7}$$

where ... denotes higher-order terms in Planck's constant $\hbar = h/2\pi$ and $A(E)$ is the phase-space area under the contour $H_{cl}(x, p) = E$. With (1) there is the immediate problem that the classical motion is not bound, so that A is infinite. Therefore the system must be regularized. The simplest regularization is to truncate x and p by extending the Planck cell with sides l_x, l_p and area $h = l_x l_p$ as in figure 1, so that A becomes the finite area indicated, which depends on $\hbar$. This makes the system quantum-mechanically quasi-one-dimensional. We cannot justify the regularization procedure, but note the analogy between this phase-space regularization and the fact that the hyperbola billiard in two dimensions is classically unbound but has a discrete quantum spectrum [27-29]. Thus

$$N(E) = \frac{1}{h}\left[E \int_{l_x}^{E/l_p} \frac{dx}{x} - l_p\left(\frac{E}{l_p} - l_x\right) \right] + \ldots$$

$$= \frac{E}{h}\left(\log\left(\frac{E}{h}\right) - 1 \right) + 1 + \ldots \tag{8}$$

The constant (sub-leading) term should be modified by the Maslov phase. To guess this, we note that for a closed phase-space contour which turns by -2π, the extra term in the counting function is $+1/2$ (cf. the harmonic oscillator with frequency ω, for which $N(E)=\mathrm{Int}(E/\hbar\omega+1/2)$). For (1) the turn is $+\pi/2$, so the extra term should be $-1/8$. Choosing units such that $\hbar=1$, (equivalent to replacing E by $\hbar E$), we now obtain

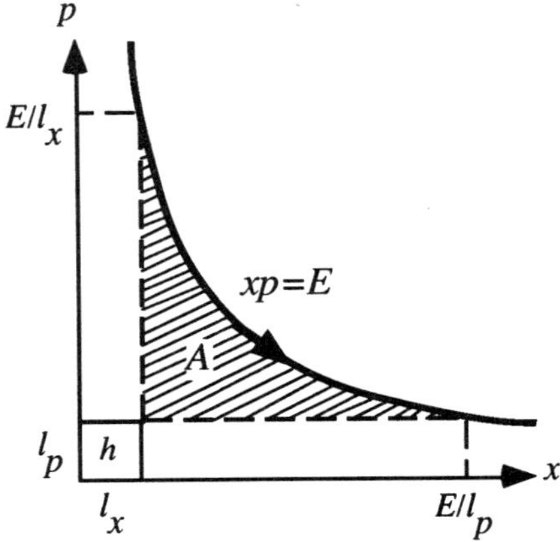

Figure 1. Phase space for $H_{cl}=xp$, with cutoffs l_x and l_p for semiclassical regularization.

$$N(E) = \frac{E}{2\pi}\left(\log\left(\frac{E}{2\pi}\right) - 1\right) + \tfrac{7}{8} + \ldots \tag{9}$$

This is precisely the asymptotic form of the smoothed counting function for the Riemann zeros, namely

$$N_{sm}(E) = \theta(E)/\pi + 1 \tag{10}$$

where

$$\theta(E) = -\frac{E}{2}\log\pi + \mathrm{Im}\log\Gamma\left(\tfrac{1}{4} + \tfrac{1}{2}iE\right) \tag{11}$$

correct to terms that do not vanish as $E\to\infty$. This is unlikely to be a coincidence.

3. CONFIGURATION AND MOMENTUM EIGENFUNCTIONS

The simplest formally hermitian operator corresponding to (1) is

$$H = \tfrac{1}{2}(xp + px) = -i\hbar\left(x\frac{\mathrm{d}}{\mathrm{d}x} + \tfrac{1}{2}\right) \tag{12}$$

The formal eigenfunctions, satisfing

$$H\psi_E(x) = E\psi_E(x) \tag{13}$$

are

$$\psi_E(x) = \frac{A}{x^{1/2 - iE/\hbar}} \tag{14}$$

We note the appearance of the power x^{-s} appearing in the Dirichlet series for $\zeta(s)$ (as integer^{-s}) and the Euler product (as prime^{-s}).

The corresponding momentum eigenfunction is

$$\phi_E(p) = \frac{1}{\sqrt{h}} \int\limits_{-\infty}^{\infty} dx \, \psi_E(x) \exp(-ipx/\hbar) \tag{15}$$

To evaluate this, we must choose a continuation of $\psi_E(x)$ across the singularity at $x=0$. The simplest choice is that the eigenfunctions are even. Then

$$\phi_E(p) = \frac{A}{|p|^{1/2+iE/\hbar} \sqrt{2\pi}} \int\limits_{-\infty}^{\infty} \frac{du}{|u|^{1/2-iE/\hbar}} \exp(-iu)$$

$$= \frac{A}{|p|^{1/2+iE/\hbar}} \left(\frac{h}{\pi}\right)^{iE/\hbar} \frac{\Gamma\left(\frac{1}{4} + \frac{iE}{2\hbar}\right)}{\Gamma\left(\frac{1}{4} - \frac{iE}{2\hbar}\right)} \tag{16}$$

where the reflection and duplication formulas for the gamma function have been used. Noting the similarity with (11), and writing x and p in terms of the sides of the Planck cell, we find

$$\psi_E(x) = \frac{\exp\{-i\theta(E/\hbar)\}}{\sqrt{l_x}\,|x/l_x|^{1/2-iE/\hbar}}$$

$$\phi_E(p) = \frac{\exp\{i\theta(E/\hbar)\}}{\sqrt{l_p}\,|p/l_p|^{1/2+iE/\hbar}} = \sqrt{\frac{l_x}{l_p}}\left[\psi_E\left(pl_x/l_p\right)\right]^* \tag{17}$$

Henceforth we set $l_x=l_p=\sqrt{(2\pi)}$, i.e. $\hbar=1$.

The meaning of this symmetry is that position and momentum eigenfunctions are each other's time-reverse (cf. figure 1): thus we have a physical interpretation of the function $\theta(E)$ at the heart of the functional equation for $\zeta(s)$ [30], which states that the function

$$Z(E) \equiv \exp\{i\theta(E)\}\zeta(1/2+iE) \tag{18}$$

is even, and from which it follows that $Z(E)$ is real when E is real.

If the hamiltonian had not been symmetrized to make it formally hermitian, we would not have obtained the results (14) and (17), containing the same combination $1/2+iE$ as occurs in $\zeta(s)$ on the critical line.

Equation (17) is a special case of a more general relation between the position and momentum eigenfunctions, obtained by allowing the multipliers A in (14) to be different for positive and negative x. The relation is

$$\psi_E(x) = \frac{\exp\{-i\theta(E)\}}{|x/2\pi|^{1/2-iE}}\left[A_+\Theta(x) + A_-\Theta(-x)\right]$$

$$\phi_E(p) = \frac{\exp\{i\theta(E)\}}{|p/2\pi|^{1/2+iE}}\left[B_+\Theta(x) + B_-\Theta(-x)\right] \tag{19}$$

where Θ denotes the unit step function, and the x and p multipliers are related by

$$\begin{pmatrix} B_+ \\ B_- \end{pmatrix} = \mathsf{M}\begin{pmatrix} A_+ \\ A_- \end{pmatrix} \tag{20a}$$

where M is the unitary matrix

$$M = \frac{(\exp(E\pi) - i)}{2\cosh(E\pi)} \begin{pmatrix} 1 & i\exp(-E\pi) \\ i\exp(-E\pi) & 1 \end{pmatrix} \tag{20b}$$

The unitarity of M implies

$$|A_+|^2 + |A_-|^2 = |B_+|^2 + |B_-|^2 \tag{21}$$

- a relation that can be interpreted in terms of phase-space currents: the total x current flowing out from the origin equals the total p current flowing into the origin (figure 2a). These currents J_x and J_p are the expectation values of the local velocity operators:

$$J_x(x) = \tfrac{1}{2} \int_{-\infty}^{\infty} dx' \psi^*(x') \left[\delta(x - x') \frac{\partial H}{\partial p} + \frac{\partial H}{\partial p} \delta(x - x') \right] \psi(x') \tag{22a}$$

$$= x|\psi(x)|^2 = 2\pi \left[|A_+|^2 \Theta(x) - |A_-|^2 \Theta(x) \right]$$

and similarly

$$J_p(p) = 2\pi \left[-|B_+|^2 \Theta(p) + |B_-|^2 \Theta(p) \right] \tag{22b}$$

Of course, the hamiltonian xp is simply a canonically rotated form of the upturned harmonic oscillator p^2-x^2, which is in turn a complexified version of the usual harmonic oscillator p^2+x^2. These connections have been noted before. Nonnemacher and Voros[31] calculate the Wigner function corresponding to xp, in a study of eigenstates near hyperbolic points. Bhaduri et al [32] and Khare [33] show that the density of scattering states of the second-order operator p^2-x^2 resembles $d\theta(E)/dE$ (the difference is a constant); Armitage[34] studies the fourth-order combinations $(p^2{\pm}x^2)^2$; and Okubo[35] studies the two-dimensional hamiltonian $p_x^2-x^2-p_y^2+y^2$. The first-order operator xp is the simplest representative of this class, with the monomials (14) avoiding the complications of the parabolic cylinder eigenfunctions of p^2-x^2. Indeed, it is possible give a very simple derivation of transmission and reflection from the potential $-x^2$, using a quantum canonical transformation of the states (14) with appropriate connections across the singularity at $x=0$.

4. x AND p CONNECTIONS

It would be desirable to replace the semiclassical regularization of xp in (section 2) with a quantum boundary condition that would generate a discrete spectrum in a natural way. We do not know how to do this, but offer some remarks.

It is likely that x and $-x$ should be identified, and also p and $-p$, as in (17). This is suggested by a consideration of the complex periodic orbits of xp. With imaginary time the orbits (6) are periodic (as in an ordinary, rather than an inverted, harmonic oscillator), but the periods are wrong: $2i\pi m$, rather than $i\pi m$ as required by property g in section 1. Note however that after odd multiples of the time $i\pi$, x evolves to $-x$ and p to $-p$, so that identification of $\pm x$ and $\pm p$, as shown in figure 2, produces the required complex periods.

Even after these identifications, the system remains open. Ways to close it, and thereby force the spectrum to be discrete, are suggested by the *symmetries* of xp. Using these, we will try to incorporate the fact that the eigenstates of a hermitian operator with

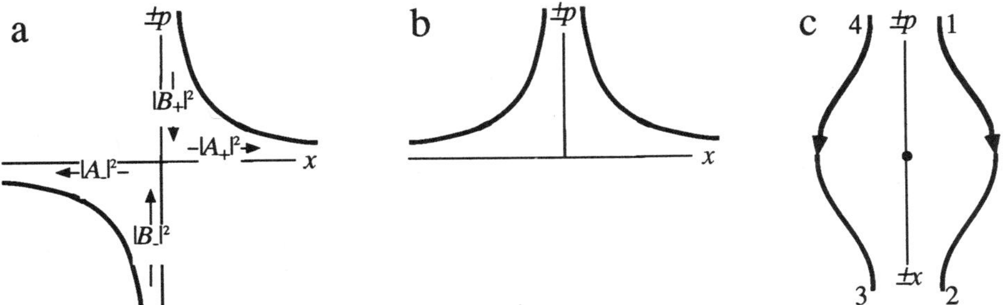

Figure 2. (a) phase space for $H_{cl}=xp$, showing positive-energy contours; (b) with p and $-p$ identified; (c) with x and $-x$ identified.

symmetry can be written as superpositions of solutions of the eigenequation acted on by operations in the symmetry group, with each solution in the superposition multiplied by the appropriate group character.

An obvious symmetry of (2) is that xp is invariant under *dilations*:

$$x \rightarrow Kx, \quad p \rightarrow p/K \tag{23}$$

From (6), K corresponds to evolution after time $\log K$. This implies that the operator (11), corresponding to xp, generates dilations, in the same way that the momentum operator generates translations; the following sequence of transformations makes this obvious:

$$f(Kx) = f\left(\exp\{\log K + \log x\}\right) = \exp\left\{(\log K)\frac{d}{d\log x}\right\}f(x)$$

$$= \exp\left\{(\log K)x\frac{d}{dx}\right\}f(x) = K^{x\frac{d}{dx}}f(x) = \frac{1}{K^{\frac{1}{2}-iH}}f(x) \tag{24}$$

It is tempting to choose the integer dilations $K=m$, corresponding to evolution times $\log m$, and the characters unity, and write

$$\psi_E(x) \rightarrow \sum_{m=1}^{\infty}\psi_E(mx) = \frac{\text{constant}}{|x|^{1/2-iE}} \times \sum_{m=1}^{\infty}\frac{1}{m^{1/2-iE}} = \frac{\text{constant}}{|x|^{1/2-iE}}\zeta\left(\tfrac{1}{2}-iE\right) \tag{25}$$

A requirement that this must vanish would, if interpreted as an eigencondition, yield the Riemann zeros E_n as eigenvalues. However, we see no reason to impose this requirement, and moreover the set of dilations $K=m$ does not form a group (the inverse multiplications $1/m$ are missing). Even worse, putting $E=E_n$ in (25) destroys the 'eigenfunction' by making it vanish for all x.

Another possibility, closely related to the ideas of Connes[3], is to use not all integers but the group of integers under multiplication (mod k). This would have two advantages. First, the group involves only integer and not fractional dilations. Second, it opens the possibility that the group characters[36] can appear as multipliers in the Dirichlet series for ζ, thereby yielding the zeros of the different Dirichlet L-functions (which are all conjectured to have zeros in the line $\text{Re}s=1/2$) as eigenvalues of different self-adjoint extensions of xp.

Another way to close the system xp could be to connect the asymptotic positions with the asymptotic momenta. Then the current flowing out at $x=\pm\infty$ would be re-injected at $p=\pm\infty$. We envisage two such connections. Referring to figure 2c, we could connect 1 with 2 and 3 with 4, thus preserving the separation of the original quadrants (opposite in figure 2a)

361

and yielding a phase space with cylindrical topology; or we can connect 1 with 3 and 2 with 4, thereby connecting the quadrants (as does the matrix M in (20)) and yielding a phase space with Möbius topology.

A way to accomplish this connection is suggested by the fact that the dilations K under which xp is invariant need not be constant but can be any function of xp. The choice $K=h/(xp)$ yields the canonical transformation

$$x \to x_1 = \frac{h}{p}, \quad p \to p_1 = \frac{xp^2}{h} \tag{26}$$

Because of the h-dependence, we call this *quantum exchange* (the simpler canonical exchange $x \to p$, $p \to -x$ does not leave xp invariant). Under quantum exchange, the hyperbolas $xp=E$ are of course invariant curves; $E=h$ is a curve of fixed points, with points on the curves $E<h$ mapping towards increasing x, and points on the curves $E>h$ mapping towards decreasing x. To see the corresponding transformation of quantum states, we represent these in Hilbert space as kets $|\psi\rangle$, and employ the notations

$$\langle x|\psi\rangle \equiv \psi(x), \qquad \langle p|\psi\rangle \equiv \phi(p),$$
$$\langle x_1|\psi\rangle \equiv \psi_1(x_1), \quad \langle p_1|\psi\rangle \equiv \phi_1(p_1) \tag{27}$$

Then the quantum implementation of exchange is

$$\psi_1(x_1) = \frac{\sqrt{h}}{|x_1|}\phi\left(\frac{h}{|x_1|}\right) \tag{28}$$

(obviously, this would preserve normalization of the state).

Superposition of states related by this exchange operation gives, after using (17)

$$\psi_E(x) \to \psi_E(x) + \frac{\sqrt{h}}{x}\phi_E\left(\frac{h}{x}\right) = \frac{2\cos\{\theta(E)\}}{\left|x/\sqrt{h}\right|^{1/2-iE}} \tag{29}$$

If we could argue that this should vanish, the resulting 'quantization condition' would be vanishing of the first term of the main sum of the Riemann-Siegel formula [2]. This would give zeros with the correct density, and it is tempting to regard it as arising from some hamiltonian operator, and seek to generate the true Riemann operator from a series of corrections.

However, this hope is unlikely to be realised, because (29) possesses complex zeros and so cannot be associated naively with a hermitian operator. To demonstrate the existence of these zeros off the critical line, we write (29) in the following form, which follows from (11):

$$g(s) = f(s) + f(1-s) = 0,$$
$$\text{where } f(s) = \frac{\pi^{s/2}}{\Gamma(s/2)} \tag{30}$$

This has zeros for s real, that is E imaginary, at

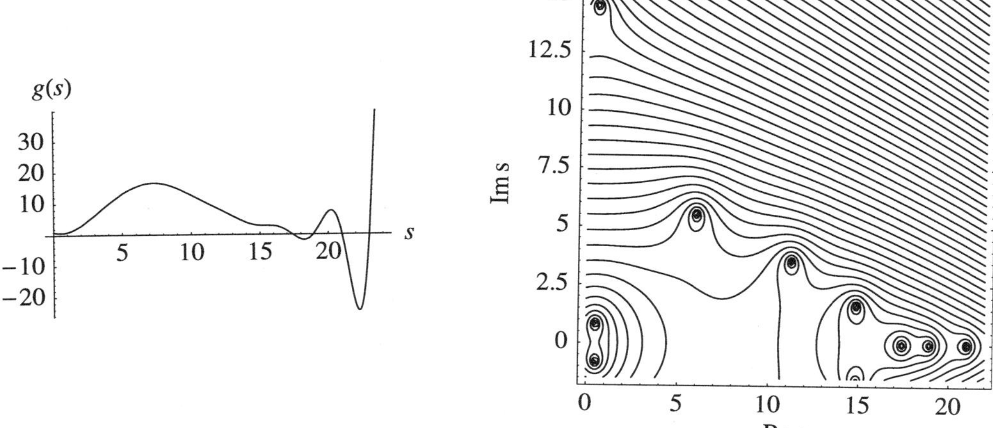

Figure 3. The function $g(s)$, whose zeros are the same as the first term of the Riemann-Siegel main sum, (a) on the real s axis, (b) contours of $|g(s)|$, showing zeros enclosed by loops.

$$s \approx 2m + 1 + \frac{(-1)^m}{\sqrt{\pi m}}\left(\frac{\pi e}{m}\right)^{2m}, \quad (m = 8,9,\ldots) \tag{31}$$

The first few are illustrated in figure 3a. There are also at least three zeros, shown in figure 3b, between the real axis and the critical line. Also visible in figure 3b are zeros of (29) that are on the line but do not correspond to Riemann zeros; these lie near $s=1/2\pm0.82i$. Similar arguments establish the existence of zeros off the line when more terms of the Riemann-Siegel main sum are added to (29). It follows that the vanishing of (29) is not a boundary condition corresponding to a hermitian operator.

Combining the two symmetries - integer dilation and quantum exchange - suggests the 'boundary condition'

$$\sum_{m=1}^{\infty} \psi_E(mx) + \frac{\sqrt{h}}{x}\sum_{k=1}^{\infty}\phi_E(mh/x) = \frac{2}{\left|x/\sqrt{h}\right|^{1/2-iE}}Z(E) = 0 \tag{32}$$

Using (24), this can be put into the intriguing form (with operators temporarily denoted by carats for clarity)

$$\langle x|\hat{x}^{\frac{1}{2}}\zeta\left(\tfrac{1}{2}-i\hat{H}\right)|\psi_E\rangle + \langle p|\hat{p}^{\frac{1}{2}}\zeta\left(\tfrac{1}{2}+i\hat{H}\right)|\psi_E\rangle = 0 \quad (xp = h) \tag{33}$$

These conditions do generate the Riemann zeros, but we see no way to interpret either of them geometrically. (With a $-$ sign, (33) would be an identity.)

5. GAUSS MAP AS A BOUNDARY CONDITION?

The relations (25) and (32) are multiplicative: they involve formal eigenfunctions of xp at multiples of any given x and the associated momentum h/x. A different relation, combining multiplication with addition, connects values of x related by the Gauss map that generates continued fractions. This involves the generalized transfer operator [37], and the requirement that this operator has eigenvalue unity [38]. The eigencondition corresponding to this map is

$$\sum_{n=1}^{\infty} \frac{1}{(n+x)^s} f_s\left(\frac{1}{n+x}\right) = f_s(x) \tag{34}$$

This was introduced[38] as a quantum map giving discrete eigenvalues associated with the modular domain. The natural exponent is then $s=1+iE$, with E real, so that factors in the sum are 'semiclassical' complexified square roots of the jacobians in the corresponding 'classical' transfer operator, which would have $s=2$. However, the Riemann zeros follow from the different association $s=1/2+iE$, with E real, This is semiclassically mysterious because the factors in (34) now correspond to $1/4$ powers of the classical jacobians. The argument, explained to us by Bogomolny (personal communication) is as follows.

Define

$$h_s(x) \equiv f_s(x-1) \tag{35}$$

and seek a formal eigenfunction of (34) in the form

$$h_s(x) = \sum_{m=1}^{\infty} \sum_{k=1}^{\infty} \frac{1}{(mx+k)^s} \tag{36}$$

where $s=1/2+iE$. The condition (34) becomes

$$
\begin{aligned}
h_s(x) &= \sum_{n=1}^{\infty} \frac{1}{(n+x-1)^s} h_s\left(\frac{n+x}{n+x-1}\right) \\
&= \sum_{n=1}^{\infty} \sum_{m=1}^{\infty} \sum_{k=0}^{\infty} \frac{1}{\left[x(m+k)+n(m+k)-k\right]^s} - \sum_{n=1}^{\infty} \sum_{m=1}^{\infty} \frac{1}{\left[m(x+n)\right]^s} \\
&= \sum_{n=1}^{\infty} \sum_{l=1}^{\infty} \sum_{k=0}^{l-1} \frac{1}{\left[xl+nl-k\right]^s} - \zeta(s) \sum_{n=1}^{\infty} \frac{1}{(x+n)^s} \\
&= h_s(x) - \zeta(s) \sum_{n=1}^{\infty} \frac{1}{(x+n)^s}
\end{aligned}
\tag{37}
$$

where the last equality follows after noticing that the sums over n and k can be conflated into a single sum over the variable $nl-k$. Obviously the condition is satisfied whenever $1/2+iE$ is a Riemann zero.

It might seem that the eigenfunctions disappear at the Riemann zeros even without the condition (34), because the summation in (36) can be taken over multiples of coprime (m,k) pairs and $\zeta(s)$ extracted as a factor:

$$h_s(x) = \sum_{l=1}^{\infty} \sum_{(m,k)=1} \frac{1}{\left[l(mx+k)\right]^s}$$
$$= \zeta(s) \sum_{(m,k)=1} \frac{1}{(mx+k)^s}$$

(38)

If this were a valid objection, the solution (36) would be empty. But it is not valid, because (36) is a formal expression that does not converge when E is real. It can be analytically continued onto the critical line, for example by

$$h_s(x) = \frac{1}{\Gamma(s)\left[\exp(2\pi i s)-1\right]} \int_C dt \frac{t^{s-1}}{\left[\exp(t)-1\right]\left[\exp(xt)-1\right]}$$

(39)

where C is a loop starting and ending at $t=+\infty$, encircling the origin positively and enclosing no other poles. The integral, when evaluated numerically, does not vanish at the Riemann zeros (figure 4).

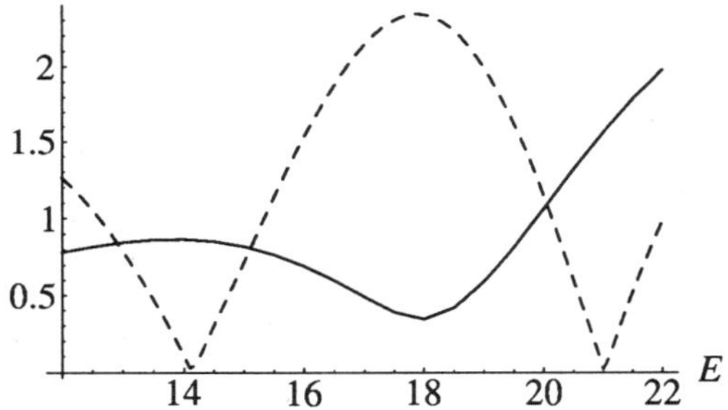

Figure 4. Absolute values of $\zeta(s)$ (dashed curve) and the Gauss map eigenfunction $h_S(0.5)$ (computed from the integral (35)) (full curve) on the critical line $s=1/2+iE$.

Now, $h_S(x)$ in (36) can be regarded as a sum of eigenfunctions (14) of xp, evaluated at positions $x+k/m$ that differ by rational numbers. Therefore the condition (34) might be interpretable as a boundary condition, relating the eigenfunction at each such position to its pre-images under the Gauss map. We do not know how to pursue this suggestion.

6. Concluding remarks

We have presented several tantalizing connections between xp and $\zeta(s)$. However, it is clear that more is required to transform our hints and guesses into an unambiguous and satisfactory construction of the Riemann operator. There are two principal unsolved problems.

First, the space on which xp acts is not known. Somehow the plane must be sewn up into a region that makes the dynamics bound, at least quantally. We have speculated that this might involve connecting x and p, or relating multiples of x or rational translations of x (to see how complicated this can get, compare the space obtained by identifying x with nx for real x and all integers n with the familiar circle obtained by identifying x with $n+x$). Perhaps the required space is a quantum graph [39, 40], with xp acting on bonds between vertices (one difficulty is that xp does not sit naturally on a general graph).

Second, we do not know how to associate the primes with the periodic orbits of the Riemann dynamics.

In terms of the properties listed in the Introduction, xp is consistent with a, part of b (xp dynamics is unstable but not bound), c, d, g, h and i. Concerning e, the appearance of times that are logarithms of integers begins to be plausible in view of the association between dilation and evolution, but primes do not appear in any obvious way. We have no explanation of f.

There are probably more connections between xp and $\zeta(s)$. Our hope is that in writing this paper we will stimulate others to uncover them.

ACKNOWLEDGEMENT

We thank Professor Alain Connes for giving us a detailed explanation of his ideas about the Riemann zeros, and Professors Zeev Rudnick and Peter Sarnak for several helpful conversations.

REFERENCES

1. Riemann, B. Über die Anzahl der Primzahlen unter einer gegebenen Grosse, *Monatsberichte d. Preuss. Akad. d. Wissens., Berlin* 671-680 (1959).
2. Edwards, H.M. *Riemann's Zeta Function* (Academic Press, New York and London, 1974).
3. Connes, A. Formule de trace en géométrie non-commutative et hypothèse de Riemann, *C.R. Acad. Sci. Paris* **323**, 1231-1236 (1996).
4. Goldfeld, D. A spectral interpretation of Weil's explicit formula, *Springer Math. Notes* **1593**, 137-152 (1994).
5. Berry, M.V. in *Quantum chaos and statistical nuclear physics* (eds. Seligman, T.H. & Nishioka, H.) 1-17 (1986).
6. Berry, M.V. Quantum chaology (The Bakerian Lecture), *Proc. Roy. Soc. Lond.* **A413**, 183-198 (1987).
7. Keating, J.P. in *Quantum Chaos* (eds. Casati, G., Guarneri, I. & Smilansky, U.) 145-185 (North-Holland, Amsterdam, 1993).
8. Gutzwiller, M.C. Periodic orbits and classical quantization conditions, *J. Math. Phys.* **12**, 343-358 (1971).
9. Gutzwiller, M.C. *Chaos in classical and quantum mechanics* (Springer, New York, 1990).
10. Montgomery, H.L. *Proc. Symp. Pure Math.* **24**, 181-193 (1973).
11. Odlyzko, A.M. Zeros of zeta functions, *Math. of Comp.* **48**, 273-308 (1987).
12. Rudnick, Z. & Sarnak, P. Zeros of principal L-functions and random-matrix theory, *Duke Math. J.* **81**, 269-322 (1996).
13. Bogomolny, E.B. & Keating, J.P. Random matrix theory and the Riemann zeros I: three- and four-point correlations, *Nonlinearity* **8**, 1115-1131 (1995).
14. Bogomolny, E.B. & Keating, J.P. Random-matrix theory and the Riemann zeros II: n-point correlations, *Nonlinearity* **9**, 911-935 (1996).
15. Bohigas, O. & Giannoni, M.J. *Chaotic Motion and Random-matrix Theories* 1-1-99 (Springer-Verlag, 1984).
16. Berry, M.V. Semiclassical theory of spectral rigidity, *Proc. Roy. Soc. Lond.* **A400**, 229-251 (1985).
17. Bogomolny, E.B. & Keating, J.P. Gutzwiller's trace formula and spectral statistics: beyond the diagonal approximation, *Phys. Rev. Lett.* **77**, 1472-1475 (1996).
18. Berry, M.V. Semiclassical formula for the number variance of the Riemann zeros, *Nonlinearity* **1**, 399-407 (1988).
19. Seligman, T.H., Verbaarschot, J.J.M. & Zirnbauer, M.R. Spectral fluctuation properties of Hamiltonian systems: the transition region between order and chaos, *J. Phys. A* **18**, 2751-2770 (1985).
20. Berry, M.V. & Robnik, M. Statistics of energy levels without time-reversal symmetry: Aharonov-Bohm chaotic billiards, *J. Phys. A* **19**, 649-668 (1986).
21. Katz, N. & Sarnak, P. Zeros of zeta functions, their spacings and their spectral nature, *preprint* (1997).

22. Sarnak, P. Quantum chaos, symmetry and zeta functions, *Curr. Dev. Math.* 84-115 (1997).

23. Robbins, J.M. Maslov indices in the Gutzwiller trace formula, *Nonlinearity* **4**, 343-363 (1991).

24. Berry, M.V. The Riemann-Siegel formula for the zeta function: high orders and remainders, *Proc.Roy.Soc.Lond.* **A450**, 439 - 462 (1995).

25. Berry, M.V. & Howls, C.J. High orders of the Weyl expansion for quantum billiards: resurgence of periodic orbits, and the Stokes phenomenon, *Proc. Roy. Soc. Lond.* **A447**, 527-555 (1994).

26. Balazs, N.L. & Voros, A. Chaos on the pseudosphere, *Physics Reports* **143**, 109-240 (1986).

27. Sieber, M. & Steiner, F. Classical and quantum mechanics of a strongly chaotic billiard, *Physica* **D44**, 248-266 (1990).

28. Simon, B. Nonclassical eigenvalue asymptotics, *J. Funct. Anal.* **53**, 84-98 (1983).

29. Sieber, M. & Steiner, F. Quantization of chaos, *Phys. Rev. Lett.* **67**, 1941-1944 (1991).

30. Titchmarsh, E.C. *The theory of the Riemann zeta-function* (Clarendon Press, Oxford, 1986).

31. Nonnemacher, S. & Voros, A. Eigenstate structures around a hyperbolic point, *J. Phys. A.* **30**, 295-315 (1997).

32. Bhaduri, R.K., Khare, A. & Law, J. Phase of the Riemann zeta function and the inverted harmonic oscillator, *Phys. Rev.* **E52**, 486- (1995).

33. Khare, A. The phase of the Riemann zeta function, *Pramana* **48**, 537-553 (1997).

34. Armitage, J.V. in *Number theory and dynamical systems* (eds. Dodson, M.M. & Vickers, J.A.G.) 153-172 (University Press, Cambridge, 1989).

35. Okubo, S. Lorentz-invariant hamiltonian and Riemann hypothesis, *Preprint from University of Rochester* (1997).

36. Apostol, T.M. *Introduction to analytic number theory* (Springer-Verlag, New York, 1976).

37. Mayer, D.H. On the Thermodynamic Formalism for the Gauss Map, *Commun. Math. Phys* **130**, 311-333 (1990).

38. Bogomolny, E.B. & Carioli, M. Quantum maps from transfer operators, *Physica* **D67**, 88-112 (1993).

39. Jakobson, D., Miller, S., Rivin, I. & Rudnick, Z. Eigenvalue spacings for regular graphs, *preprint* (1996).

40. Kottos, T. & Smilansky, U. Quantum chaos on graphs, *preprint from Weizmann Institute, Israel* (1997).

PARAMETRIC RANDOM MATRICES: STATIC AND DYNAMIC APPLICATIONS

Michael Wilkinson

Department of Physics and Applied Physics, John Anderson Building, University of Strathclyde, Glasgow, G4 0NG, U.K.

Abstract The random matrix approach can be extended to parameter dependent Hamiltonians, by hypothesising that parametric random matrices provide a good description of these systems. The hypothesis can be applied to both 'static' properties (pertaining to solutions of the time independent Schrödinger equation), and 'dynamic' properties, where the parameter is time dependent, and the objective is to understand properties of the time dependent Schrödinger equation.

This article reviews the formulation of parametric random matrix models, and briefly discusses their use to estimate densities of singularities in the energy levels, which are important in asymptotic theories for the dynamical properties. The existing knowledge about the dynamic properties of parametric random matrix models is then discussed.

1. Introduction

The central hypothesis of random matrix theory can be summarised as follows: there is a broad category of systems, which have been termed 'complex quantum systems', for which many of the statistical properties of energy levels are indistinguishable from those of typical samples from ensembles (probability distributions) of random matrices of large dimension. The properties of the random matrix model are said to be 'universal'. This idea was introduced by Wigner, who took an interest in the statistical description of nuclear energy levels. The theory was greatly extended by Dyson, Mehta, Porter and others, who identified the most natural random matrix models, and showed that many of the relevant statistical properties of these matrices could be calculated analytically. Most of the important early papers on random matrix theory are collected in [1], and the principal mathematical results on random matrix theory are described in [2]. The physical systems for which random matrix theories provide a good model are mainly of three types: systems with chaotic classical motion and a small number of degrees of freedom, metallic systems with diffusive electron motion, and many-body systems (treated without the use of one-body effective approximations).

The random matrix approach has been extended in several directions. In this article I will be concerned with the extension of the random matrix approach to deal with systems where the Hamiltonian depends upon a parameter, which will usually be

Supersymmetry and Trace Formulae: Chaos and Disorder
Edited by Lerner *et al.*, Kluwer Academic / Plenum Publishers, New York, 1999

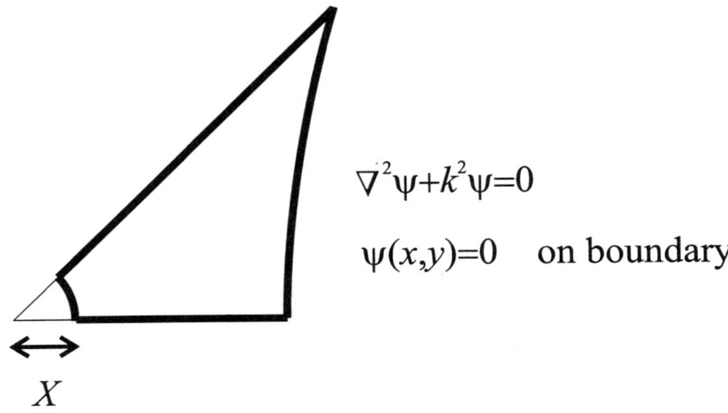

Figure 1. An example of a parametric Hamiltonian, to which the random matrix model is applicable: the system is a 'quantum billiard', with boundary defined by the bold line. The shape depends on a parameter X, which is the radius of a circular section of the boundary.

denoted by X: an example would be a parameter describing the change in shape of a 'quantum billiard', such as that illustrated in figure 1. An immediate consequence of having a parameter in the Hamiltonian is that the energy levels E_n become functions of the parameter: the functions $E_n(X)$ are shown for this quantum billiard in figure 2(a) (the plot is taken from [3], which gives a full specification of the system). The plot exhibits some striking features, for example the energy level curves never cross, but do approach each other at events termed avoided crossings, which have a characteristic structure. It is natural to ask whether these structures have the same degree of universality as the spectrum, and whether random matrix methods can give a description of these structures. Figure 2(b) is a plot of the eigenvalues of a matrix with elements which are randomly generated but which depend smoothly on the parameter X: it clearly forms the basis for a reasonable model for statistics of the parameter dependence of the energy levels. In section 2 I will describe the appropriate random matrix models for parametric dependence of energy levels and other quantities, and review some of the results obtained in the statistical characterisation of these pictures.

Sometimes parametric dependence is of interest because an experimenter is able to vary some variable smoothly and observe the results, but there are deeper reasons why parametric dependences are of importance. There are many problems where the parameter would be a function of time, $X(t)$. An example is illustrated in figure 3(a): a particle is trapped inside an enclosure, the shape of which can be varied by moving a piston, which has coordinate $X(t)$. The energy levels of the particle may resemble those plotted in figure 2: it is natural to ask whether the form of the energy level curves have any implications for the manner in which the particle responds to the motion of the piston. In some regimes the response is equivalent to that of a classical particle, and the detailed structure of the energy levels is irrelevant. In other regimes however the dynamics of the trapped particle does depend on the structure of the energy levels, and an understanding of parametric random matrix theory is essential to understanding this dynamical problem.

The piston model illustrated in figure 3(a) is a prototype for a rather broad class of problems, in which $X(t)$ represents the coordinate of a degree of freedom which is weakly perturbed by the remainder of the systems: for example, if the piston is very massive compared to the particle, the coordinate of the piston can be regarded as having a specified time dependence. Below I list some systems for which this a model:

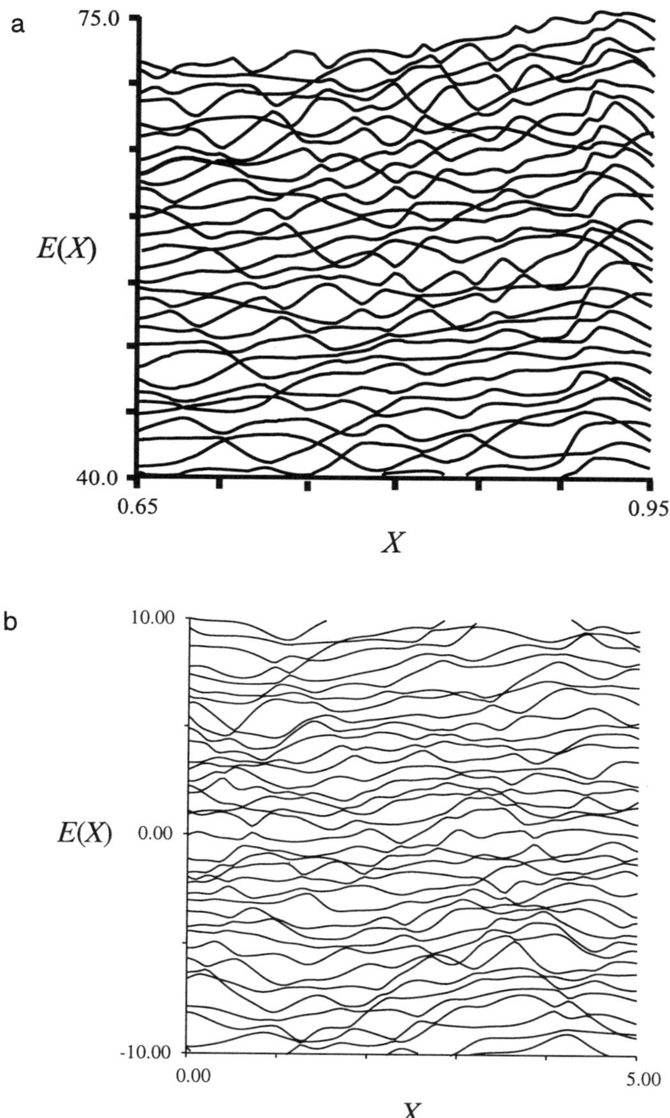

Figure 2. Energy levels E_n plotted as a function of a parameter X: a) Quantum billiard illustrated in figure 1, with its area scaled so that the mean level spacing is unity. b) The parameter dependent random matrix model, (1.1), with dimension $N = 40$.

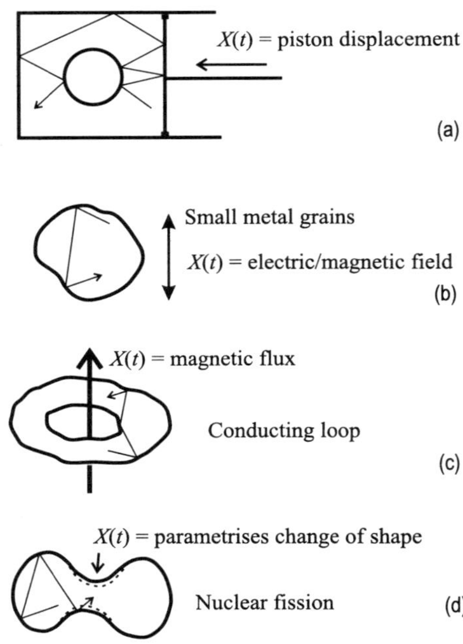

Figure 3. Some examples of applications of parametric Hamiltonians: a) The 'piston model'. b) Electron in a small metal particle interacting with an electromagnetic field. c) Electron in a conducting loop threaded by a flux. d) Nucleon in the collective model for nuclear fission.

a) The coordinate $X(t)$ might represent the motion of the nuclei in a molecule, and the Hamiltonian $\hat{H}(X)$ describes the dynamics of the electronic degrees of freedom for a fixed configuration of the nuclei. Because the nuclei are thousands of times heavier than the electrons, the motion of the coordinate $X(t)$ is slowly varying. The electronic system can respond to the motion of the nuclei by undergoing radiationless transitions [4].

b) The coordinate $X(t)$ might represent an electric or magnetic field perturbing a small metallic particle (figure 3(b)). The electrons respond to the perturbation by polarisation, which may be in phase with the perturbation, or out of phase. The out of phase response corresponds to the dissipative absorption of radiation [5].

c) If the coordinate $X(t)$ represents the magnetic flux passing through a conducting loop, then rate $\dot{X}$ of variation of the flux equals the electromotive force acting around the loop (figure 3(c)). This e.m.f. causes a current to flow around the loop. This is a very convenient model for electrical conduction, because it is possible to discuss DC conduction without either dealing with an open system, or considering the limit of AC conduction where the frequency $\omega \to 0$ [6].

d) In the collective model of the nucleus, the nucleons are considered to have almost independent motion in inside a droplet of 'nuclear fluid' [7]. The coordinate $X(t)$ could represent a parametrisation of a family of shapes for the nuclear droplet, whilst it undergoes processes such as fission, or a heavy-ion collision (figure 3(d)). The variation of the shape of the droplet parametrised by $X(t)$ results in excitation of the gas of nucleons: these excitations are irreversible, and can be thought of as representing viscosity of the nuclear fluid [8].

Random matrix models could give a useful statistical description of any of these processes. In all of the examples above, the Hamiltonian $H(X)$ describes a large number of fermions. Low lying excitations of many fermion systems can often be described by weakly interacting fermionic quasiparticles: for many purposes, the system can be modeled adequately by a set of non-interacting fermions [9]. Most treatments of electrons in metals and of the collective model of the nucleus make use of this simplification, and I will emphasise the non-interacting case where $H(X)$ represents the single particle Hamiltonian for a gas of non-interacting fermions.

Another significant reason to consider parametric dependences is that there can be singular structures in the parameter space: for example if parameters are varied the energy levels of a system may become degenerate. The singularities are often of physical significance, for example a symmetry class or topological index associated with a state of a system can change if the state becomes degenerate [10].

I will therefore discuss work based upon two distinct extensions of the random matrix hypothesis, relating respectively to 'static' and 'dynamic' applications of parametric Hamiltonians

I) *Static parametric random matrices*

Random matrices can model the parametric dependence of energy levels, wavefunctions and matrix elements. It will be argued that a natural parametrisation of the standard random matrix ensembles is

$$\hat{H}(X) = \cos(X)\hat{H}_1 + \sin(X)\hat{H}_2 \tag{1.1}$$

where $\hat{H}_1$ and $\hat{H}_2$ are two independent realisations of the random matrix ensemble, with large dimension N. This parametrisation, and its application as a model for parametric energy levels of complex systems, was originally proposed in [3].

II) *Dynamic parametric random matrix theory*

Random matrix models can give a 'universal' description of the dynamics of a parametric Hamiltonian under a specified time dependence of the parameter $X(t)$. Specifically, many aspects of the dynamical behaviour of a complex quantum system under time dependent variation of a parameter $X(t)$ are independent of the detailed structure of the energy levels and wavefunctions, and are the same as those of the solutions of the Schrödinger equation

$$i\hbar\frac{\partial|\psi\rangle}{\partial t} = \hat{H}(X(t))|\psi\rangle \tag{1.2}$$

where the Hamiltonian is given by (1.1), and $X(t)$ has some specified time dependence. These properties should, after suitable scaling, be identical to the dynamics of (1.1) under a time dependent variation of X. This hypothesis was originally proposed in [11].

Section 2 will review the necessary background in random matrix theory, before discussing the properties of the parametric model (1.1) and some selected results on the statistics of parametric energy levels. Many results have been obtained in this area using a variety of methods, which are well represented by earlier articles in this volume. I will concentrate on results related to singularities of energy levels, which are most relevant to the discussion of the dynamical behaviour in later sections. Section 3 discusses the dynamics of complex systems, the concept of energy diffusion and its relation to dissipation, and introduces the dimensionless parameters which define the different regimes of the dynamics. Section 4 discusses the different theoretical approaches which have been applied to give quantum theories for energy diffusion. Finally section 5 compares perturbative and non-perturbative approaches to the dynamics of complex systems, and discusses an open problem.

Finally, a note about notation. Many different probability distributions will be introduced: in order to avoid naming many different probability density functions, the probability element for the variable X lying in the interval $[X, X + dX]$ will usually be written $dP = P[X]dX$.

2. Static parametric random matrix theory

2.1 Parametric random matrix models

Complex quantum systems are expected to have universal spectral statistics. The spectral statistics are most conveniently calculated using suitably chosen random matrix ensembles, of which the gaussian invariant ensembles are most convenient. These are ensembles of hermitian matrices, with either real, complex or quarternion elements: these cases are referred to as the orthogonal, unitary, and symplectic ensembles respectively [12]. The number of real numbers required to represent each element is β.

The components of the matrix be written $H_{ij} = \sum_{k=0}^{\beta-1} H_{ij,k}\mathbf{e}_k$, where $\{ij\}$ are the indices of the matrix element, $\mathbf{e}_0 = 1$, $\mathbf{e}_1 = i$, and the other $\mathbf{e}_k$ are the other two bases for quarternion arithmetic. Each component of each independent element $H_{ij,k}$ of the gaussian matrices are independently gaussian distributed. The ensemble is therefore specified by the probability density in the space of matrices, with volume element dV defined in the natural way:

$$dP = P(\{H_{ij,k}\})dV, \quad dV = \prod_{i,j \geq i} dH_{ij,0} \prod_{i,j > i} \prod_{k=1}^{\beta-1} dH_{ij,k}$$

$$P(\{H_{ij,k}\}) = A \prod_{i,j > i} \exp[-a\,H_{ij,0}^2] \prod_{k=0}^{\beta-1} \prod_{i,j \geq i} \exp[-a\,H_{ij,k}^2] = A \exp(-a\,\mathrm{tr}\hat{H}^2) \tag{2.1}$$

for some constants a (arbitrary, often taken to be 1 or $\beta/2$) and A (chosen to normalise the distribution). This expression has some very convenient features: both the volume element and the probability density $P(\{H_{ij,k}\})$ are invariant under appropriate classes of unitary transformations of the matrix $\hat{H}$, with elements of the same type as those of the matrix. In the case $\beta = 2$ these are general unitary transformations, when $\beta = 1$ these are orthogonal transformations, and when $\beta = 4$ these are termed symplectic transformations [2,12]. The three ensembles are termed GUE, GOE and GSE respectively, and in this review they will be referred to generically as the GXE.

The joint probability distribution of eigenvalues of the gaussian invariant matrix ensembles can be obtained analytically, and many marginal distributions pertaining to finite numbers of levels have been calculated [2].

There are natural extensions of the gaussian invariant ensembles to parameter dependent models. The simplest one is (1.1). This model has some very convenient features [3]:

a) Both $\hat{H}$ and $d\hat{H}/dX$ are statistically stationary (the statistical properties are independent of X). This is a simple consequence of the fact that $\hat{H}_1$ and $\hat{H}_2$ are statistically independent.

b) All of the elements of the matrices $\hat{H}$ and $d\hat{H}/dX$ are statistically independent of each other.

These two results are used to demonstrate important properties of the matrix elements

$$\partial_X H_{nm} = \left\langle \phi_n \left| \frac{\partial \hat{H}}{\partial X} \right| \phi_m \right\rangle \tag{2.2}$$

where the states $|\phi_n\rangle$ are the eigenfunctions of $\hat{H}(X)$, i.e.

$$\hat{H}(X)|\phi_n(X)\rangle = E_n(X)|\phi_n(X)\rangle . \tag{2.3}$$

In the case where the parameter X is time-dependent, this basis is time-dependent, and is termed the *adiabatic basis*. These matrix elements are particularly significant for further development, because they occur in perturbation theory. They have the following properties

c) The matrix $\partial_X \tilde{H} = \{\partial_X H_{nm}\}$ has statistics representative of a gaussian invariant ensemble. This is a consequence of the fact that because of property (b), the transformation which diagonalises $\hat{H}$ has no significance when applied to $d\hat{H}/dX$: the matrix $\partial_X \tilde{H}$ is therefore the result of applying an arbitrary unitary transformation to $d\hat{H}/dX$, and unitary invariance of the ensemble implies that this matrix has GXE distributed elements.

d) The matrix elements $\partial_X H_{nm}$ are statistically independent of the energy levels E_n. This also follows from property b) above.

This method for constructing parametric random matrices can be extended in various ways. For numerical work, the fact that the model (1.1) is periodic in X may be inconvenient. This can be overcome by using generating the matrix elements by smoothing white noise:

$$H_{ij,k}(X) = \int_{-\infty}^{\infty} dX' \, g(X - X')W_{ij,k}(X') \tag{2.4}$$

where $g(x)$ is a smooth function, decaying to zero as $|x| \to \pm\infty$, and the $W_{ij,k}$ are independent white noise functions: the $W_{ij,0}$ is symmetric in i,j, and for $k \neq 0$ they are antisymmetric. For $j > i$, $j' > i'$ they satisfy

$$\langle W_{ij,k}(X)\rangle = 0, \quad \langle W_{ij,k}(X)W_{i'j',k'}(X')\rangle = \delta(X - X')\delta_{ii'}\delta_{jj'}\delta_{kk'} . \tag{2.5}$$

This version of the model was introduced in [13]. Also, there are contexts in which it is necessary to consider more than one parameter. A suitable extension of (1.1) containing $\mathcal{N}$ parameters $\mathbf{X} = (X_1, .., X_\mathcal{N})$ is

$$\hat{H}(\mathbf{X}) = \sum_{i=1}^{\mathcal{N}} \cos(X_i)\hat{H}_{2i-1} + \sin(X_i)\hat{H}_{2i} . \tag{2.6}$$

2.2 Application of parametric random matrix models

Figure 1 compared the energy levels of a 'real' complex quantum system with those of the parametric random matrix models. In order for the random matrix eigenvalues to simulate the energy levels of the real system, three different scaling transformations must be applied:

a) The energy levels are scaled so that the mean level spacings are the same. This requires an estimate for the mean level spacing, Δ, for the physical system of interest (The level spacing for the random matrix ensemble is given by the Wigner 'semicircle law', discussed in [1,2]). The level spacing Δ is estimated using the density of states:

$$\rho(E) = \sum_n \delta_\epsilon(E - E_n) \tag{2.7}$$

where $\delta_\epsilon(x)$ is a 'smoothed delta function', meaning a function with unit weight, with its support localised in a region of width ϵ centred on $x = 0$ (an example is the gaussian function $\delta_\epsilon(x) = \exp(-x^2/2\epsilon^2)/\sqrt{2\pi}\epsilon$. The mean level spacing at energy E is estimated as $\Delta = 1/\rho(E)$. The smoothing parameter ϵ is chosen to be large compared to Δ, but small compared to all other important energy scales in the problem: this separation of scales is necessary for the application of random matrix models, and is achievable for all of the types of complex systems discussed in the introduction. If the system has a meaningful semiclassical limit ($\hbar \to 0$, holding all classically defined quantities fixed), the density of states can be estimated using the Weyl formula [14]

$$\rho(E) \sim \frac{1}{(2\pi\hbar)^d} \int d\alpha\, \delta(E - H(\alpha)) = \frac{1}{(2\pi\hbar)^d} \frac{d\Omega}{dE} \tag{2.8}$$

where α stands for the canonical coordinates of phase space $(\mathbf{q}, \mathbf{p})$, and the measure is $d\alpha = \prod_{i=1}^d dq_i dp_i$ (d being the number of degrees of freedom); $\Omega(E)$ is the volume of phase space enclosed by the surface $H(\alpha) = E$.

b) The parameters must be scaled so that sensitivity of the energy levels to a perturbation are in agreement. Both the mean and the variance of the slopes E_n/dX of the energy levels must be adjusted so that they are in agreement. The mean slope at energy E is defined in a similar way to the density of states:

$$\left\langle \frac{dE_n}{dX} \right\rangle = \frac{1}{\rho(E)} \sum_n \frac{dE_n}{dX} \delta_\epsilon(E - E_n) \ . \tag{2.9}$$

For semiclassical systems, this can be estimated from classical objects using a formula analogous to the Weyl estimate [15]

$$\left\langle \frac{dE_n}{dX} \right\rangle \sim \frac{1}{\Omega'(E)} \int d\alpha\, \frac{\partial H}{\partial X}(\alpha)\, \delta(E - H(\alpha)) \equiv \left\langle \frac{\partial H}{\partial X} \right\rangle_E \tag{2.10}$$

where the last equality defines the microcanonical average $\langle .. \rangle_E$, and $\Omega' = d\Omega/dE$. For the parametrised random matrix models, $\langle dE/dX \rangle = 0$, and it is necessary to add a slope to the random matrix eigenvalues: the value of the slope can be estimated using (2.10).

c) The variance of the slopes of the energy levels is defined using a similar relation to (2.9):

$$\mathrm{Var}(dE_n/dX)_E = \frac{1}{\rho(E)} \sum_n \left[\left(\frac{dE_n}{dX}\right) - \left\langle\frac{dE_n}{dX}\right\rangle\right]^2 \delta_\epsilon(E - E_n) \ . \tag{2.11}$$

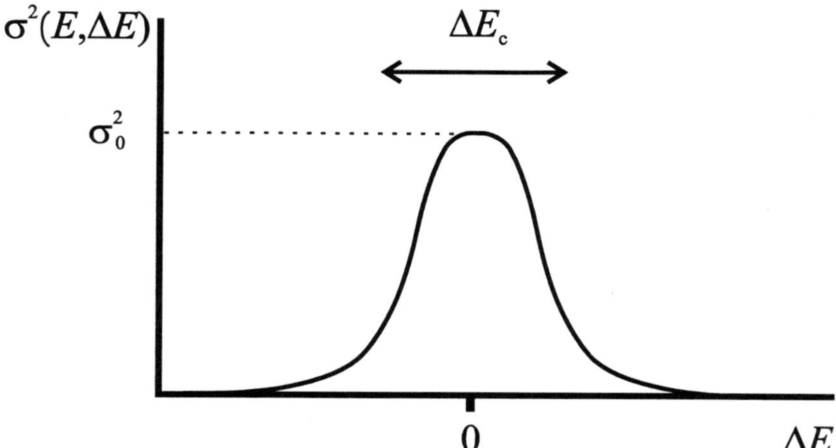

Figure 4. Schematic illustration of the form of the function $\sigma^2(E, \Delta E)$, illustrating the definitions of the quantities σ_0^2 and ΔE_c (the latter is termed the Thouless energy).

This quantity can be estimated directly using by a numerical experiment, and it can also be estimated semiclassically. In this case the semiclassical estimate is indirect: the variance is related to a statistic describing the matrix elements introduced in (2.2):

$$\sigma^2(E, \Delta E) = \sum_{n \neq m} |\partial_X H_{nm}|^2 \delta_\epsilon(E - \tfrac{1}{2}(E_n + E_m)) \delta_\epsilon(\delta E - (E_n - E_m)) \ . \tag{2.12}$$

Two different lines of argument can be used to suggest that

$$\mathrm{Var}\left(\frac{dE_n}{dX}\right) = \frac{2}{\beta}\sigma^2(E, 0) \equiv \frac{2}{\beta}\sigma_0^2 \tag{2.13}$$

where the final equality defines σ_0^2. One argument [16] is based upon a random matrix approach, and uses the fact that the wavefunctions of a complex system can be modeled by random functions. Another approach [17] uses the periodic orbit corrections to (2.10), and shows that these are consistent with the hypothesis that the diagonal matrix elements $\partial_X H_{nn} = dE_n/dX$ are independently gaussian distributed, with a variance which satisfies (2.13). Equation (2.13) has been verified numerically [3]. It gives a relation between properties of eigenvalues and the matrix elements which describe the response of the system to a perturbation: it can be viewed as a quantitatively accurate version of the Thouless relation [18], which connects conductivity with the curvatures $d^2 E_n/dX^2$ (as opposed to the slopes dE_n/dX) of energy levels.

A practical advantage of relating (2.11) to the statistic $\sigma^2(E, \Delta E)$ is that the latter is more easily estimated, both numerically (because there are more off-diagonal than diagonal elements, leading to better statistics), and semiclassically: the semiclassical formula for σ^2 is [19,20]

$$\sigma^2(E, \Delta E) \sim \frac{1}{2\pi\hbar\rho} \int d\tau \ \exp[\mathrm{i}\Delta E\tau]\left\langle \frac{\partial H}{\partial X}(\tau)\frac{\partial H}{\partial X}(0)\right\rangle_E \tag{2.14}$$

where the correlation function of $\partial H/\partial X$ is defined using the microcanonical average. The function $\sigma^2(E, \Delta E)$ usually decays rapidly to zero for sufficiently large values of ΔE, as shown schematically in figure 4 (the case where the parameter describes variation of the shape of the boundary of a quantum billiard is an important exception

and ΔE_c: the latter quantity is often termed the Thouless energy, particularly in the theory of disordered metals [18].

To summarise: the random matrix model (1.1) is expected to provide a good model for the parametric spectra of real systems after re-scaling the energy and parameter so that the level density ρ and $\mathrm{Var}(dE_n/dX)$ correspond to those of the real system, and after adding a drift $\langle dE_n/dX \rangle$ to the resulting set of levels. Semiclassical formulae have been given which can be used to estimate all of these scaling factors.

2.3 Singularities of energy levels

The parametric dependence of energy levels such as those plotted in figure 2 can be analysed statistically. One approach involves calculating correlation functions as functions of the parameter, for example the correlation function of the slope, dE_n/dX [21]. Another approach is to concentrate on statistics of singularities of the energy level functions $E_n(X)$: an elementary example of a singularity is a local minimum of the energy function, and other examples will be discussed shortly. A motivation for examining singularities is that they play an important rôle in asymptotic theories. The simplest type of statistic characterising a singularity is its density in parameter space, and most of the results which have been obtained are of this type. Before discussing the method used for counting singularities, I will discuss the important singularities of energy levels and their physical significance.

a) *Minima.* In the context of molecular physics, local minima of the electronic energy levels determine stable configurations of the nuclei.

b) *Degeneracies.* Degeneracies between pairs of neighbouring energy levels, for real values of the parameters, have a characteristic structure which can be understood using degenerate perturbation theory. In the neighbourhood of a degeneracy, the eigenfunctions are linear combinations of two nearly degenerate states, and the energy levels are eigenvalues of the a 2×2 matrix, with elements

$$H_{nm}(X) = \langle \phi_n(X_0)|\hat{H}(X)|\phi_m(X_0)\rangle = \delta_{nm}E_n(X_0) + \partial_X H_{nm}(X_0)\Delta X + O(\Delta X^2) \quad (2.15)$$

where X_0 is a point close to the degeneracy, $\Delta X = X - X_0$, and m takes the values n or $n+1$. The eigenvalues of this 2×2 matrix are

$$E_\pm = \tfrac{1}{2}(H_{n,n} + H_{n+1,n+1}) \pm \tfrac{1}{2}\sqrt{(H_{n+1,n+1} - H_{n,n})^2 + 4|H_{n,n+1}|^2} \quad (2.16)$$

Equation (2.15) has an obvious extension to the case in which the parameter space has several dimensions, in which case the point $\mathbf{X}_0$ about which the matrix elements are Taylor expanded can be chosen to be the point of degeneracy itself. The two nearly degenerate eigenvalues are then of the form

$$E(\mathbf{X}) \sim E_n(\mathbf{X}_0) + \tfrac{1}{2}[(\partial_{X_i} H_{nn}(\mathbf{X}_0) + \partial_{X_i} H_{n+1,n+1}(\mathbf{X}_0))\Delta X_i]$$

$$\pm \tfrac{1}{2}\left\{[(\partial_{X_i} H_{n,n}(\mathbf{X}_0) - \partial_{X_i} H_{n+1,n+1}(\mathbf{X}_0))\Delta X_i]^2 + 4|\partial_{X_i} H_{n,n+1}(\mathbf{X}_0)\Delta X_i|^2\right\}^{1/2} \quad (2.17)$$

where ΔX_i are the components of $\Delta \mathbf{X} = \mathbf{X} - \mathbf{X}_0$, and repetitions of the index i are summed over. The off-diagonal matrix element $\partial_{X_i} H_{n,n+1}(\mathbf{X}_0)\Delta X_i$ has β independent components. In order to create a degeneracy, the argument of the square root in (2.17) must be made to vanish. This requires that $\beta + 1$ independent numbers must vanish: it follows that $\beta + 1$ parameters must be varied in order to create degeneracies [22]. In cases where the Hamiltonian has real-valued matrix elements, the degeneracies occur as isolated points in a space with two parameters, and they have the double-cone structure illustrated in figure 5(a).

c) *Avoided crossings.* If only one parameter is varied, the energy levels approach each other at events called avoided crossings. In cases where the gap of the avoided crossing is small compared to the mean level spacing, and where the nearly degenerate levels are well separated from all of their neighbours, the combination of degenerate perturbation theory and linearisation of the parameter dependence of the Hamiltonian gives a good description of the structure of the avoided crossing. Equation (2.16) then implies that the energy levels in the vicinity of the avoided crossing are of the following form

$$E = E_0 + \tfrac{1}{2}B(X - X_0) \pm \tfrac{1}{2}[\epsilon^2 + A^2(X - X_0)^2]^{1/2} \tag{2.18}$$

where the parameters ϵ, A, B, are termed respectively the gap, the slope difference, and the mean slope of the avoided crossing; X_0 and E_0 describe the position of the singularity. The form of the avoided crossing is illustrated in figure 5(b).

d) *Complex branch points.* The eigenvalues may become degenerate for complex values of a single parameter: this is illustrated by the functional form of (2.18), where the energy levels are degenerate at the complex branch points $X_0 \pm i\epsilon/A$. These energy levels have a square root singularity at the branch point: application of degenerate perturbation theory with complex non-hermitian matrices shows that this is the typical behaviour at complex-parameter degeneracies.

It is useful to think of the complex branch point degeneracies as connecting distinct sheets of a Riemann surface for the energy level function. Because the singularity is of square root type, branch cuts must be inserted to make the energy $E(z)$ a single valued function of the complex parameter z. Consider what happens if the branch cuts cross the real axis, as illustrated in figure 5(c). If $E(z)$ is followed along the path γ, the energy undergoes a smooth transition from one real-axis energy level to another. These paths are important when considering extensions of the adiabatic theorem which describe amplitudes for transitions between quantum states: this is discussed in section 4.

2.4 Densities of singularities

The density in parameter space of the singularities discussed above can all be calculated by an extension of a result which is sometimes known as the Kac-Rice formula, which gives the frequency of zero-crossings of a statistically stationary random function. The formula assumes that the joint probability density for the random function $f(x)$, and its derivative $f'(x)$ evaluated at the same point, are known: the probability element is denoted by $P[f, f']df\,df'$. The approach is to consider a randomly chosen point x_0, and to estimate the probability of finding a zero crossing in a small interval of width $[x_0, x_0 + \delta x]$: this probability is $\delta P = \mathcal{D}\delta x + O(\delta x^2)$, where $\mathcal{D}$ is the density of zero crossings. Provided δx is sufficiently small, the distance to the zero crossing may be approximated by $-f/f'$. The probability δP is estimated as an integral over f and f' of the probability density $P[f, f']$, multiplied by a characteristic function which selects for the condition $0 < f/f' < \delta x$. Equating this estimate with $\mathcal{D}\delta x$ leads to the equation

$$\mathcal{D} = \int_{-\infty}^{\infty} df' \, |f'|P[0, f'] \tag{2.19}$$

which is often known as the Kac-Rice formula [23,24].

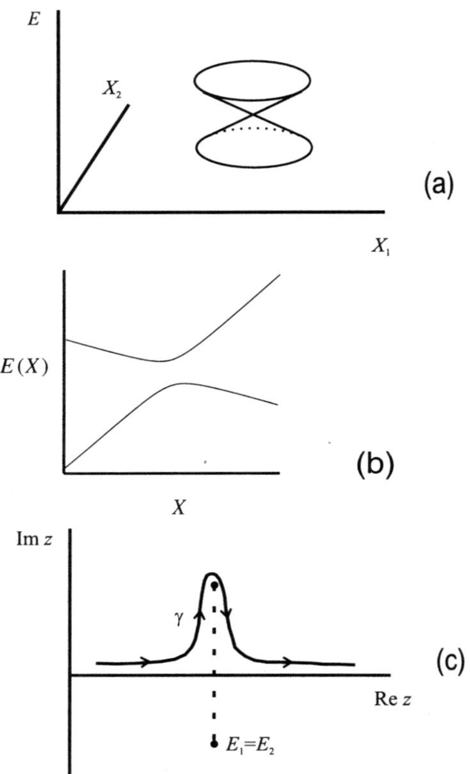

Figure 5. Structures associated with degeneracies between energy levels: a) Double-cone (diabolo) structure of energy level surfaces in neighbourhood of degeneracy for system with real-valued Hamiltonian. b) Avoided crossing, seen when only one parameter is varied. c) Branch points, where $E_1 = E_2$, and branch cuts in complex parameter space associated with an avoided crossing.

The same approach can be extended to estimate the density of all of the structures described above: the approach is always to calculate the probability of the singularity lying inside a small volume element in the parameter space, centred on a randomly chosen test point, and to equate this with $\mathcal{D}\delta V$, where $\mathcal{D}$ is the density of singularities, and δV the measure of the volume element.

The densities of all of the singularities listed above can be expressed in terms of two of the scaling parameters, discussed in section 2.2, namely the density of states ρ, and the parameter σ_0 describing the sensitivity of energy levels to variation of the parameters. To estimate the density of singularities for any 'real' system, it is only necessary to estimate these two parameters.

When there is more than one parameter, the quantity σ_0 is defined by

$$\sigma_0^2 = \det(\tilde{\Sigma}), \qquad \tilde{\Sigma} = \{\sigma_{ij}^2\}$$

$$\sigma_{ij}^2 = \langle \partial_{X_i} H_{nm} \partial_{X_j} H_{nm}^* \rangle \tag{2.20}$$

where the average runs over all states with E_n close to E_m: the correlation σ_{ij} can be estimated semiclassically using an obvious adaptation of equation (2.14).

As an example of the use of the Kac-Rice approach, consider how to calculate the density of degeneracies for the GOE version of the parametrised random matrix model (2.6), with $d = 2$. Assume that the test point $\mathbf{X}_0$ is sufficiently close to an exact degeneracy that (2.16) gives a good approximation. The displacement ΔX_i from the test point to the exact degeneracy is seen to be given by the solution of the following linear equations

$$\begin{pmatrix} \partial_{X_1} H_{n+1,n+1} - \partial_{X_1} H_{n,n} & \partial_{X_2} H_{n+1,n+1} - \partial_{X_2} H_{n,n} \\ \partial_{X_1} H_{n,n+1} & \partial_{X_2} H_{n,n+1} \end{pmatrix} \begin{pmatrix} \Delta X_1 \\ \Delta X_2 \end{pmatrix} = \begin{pmatrix} \Delta E \\ 0 \end{pmatrix} \tag{2.21}$$

where $\Delta E = E_{n+1}(\mathbf{X}_0) - E_n(\mathbf{X}_0)$. The distance from the test point to the degeneracy is $R = \sqrt{\Delta X_1^2 + \Delta X_2^2}$ can be written in the form $R = f \Delta E$, where f is a quantity which is constructed from the matrix elements $\partial_{X_i} H_{nm}$. In the case of the parametric random matrix Hamiltonian, the energy levels E_n were shown to be statistically independent of the matrix elements $\partial_{X_i} H_{nm}$ of the perturbation operators, and the joint probability density for these quantities may be written as a product, $P[\Delta E]P[f]$. The probability for finding a degeneracy in a small annulus of radius R and width dR centred on the test point is then $P[R]dR$, where

$$P[R] = \int_0^\infty df \int_0^\infty d\Delta E\, P[f]\, P[\Delta E]\, \delta(R - f \Delta E) \tag{2.22}$$

For small R, $P[R]dR = 2\pi \mathcal{D} R\, dR$, where $\mathcal{D}$ is the density of degeneracies, and the integral is dominated by the region where ΔE is small. The distribution of ΔE is the much studied level spacing distribution, and its behaviour for small ΔE is known for all three GXE ensembles [2]: for the GOE matrices of large dimension, $P[\Delta E] \sim \frac{1}{6}\pi^2 \rho^2 \Delta E$ for $\Delta E \rho \ll 1$. Using this in (2.21) gives

$$P[R] = \tfrac{1}{6}\pi^2 \rho^2 R \langle f^{-2} \rangle = 2\pi \mathcal{D} R\,. \tag{2.23}$$

The quantity f is a combination of the matrix elements $\partial_{X_i} H_{nm}$: using the fact that these are independent gaussian variables, the average $\langle f^{-2} \rangle$ can be calculated, and is found to be $4\sigma_0^2$, which gives an exact asymptotic result for the density of degeneracies [25], for GOE matrices with large dimension.

2.5 Summary of results on densities of singularities

The calculation for the density of degeneracies can be carried out for all three of the standard ensembles: the density of degeneracies $\mathcal{D}_\beta$, in a space of $\beta+1$ parameters, of one level with the level above is [26]

$$\mathcal{D}_1 = \frac{\pi}{3}\rho^2\sigma_0^2 \quad \text{(GOE)} ,$$

$$\mathcal{D}_2 = \frac{2\sqrt{\pi}}{3}\rho^3\sigma_0^3 \quad \text{(GUE)} ,$$

$$\mathcal{D}_4 = \frac{16\sqrt{2}\pi^{3/2}}{45}\rho^5\sigma_0^5 \quad \text{(GSE)} . \tag{2.24}$$

These expressions have been derived for the case where the Hamiltonian is the parametric GXE model, but if the 'static' parametric random matrix hypothesis is true, this estimate should work for a typical complex quantum system.

The densities of avoided crossings with small gap sizes are also known. Avoided crossings with narrow gaps, $\epsilon \ll \Delta$, have a structure which is approximated by (2.16). In order to characterise the avoided crossings it is natural to define a density of avoided crossings, $\mathcal{D}_\beta(\epsilon, A, B)$. The number of avoided crossings encountered by a given level with the level above, in unit interval of the parameter X, and with parameters in the intervals $[\epsilon, \epsilon + d\epsilon]$, $[A, A + dA]$, $[B, B + dB]$, is $\mathcal{D}(\epsilon, A, B) d\epsilon \, dA \, dB$. The exact densities for the three canonical ensembles are [13,3]

$$\mathcal{D}_1(\epsilon, A, B) = P[B]\frac{\pi\rho^2}{24\sigma^2}A^2 \exp[-A^2/8\sigma^2] \quad \text{(GOE)} ,$$

$$\mathcal{D}_2(\epsilon, A, B) = P[B]\frac{\pi^{3/2}\rho^3}{12\sigma^3}\epsilon A^3 \exp[-A^2/4\sigma^2] \quad \text{(GUE)} ,$$

$$\mathcal{D}_4(\epsilon, A, B) = P[B]\frac{8\pi^{7/2}\rho^5}{135\sqrt{2}\sigma^5}\epsilon^3 A^5 \exp[-A^2/2\sigma^2] \quad \text{(GSE)} . \tag{2.25}$$

where $P[B]$ is the probability density for the mean slope, which is gaussian, with variance σ^2/β. Some other exact results on densities of degeneracies are known, and are reviewed in [26].

3. Dynamic parametric random matrix theory

3.1 Characterisation of dynamics of complex systems

The natural way to characterise the dynamics of the time-dependent parametric random matrix Hamiltonian is by means of statistical properties of its evolution operator. The matrix elements of the evolution operator are basis-dependent, but there is one basis, the adiabatic basis (defined in (2.3)), which is particularly natural. The evolution operator for a process in which the parameter evolves from $X_i = X(t_i)$ to $X_f = X(t_f)$ as the time increases from t_i to t_f is characterised by the following matrix elements

$$U_{nm}(t_i, t_f) = \langle \phi_n(X_f)|\hat{U}(t_f, t_i)|\phi_m(X_i)\rangle \tag{3.1}$$

which is the amplitude to make a transition from the state $|\phi_m\rangle$ at time t_i, to the state $|\phi_n\rangle$ at time t_f. Here $\hat{U}(t_1, t_2)$ is the evolution operator, satisfying the Schrödinger equation $i\hbar\partial_t\hat{U}(t, t_0) = \hat{H}(t)\hat{U}(t, t_0)$, with boundary condition $\hat{U}(t_0, t_0) = \hat{I}$. The transition

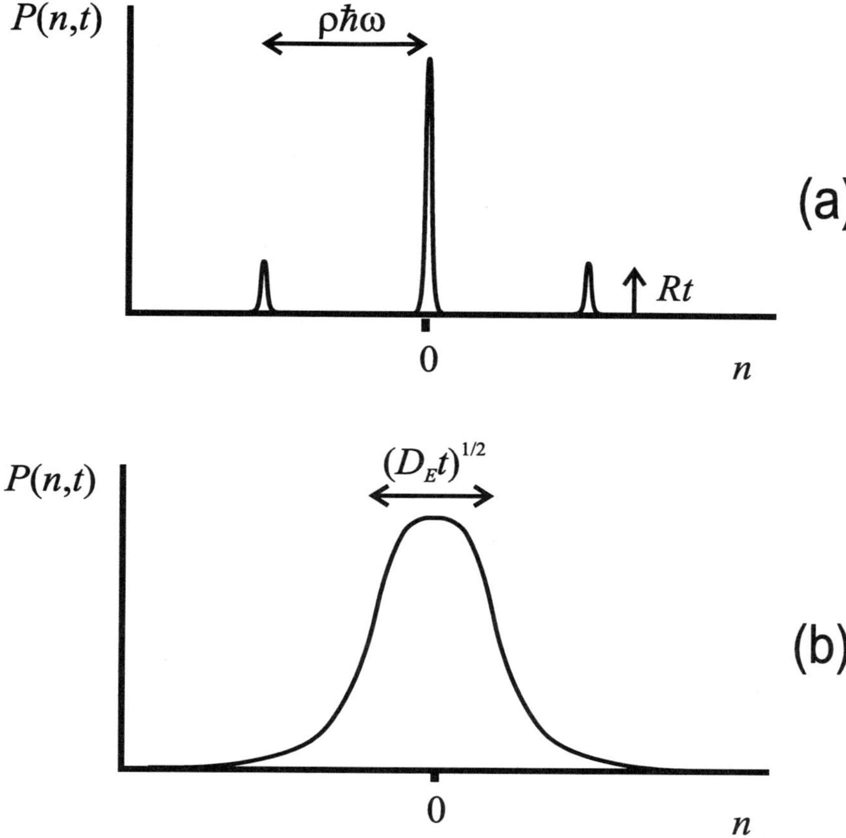

Figure 6. Schematic illustration of the two limiting behaviours of $P(n, t)$, the probability for making a transition through n levels: a) Resonant response, b) Energy diffusion.

probability $P(n, t)$ is defined by

$$P(n, t) = \langle |U_{n_0, n_0+n}(T, T+t)|^2 \rangle \tag{3.2}$$

where the average denoted by $\langle .. \rangle$ could be either an average over the initial state n_0, or an ensemble average in a random matrix model. This quantity gives a very natural physical characterisation of the dynamics of the system: it represents the probability for making a transition through n states after time t. The function $P(n, t)$ will depend on the energy E of the initial state $|\phi_{n_0}\rangle$.

There are two distinct physical processes which could determine the form of $P(n, t)$: they will be illustrated by considering the effect of a periodic perturbation $X(t) = X_0 \cos(\omega t)$. These two behaviours are illustrated in figure 6: they are respectively a resonant response, and diffusion in energy. I will argue that both are realised: one characterises situations in which some measure of the perturbation is small compared to the typical level separation Δ, the other characterises situations in which the perturbation is large.

The resonant and diffusive dynamics can be characterised as follows. In the resonant regime, a significant amplitude for transition only occurs when energy levels are separated by an amount close to $\pm\hbar\omega$, and the probability for making an upward or downward transition is characterised by a rate constant R. For short times, satisfying

$Rt \ll 1$, the function $P(n)$ may be approximated by

$$P(n, t) \sim (1 - 2Rt)\,\delta_\epsilon(n) + Rt\,\delta_\epsilon(n - \rho\hbar\omega) + Rt\,\delta_\epsilon(n + \rho\hbar\omega) \,. \qquad (3.3)$$

The broadening ϵ of the delta functions satisfies $\epsilon t \sim \hbar$, and this expression remains valid provided t is small enough that $t \ll \rho\hbar$ and $Rt \ll 1$. The rate constant is well known: it is given by a Fermi golden rule [4], which shows that R is proportional to the mean-square matrix element of the perturbation, $\langle|\partial_X H_{nm}|^2\rangle$, where the average runs over final states separated by $\sim \hbar\omega$ from the initial state. For frequencies small enough that $\hbar\omega$ is small compared to the scale on which $\sigma^2(E, \Delta E)$ depends on E, R may be written in terms of the statistic $\sigma^2(E, \Delta E)$:

$$R = \frac{\pi\rho}{2\hbar}\sigma^2(E, \hbar\omega)X_0^2 \qquad (3.4)$$

where $\sigma^2(E, \Delta E)$ is defined in (2.12).

In the case of diffusive dynamics, the function $P(n, t)$ is a approximated by a gaussian, and is characterised by a diffusion constant D_E:

$$P(n, t) \sim \frac{1}{\rho\sqrt{4\pi D_E t}}\exp(-n^2/4\rho^2 D_E t) \,. \qquad (3.5)$$

In both the resonant and the diffusive cases the second moment of the distribution $P(n, t)$ grows linearly in time, and may be characterised by a diffusion constant D_E:

$$\langle\Delta E^2\rangle = \rho^{-2}\int_{-\infty}^{\infty} dn\; n^2 P(n, t) \sim 2D_E t \,. \qquad (3.6)$$

3.2 Diffusion in energy

The process of resonant absorption is very familiar, but the dynamical response characterised by diffusion in energy is less well known. Here it will be introduced using a classical mechanical model: it must also be realised in the corresponding quantum system in the semiclassical limit ($\hbar \to 0$, with all classical quantities held fixed). This classical calculation is a simplified form of an argument given by Ott [27].

The classical model is a Hamiltonian system with chaotic dynamics, and a parameter which can be varied: the piston model in figure 3(a) is a good example. The change in the energy of the particle upon varying the parameter $X(t)$ is

$$\Delta E(t) = \int^t dt'\; \dot{X}(t')\frac{\partial H}{\partial X}(\mathbf{q}(t'), \mathbf{p}(t'), t') \qquad (3.7)$$

where $\mathbf{q}(t'), \mathbf{p}(t')$ is the phase space trajectory. Averaging gives:

$$\langle\Delta E(t)\rangle = \int^t dt'\; \dot{X}(t')\left\langle\frac{\partial H}{\partial X}(t')\right\rangle_E \qquad (3.8)$$

$$\langle\Delta E^2(t)\rangle = \int^t dt_1 \int^t dt_2\; \dot{X}(t_1)\dot{X}(t_2)\left\langle\frac{\partial H}{\partial X}(t_1)\frac{\partial H}{\partial X}(t_2)\right\rangle_E \qquad (3.9)$$

where the averages on the l.h.s. are microcanonical averages, defined by (2.10). These expressions will be simplified by considering the case in which $\dot{X}$ is independent of time, and where the microcanonical average of $\partial H/\partial X$ is zero. The correlation function appearing in (3.8) is also assumed to decay faster than $1/|t_2 - t_1|$, on timescales greater than some characteristic time τ_c (this assumption is realised in fully chaotic systems). For $t \gg \tau_c$, (3.8) can be approximated as follows [27]:

$$\langle \Delta E^2 \rangle \sim t\dot{X}^2 \int_{-\infty}^{\infty} d\tau \left\langle \frac{\partial H}{\partial X}(\tau) \frac{\partial H}{\partial X}(0) \right\rangle_E = 2D_E t \ . \tag{3.10}$$

The conclusion is that $\langle \Delta E^2(t) \rangle$ is proportional to t, and to an energy diffusion constant D_E. A simple adaptation to the case where $X(t) = X_0 \sin(\omega t)$ gives the diffusion constant

$$D_E = \tfrac{1}{2} X_0 \omega^2 \int_{-\infty}^{\infty} d\tau \ \exp[i\omega t] \left\langle \frac{\partial H}{\partial X}(\tau) \frac{\partial H}{\partial X}(0) \right\rangle_E \tag{3.11}$$

which agrees with (3.10) in the limit $\omega \to 0$, after identifying the time average $\langle \dot{X}^2 \rangle = \tfrac{1}{2} X_0^2 \omega^2$. The correlation function appearing in (3.10) and (3.11) is exactly the same as that in (2.14), and in order to facilitate comparison with quantum mechanical formulae, these results can be expressed in terms of the matrix element statistic σ_0^2 via (2.14): the 'semiclassical' expression for the diffusion constant, corresponding to (3.10), is

$$D_E^{(sc)} = \pi \hbar \rho \sigma_0^2 \dot{X}^2 \ . \tag{3.12}$$

The situation is very different when the classical motion is integrable, because the Fourier transform of the correlation function behaves very differently from ergodic systems. In the case of integrable motion with no separatrices, the fluctuations of the energy change $\Delta E(t)$ are exponentially small: $\Delta E \sim \exp(-K/\dot{X})$ in the limit $\dot{X} \to 0$ [28]. When there are separatrices, ΔE is much larger [29].

3.3 Relation between energy diffusion and dissipation

Diffusion is an archetypical irreversible process, and is natural to ask whether the diffusion of energy is related to dissipation, the irreversible transfer of energy. It will be shown that, for an important model, there is a direct connection.

Dissipation is the irreversible transfer of energy from an observed, usually macroscopic, degree of freedom into a large number of degrees of freedom (a bath) describing un-observed microscopic motion. The term 'irreversible' means that the energy transfer is not reversed when the trajectory $X(t)$ of the observed coordinate is reversed, except for exceptional choices of initial conditions.

At sufficiently low temperatures, most situations are described by one of two types of bath. In the case where the bath degrees of freedom are bosons, they are able to relax into their ground state at zero temperature. Small displacements are described by approximately linear equations of motion, and the bath is modeled by a set of harmonic oscillators. Examples are when the dissipation is caused by coupling to lattice vibrations (phonons), or the electromagnetic field (photons). This situation has been discussed by many authors, notably Feynmann and Vernon [30], Leggett and co-workers [31,32]).

The second situation is when the dissipation is effected by coupling to a bath of fermions. The bath of particles may be electrons (in solid state or molecular physics) or nucleons (in nuclear physics applications). At sufficiently low temperatures and high densities systems of fermions can be approximated as systems of weakly interacting fermions, which can be modeled by non-interacting fermions which satisfy an effective Schrödinger equation [9]. This is the standard model for dissipation in electronic systems: this review considers an unfamiliar application of the model, involving diffusion of single particle energies.

Consider a system of non-interacting fermions, with motion described by an effective single particle Hamiltonian $\hat{H}(X)$. If there are many fermions, the ground state configuration of the system includes particles which are in highly excited states of the single particle Hamiltonian, and it is these particles which require the least additional energy to be excited by a perturbation of the system. If the Hamiltonian $\hat{H}(X)$ corresponds to a chaotic or diffusive single particle motion, it is appropriate to use random matrix models.

Consider the occupation probability $f(E,t)$ for single particle states at energy E and time t. The total energy of the system of fermions is

$$E_T(t) = \int_{-\infty}^{\infty} dE \, f(E,t) \, E \, \rho(E) \, . \tag{3.13}$$

If the single particle energies evolve diffusively, the occupation probability $f(E,t)$ satisfies a diffusion equation. Accounting for the drift of energy levels considered in section 2.2, and allowing for an energy dependence of the density of states and the diffusion constant, continuity considerations imply that this equation must be of the form

$$\frac{\partial(\rho f)}{\partial t} + \dot{X}\frac{\partial}{\partial E}\left[\left\langle\frac{dE_n}{dX}\right\rangle\rho f\right] = \frac{\partial}{\partial E}\left[D_E\rho\frac{\partial f}{\partial E}\right] \, . \tag{3.14}$$

Differentiating (3.13), using this differential equation, and integrating by parts shows that the diffusion of single particle energies implies that E_T changes at a rate

$$\frac{dE_T}{dt} = \dot{X}\int_{-\infty}^{\infty} dE \, \rho f\left\langle\frac{dE_n}{dX}\right\rangle + \int_{-\infty}^{\infty} dE \, \rho D_E\frac{\partial f}{\partial E} \, . \tag{3.15}$$

The first term represents a reversible transfer of energy to the system of fermions: it changes sign when $\dot{X}$ is reversed. Equation (3.12) shows that the second term is proportional to $\dot{X}^2$, and is therefore always positive. This term therefore represents an irreversible transfer of energy to the system. In the important case where $f(E,t)$ is a low temperature Fermi-Dirac distribution, the rate of dissipation takes a particularly simple form:

$$\left.\frac{dE_T}{dt}\right|_{\text{irrev}} = \rho D_E \tag{3.16}$$

with both ρ and D_E evaluated at the Fermi energy. This equation was derived in [16], following a less rigorous calculation in [33].

In the case of integrable classical motion, the rate of dissipation is greatly reduced in the low-frequency limit relative to comparable ergodic systems [34]. This model for dissipation with a fermionic bath therefore shows that interactions between particles in the bath are not necessary for dissipation, but that ergodic motion is required.

It is instructive to discuss the origin of the irreversibility from a quantum mechanical point of view, by considering the effect of reversing the motion of the parameter $X(t)$ on the matrix elements of the evolution operator. Let $\tilde{U}(T,0)$ be the matrix formed from the matrix elements $U_{nm}(T,0)$ of the evolution operator in the adiabatic basis when $X(t) = F(t)$ changes from X_i to X_f as time increases from $t_i = 0$ to $t_f = T$, and let $\tilde{U}(2T,T)$ be the matrix corresponding to reversing the motion of $X(t)$, i.e. for $2T > t > T$, $X(t) = F(2T-t)$. Careful consideration of the Schrödinger equation shows that $\tilde{U}(2T,T) = \tilde{U}^T(T,0)$. Only in exceptional circumstances will $\tilde{U}(2T,T)\tilde{U}(T,0)$ be diagonal, corresponding to reversible behaviour.

3.4 Comparison between energy diffusion and perturbation theory

Two approaches, perturbation theory and energy diffusion, have been applied to the same model, namely the dynamics of complex systems. It is important to determine when each approach is applicable, and the extent to which the results are equivalent.

The transition probability $P(n, t)$ is clearly different, but the rate of dissipation could be comparable.

In the case where the excursion of the parameter $X(t)$ is small, the dissipative response of the system may be analysed using perturbation theory. If $X(t)$ oscillates sinusoidally with amplitude X_0, the transition rate is estimated using the Fermi golden rule, with the rate constant given by (3.4). For the system of non-interacting fermions, initially in the ground state, there are $\rho \hbar \omega$ electrons which can be excited, and upon excitation each of these fermions gains an energy $\hbar \omega$: the rate of absorption (dissipation) of energy is then

$$\frac{dE_T}{dt} = R \rho \hbar^2 \omega^2 = \tfrac{1}{2} \pi \hbar \rho^2 \sigma^2 (E_F, \hbar \omega) X_0^2 \omega^2$$

$$= \pi \hbar \rho^2 \sigma^2 (E_F, \hbar \omega) \langle \dot{X}^2 \rangle . \tag{3.17}$$

This simple formula is equivalent to that obtained by the rather less direct procedure of applying quantum mechanical linear response theory, as discussed by Kubo [35]. It is particularly close to the formula given by Greenwood [6] (known as the Kubo-Greenwood formula) expressing the DC conductivity in terms of matrix elements.

Now consider the comparison between (3.17) and the energy diffusion formula, (3.16), for a semiclassical system. Using equation (2.14) the matrix elements statistic σ^2 can be expressed in terms of an integral over the classical microcanonical correlation function of $\partial H / \partial X$. This integral is precisely the same as that which appears in the expression (3.11) for the energy diffusion rate. It follows for systems where semiclassical estimates are valid, the classical energy diffusion model and the Kubo-Greenwood formula are equivalent.

It is important to clarify when the perturbative approach is justified. Recall the arguments used to derive the Fermi golden rule: a perturbation of frequency ω is applied, to a system which is originally prepared in an eigenstate $|\phi_{n_0}\rangle$ of the unperturbed Hamiltonian. The amplitude $a_n(t)$ for transitions into other levels is estimated by expressing the Schrödinger equation in the basis of unperturbed states, multiplied by phase factors $\exp[-iE_n t/\hbar]$, and integrating to obtain

$$a_n(t) = a_n(0) - \frac{i}{\hbar} \sum_m \int_0^t dt' \; \exp[i(E_n - E_m)/\hbar] \cos(\omega t') \, V_{nm} \, a_m(t') \tag{3.18}$$

where $V_{nm} = X_0 \partial_X H_{nm}$ are the matrix elements of the perturbation. The transition rate R is deduced by considering the behaviour of the $a_n(t)$ for sufficiently short times, by replacing the amplitudes $a_m(t)$ in the integral with their values at $t = 0$. If initially only the state with index n_0 is occupied (i.e. $a_n(0) = \delta_{n,n_0}$), then only the term $m = n_0$ then contributes to the sum, and the exponential factor enhances the transition probability when $E_n - E_m = \pm \hbar \omega$. Examination of the resulting expression shows that the probability for transitions to states close to $E_{n_0} + \hbar \omega$ is proportional to time, implying that the transition probability may be written in the form (3.3). This calculation becomes meaningless if the total transition rate R is sufficiently large that the amplitude a_{n_0} to remain in the initial state is significantly reduced for times shorter than one period of the perturbation, i.e. when $R \gg \omega$. Using the expression (3.4) for R, this condition for the breakdown of the perturbative solution can also be written

$$(\rho \sigma X_0)^2 \gg \rho \hbar \omega \tag{3.19}$$

i.e. the square of the dimensionless size of the perturbation should be large compared to the number of level spacings per photon energy. It will be instructive to consider when this condition for breakdown of perturbation theory may be satisfied. Two cases will be considered: first the case of electrical conduction in metals, then later the case of the semiclassical limit ($\hbar \to 0$, holding all classical quantities fixed).

Consider the rate of dissipation in a cubic sample, in d-dimensions, of side L, with internal electric field $\mathcal{E}$, and bulk conductivity Σ. The density of states is $\rho = nV$, where n is the density of states per unit volume, $V = L^d$, and the conductivity can be expressed in terms of the spatial diffusion constant D using the 'zero temperature Einstein relation', $\Sigma = ne^2 D$. To introduce the electric field, two opposite edges of the sample may be regarded as being connected by a resistanceless conductor, with a flux X passing through the loop, so that $\mathcal{E} = \dot{X}/L$; it will be assumed that the flux oscillates sinusoidally with amplitude X_0 and frequency ω.

The mean rate of dissipation may be written $\langle dE_T/dt \rangle = \Sigma L^d \langle \mathcal{E}^2 \rangle$. Comparison of (3.17) and the zero-temperature Einstein relation shows that $\sigma^2 = e^2 D L^{-(d+2)}/\pi\hbar n$. The condition (3.19) for perturbation theory to fail can then be estimated as follows:

$$\frac{R}{\omega} \sim \frac{e^2 D \mathcal{E}^2}{\hbar^2 \omega^3} \sim \frac{E_{\text{coll}}^2}{(\hbar\omega)^2} \frac{\omega_s}{\omega} \gg 1 \tag{3.20}$$

where $E_{\text{coll}} = e\lambda\mathcal{E}$ is the typical amount of energy transferred from the electromagnetic field to the electron between collisions; λ is the mean free path and ω_s is the scattering rate. Clearly, the perturbative approach only applies for sufficiently high frequencies: it is an unproven assumption that the predictions of perturbation theory or linear response theory remain valid in the limit $\omega \to 0$ with the amplitude of the electric field held constant. A similar point about the use of perturbation theory to calculate electrical conductivity has been made by van Kampen [36], who uses a classical argument. The energy diffusion model applies in the non-perturbative situation, and the classical calculation of section 3.2 yields a diffusion constant that gives a rate of dissipation equal to that of the perturbative quantum calculation. This lends strong support to the use of the Kubo-Greenwood formula, but not all systems exhibit classical behaviour, and the classical diffusion constant may not be valid for all regimes of the corresponding quantum system. It is important, therefore, to develop purely quantum models for the energy diffusion constant D_E. This will be done in section 4: first it will be useful to consider the dimensionless parameters which distinguish the different regimes of the model.

3.5 Dimensionless parameters

The following dimensional quantities might be relevant to a universal theory for dynamics:

$$\hbar, \quad \rho, \quad \sigma_0, \quad \dot{X}, \quad X_0, \quad \Delta E_c \ . \tag{3.21}$$

The final quantity is the energy scale at which the statistic $\sigma^2(E, \Delta E)$ characterising the matrix elements $\partial_X H_{nm}$ becomes small: it will be termed the Thouless energy (see Fig. 4). These quantities can be formed into three independent dimensionless combinations, which may be written:

$$\chi = \rho\sigma_0 X_0, \quad \kappa = \rho^2 \sigma_0 \hbar \dot{X}, \quad \eta = \rho\sigma_0 \hbar \dot{X}/\Delta E_c \ . \tag{3.22}$$

The parameter χ is a dimensionless measure of the strength of the perturbation: the condition for Rayleigh- Schrödinger perturbation theory to be valid is $\chi \ll 1$. The parameters κ and η are dimensionless measures of the velocity of the perturbation: the velocity $\dot{X}$ is converted to an energy scale $\hbar\omega_q$, where $\omega_q = \rho\sigma_0\dot{X}$ is the rate at which the adiabatic eigenfunctions decorrelate, and this energy scale is compared with the mean level spacing or the Thouless energy respectively. Because $\Delta E_c \rho \gg 1$, it follows that $\kappa \gg \eta$. The significance of the parameters κ and η will become clear in section 4.

If the time-dependence of the coordinate is sinusoidal, with amplitude X_0 and frequency ω, then $\dot{X}$ can be identified with $X_0\omega$. Another significant dimensionless parameter is $\nu = \rho\hbar\omega$, which is the ratio of the 'photon energy' to the mean level spacing. When $\nu \ll 1$ and $\chi \ll 1$, the repulsion between energy levels which is predicted by random matrix theory implies that the rate of dissipation will be reduced, because there is less likely to be a state in resonance with the initial state. When the thermal energy kT is large compared to Δ, Gorkov and Eliashberg [37] showed, within a perturbative calculation, that this reduction is by the factor $R_\beta^{(2)}(\nu)$, where $R_\beta^{(2)}$ is universal the two-level correlation function of energy levels, satisfying $R_\beta^{(2)}(\nu) \sim \nu^\beta$ for $\nu \ll 1$ [2]. The use of a perturbative model for absorption in this case appears questionable, because the model is applied to a regime in which there are pairs of levels isolated from other states: unless the model is further elaborated by introducing coupling to phonons, these two-level systems will exhibit Rabi oscillations rather than continuous absorption.

It is useful to consider how these quantities scale in the semiclassical limit. Using (2.8) and (2.14) respectively gives $\rho \sim \hbar^{-d}$, $\sigma \sim \hbar^{(d-1)/2}$. Also, equation (2.14) shows that $\Delta E_c \sim \hbar/\tau_c$, where τ_c is the timescale for decay of classical autocorrelation functions. From these it follows that the scaling of the dimensionless parameters is $\chi \sim \hbar^{-(d+1)/2}$, $\kappa \sim \hbar^{-(3d-1)/2}$, $\eta \sim \hbar^{-(d+1)/2}$, and $\nu \sim \hbar^{-d+1}$. It follows that in the semiclassical limit all of these dimensionless parameters diverge, for $d > 1$. The dimensionless perturbation parameter which determines whether the Fermi golden rule is applicable is seen, by comparison with (3.19), to be $\chi^2/\nu \sim \hbar^{-2}$: this also diverges in the semiclassical limit, as might be anticipated from the fact that the energy diffusion mechanism has been demonstrated in the classical case.

4. Quantum theories for diffusion in energy

4.1 The adiabatic Schrödinger equation

This section will consider the dynamics of the parametric Hamiltonian (1.1) when the excursion of the parameter $X(t)$ is non-perturbative. In this case it is natural to expand the solution in terms of the adiabatic basis, defined by equation (2.3): the wavefunction is written

$$|\psi(t)\rangle = \sum_n a_n(t) \exp[-i\theta_n(t)] \, |\phi_n(X(t))\rangle \tag{4.1}$$

where $\theta_n(t)$ is a phase integral, and the phases of the adiabatic eigenfunctions are gauged to satisfy the 'Berry phase' connection rule:

$$\theta_n(t) = \int_{t_0}^{t} dt' \, E_n(t') , \quad \left\langle \phi_n \left| \frac{d\phi_n}{dt} \right. \right\rangle = 0 . \tag{4.2}$$

Substituting (4.1) into the time dependent Schrödinger equation, and using (4.2) leads to the following equation of motion for the coefficients $a_n(t)$, which will be termed the adiabatic Schrödinger equation:

$$\frac{da_n}{dt} = \dot{X} \sum_{m \neq n} \frac{\partial_X H_{nm}}{E_m - E_n} \exp[i(\theta_n - \theta_m)] \, a_m . \tag{4.3}$$

This equation has been used by many authors in discussions of the quantum adiabatic theorem (e.g. [38]). Most of the discussion in this section will be based upon this natural representation of the Schrödinger equation.

Equation (4.3) can be integrated immediately to give

$$a_n(t) = a_n(t_0) + \dot{X} \int_{t_0}^{t} dt' \left. \frac{\partial_X H_{nm}}{E_n - E_m} \right|_{X(t')} \exp[\mathrm{i}(\theta_n(t') - \theta_m(t'))]\, a_m(t') \qquad (4.4)$$

which will be useful for subsequent discussions.

4.2 Non-adiabatic transitions

When $\dot{X}$ is small, the phase factor in (4.4) ensures that the integral is very small, implying that the coefficients $a_n(t)$ are approximately constant: this is an expression of the quantum adiabatic theorem. In this limit the dynamical behaviour is reversible: the original state is recovered (apart from a phase factor) when $X(t)$ is returned to its original value.

For small $\dot{X}$, in cases where the perturbation is turned on and off sufficiently smoothly, the transition probabilities are exponentially small. The calculation of these transition probabilities is a difficult problem, which is still far from completely solved [38]. For a two level system, in which the adiabatic states asymptotically approach constant state vectors as $|t| \to \pm\infty$, the probability P_{12} for transition between an initial state $|\phi_1\rangle$ for $t \to -\infty$ to a final state $|\phi_2\rangle$ for $t \to \infty$ satisfies

$$\lim_{\dot{X} \to 0} P_{12} \exp[2|\mathrm{Im} S_{12}|/\hbar] = 1 \qquad (4.5)$$

where S_{12} is a phase integral

$$S_{12} = \int_\gamma dz\, E(z) \qquad (4.6)$$

in which the energy is integrated along a path in a complex time variable z, which connects levels E_1 and E_2 by looping around a branch point, as illustrated in figure 5(c). This result is known as the Dykhne formula [39]. Note that in cases where the Hamiltonian is a function of a parameter X, the integral S_{12} is inversely proportional to $\dot{X}$, so that the transition probability is expected to vanish exponentially in the limit $\dot{X} \to 0$. If there are several branch points, the one which gives the largest transition probability is relevant. When there are more than two levels, the Dykhne formula applies to adjacent energy levels. For pairs of levels which are not adjacent, the exponent is given by a sum of contributions of the form (4.6), with only certain combinations of branch points allowed to contribute [38]: the general solution for the $\dot{X} \to 0$ asymptotic behaviour of the transition probabilities between states which are not nearest neighbours is still an open problem.

In the case where energy levels exhibit a narrow avoided crossing, with the separation of energy levels given by (2.18), the integral (4.6) takes the value $S = \pi\epsilon^2/2A\dot{X}$, and the transition probability predicted by (4.5) is

$$P_{12} = \exp[-\pi\epsilon^2/2A\dot{X}\hbar] \qquad (4.7)$$

which is known as the Landau-Zener formula [40,41,42]. As well as being a special case of the Dykhne formula, the Landau-Zener formula is the exact transition probability for a two state model with Hamiltonian

$$\hat{H}(t) = \tfrac{1}{2} \begin{pmatrix} A\dot{X}t & \epsilon \\ \epsilon & -A\dot{X}t \end{pmatrix}. \qquad (4.8)$$

Note that, for this model, equation (4.7) gives an accurate result for large as well as small values of $\dot{X}$. In a many-level system, the Landau-Zener formula gives an accurate expression for the transition probability at narrowly avoided crossings: this has been verified numerically [16], and the current state of theoretical knowledge is discussed in [43].

4.3 Diffusion by the Landau-Zener mechanism

Figure 2 shows that energy levels of complex quantum systems may exhibit many avoided crossings as a parameter is varied. The Landau-Zener mechanism allows the diffusion of occupation probability as a parameter is varied, by transitions between neighbouring levels at those avoided crossing which have sufficiently narrow gaps to give a significant transition probability.

When $\dot{X}$ is small, only avoided crossings with very small gaps give a significant transition probabilities. These narrowly avoided crossings are rare events, and when writing down an asymptotic theory valid in the limit $\dot{X} \to 0$, they may be assumed to be completely uncorrelated.

The energy diffusion constant may be estimated approximately as follows:

$$D_E \sim R_{\mathrm{LZ}}/\rho^2 \sim \frac{\rho\sigma_0\dot{X}}{\rho^2}\langle P_t\rangle \tag{4.9}$$

where R_{LZ} is the rate of Landau-Zener transitions between adjacent levels, $\langle P_t\rangle$ is the mean probability of transition at an avoided crossing, and $\rho\sigma_0\dot{X}$ is an estimate of the rate at which avoided crossings are encountered. Using (2.25), the probability distribution of gap sizes ϵ is of the form $P[\epsilon]d\epsilon \sim \rho^\beta\epsilon^{\beta-1}d\epsilon$: this enables $\langle P_t\rangle$ to be estimated crudely, by averaging the Landau-Zener formula over the distribution of gap sizes ϵ:

$$\langle P_t\rangle \sim \rho^\beta \int_0^\infty d\epsilon\ \epsilon^{\beta-1}\exp[-\pi\epsilon^2/2\hbar A\dot{X}] \sim \kappa^{\beta/2}\ . \tag{4.10}$$

This leads to the estimate that $D_E \sim \kappa^{(\beta+2)/2}/\rho^3\hbar \propto \dot{X}^{(\beta+2)/2}$.

Using the exact result (2.25) for the density of avoided crossings, and integrating over the distribution of slopes A as well as that of the gap sizes ϵ, the calculation sketched above can be refined into a precise asymptotic formula for the diffusion constant, valid in the limit $\dot{X} \to 0$: for the orthogonal and unitary ensembles, the results are [33]

$$D_E = 2^{-5/4}\pi\Gamma(\tfrac{3}{4})\hbar^{1/2}\sigma_0^{3/2}\dot{X}^{3/2} \quad \text{(GOE)} ,$$

$$D_E = \pi\hbar\rho\sigma_0^2\dot{X}^2 \quad \text{(GUE)} . \tag{4.11}$$

The result for the unitary ensemble is exactly the classical result (3.12): this appears to be a coincidence.

It is worth commenting on two points about the Landau-Zener mechanism for energy diffusion. The first concerns the observability of the predictions of (4.11). These expressions are valid in the limit where Landau-Zener transitions only have significant probability at very narrowly avoided crossings, which are rare events: if a single parameter $X(t)$ varies over a finite range, there will be a narrowest avoided crossing between each pair of levels, and in cases where this is large, there will be a barrier to diffusion. This difficulty is avoided if the Hamiltonian has at least $\beta + 1$ parameters, and if $\mathbf{X}(t)$ explores the parameter space ergodically: in this case, the trajectory $\mathbf{X}(t)$ passes arbitrarily close to the exact degeneracies which exist in parameter spaces with $\beta + 1$ dimensions. Provided exact degeneracies exist between each pair of levels in the region explored by $\mathbf{X}(t)$, arbitrarily narrow avoided crossings are encountered between all pairs of levels, and the prediction (4.11) holds. An example of a physical realisation for a system with GOE statistics would be to consider a perturbation consisting of two perpendicular electric fields, oscillating at different frequencies.

A second point concerns the relation between the energy diffusion mechanism and the Pauli exclusion principle. The exclusion principle does not result in an inhibition of diffusion because the electrons are treated as independent particles: a Slater determinant remains a valid antisymmetric state when all of the single particle states are mapped under the evolution operator of the same one-body Hamiltonian [16].

4.4 Energy diffusion driven by matrix element fluctuations

In the case where transitions between states are effected by the Landau-Zener mechanism, the solutions of the differential equation (4.3) are determined by the fact that the phase factors fluctuate much more rapidly than the other quantities, except in the neighbourhood of close avoided crossings. For larger values of $\dot{X}$, the matrix elements $\partial_X H_{nm}$ and the energy levels E_n fluctuate as rapidly as the phase factors: this requires a different theory, which is given in this sub-section.

The exact equation of motion (4.3) is very difficult to solve when the fluctuations of the matrix elements must be accounted for. Instead, this equation will be replaced by a model system of the form

$$\frac{da_n}{dt} = \epsilon \sum_{m \neq n} Z_{nm}(t) \exp[i\lambda(n - m)t]\, a_m \; . \tag{4.12}$$

In this equation the coefficients $Z_{nm}(t) = -Z_{mn}^*(t)$ are gaussian random functions, with statistics defined (for $m > n$ and $m' > n'$) by

$$\langle Z_{nm}(t) \rangle = 0$$

$$\langle Z_{nm}(t) Z_{n'm'}^*(t') \rangle = \delta_{nn'} \delta_{mm'}\, f(n - m)\, C(t - t') \; . \tag{4.13}$$

The correlation function satisfies $C(0) = 1$, and decays to asymptotically approach zero on a timescale τ_q. The stochastic differential system described by (4.12) and (4.13) was introduced in [44]. It is a reasonable model for the adiabatic form of the Schrödinger equation (4.3) if the following identifications are made (the terms on the left refer to the adiabatic Schrödinger equation, those on the right refer to the stochastic model):

$$\dot{X} \leftrightarrow \epsilon, \qquad \frac{\sigma^2(\Delta n/\rho)}{\Delta n^2/\rho^2} \leftrightarrow f(\Delta n), \qquad \frac{1}{\rho\hbar} \leftrightarrow \lambda \; . \tag{4.14}$$

In making the second identification, equation (4.4) is treated as if $\Delta E = \Delta n/\rho$, i.e. fluctuations in the energy levels are ignored, and only fluctuations of the matrix elements are considered. This is reasonable, because random matrix spectra are highly 'rigid' [2].

The objective is now to show how the model (4.12), (4.13) exhibits energy diffusion, and to determine the diffusion constant. The equation (4.12) can be integrated immediately, to give an expression analogous to (4.4). Squaring this expression gives the exact result

$$|a_n(t)|^2 = |a_n(0)|^2 + 2\mathrm{Re}\,\epsilon \sum_m \int_0^t dt'\, Z_{nm}(t') \exp[i\lambda(n - m)t']\, a_n^*(0)\, a_m(0)$$

$$+ 2\mathrm{Re}\,\epsilon^2 \sum_{m_1} \sum_{m_2} \int_0^t dt_1 \int_0^{t_1} dt_2\, Z_{nm_1}(t_1) Z_{m_1 m_2}(t_2)$$

$$\times \exp[i\lambda\{(n - m_1)t_1 + (m_1 - m_2)t_2\}]\, a_n^*(0)\, a_{m_2}(t_2)$$

$$+\epsilon^2 \sum_{m_1} \sum_{m_2} \int_0^t dt_1 \int_0^t dt_2 \ Z^*_{nm_1}(t_1) \, Z_{nm_2}(t_2)$$

$$\times \exp[i\lambda\{(n-m_2)t_2 - (n-m_1)t_1\}] \, a^*_{m_1}(t_1) \, a_{m_2}(t_2) \ . \tag{4.15}$$

Ensemble averaging gives an expression for the occupation probability for the nth state, $P_n = \langle |a_n|^2 \rangle$. This expression will be treated in the leading order of perturbation theory in the parameter ϵ. The linear terms vanish upon averaging, and the terms at $O(\epsilon^2)$, are simplified by noting that in the limit $\epsilon \to 0$ the amplitudes $a_n(t)$ vary slowly, so that the they may be approximated by constants in the integrals. If the initial phases of the amplitudes are random, such that

$$\langle a_n(t) a^*_{n'}(t') \rangle = \delta_{nn'} P_n(0) \tag{4.16}$$

equation (4.15) can be approximated as follows

$$P_n(t) \sim P_n(0) + 2 \,\mathrm{Re}\, \epsilon^2 \sum_m \int_0^t dt_1 \int_0^{t_1} dt_2 \ \langle Z_{nm}(t_1) \, Z_{mn}(t_2) \rangle \, \exp[i\lambda(n-m)(t_1-t_2)] \, P_n(0)$$

$$+\epsilon^2 \sum_m \int_0^t dt_1 \int_0^t dt_2 \ \langle Z^*_{nm}(t_1) \, Z_{nm}(t_2) \rangle \, \exp[i\lambda(n-m)(t_2-t_1)] \, P_m(0) \ . \tag{4.17}$$

If $t \gg \tau_q$, where τ_q is the decorrelation timescale for the matrix elements, the integrals are dominated by contributions from regions where $|t_1 - t_2|/\tau_q$ is not large. Also, because the correlation function depends only upon $t_2 - t_1$, (4.17) may be approximated as follows

$$P_n(t) \sim P_n(0) + \epsilon^2 t \sum_m \int_{-\infty}^{\infty} d\tau \ \langle Z_{nm}(\tau) \, Z_{nm}(0) \rangle \, \exp[i\lambda(n-m)\tau] \, (P_m(0) - P_n(0)) \ . \tag{4.18}$$

Differentiating with respect to t gives a differential equation for the probabilities P_n:

$$\frac{dP_n}{dt} = \sum_m R_{nm}(P_m - P_n) \tag{4.19}$$

where, using (4.18) and (4.13)

$$R_{nm} = \epsilon^2 f(n-m) \int_{-\infty}^{\infty} d\tau \ \exp[i\lambda(n-m)\tau] \, C(\tau) \ . \tag{4.20}$$

At large times the averaged occupation probability P_n is expected to be distributed over many states, and to vary slowly with respect to the level index n. In this case equation (4.19) can be approximated by a diffusion equation $\partial_t P = D_n \partial_n^2 P$, with diffusion constant

$$D_n = \tfrac{1}{2} \sum_m R_{nm}(n-m)^2 \ . \tag{4.21}$$

Equations (4.19) and (4.21) describe the behaviour of the model (4.12), in the limit $\epsilon \to 0$. It is important to convert this expression into the notation appropriate to the original model, (1.1), using the associations given in (4.14). The expression for the rate constant becomes

$$R_{nm} = \dot{X}^2 \frac{\sigma^2(\Delta E)}{\Delta E^2} \int_{-\infty}^{\infty} d\tau \ \exp[i\Delta E \tau/\hbar] \, C(\tau) \equiv \dot{X}^2 \frac{\sigma^2(\Delta E)}{\Delta E^2} F(\Delta E) \tag{4.22}$$

where the last equality defines $F(\Delta E)$, and the first argument of $\sigma^2(E, \Delta E)$ has been suppressed. Approximating the sum in (4.21) by an integral, the energy diffusion constant $D_E = D_n/\rho^2$ is then

$$D_E = \tfrac{1}{2}\rho \dot{X}^2 \int_{-\infty}^{\infty} d\Delta E \ \sigma^2(\Delta E) \, F(\Delta E) \ . \tag{4.23}$$

The value of the diffusion constant depends upon the relative width of the support of the functions $\sigma^2(\Delta E)$ and $F(\Delta E)$. The support of the former function is (by definition) ΔE_c. The support of $F(\Delta E)$ is $\hbar/\tau_q$, where τ_q is the timescale for the decay of correlations of matrix elements. This timescale is determined by the decorrelation of the adiabatic eigenfunctions: Rayleigh-Schrödinger perturbation theory shows that this occurs when the parameter X changes by $\Delta X_q \sim (\rho\sigma_0)^{-1}$, implying that $\tau_q \sim \Delta X_q/\dot{X} \sim (\rho\sigma_0\dot{X})^{-1}$. When $\Delta E_c\tau_q/\hbar \ll 1$, the integral in (4.23) can be approximated as follows

$$D_E = \tfrac{1}{2}\rho\dot{X}^2\sigma^2(0) \int_{-\infty}^{\infty} d\Delta E\; F(\Delta E)$$

$$= \tfrac{1}{2}\rho\dot{X}^2\sigma_0^2 \int_{-\infty}^{\infty} d\tau\; C(\tau) \int_{-\infty}^{\infty} d\Delta E\; \exp[\mathrm{i}\Delta E\tau/\hbar]$$

$$= \pi\hbar\rho\dot{X}^2\sigma_0^2 \int_{-\infty}^{\infty} d\tau\; C(\tau)\,\delta(\tau)\;. \tag{4.24}$$

This expression can be further simplified using the fact that $C(0) = 1$, giving a simple formula for D_E under the condition $\Delta E_c\tau_q/\hbar \gg 1$:

$$D_E = \pi\hbar\rho\sigma_0^2\dot{X}^2, \qquad \hbar\rho\sigma_0\dot{X}/\Delta E_c = \eta \ll 1\;. \tag{4.25}$$

This result is exactly that predicted by the classical diffusion model, and is also consistent with the Kubo-Greenwood formula. When $\eta \gg 1$ the support of $\sigma^2(\Delta E)$ is broader than that of $F(\Delta E)$, and it is clear that the diffusion constant will be smaller than that predicted by (4.25). Also, in [44] it was argued that the assumptions underlying the use of (4.12) as a model for Schrödinger equation of real systems become questionable when $\eta \gg 1$, and that these deficiencies of the model suggest that it may overestimate the energy diffusion constant when $\eta \gg 1$.

This discussion of the adiabatic Schrödinger equation therefore leads to the following conclusions. The model discussed above becomes relevant when $\dot{X}$ is sufficiently large that the Landau-Zener mechanism ceases to be relevant, i.e. when $\kappa = \hbar\rho^2\sigma_0\dot{X} \gg 1$. Provided $\eta = \hbar\rho\sigma_0\dot{X}/\Delta E_c \ll 1$, the perturbative analysis of the model presented above predicts an energy diffusion constant which is exactly in agreement with the classical model, and with the low frequency limit of the Kubo formula. When $\eta \gg 1$, the theory strongly suggests that the diffusion constant is smaller than that predicted by the classical model.

4.5 Numerical studies

This section has discussed two approaches to calculating the energy diffusion constant, one based upon the Landau-Zener mechanism, the other based upon a system of stochastic differential equations which resemble the adiabatic form of the Schrödinger equation. Neither of these models is mathematically rigorous, and their predictions need to be tested numerically.

The relevant numerical tests have been performed using parametric random matrices as the Hamiltonian. Results for the Landau-Zener model are reported in [16], for the parametrised GOE model. These tests were performed using the version of the model described by (2.4) and (2.5), in which the Hamiltonian is obtained by smoothing white noise. It was necessary to use this version of the model to see clear agreement with the theoretical prediction at small values of $\dot{X}$, because if the model (1.1) were used there would be a smallest avoided crossing between each pair of levels, as discussed at the end of section 4.3. The numerical experiments showed that the prediction of (4.11) is verified (for the GOE case), and that as $\dot{X}$ is increased to the point where $\kappa \sim 1$, there is a crossover to a regime in which the diffusion constant is given by the semiclassical

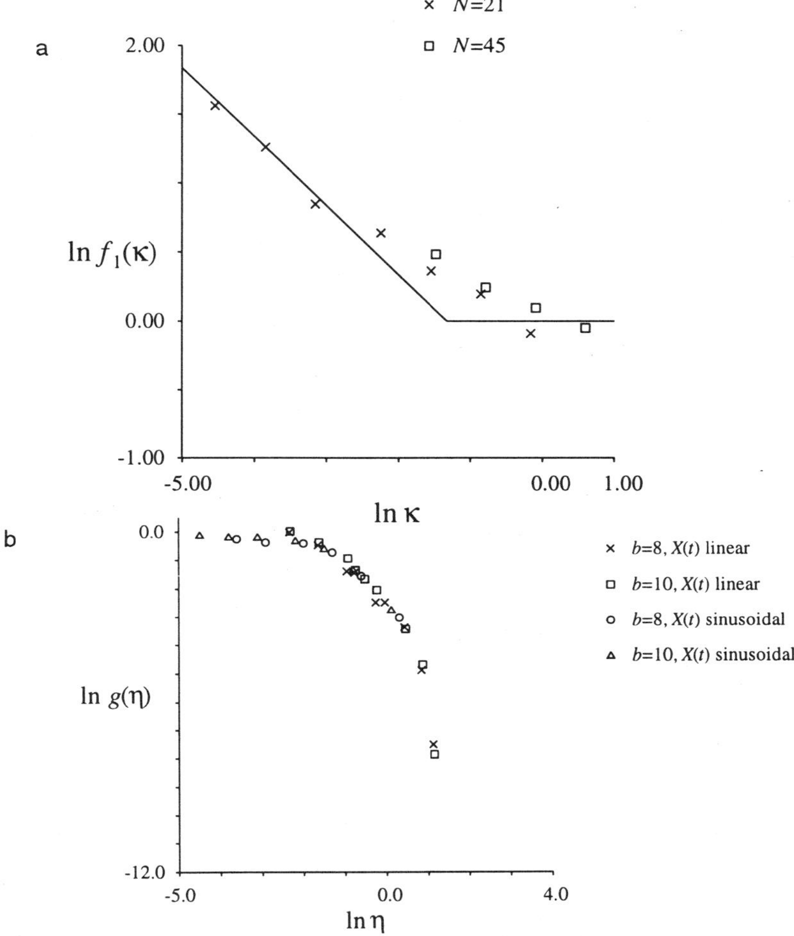

Figure 7. Numerical results showing the ratio of the energy diffusion constant D_E, to the semiclassical prediction $D_E^{(cl)}$. a) Results for the parametrised GOE model, showing crossover from the Landau-Zener mechanism to the semiclassical behaviour. b) Results showing the behaviour of the banded random matrix model (4.27), parametrised in a similar manner to (1.1), showing a sharp fall below the semiclassical prediction when $\eta \gg 1$.

value, $D_E^{(sc)}$, given by equation (3.12). The numerical results are consistent with the following scaling relation

$$\frac{D_E(\dot{X})}{D_E^{(sc)}} = f_\beta(\kappa) \tag{4.26}$$

where the scaling function f_β is expected to be universal, within a given symmetry class, with $f_\beta(x)$ approaching unity in the limit $x \to \infty$. The data from [16] are re-plotted in figure 7(a), showing the scaling relation in this form.

To test the predictions of the model described in section 4.4, it is necessary to introduce a model for which the Thouless energy is well defined. The standard parametric model, (1.1) has $\sigma^2(E, \Delta E)$ independent of both E and ΔE, so that E_c must be taken to infinity, and this model always corresponds to the $\eta \to 0$ limit. A parameter dependent random matrix model with a finite Thouless energy can be constructed from

the Wigner ensemble of banded random matrices, which are real symmetric matrices
with independently gaussian distributed elements satisfying

$$H_{ij} = \alpha\, i\, \delta_{ij} + V_{ij}$$

$$\langle V_{ij}\rangle = 0 \,, \qquad \langle V_{ij}^2\rangle = \begin{cases} 1 + \delta_{ij} \,, & |i - j| \le b \\ 0 \,, & |i - j| > b \end{cases} . \qquad (4.27)$$

The parameter dependence can be built into this model in a way analogous to either
(1.1), or to (2.4) and (2.5). The Thouless energy for these models is calculated in [44],
where it is found that $\Delta E_c \sim \alpha b^2$. The time-dependent Schrödinger equation for this
random matrix model was integrated numerically: the evolution was shown to be well
described by an energy diffusion process, and the diffusion constant was evaluated as a
function of $\dot X$. It was found that D_E is given by (4.25) for $\eta \ll 1$, and that D_E is much
smaller than the value predicted by this theory for $\eta \gg 1$. The in the region where η
is not too large or too small, the data fit a scaling relation of the form

$$\frac{D_E(\dot X)}{D_E^{(\mathrm{sc})}} = g(\eta) \qquad (4.28)$$

with $g(x) \to 1$ as $x \to 0$, and $g(x)$ decreasing as $x \to \infty$. In this case the scaling
function was found [44] to be non-universal: $g(x)$ was different for the versions of the
model based upon the different parametrisations, (1.1) and (2.4). The data for the
parametrisation analogous to (1.1) are plotted in figure 7(b), showing the crossover
phenomenon.

4.6 Dynamical localisation

No discussion of the dynamics of complex quantum systems would be complete
without a mention of dynamical localisation. This phenomenon occurs when a chaotic
system is driven by a strictly periodic perturbation: initially diffusion of energy (or
other dynamical variables) proceeds as might be expected, but after some timescale
t_{loc} diffusion ceases, and the dispersion of the variable remains approximately constant.
This effect was originally observed and explained [45,46] in numerical experiments on
the kicked quantum rotor, and has also been observed and analysed [11] for the 'universal' model Hamiltonian, (1.1).

The general explanation is based upon the following observations. Firstly, if the
perturbation is strictly periodic with period T, the evolution operator at NT is obtained
from a product of the Floquet operator: $\hat U(NT, 0) = [\hat U(T, 0)]^N$. When the Floquet
operator is expressed in the adiabatic basis, it is approximately banded: the matrix
elements $U_{nm}(T, 0)$ are very small for $|n - m| \gg \rho\sqrt{D_E T}$. Banded random matrices, of
bandwidth b, are a reasonable model for the matrix elements of the Floquet operator:
these matrices have localised eigenfunctions, with localisation length $L \sim b^2$. The long-
time evolution operator can be written in terms of localised eigenfunctions $|\chi_i\rangle$ of the
Floquet operator, with corresponding eigenvalues $\exp[i\Theta_i]$:

$$U_{nm}(NT, 0) = \sum_i \exp[iN\Theta_i]\langle\phi_n|\chi_i\rangle\langle\chi_i|\phi_m\rangle \,. \qquad (4.29)$$

Localisation implies that $\langle\chi_i|\phi_n\rangle$ is negligible for $|i - n| \gg L$. The matrix element
$U_{nm}(NT, 0)$ remains negligible for all N if there are no Floquet eigenfunctions $|\chi_i\rangle$
with significant amplitude in both the states n and m, i.e. if $|n - m| \gg L$. It follows
that the dispersion $\langle\Delta E^2\rangle$ of the energy saturates at a time

$$t_{\mathrm{loc}} \sim \rho^2 D_E T^2 \,. \qquad (4.30)$$

This estimate has been verified [11] in numerical experiments on the Hamiltonian (1.1).

Of all the effects described in this review, dynamical localisation is probably the most difficult to observe experimentally, because it is a quantum coherence effect which depends upon the Hamiltonian of the electron being precisely periodic: it would be very easily destroyed by noise from sources such as thermally excited phonons.

5. Concluding remarks

5.1 Summary of the dynamical regimes

The dynamical behaviour of the model (1.1) is surprisingly complex, and it is helpful to present a brief summary of the different regimes. These are conveniently divided into perturbative and non-perturbative cases, defined by whether the dimensionless perturbation parameter $\chi = \sigma_0 \rho X_0$ is large or small. To simplify the discussion it will be assumed that the excursion of the parameter is sinusoidal, $X(t) = X_0 \sin(\omega t)$.

The dynamical behaviour will be discussed in terms of the behaviour of the second moment of the transition probability $P(n)$: equation (3.6) implies that this quantity has linear growth for sufficiently short times, characterised by a coefficient D_E. The quantity D_E is convenient because it can be calculated for both the perturbative resonant response and the non-perturbative diffusive response, so that these regimes may be compared directly. It is also important physically because it is related to the energy absorbed irreversibly in the independent fermion model, where it is proportional to the energy absorbed.

The behaviour of (1.1) and the other closely related models which have been discussed are summarised schematically in figure 8, which compares three different situations: (a) the classical response, (b) the response predicted by perturbation theory, and (c), the behaviour predicted by the non-perturbative models discussed in section 4. In all three cases, the ratio $F(\omega) = D_E / D_E^{(\mathrm{sc})}$ (where $D_E^{(\mathrm{sc})}$ is given by (3.12)) is plotted as a function of frequency, on a logarithmic scale. The classical plot just shows a crossover behaviour at high frequencies: equation (3.11) predicts that $F(\omega)$ is reduced when the frequency exceeds the inverse of the classical de-correlation timescale, i.e. when $\omega \gg \omega_c \equiv \Delta E_c / \hbar \sim \eta / \chi$. Equation (3.4) shows that the quantum perturbative case shows the same crossover behaviour at ω_c, and in addition the theory of Gorkov and Eliashberg [37] implies that there is another crossover when $\rho \hbar \omega = \nu \sim 1$. The non-perturbative case also exhibits two crossovers, but these occur at frequencies where $\kappa = 1$ and $\eta = 1$: these are both lower than the crossover frequencies for the perturbative case by a factor of χ.

5.2 Discussion

The results for the non-perturbative case are surprising, because the crossover between different regimes is determined by the value of the velocity, not the frequency. The theoretical arguments and numerical results discussed in section 4 show that the classical estimate $D_E^{(\mathrm{sc})}$ can apply even when the photon energy is small compared to the mean level spacing: this is at variance with the perturbative prediction by Gorkov and Eliashberg [37]. This prediction has been verified for the random matrix model ([16], results illustrated in figure 7(a)), but has not yet been tested by simulations on real systems. There is however no logical objection to this result, and random matrix models have always been found to work well for predictions on the smallest energy scales.

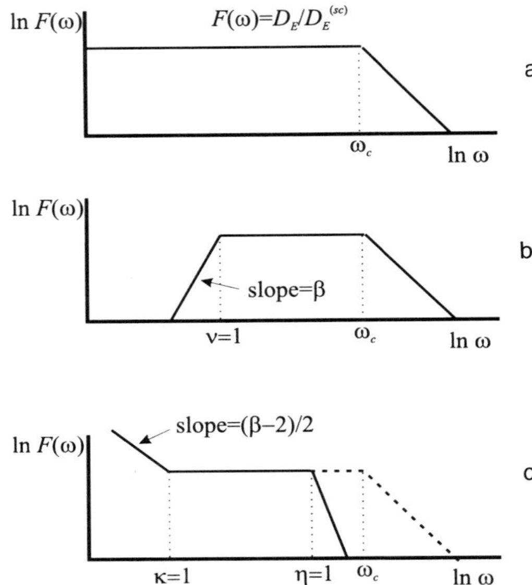

Figure 8. Schematic plots of the diffusion constant D_E, divided by the semiclassical prediction $D_E^{(sc)} = \pi \hbar \rho \sigma_0^2 \langle \dot{X}^2 \rangle$, for three different cases: a) Classical. b) Perturbative, quantum. c) Non-perturbative quantum: the bold line is the random matrix prediction, the dotted line the expected behaviour of semiclassical systems.

The case of the second crossover at higher velocities is more problematic. The reduction in the diffusion constant relative to the classical prediction has been verified numerically for a specific random matrix model (some results were shown in figure 7(b)). It will now be argued that this feature of the random matrix model will not always be realised in physical applications. The regime where both $\eta \gg 1$ and $\chi \gg 1$ was shown in section 3.5 to correspond to the semiclassical ($\hbar \to 0$) limit, and it can also be shown to correspond to the infinite volume limit in the problem of electrical conduction, which was discussed in section 3.4. In the case of the semiclassical limit, the classical behaviour must be realised, for times below some scale t^* which diverges in the limit $\hbar \to 0$. The reduction of the diffusion constant where $\eta \gg 1$ would be observable on short timescales, and is therefore inconsistent with the semiclassical prediction. The conclusion is that there is some aspect of the random matrix models used in section 4.4 which is inconsistent with the properties of semiclassical systems. In the case of the $V \to \infty$ limit for electrical conduction, the argument is that the behaviour predicted in section 4, and observed for random matrix models, is apparently never observed experimentally. Possibly this is because most observations of DC electrical conduction correspond to regimes in which the electron dynamics is semiclassical.

To conclude this discussion: the results for the random matrix model of section 4.4 are significant for two reasons:

a) Because they show that the usual linear response approach to calculating a dissipative response is not *universally* valid in parameter regimes for which measurements are made. The existence of one model for which there is an anomalous behaviour indicates that the assumptions upon which linear response theory is based must be treated with caution.

b) Although the random matrix model discussed in sections *4.4* and *4.5* cannot correspond to either the semiclassical limit, or the infinite volume limit of electrical conduction, it has no obvious deficiencies. It may therefore give a good description of other systems: this is a subject for continuing research.

REFERENCES

1. C. E. Porter (ed.), *Statistical Theories of Spectra: Fluctuations*, New York: Academic, (1965).
2. M. L. Mehta, *Random Matrices*, 2nd ed., New York: Academic, (1991).
3. E. J. Austin and M. Wilkinson, *Nonlinearity*, **5**, 1137-50, (1992).
4. L. D. Landau and E. M. Lifshitz, *Quantum Mechanics* (Course in Theoretical Physics, volume 3), Oxford: Pergamon, (1958).
5. L. D. Landau and E. M. Lifshitz, *Electrodynamics of Continuous Media* (Course of Theoretical Physics, volume 8), Oxford: Pergamon, (1960).
6. D. A. Greenwood, *Proc. Phys. Soc.*, **A68**, 585-96, (1958).
7. D. L. Hill and J. A. Wheeler, *Phys. Rev.*, **89**, 1102-45, (1952).
8. J. Błocki, Y. Boneh, J. R. Nix, J. Randrup, M. Robel, A. J. Sierck and W. J. Świątecki, *Ann. Phys. NY*, **113**, 130, (1978).
9. D. Pines and P. Nozieres, *The Theory of Quantum Liquids*, Menlo Park: Benjamin, (1966).
10. D. J. Thouless, M. Kohmoto, M. P. Nighingale and M. den Nijs, *Phys. Rev. Lett.*, **49**, 405-8, (1982).
11. M. Wilkinson and E. J. Austin, *Phys. Rev.*, **A46**, 64-74, (1992).
12. F. J. Dyson, *J. Math. Phys.*, **3**, 1199-1215, (1962).
13. M. Wilkinson, *J. Phys.*, **A22**, 2795-2805, (1989).
14. M. C. Gutzwiller, *Chaos in Classical and Quantum Mechanics*, New York: Springer, (1990).
15. A. I. Schnirel'man, *Usp. Math. Nauk*, **29**, 181, (1974).
16. M. Wilkinson, *Phys. Rev.*, **A41**, 4645-52, (1990).
17. M. Wilkinson, *J. Phys.*, **A21**, 1173-90, (1988).
18. D. J. Thouless, *Phys. Rep.*, **13**, 93, (1974).
19. M. Feingold and A. Peres, *Phys. Rev.*, **A34**, 591-5, (1986).
20. M. Wilkinson, *J. Phys.*, **A20**, 2415-23, (1987).
21. B. L. Altshuler and B. D. Simons, *Phys. Rev. Lett.*, **70**, 4063, (1993).
22. V. I. Arnold, *Mathematical Methods of Classical Mechanics*, Springer: New York, (1978).
23. M. Kac, *Bull. Amer. Math. Soc.*, **49**, 314-20, (1943).
24. S. O. Rice, *Bell. Sys. Tech. J.*, **24**, 46, (1945).
25. M. Wilkinson and E. J. Austin, *Phys. Rev.*, **A47**, 2601-9, (1993).
26. P. N. Walker, M. J. Sánchez and M. Wilkinson, *J. Math. Phys.*, **37**, 5019-31, (1996).
27. E. Ott, *Phys. Rev. Lett.*, **42**, 1628-31, (1979).
28. A. Lenard, *Ann. Phys. NY*, **6**, 261, (1959).
29. J. H. Hannay, *J. Phys.*, **A19**, L1067-72, (1986).
30. R. P. Feynman and F. L Vernon Jr., *Ann. Phys. NY*, **24**, 118, (1963).
31. A. O. Caldeira and A. J. Leggett, *Ann. Phys. NY*, **149**, 374, (1983).
32. A. J. Leggett, S. Chakravarty, A. T. Dorsey, M. P. A. Fisher, A. Garg and W. Zerger, *Rev. Mod. Phys.*, **59**, 1-85, (1987).
33. M. Wilkinson, *J. Phys.*, **A21**, 4021-37, (1988).
34. M. Wilkinson, *J. Phys.*, **A23**, 3603-11, (1990).
35. R. Kubo, *Canad. J. Phys.*, **34**, 1274-7, (1956).
36. N. G. van Kampen, *Acta Phys. Norwegica*, **5**, 279-84, (1971).
37. L. P. Gorkov and G. M. Eliashberg, *Zh. Eksp. Teor. Fiz.*, **48**, 1407, (1965) (Engl. transl. *Sov. Phys. JETP*, **21**, 940, (1965).
38. J. T. Hwang and P. Pechukas, *J. Chem. Phys.*, **67**, 4640-53, (1977).
39. A. M. Dykhne, *Sov. Phys. JETP*, **14**, 941-43, (1962).
40. C. E. Zener, *Proc. Roy. Soc. Lond.*, **A137**, 696-702, (1932).
41. L. D. Landau, *Phys. Z. USSR*, **1**, 426, (1932).
42. E. Majorana, *Nuovo Cimento*, **9**, 43-50, (1932).
43. A. Joye, H. Kunz and C-E. Pfister, *Ann. Phys. NY*, **208**, 299-332, (1990).
44. M. Wilkinson and E. J. Austin, *J. Phys.*, **A28**, 2277-96, (1995).
45. B. Chirikov, D. L. Shepelyansky and F. M. Izraelev, *Sov. Sci. Rev.*, **C2**, 209, (1981).
46. D. R. Grempel, S. Fishman and R. Prange, *Phys. Rev. Lett.*, **49**, 509-12, (1982).

INDEX

Drude formula, 26
Dykhne formula, 390
Dyson's Brownian motion, 59, 62, 67, 71

Edge state, 75, 79, 80
Eigenfunction, 31, 49, 50, 52, 91, 103, 105, 132,
    147–149, 166, 167, 192–196, 201, 211–213,
    217, 219, 220, 229, 245–254, 256–258, 261,
    262, 265, 266, 269, 270, 283, 284, 288, 290,
    317, 355, 356, 375, 378, 388, 389, 394, 396
  correlations, 245, 246, 250, 252, 253, 256, 290
  distribution, 245
Eigenstates, 17, 20, 22, 23, 103–106, 108–111, 114,
    116, 118–120, 122, 123, 126, 127, 131, 136,
    139, 194, 212, 227–229, 252, 253, 258, 262,
    264, 270, 271, 283, 339, 356, 387
  correlations, 131, 252
Eilenberger equation, 329
Einstein relation, 26, 388
Ergodic regime, 28, 29, 52
Ergodic time, 30, 50
Ergodicity, 7–9, 19, 60, 105, 178, 228, 350
Evolution operator, 18, 31, 45, 49, 50, 85, 86, 89–92,
    96, 97, 99–101, 111, 119, 174, 181, 190, 194–
    199, 201, 382, 386, 392, 396, 400
Expectation value, 86–91, 356, 401

Floquet multipliers, 92
Floquet operators, 173, 396
Fock space, 154, 157–159, 163, 164
Fredholm alternative, 202
Fredholm determinant, 101, 202–204
Fredholm integral equation, 202
Fredholm theory, 201, 202, 204
Fundamental domains, 208

Generating functions, 63, 67, 70, 72
Geodesic, 1, 3, 7, 207, 208, 210, 211, 356
Ginzburg–Landau functional, 34
Gradient expansion, 37
Grassmann algebra, 32, 159, 160
Grassmann variables, 32, 159, 161, 163, 231, 277
Green function, 4, 23, 24, 27, 32–37, 42, 43, 45, 46,
    63, 147, 148, 193–195, 201, 202, 204, 206,
    208–210, 230, 231, 246, 262, 335

Hamiltonian vector fields, 156, 167
Hardy–Littlewood conjecture, 13
Heat kernel, 95, 316, 318, 321–324
Heisenberg time, 5, 9, 29, 31, 42, 52, 107, 111, 116,
    118, 122, 133, 153, 186, 216, 264, 265
Hessian, 154, 166, 168, 169, 171
Hierarchic equations, 68, 71
Homoclinic recurrences, 104, 124, 126, 131
Hubbard–Stratonovich transformation, 34, 35, 43, 46,
    335
Husimi phase space theory, 104
Husimi projection, 109, 110
Hyperbolicity, 50, 51, 71, 87, 92, 93, 95, 96, 101,
    108, 119, 167, 178–180, 189, 356
Hyperbolicity assumption, 93

Insulator, 76, 79, 80, 269
Intensity statistics, 126
Interaction energy, 134, 137, 139, 140, 152
Invariance, 36, 48, 59, 60, 62, 67, 71, 72, 154, 164,
    170, 228, 266, 270, 271, 274, 349, 350, 375
Inverse participation ratio, 117, 120, 121, 127, 131,
    248, 250, 252, 253
Irreversibility, 53, 54, 386
Isometries, 170, 207–209, 212

Jacobian, 65, 66, 70, 92–95, 98, 99, 162, 234, 238,
    265, 275, 356

Kac–Rice formula, 379
Kubo formula, 394

Landau levels, 75–77, 79, 323
Landau–Zener transition, 391
Landauer formula, 291
Legendre function, 208, 209, 222
Level correlations, 4, 18, 19, 21, 153, 156, 193, 252
Level statistics, 6, 7, 27, 28, 39, 44, 52, 233, 253,
    254, 270, 339
Linearized dynamics, 30, 104, 126
Liouville equation, 238, 279–281
Liouville measure, 155, 264
Localised state, 75–77, 269, 271, 277, 290, 331
Localization, 18, 19, 22, 25–28, 30, 31, 39–41, 51–
    53, 56, 75–80, 82, 104, 108–111, 170, 171,
    191, 192, 227, 229, 236–238, 241, 246–248,
    252, 253, 270, 271, 277, 290, 331, 334, 340,
    396, 397
  dynamical, 396, 397
  of superintegrals, 170
  weak, 18, 25, 26, 30, 31, 39–41, 51–53, 56, 227,
    237, 246, 253, 270
Long range correlations, 24, 288, 356
Lyapunov exponent, 52, 79–81, 86, 93, 104, 105,
    109, 117, 118, 120, 122, 124, 213, 217

Maps, 1–4, 7, 9, 11, 14, 47, 82, 86, 90–95, 97, 98,
    100–102, 105, 106, 108, 110, 112, 114, 117–
    122, 125, 127, 129, 153–157, 159, 161, 163,
    164, 166–174, 177–183, 185, 189, 190, 192,
    195–199, 220, 320, 356, 392
  area preserving, 155, 173, 177, 196
  baker's, 118–122, 125, 129, 189, 192, 195, 198,
    199, 220
  cat, 1, 3, 7, 154, 155, 157, 189
  quantized symplectic, 153, 163
  quantum maps, 2, 3, 14, 173, 177, 192, 356
Markov graph, 85, 91
Maslov index, 3, 187, 204
Matrix elements fluctuations, 245, 250
Mean-field approximation, 34, 35, 43, 134, 141, 142,
    252, 345
Mesoscopic fluctuations, 18, 28, 144
Mesoscopic systems, 18, 20, 26, 193, 228, 237, 242,
    250, 255, 272, 300, 327, 331, 343
Monodromy matrix, 3, 30, 92, 179, 205, 206, 221
Multifractality, 241, 252, 253, 271, 283, 284, 288, 315